HANDBOOK OF GEOPHYSICAL EXPLORATION

SEISMIC EXPLORATION

VOLUME 35

SEISMIC WHILE DRILLING

FUNDAMENTALS OF DRILL-BIT SEISMIC FOR EXPLORATION

HANDBOOK OF GEOPHYSICAL EXPLORATION

SEISMIC EXPLORATION

Editors: Klaus Helbig and Sven Treitel

PUBLISHED VOLUMES

1984 – Mathematical Aspects of Seismology. 2nd Enlarged Edition (M. Båth and A.J. Berkhout)*
1984 – Seismic Instrumentation (M. Pieuchot) ISBN 0-08-036944-8
1984 – Seismic Inversion and Deconvolution (a) Classical Methods (E.A. Robinson)*
1985 – Vertical Seismic Profiling (a) Principles. 2nd Enlarged Edition (B.A. Hardage)*
1987 – Pattern Recognition & Image Processing (F. Aminzadeh)*
1987 – Seismic Stratigraphy (B.A. Hardage)*
1987 – Production Seismology (J.E. White and R.L. Sengbush)*
1989 – Supercomputers in Seismic Exploration (E. Eisner)*
1994 – Seismic Coal Exploration (b) In-Seam Seismics (L. Dresen and H. Rüter)*
1994 – Foundations of Anisotropy for Exploration Seismics (K. Helbig) ISBN 0-08-037224-4
1998 – Physical Properties of Rocks: Fundamentals and Principles of Petrophysics (J.H. Schön) ISBN 0-08-041008-1
1998 – Shallow High-Resolution Reflection Seismics (J. Brouwer and K. Helbig) ISBN 0-08-043197-6
1999 – Seismic Inversion and Deconvolution (b) Dual-Sensor Technology (E.A. Robinson) ISBN 0-08-043627-7
2000 – Vertical Seismic Profiling: Principles. 3d Updated and Revised Edition (B.A. Hardage) ISBN 0-08-043518-1
2001 – Seismic Signatures and Analysis of Reflection Data in Anisotropic Media (I. Tsvankin) ISBN 0-08-043649-8
2001 – Computational Neural Networks for Geophysical Data Processing (M.M. Poulton) ISBN 0-08-043986-1
2001 – Wave Fields in Real Media: Wave Propagation in Anisotropic, Anelastic and Porous Media (J.M. Carcione) ISBN 0-08-043929-2
2002 – Multi-Component VSP Analysis for Applied Seismic Anisotropy (C. MacBeth) ISBN 0-08-0424439-2
2002 – Nuclear Magnetic Resonance. Petrophysical and Logging Applications (K.J. Dunn, D.J. Bergman and G.A. LaTorraca) ISBN 0-08-043880-6
2003 – Seismic Amplitude Inversion in Reflection Tomography (Y. Wang) ISBN 0-08-044243-9
2003 – Seismic Waves and Rays in Elastic Media (M.A. Slawinski) ISBN 0-08-043930-6
2004 – Quantitative Borehole Acoustic Methods (X. Tang and A. Cheng) ISBN 0-08-044051-7

* *Book out of print.*

SEISMIC EXPLORATION

Volume 35

SEISMIC WHILE DRILLING

FUNDAMENTALS OF DRILL-BIT SEISMIC FOR EXPLORATION

by

Flavio Poletto
Istituto Nazionale di Oceanografia e di
Geofisica Sperimentale (OGS)
Sgonico (Trieste)
Italy

and

Francesco Miranda
ENI E&P Division
Milan
Italy

2004

ELSEVIER

Amsterdam – Boston – Heidelberg – London – New York – Oxford
Paris – San Diego – San Francisco – Singapore – Sydney – Tokyo

ELSEVIER B.V.
Sara Burgerhartstraat 25
P.O. Box 211, 1000 AE
Amsterdam, The Netherlands

ELSEVIER Inc.
525 B Street
Suite 1900, San Diego
CA 92101-4495, USA

ELSEVIER Ltd
The Boulevard
Langford Lane, Kidlington,
Oxford OX5 1GB, UK

ELSEVIER Ltd
84 Theobalds Road
London WC1X 8RR
UK

First edition 2004

Library of Congress Cataloging in Publication Data
A catalog record is available from the Library of Congress.

British Library Cataloguing in Publication Data
A catalogue record is available from the British Library.

ISBN: 0-080-43928-4
ISSN: 0950-1401 (Series)

The paper used in this publication meets the requirements of ANSI/NISO Z39.48-1992 (Permanence of Paper).
Printed in The Netherlands.

to Paola and my family,
and all who encouraged my studies

Flavio Poletto

to Barbara

Francesco Miranda

Contents

When you can measure what you are speaking about, and express it in numbers, you know something about it...and you have, in your thoughts, advanced to the stage of science.

William Thomson, Lord Kelvin, British Mathematician and Physicist, 1894.

Alike in the experimental knowledge obtained, and in the analytical methods and results, nothing that has once been discovered ever loses its value ...

A. E. H. Love (A Treatise on the Mathematical Theory of Elasticity, 1952)

Preface

The use of "opportunity sources" for remote sensing has a long history; we could start from the use of the sun to illuminate objects, and then extract an image of the surroundings. Nothing new under the sun, then. However, this corresponds to an incoherent use of the source. The observer just maps the distribution of the spectrum of the reflected energy, in a relatively very narrow band, maybe using the binocular view as a tool to gather 3D information. Electronic technology permits much more than that in that it allows for the store of a replica both of the source waveform and of the reflected waveforms. Their cross correlation yields the travel times to the objects (and more), thus allowing the extraction of a possibly three-dimensional image of the surroundings. This coherent use of the illumination source is recent. It started with radar, where the use of, say, television transmitters as illuminators allowed the detection of incoming airplanes using only passive receivers, and therefore avoided giving up the existence of the observers. This "stealth" connotation was the first and foremost motivation for the study of these techniques, and was in effect rather simple, at least from the theoretical point of view. In this case, the synthesis of a directive antenna was possible, due to the motion of the target. In other cases, it is possible to use the motion of the opportunity source to synthesize the antenna: this happens for instance using geo synchronous satellites for the illumination of targets and then recording the backscattered waves to form images.

The applications to geophysical surveys using acoustical waves had different motivations. Here the "opportunity sources" could be the propellers of a ship, illuminating the subsurface: the synthesis of the antenna comes from the motion of the ship. Another interesting source was indeed the drill bit (Seismic While Drilling, SWD). Here, the progression of the bit towards greater depths creates an array of sources that can be used to focus the beam of the illuminating waves.

This application resulted in outstanding interest, since among the several problems to be faced during the drilling of an oil-and-gas or geothermal well is that of properly predicting the kind of strata to be encountered ahead of the drill bit. This provides for the possibility of making proper decision while drilling and preventing unwanted risky situations.

Geophysicists use Surface Reflection seismology for that purpose. However, if reflection seismic provides a proper "time image" of the distribution of the acoustic interfaces in the underground, it is often not sufficiently detailed to be properly matched with the data while drilling, and the lack of real time check shot make the exercise of getting a proper correlation difficult.

To meet this challenge a really interesting "opportunity" source was indeed the drill bit commonly used for drilling oil-and-gas or geothermal wells. The advantages of such a "weak" "opportunity" power source were clear from the beginning since it was evident that it should have the possibility of high-quality images. The difficulties come from the non-availability of the source at the surface and, therefore, the need to estimate it, either with surface sensors or by direct recording. However, the potential advantages of the technique were also very clear:

- Positioning the bit on the seismic section, in time and not just in depth;
- Positioning the bit closer to the target, and, therefore, generating the hope of getting higher frequency signals.
- Obtaining the equivalent of the check shots, while drilling;
- Obtaining a better tomographic representation of the reservoir using direct arrivals from the deep besides reflections on the buried layers. This allows for a better angular coverage and, therefore, better spatial resolution.
- Deriving the travel times to the surface during the drilling of the well while using seismic frequencies, whereas the tools that could measure the velocities of the formations after the completion of the bore use sources of much higher frequencies, and after the execution of the bore. The dispersive character of the finely layered medium entails discrepancies between the synthetic seismogram obtained using sources at several kHz and the seismic data, making more difficult the combination of the data sets and the comparison of the synthetic seismogram thus obtained with the surface seismics.
- Exploiting the real time possibility. That, however, requires a complex organization to be able to process the data, digest them and then make operational decisions that have to be all in real time, to justify the entire procedure. Even decades after the first experiments, the growing processing appetites like the interest in 3D depth migration of the data, makes this goal more interesting than ever, but still partially elusive.

A quick scan for SWD on the web shows that after more than 16 years, many, if not all, these advantages have been systematically realized.

All these characteristics made and make SWD an interesting project. The first studies in Italy on that, scientifically coordinated by the Politecnico di Milano, began in AGIP and at the Osservatorio Geofisico Sperimentale in Trieste (OGS)[1] in the late eighties, leading in 1987 to a research proposal to the EU (then the CEE). Among the proposers were Sergio Persoglia (OGS) and one of us (Fabio Rocca). From the AGIP side, the partner was Gian Piero Angeleri. The acquisitions of data for the first three wells were carried out in 1987, 1988, 1989. Other participants to the studies, from Politecnico side, were Luigi Zanzi and Giancarlo Bernasconi: part of this work was also done at Stanford in cooperation with Clement Kostov, who continued working in this area, at Schlumberger. Then (mid 1988) came in the project Flavio Poletto (OGS) who began his activity processing the data.

[1]Now, Istituto Nazionale di Oceanografia e di Geofisica Sperimentale.

In 1990, Francesco Miranda (AGIP), a PhD from the UC Swansea, then graduated with a thesis on travel time tomography (one of the very first application of SWD), joined Geobit. Both Flavio Poletto and Francesco Miranda have been continuously active on this project since and much of their experience in SWD is found in this very useful book.

From this moment on, they became deeply involved with the project, named Geobit at the beginning, to assume later – with one of us (Luca Bertelli joining the Agip Management) and Luca Aleotti – the name of SEISBIT ®[2].

Thanks to the challenging ground testing opportunities offered by Agip and the support offered by Agip Management, the SEISBIT ® technology has been gradually refined in the OGS laboratories and in the field tests, and has been utilized as an industrial tool in several challenging deep wells as a useful support for steering decisions while drilling.

The material in this book is noticeable both for the quantity as well as for the quality; the character is multidisciplinary, starting from the technology and the mechanics of well drilling, the generation of seismic waves, to the data acquisition and processing.

The material is presented fluently, following a subtle but very clear line of thought. The quantity of material proves the generosity of the authors in making available results that would be otherwise dispersed in myriads of reports and papers in different journals, which would make them hardly available to the general reader. This spurs the reader to further studies of this extremely interesting material. The wealth of citations, and the relevant formulas that are always available, make the material contained in the book very rich and useful and give to the book great vitality and survivability. The material is presented in a lively fashion, and the results are always commented on so that the reader is supported in the analysis of the past work.

Fabio Rocca[3]

Luca Bertelli[4]

[2]Registered mark of ENI E&P and OGS

[3]Politecnico di Milano

[4]ENI E&P

Preface by authors:

With this book we hope to give a theoretical and practical introduction to seismic while drilling by using the drill-bit noise. This recent technology offers important products for geophysical control of drilling. It involves aspects typical of borehole seismics and of the drilling control surveying, hitherto the sole domain of mudlogging. Therefore, especially for aspects related to the drill-bit source performance and borehole acoustics, the book attempts to provide a connection between experts working in geophysics and in drilling. There are different ways of thinking related to basic knowledge, operational procedures and precision in the observation of the physical quantities. Thus, we could say that the goal of the book is to help "build a bridge" between geophysicists involved in seismic while drilling – who may need to familiarize themselves with methods and procedures of drilling and drilling-rock mechanics – and drillers involved in geosteering and drilling of "smart wells" – who may have to familiarize themselves with seismic signals, wave resolution and radiation. For instance, an argument of common interest for drilling and seismic while drilling studies is the monitoring of the drill-string and bit vibrations.

The book develops the main concepts of drill-bit seismic while drilling, presents an overall review of the status of the art, and shows practical applications. A brief description of the main concepts discussed in this book follows.

Chapter 1 (Introduction and overview) – introduces seismic-while-drilling with an overview of the conventional borehole seismic methods. This chapter gives a short description of the history of seismic while drilling. It summarizes the results of this method, with a review of the existing approaches and perspectives.

Chapter 2 (Principles of drilling) – is the first part of the general theory. It describes the drilling principles, the wellsite environment, the main drilling components, the well planning and the drilling modes. The drilling equations relevant to the seismic-while-drilling theory are summarized.

Chapter 3 (General theory: drill-bit seismic waves) – is the second part of the general theory. It describes the drilling energy in terms of drilling parameters and the drill-bit signal production. Here, we develop the analysis of the energy balance in the drilling process and calculate the vibrations produced by drilling bits and the radiated wavefields. The analysis is done for roller-cone and polycrystalline diamond compact (PDC) bits and includes anomalous drilling conditions. Performances of drill-bit and conventional seismic sources are compared.

Chapter 4 (General theory: drill-string waves and noise fields) – analyzes the signal and noise waves in drill strings. The wave-propagation equations are given for homogeneous and non-homogeneous strings, including passband and stopband effects for extensional and torsional waves – as well as for guided waves in the borehole. The analysis includes attenuation, sources of noise in the drill string, at the wellsite, and the radiated noise fields.

Chapter 5 (Acquisition of SWD data) – describes the acquisition of drill-bit while-drilling data. Sensor technology and specifications, acquisition systems, surface acquisition lines, acquisition procedures by drilling-parameter control are described here. The discussion includes the use of acceleration and strain dual sensors in the drill string. Use of seismic arrays and sampling in depth of seismic-while-drilling data are a part of the arguments of this chapter.

Chapter 6 (Preprocessing of SWD data) – describes preprocessing of the seismic-while-drilling data, i.e., the production of interpretable seismograms. Preprocessing includes correlation, stacking, time correction, phase correction and deconvolution. Modeling of drill-string waves is used to interpret the delays of the pilot signals in the correlations.

Chapter 7 (Processing of signal and noise RVSP fields) – describes the processing of signal and noise wavefields. Repeatability of the source and selective stacking are analyzed by drilling parameters. Noise removal is discussed using orthogonalization analysis or independent pilot-signal separation. Main steps of wavefield processing of reverse VSP data, data quality control and cross-correlogram migration are further subjects of this chapter.

Chapter 8 (Applications) – describes the seismic-while-drilling products, with geophysical applications and perspectives. Results of while-drilling monitoring of the well while drilling with case histories obtained in different lithologies, in variable drilling conditions, and in complex areas of poor seismic response are covered in this part. The analysis includes crosshole geometry and 3D onshore reverse VSP data. Use of seismic-while-drilling data for pressure assessment and new trends are discussed.

The book is a theoretical support intended for use by researchers, graduate and post-graduate students in geophysics and drilling engineering. It includes examples of practical applications for use by people working in the field of borehole geophysics and drilling. Moreover, the book can be of help as a reference text for drill-bit data acquisition as well as for while-drilling data processing and interpretation.

Acknowledgments

We are grateful to Klaus Helbig for reading through the early draft of the book and offering valuable suggestions and corrections. A valuable contribution to the seismic-while-drilling technology described in the book was given by Fabio Rocca. He proposed the initial drill-bit seismic research project in Italy. Thanks also to Luca Bertelli who contributed to the growth of seismic while drilling by pushing the technology in ENI E&P. José M. Carcione has given important suggestions and discussions for analysis of borehole waves and overpressure.

We wish also to express our thanks to friends and colleagues who contributed to the development of the SWD technology at OGS and ENI E&P and helped us in the preparation of this book. In particular, Massimo Malusa, Piero Corubolo, Lorenzo Petronio, Giuliano Dordolo, Andrea Schleifer, Cinzia Bellezza, Roberto Miandro, Enrico Visigalli and Aronne Craglietto. Giorgia Pinna and Pierzaverio Marchetti helped us in the discussion of the drilling technology. Finally, we thank Anthony Gangi for his valuable contribution in reading the book, making comments and giving important suggestions to improve its clarity.

Thanks are given to ENI E&P Division and to the Istituto Nazionale di Oceanografia e di Geofisica Sperimentale (OGS) for the support of the seismic-while-drilling research and development projects and for the permission to publish the results contained in this work. This technology was partially funded by the European Community.

ABOUT THE AUTHORS

Flavio Poletto was born in Italy in 1952. He received the degree "Dottore in Fisica" from the University of Trieste in 1988. He works at the Istituto Nazionale di Oceanografia e di Geofisica Sperimentale – OGS (former Osservatorio Geofisico Sperimentale), Italy. Here, he has been working in seismic processing from 1984 to 1987. Since 1988, he has been working as a geophysicist on seismic-while-drilling research by using the drilling noise to obtain reverse VSP's.

He is the author of many papers on SWD, drill-string acoustics, and the co-author of patents regarding the drill-bit-noise separation for RVSP by using statistical independence and drilling parameters.

In 2001 and 2003, he received the Honorable Mention for two papers selected in the category of the Best Paper in GEOPHYSICS. Currently, he works as senior research geophysicist and is coordinator of the Seismic-while-drilling Project at OGS. His research activity includes borehole geophysics, acoustic and seismic while drilling, with applications extended to drilling diagnostics and geosteering.

Francesco Miranda was born in Belgium in 1962. He received a BSc degree in Geology in 1984 and a PhD degree in Geophysics 1989 from UC Swansea (UK) with a thesis on Travel-time and Amplitude Tomography. He worked at DMT (Bochum, Germany) where he was active in research in seismic tomography, borehole seismic data acquisition and processing, processing of in seam seismic for coal fields, and special processing and acquisition techniques for surface 2 D seismic.

Since 1990 he works in ENI E&P Division in Milano in the borehole seismic department. He has been following acquisition, processing, interpretation and special studies on borehole seismic. He has been actively involved in research in the field of Seismic While Drilling in which he is the author of many papers and of one patent and in the field of cross-hole seismic. Geosteering is one of the fields in which he has been involved in the last years.

Glossary of main symbols

Y	Young's modulus
μ	shear modulus
$\mathcal{Z}$	impedance
ρ	mass density
V, c	wave velocity
k	wavenumber, integer index
Q	quality factor
c_i, r_i, R	reflection coefficients
p	pressure
W	power, wave Fourier transform
E	energy
I	moment of inertia
A	cross section
u	axial displacement
θ	angular displacement
s, ε	strain
a	acceleration
z	axial coordinate
R	radius of curvature
F	force
g	acceleration of gravity
T	tension, period
σ	stress, rock strength
SS	synthetic seismogram
AI	acoustic impedance (log)

WOB	weight on bit
ROP	rate of penetration.
RPM	revolution per minute
FLOW	mud flow
TOB	torque on bit
E_S	specific energy
KB	kelly bush
TD	total drilled depth
TVD	true vertical depth
BHA	bottom hole assembly
DC	drill collars
HWDP	heavy weight drill pipes
DP	drill pipe
ID	pipe inner diameter
OD, D	pipe outer diameter
r_I	pipe inner radius
r_O	pipe outer radius
SPM	strokes per minute
DH	downhole
SUB	special DH tool
WD	while drilling
FE	formation evaluation
LWD	logging while drilling
MWD	measurement while drilling
GR	gamma ray (log)

Conversion factors for the main physical quantities

Quantity	Unit to be multiplied	by	to obtain
Distance	meter [m]	3.281	foot [ft]
Distance	centimeter [cm]	0.3937	inch [in]
Velocity	meter/second [m/s]	1.944	knot
Area	cm^2	0.155	$inch^2$
Volume	liter [l]	33.81	ounce [oz]
Volume	cubic centimeter [cm^3]	0.06102	cubic inch
Mass	kilogram [kg]	2.205	pound [lb]
Mass	pound [lb]	16	ounce [oz]
Density	kg/m^3	0.06243	$pound/foot^3$
Density	kg/m^3	0.008345	pound/gallon [ppg]
Force	newton [N]	0.2248	pund-force [lbf]
Force	kilonewton [kN]	0.102	ton force [tonf]
Torque	newton meter [N m]	0.7376	pound-force foot [lbf ft]
Torque	kilonewton meter [kN m]	102	kilogram-force meter [kgf m]
Pressure	megapascal [MPa]	10	bar
Pressure	kilopascal [kPa]	0.145	$pund/inch^2$ [psi]
Power	kilowatt [kW]	1.341	horsepower [hp]
Power	watt [W]	0.239	calorie/second [cal/s]
Energy	joule [J]	8.851	pound-force inch
Energy	kilojoule [kJ]	0.239	kilocalorie [Cal]

Chapter 1

Introduction and overview

1.1 Geophysics for exploration and drilling

Oil and gas exploration consists of three main phases: geophysical exploration with the geological interpretation of processed data, drilling, and production. This process develops as follows:

◇ Geophysical exploration and geological interpretation identify – by using seismic, gravimetric, and magnetic measurements – a geological region in which to drill oil wells and, after a reservoir as been found, in which to develop the reservoir.

◇ Explorative wells are drilled firsts, followed by appraisal, development, and production wells which are, in many cases, grouped in clusters of wells drilled from the same platform.

◇ After drilling, the goal is to achieve optimal production. This goal requires knowledge of the reservoir conditions, in particular the oil distribution and the position of the oil-water contact.

Geophysical and acoustic measurements play a key role in the first phase. Surface reflection seismics is one of the geophysical tools most widely used. The drilling location of a well is usually determined using surface seismic sections; the time seismic sections – both 2D and 3D – give the interpreter the travel time information of the structures and reflectors encountered during drilling.

The role of geophysics during the second phase of oil and gas exploration – the drilling phase – is also important. This is the central argument of this book. In the process of geophysical monitoring of drilling, it is necessary to correlate the data of the surface seismic lines with the well-log data measured in depth, either during or after drilling. This correlation can be done by conventional wireline vertical seismic profiles (VSP), where the receivers, deployed downhole to record the signal emitted by a source placed at the surface near the well, provide the necessary information to convert borehole data from depth to seismic-time domain. However, due to limitations of this conventional approach, seismic-while-drilling measurements were recently introduced. These techniques are described in this book.

The production phase is also supported by geophysical and acoustic measurements. Permanent sensors are used in instrumented wells to detect induced microseismicity (microcracks) and to determine the production conditions. Permanent acoustic sensors can be located in the well to monitor the oil flow between different branches. In addition, seismic surveys repeated at intervals during production (4D) are used to monitor the evolution of a reservoir. This information allows drilling engineers to reduce risk during production and optimize the drilling of new wells.

Geophysical monitoring of drilling

Drilling a well to reach a prefixed target is a complex process that involves high cost and risk. These can be reduced if while-drilling measurements are employed. Wells are normally drilled on the basis of seismic information and of information coming from other wells in the same area. Correlation among the well data is done using seismic lines; however, uncertainties can be present in the inter-well area and can lead to the mispositioning of new well locations or to wrongly locating the well's target. The geological model for the to-be-drilled well is given in a vertical depth scale. This is usually calculated by the time-to-depth conversion of the interpreted seismic time sections or by the interpretation done on the depth-migrated sections. A correct definition of the velocity model used for the time-to-depth conversion is one of the most critical issues. Therefore, especially in geologically complex areas, it is important to "follow seismically" the drilling of the well, and to monitor continuously its evolution from a geological and geophysical point of view. If the initial geological model is incorrect, the prognosis of the new well will be incorrect too, and actions must be taken, while drilling, to rectify it when possible.

To monitor a well from a geological and geophysical point of view, explorationists obtain information about a selected property of the rock, using several methods. This information is then used to make drilling decisions. The geological and geophysical data existing prior to drilling give information about the local geology of the area. The mudlogging service – which is typically present on the well site to measure the drilling parameters for quality control of the drilling activity (Chapter 2) – gives lithological information about the drilled rocks by analyzing the cuttings that come out of the well. High-resolution borehole measurements (logs) of the acoustic, density and electromagnetic properties of the rock are measured in the well during interruptions of drilling, or by using logging-while-drilling (LWD) and measurement-while-drilling (MWD) technologies based on mud-pulse or cabled telemetry to send data to the surface in real time (Chapter 2). Well logs give the rock properties only in the immediate vicinity of the well, with a distance of the order of a few meters.

Finally, the borehole seismic data identify the reflections caused by the geological structures, and give information about the acoustic and elastic properties of the drilled and to-be-drilled rocks, with an investigation distance of the order of several hundred meters. This is the vertical seismic profiling (VSP). The technology now exists to provide VSP seismic information while drilling and steer the drilling of a well from the geophysical point of view. This technology uses the drill-bit noise as the seismic source.

1.2 Conventional borehole seismic methods

Before introducing the reverse-VSP seismic-while-drilling concept, we give a short overview of the conventional vertical seismic profile (VSP) and discuss the advantages of borehole seismic methods. The conventional VSP uses geophones lowered into a well by a wireline unit and seismic sources at the surface (Figure 1.1). This well-known acquisition technology is described by several authors (for example: Hardage, 2000; Balch and Lee, 1984; Gal'perin, 1974).

The main products of the processing of seismic well data are the seismic velocities, the reflection sections (interesting from the point of view of interpretation) and the synthetic seismograms obtained with the joint use of well logs.

1.2.1 Vertical Seismic Profile

Surface reflection seismics is one of the geophysical tools most widely used in oil exploration and oil field development. Surface seismic data are used to map the subsurface geology, determine structures and stratigraphic sequences, and to obtain velocity information. However, the information in the surface seismic data is affected by absorption, attenuation and diffraction effects. The corresponding degradation can be partially eliminated by stacking the data, i.e., by using the statistical properties of the signal and noise.

In the VSP geometry, we have the advantage of measuring the seismic signals reflected both from reflectors located above and below the observation point in the borehole. This geometry makes it possible to obtain improved results with respect to surface seismics. The reflections coming from above are contained in the *downgoing* wavefield (Hardage, 2000). The downgoing wavefield is, in turn, reflected upward from reflectors located below the observation point in the well. Hence, the same downgoing wavefield is contained also in the *upgoing* wavefield (Hardage, 2000). Thus, the use of the downgoing wavefields allows the deterministic elimination of the downgoing filter from the upgoing waves since the propagation of the two wavefields is, at least in part, essentially the same (if acquisition is performed with the source at small offset). After downgoing deconvolution, the VSP reflections are greatly improved. VSP surveys allow the identification of multiple events and phase problems of surface seismics, and can be used to estimate operators for deconvolving surface seismic data.

The VSP is used to obtain the location in space of the reflectors and diffractors that cause the reflections and diffractions observed on the surface seismic sections. These data can be inverted to predict the acoustic impedance contrasts beneath the well bottom. All this information, together with synthetic seismograms, makes it possible to better correlate the structural and stratigraphic characteristics of the well profile with the surface seismics. The result is a considerable improvement in the interpretation of the seismic and well data. The study of the arrival times of the S-waves separated into two polarizations makes it possible to determine possible anisotropy and the trend of fractures (MacBeth, 2002). Moreover, the VSP offers greater resolution – due to the reduction of the length of the ray path – than surface seismics. Indeed, some formations in some wells can only be observed with the resolution of well seismics.

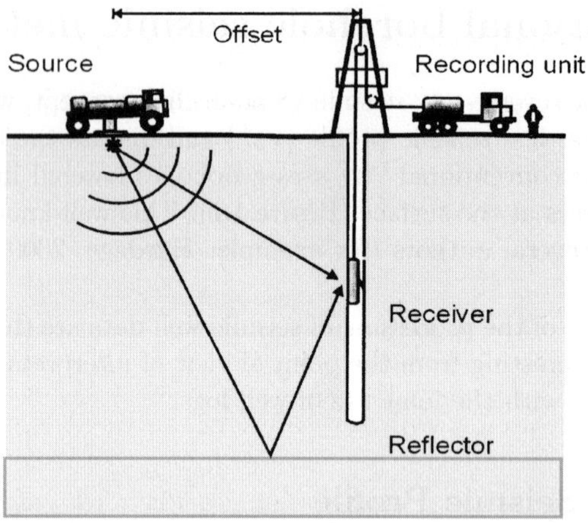

Figure 1.1: Conventional wireline VSP uses the receivers in the well and surface source. The distance of the source from the well is the offset (which is equal to the source-receiver offset if the well is vertical).

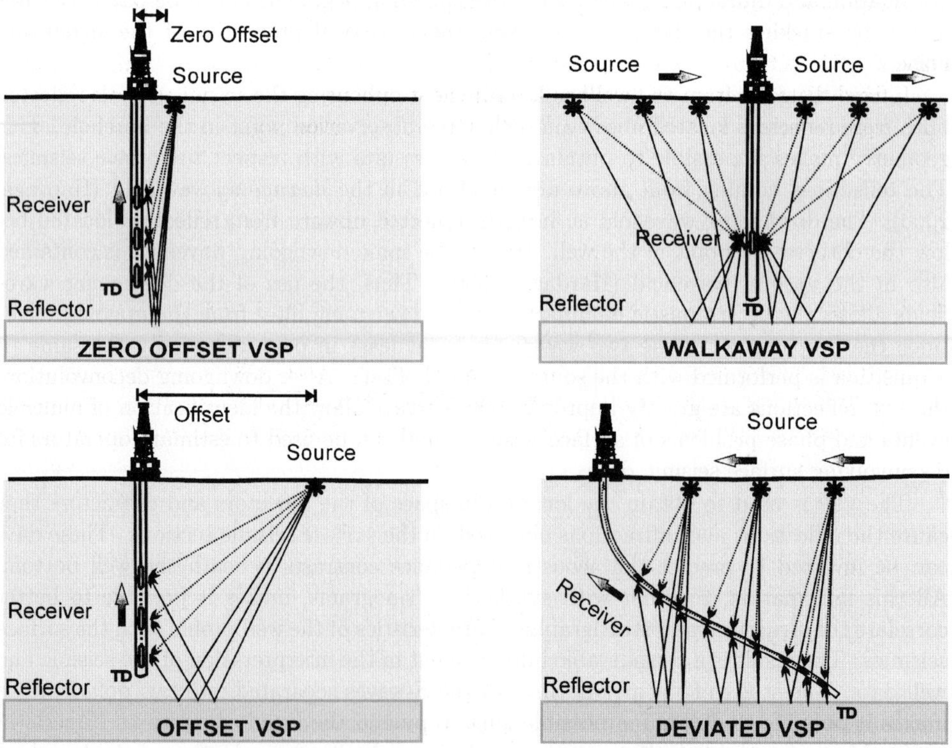

Figure 1.2: Schematic representation of main types of conventional onshore VSP acquisition geometries. A similar geometry holds for offshore applications.

1.2.2 Typical acquisition geometries of conventional VSP

Most of the data and geophysical information acquired from conventional VSP methods discussed below are obtained only after the drilling process has been completed, rather than during the drilling process, i.e., while drilling.

VSPs are characterized according to the acquisition geometry and the well in which they are acquired. Figure 1.2 schematically summarizes the following different types of conventional VSPs.

Standard VSP (zero offset) – Standard VSP refers to zero-offset acquisition of VSP in vertical wells. A seismic source is placed on the surface near the wellhead, and a geophone (or a group of geophones) is lowered to the bottom. Then, the geophone is raised to record data with regular depth acquisition intervals. Standard VSP provides the time-to-depth conversion and the identification of the primary reflected events in two-way time (TWT) on the seismic sections, with a lateral resolution determined by the Fresnel zone of the observed reflections (Hardage, 2000). The standard VSP data make it possible to estimate the acoustic impedance contrasts below the total depth (TD) and provide information regarding the attenuation of the signal as a function of the encountered lithology. Figure 1.3 shows a zero-offset VSP after processing, compared to well logs.

Offset VSP (OVSP) – The geometry of the OVSP involves placing the source at a pre-fixed distance from the wellhead, with the geophone anchored at regularly decreasing levels in a normally vertical well. Recording waves, which are propagated with oblique angles against the vertical, introduce normal moveout (NMO) effects, which are corrected during the processing phase. The final result of an OVSP is compared to the surface seismic section and allows geophysicists to extend laterally from the well the structural survey with VSP data (Figure 1.4).

Walkaway SP (WSP) – The WSP is recorded by running the geophone into the well in a fixed position and recording the seismic events produced by a source which moves on the surface along a straight line intersecting the wellhead. Unlike the OVSP, the WSP illuminates the zone beneath the sensor in more detail. Also in this case, after processing, the final result can be superimposed to the conventional seismic section.

VSP for a deviated well (DVSP) – A deviated well is often drilled to accommodate particular operating requirements. In such a case, the acquisition geometry of the DVSP involves moving the energy source to different points along the well's trajectory in the surface plane so that the source lies vertically (or in points close to the vertical position) above the geophone. A seismic pseudo-section with greater resolution than surface seismic is obtained after data processing is performed. This result allows local structural reconstruction.

The OVSP and WSP are used for the processing of structural sections, the analysis of dips and fractures and the determination of the borders of a reservoir. Due to the limitations of using tools with few geophones in the well, it is expensive – and implies higher risk – to acquire the conventional VSP with both many receiver depth levels and many shot offset positions. Thus, the conventional offset VSPs are typically acquired with few offset traces and many depth levels, while the walkaway VSPs are typically acquired with many surface channels and few depth levels in the well.

3D VSP – In recent years the new 3D VSP technology has become an area of great interest. The strategic importance of 3D seismic is now fully confirmed, and, since 1987, well-seismic applications have been studied with 3D VSP configurations. A standard 3D VSP consists of many walkaway source lines recorded while receivers are set downhole. The 3D VSP has a higher resolution than surface 3D surveys and makes it possible to obtain images beneath surface obstacles, such as platforms, and beneath near-surface obstructions, such as shallow gas zones.

Azimuth VSP – This is a particular 3D VSP with a circular line of surface acquisition points centered on the well. Azimuth VSPs are used to study anisotropy and for S-waves splitting analysis.

The main aspects concerning the use of the conventional 3D-VSP and while-drilling 3D-VSP are discussed in Chapters 5 and 8. Actually, 3-component tools with 80 sondes per array are available by industry and allow acquisition of multicomponent–multilevel reservoir data and cost-effective 3D VSP also onshore (Input/Output INC, 2001; PGS, Technical presentation, 2002).

1.2.3 Conventional processing of VSP data

After some preprocessing, including editing, first-arrival picking, stacking, filtering and recovery of the geometrical spreading, the conventional processing of the VSP data takes place in the following phases:

Separation of downgoing and upgoing wavefields – In this phase, the upgoing wavefield, which includes primary reflections and multiples, is separated from the downgoing wavefield, which includes the direct arrivals and downgoing reflections. This basic VSP concept (Hardage, 2000) is well known to geophysicists. In an ideal zero-offset VSP, upgoing and downgoing wavefields have opposite slopes in the depth-time domain. Picking of the direct arrivals is critical in order to obtain effective separation and to correctly position the primary reflections in two-way time. The separation process can take place by calculating the total field spectrum in the frequency-wavenumber (fk) domain, filtering the downgoing field with an adequate passband filter, inversely transforming and then subtracting the downgoing field from the total – assuming that the total field equals the downgoing field plus the upgoing field. Another approach is to use median filters: by aligning the events of the downgoing field to a constant-time datum, the median filter enhances them and attenuates instead the events of the upgoing field, which have a doubled slope. The downgoing field obtained in this way is then subtracted from the total field to obtain the upgoing field. However, in reality, the downgoing field is much more energetic and the residual errors after separation may mask the upgoing reflections. Thus, a basic quality requirement for VSP acquisition is to have good source repeatability in

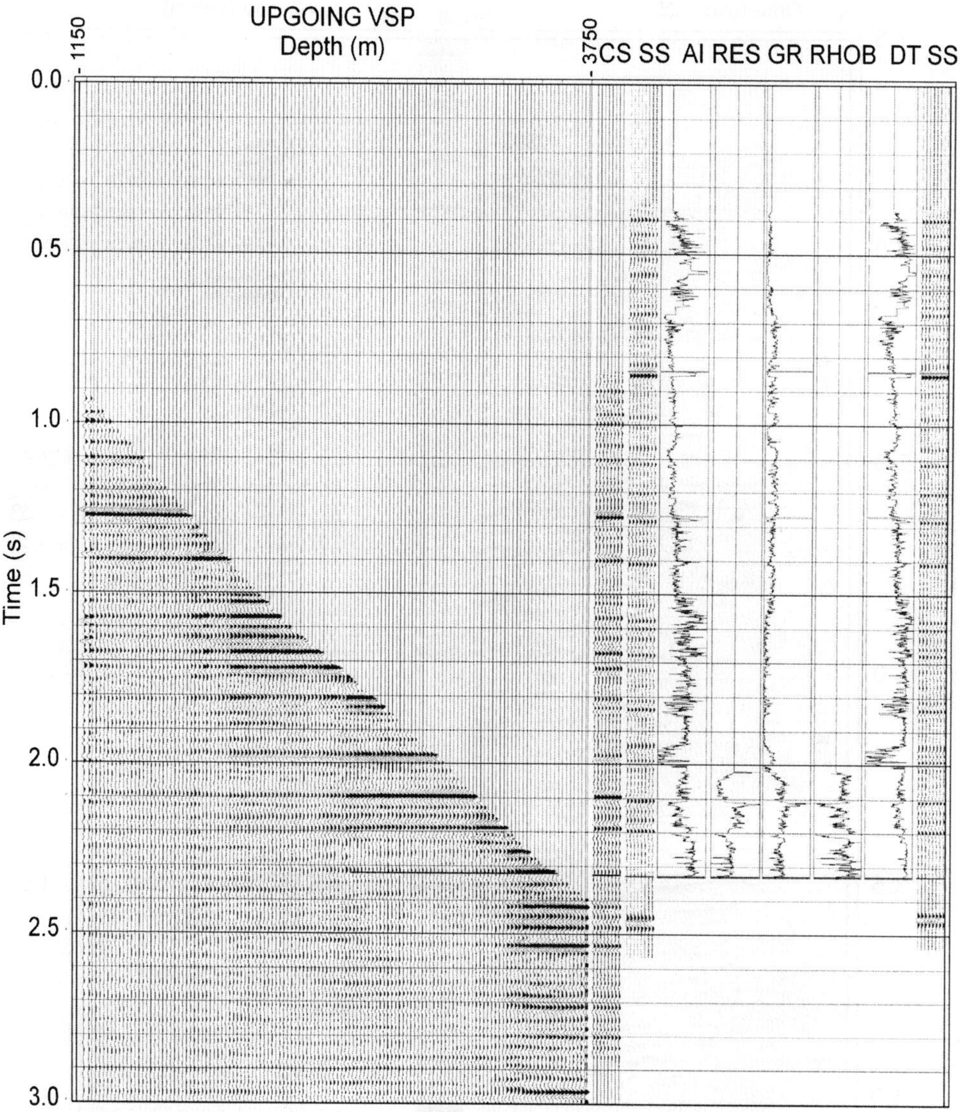

Figure 1.3: Example of final composite display for a standard conventional zero-offset VSP. From right to left the synthetic seismogram (SS), the gamma ray (GR) and acoustic impedance (AI) logs, corridor stack (CS) and the two-way-time (TWT) upgoing wavefield are displayed. The good correlation of the main reflections in the VSP and the logs can be recognized. Depth is referred to kelly-bush level (KB) (Chapter 2).

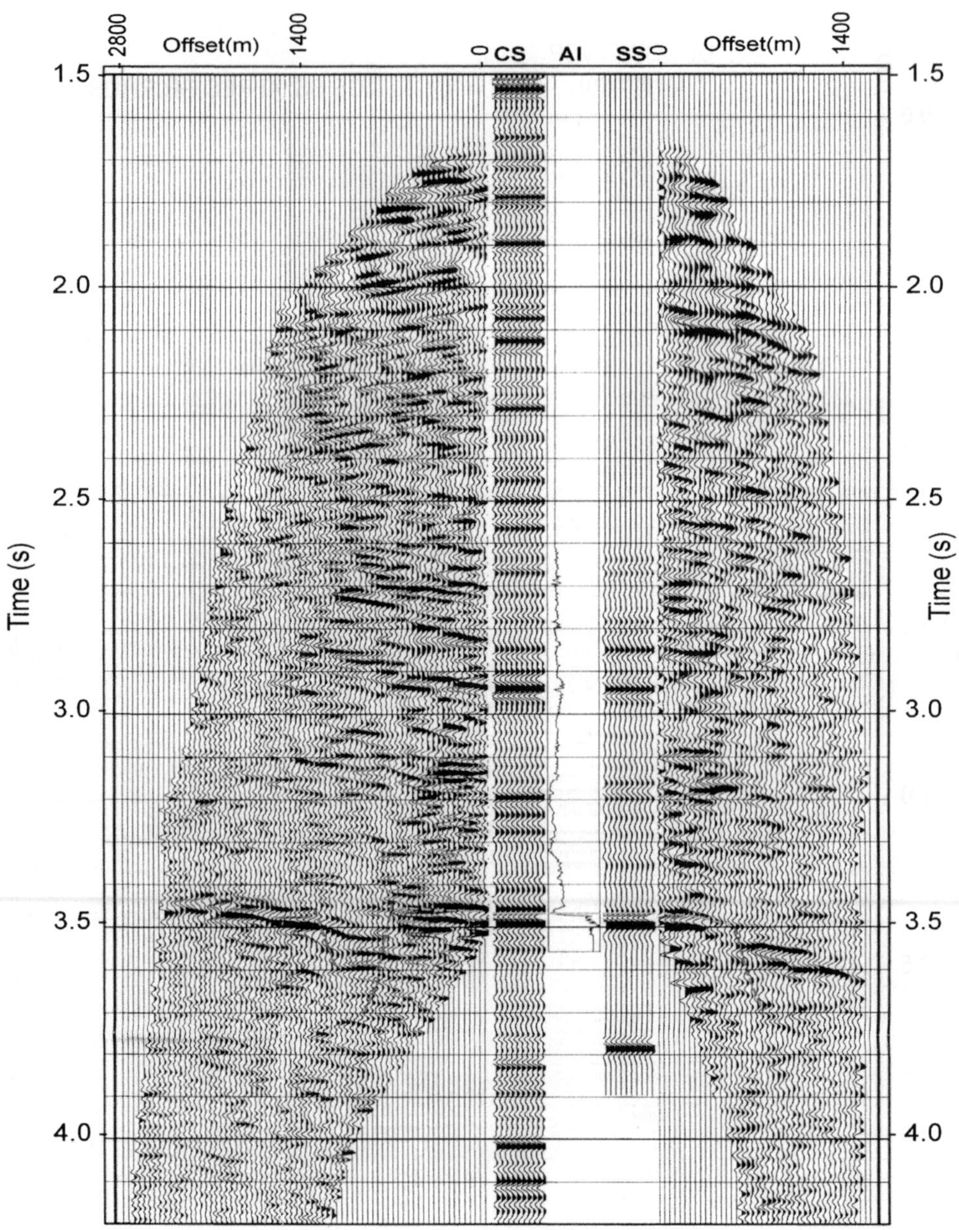

Figure 1.4: Example of final composite display for a conventional offset VSP. The offset VSP is tied to the corridor stack (CS), acoustic impedance log (AI) and synthetic seismogram (SS). The image shows the reflection structure laterally from the well position.

order to ensure the stability of the downgoing waveform. This improves separation of downgoing and upgoing fields and the subsequent waveshaping deconvolution and analysis of reflections. This means that it is desirable to have the same source waveform at the different depths in the well. In general, this condition is satisfied for conventional VSP (Hardage, 2000).

Predictive deconvolution – The process of predictive deconvolution consists of the trace by trace construction of an operator to deconvolve the upgoing field. The operator can be either deterministic or statistic. A deterministic operator is calculated in the downgoing field, which is more energetic and is considered as the result of an input to the geological structure from which the upgoing field is also derived. This operator is capable of annulling the events following the first arrival in the downgoing field. The importance of the process lies in the suppression of the multiple events and the noise, which make it difficult to identify the primary reflections in the upgoing field.

Wave shaping deconvolution – A deterministic operator that makes it possible to pass from the available mixed-phase signal to a zero- or minimum-phase signal is calculated trace by trace on the downgoing field and applied to the upgoing field. This makes it possible to eliminate the distortions due to propagation in a dispersive medium as well as those due to almost random-phase variations in the signal emitted by the source. Waveshape deconvolution allows comparison of VSP data with seismic data of homogeneous phase obtained after wavelet processing and also improves interpretation.

Corridor stack – This processing step is carried out in the case of a standard VSP and involves the sum at constant time, for selected geophone depth levels, of the events present inside a time window after the direct arrival. The trace obtained is graphically repeated several times, giving rise to a strip, a sort of local seismic section, which represents the joining element between VSP – representative of the encountered stratigraphic sequence – and the surface seismic data.

VSP-CDP stack transformation – In this processing step, the NMO effects, due to the oblique path of the seismic rays, are corrected in the data of the OVSP. Furthermore, data are moved from a time/depth representation to a time/offset representation, in order to make the print-outs comparable with conventional time and depth seismic sections. The VSP-CDP stack transformation – once the acquisition geometry and preliminary velocity model are known – carries out the time correction and the positioning of the reflected events at the correct offset (Dillon and Thomson, 1984). This method has been developed for structures with plane and parallel layers, at least in the immediate vicinity of the well.

VSP migration – The migration technique used for VSPs allows the repositioning of events and focusing of energy to their appropriate positions in space (imaging). Possible raypaths and travel times are computed by tracing rays through a velocity model of the subsurface. The 3D velocity model used for migration may be defined using structural complexity, anisotropy and velocity heterogeneity. Then all source-receiver pairs in the seismic-data volume are checked for energy that could satisfy such path and time constraints. The VSP migration algorithm is suited to handle the various source and downhole receiver geometries. The migration applied to 3D VSPs is a full 3D migration, i.e., it considers the positions of the reflectors in the 3D imaged volume, rather than in a series of 2D slices. The migration algorithm is based on the seismic wave equation or on Kirchhoff algorithms, and applied prestack, to preserve true reflection amplitudes. For data sampled not regularly in space – as is often the case for drill-bit data – Kirchhoff migration algorithms are preferred (Yilmaz, 2001).

1.2.4 Other uses of conventional VSP

As shown in the previous subsection, the main uses of conventional VSP data are stratigraphic and structural interpretation. Some other uses are listed below.

Prediction below the bit – This technique is very important when the well being drilled is located in an area where the surface seismics cannot be easily interpreted. The upgoing wavefield can be used to determine the seismic sequence below the bit, which is interpreted in terms of a geological sequence. Figure 1.5 shows an example of prediction of the top of a formation performed by using a VSP acquired during the drilling of a well before the drilling had reached the target. This technique has a high cost – due to well standby for acquisition in borehole – and, in some cases, risk – due to the uncontrolled condition of being without drill pipes in the well and to the presence of the tool in the borehole. These reasons may limit the use of repeated conventional VSP for prediction ahead of the bit.

Deconvolution of surface seismic data – Analysis of upgoing wavefields as well as of downgoing wavefields before and after deconvolution can be used to identify the multiple events. The VSP multiples may be used to identify residual multiple events, i.e., those present in the surface seismic data because they were not properly removed by the deconvolution. In particular, this is true for events with high energy and with time delay too long to be effectively filtered by predictive deconvolution.

The use of deconvolution operators derived from the VSP and applied to the surface seismic data has proved to be very useful in improving the primary reflections, even in zones at great distances (in the order of 1000 m) from the well in which the VSP was recorded.

Identification of reflections above the bit – Thanks to the symmetry between the downgoing and the upgoing wavefields, it is possible to identify the location of reflectors above the bit. This may be useful for steering the "navigation" in a layer as in the case of deviated wells.

Absorption – The energy of a seismic wave is attenuated and partially transformed into heat as it propagates through a real medium. The downgoing wavefields can be used to identify zones of greater or lesser attenuation (for a discussion about attenuation and dissipation see also Carcione, 2001).

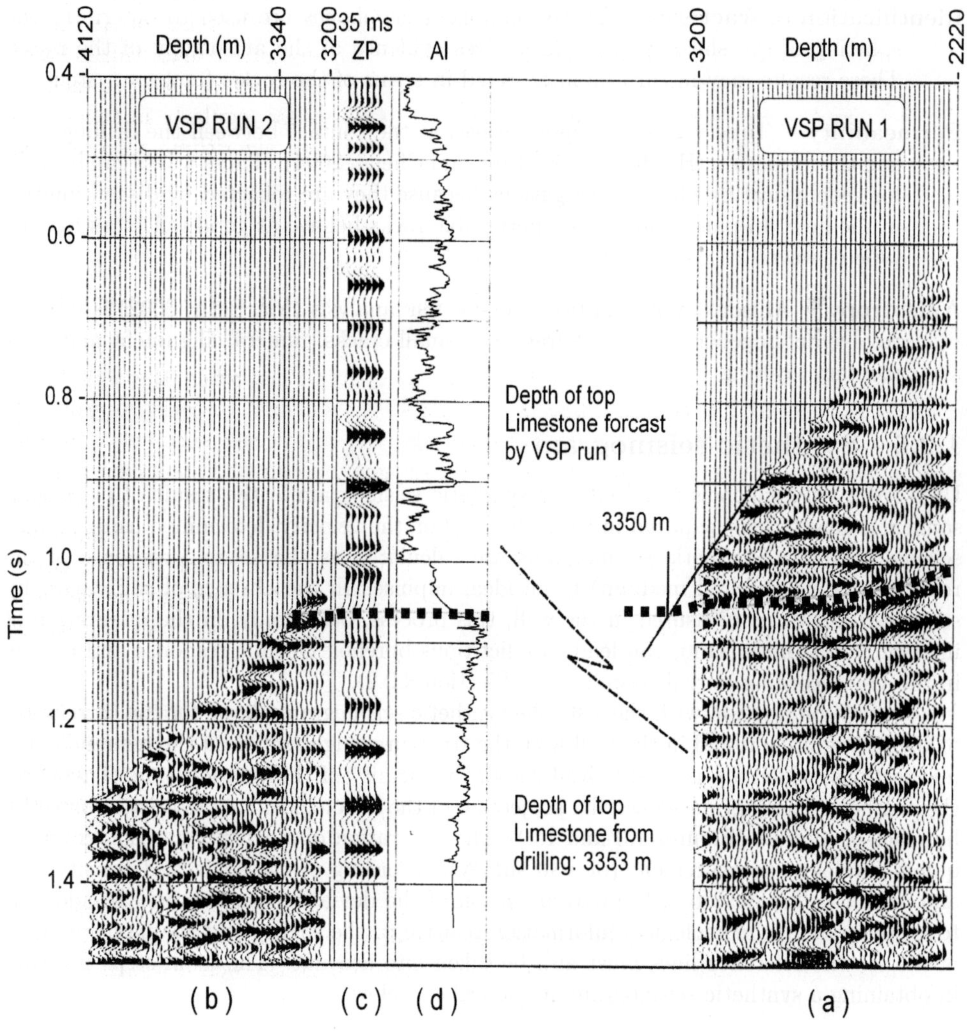

Figure 1.5: Example of prediction ahead with a VSP: a) A first run VSP was acquired up to 3200 m, before the top of a limestone formation was reached. The predicted depth of the top of carbonate was 3350 m, and was confirmed in the subsequent drilling phase. b) This event is observable in the second VSP run and c) synthetic seismogram, calculated by d) acoustic impedance. Depths are referred to kelly-bush level (KB).

Wave polarization – Using three-component geophones in the well makes it possible to study the effect of wave polarization by determining the particle motion (hodograms).

Identification of fractures – Three-component geophones are used to determine the splitting of the shear waves, which gives evidence of the anisotropy of the rocks. These results may in turn be interpreted in terms of the rock's fracture system.

Permeability – Tube waves are guided waves at the contact between the drilling mud and the formation (Hardage, 2000; Tang and Cheng, 2004). These waves can be used for geophysical borehole investigations because they are reflected by discontinuities in the hole. In particular, highly permeable rocks behave as good reflectors for this type of waves.

Anisotropy – Anisotropy by multicomponent VSP analysis (MacBeth, 2002) is used to resolve heterogeneities such as fractures, small cracks, mineral alignment and sedimentary layering.

1.2.5 Synthetic seismogram

Unlike the VSP, which is acquired, the synthetic seismogram is the result of processing sonic and density logs acquired in the well based on the velocity determined using surface seismic surveys. Synthetic seismograms are calculated as the seismic response of the reflector system (layered medium) to an ideal impulse. In the absence of calibration by seismic travel times measured in the well, this process, based only on the integration of interval travel times (ITT), can lead to a fictitious borehole seismic response and not, as in the VSP recording, a real seismic trace (Section 8.2.2).

Used in conjunction with the VSP, the synthetic seismogram allows for the correlation of the stratigraphy found in the well with the structures highlighted on the reference seismic sections, which are typically available in the vicinity of the well location. Therefore, the synthetic seismogram allows the projection of the lithological characteristics onto the horizons of interest (Figures 1.3 and 1.5). The generation of a synthetic seismogram requires a velocity measurement (or VSP survey), recording of the sonic and density logs and, if the well is deviated, the determination of the deviation parameters. In addition to all these data, the geological information relative to the survey and the interpretation of the seismic reference lines must also be taken into account. The main steps involved in obtaining a synthetic seismogram are described below.

1. **Sonic log quality control** – First, a quality control is carried out on the first arrival times derived from the available VSP velocity measurements; the time-depth values (checkshot) obtained from the best data are then selected. Then, any inconsistencies due to tool failures or to poor hole conditions are eliminated from the logs. At this point, the logs and times of the checkshots are converted to the vertical path if the well is deviated.

2. **Calibration of the sonic log** – This step allows the production of a synthetic seismogram that is in better agreement with the surface seismic surveys. The sonic

log itself is corrected on the basis of the calculation of the "drift curve", i.e., the displacements as a function of the depth between the integrated sonic times and the checkshot times.

Let $\Delta t(z)$ be the interval-transit-time (ITT) measured in the velocity-log interval Δz. We obtain the slowness

$$V^{-1}(z) = \frac{\Delta t(z)}{\Delta z}, \tag{1.1}$$

which is reciprocal of the interval seismic-log velocity. The integral transit time between two depths z_1 and z_2 is calculated as

$$t(z_1, z_2) = t(z_2) - t(z_1) = \int_{z_1}^{z_2} V^{-1}(z) \mathrm{d}z. \tag{1.2}$$

In general, the acoustic velocities determined from the sonic log differ significantly from the seismic velocities. This effect is due to dispersion (the sonic log acquires data with frequencies of 10–20 kHz while the seismic data have a typical bandwidth of 50–100 Hz), and is due to the fact that the sound measured by the log is forced to travel along the borehole wall, which is a path quite different from that of the signal generated during a seismic reflection survey. Moreover, due to errors in the interval transit time measurements, the integrated log time of equation (1.2) can have significant errors in large depth intervals ("integration drift").

The integrated transit times derived from the sonic log are corrected to match the checkshot transit times derived from the seismic velocity survey. Calibration, that is, the correction of the integrated time, is obtained using $t(z_1)$ and $t(z_2)$ measured by seismic methods. Normally, the correction is smaller than 10%.

3. **Depth-to-time conversion** – In general, the depth-to-time correspondence is calculated by adjusting the integrated-calibrated sonic log with a reference time obtained by selecting the first useful checkshot. The logs are then resampled in time.

4. **Calculation of a reflectivity trace** – Acoustic impedance is calculated from the velocities derived from the sonic log and from the densities of the density log, or given by the empirical relationship (Gardner, Gardner and Gregory, 1974)

$$\frac{\rho}{\rho_0} = \left(\frac{V}{V_0}\right)^{\alpha}, \tag{1.3}$$

where ρ is the density, V is the P-wave velocity, and $\alpha \approx 0.2$–0.3. A trace of events with reflection coefficients r_i (reflectivity trace) is obtained from the acoustic impedance $\mathcal{Z} = \rho V$ as

$$r_i = \frac{\mathcal{Z}_i - \mathcal{Z}_{i+1}}{\mathcal{Z}_i + \mathcal{Z}_{i+1}} \approx -\frac{\Delta \mathcal{Z}_i}{2\mathcal{Z}_i} \approx -\frac{1}{2} d\left(\log \frac{\mathcal{Z}}{\mathcal{Z}_0}\right) = -\frac{1}{2}(1+\alpha) d\left(\log \frac{V}{V_0}\right). \tag{1.4}$$

5. **Synthetic seismogram processing** – The reflectivity trace is the δ-pulse response of a system of horizontal layers, with acoustic discontinuities in the vertical direction (1D). To take into account the changes incurred by the fictitious signal during its propagation within a layered medium, a Ricker wavelet is convolved with the reflection-coefficient trace. The resulting seismogram has frequency and phase characteristics that make it comparable to a wide range of surface seismic data, thus facilitating interpretation.

1.2.6 Integrated interpretation of well seismic data

As mentioned in the previous paragraphs, the well seismic data are very helpful in interpreting surface seismics. Figure 1.6 shows, from left to right: four logs converted to two-way time, the synthetic seismogram, the corridor stack and the upgoing field of the conventional VSP. Sonic and gamma ray logs "in depth" are shown at top right, using the depth scale of the VSP. The image thus obtained may be considered as a sort of "seismic identity card" of the well. This generally provides the interpretative key to the reflectors that the drill has penetrated and to those immediately underneath the well bottom. In the example shown, the high level of detail available has made it possible to identify the top and bottom of the reservoir on the resistivity (ILD) log. This would not have been possible with surface seismic alone.

1.3 Motivation for seismic while drilling

The conventional VSP surveying technique uses wireline systems with recording tools in the borehole. This method – for which drilling has to be interrupted – has high costs and some operational risk. Costs are related significantly to the stand by of the drilling equipment, which rises drastically with the number of offset sources at the surface. Risks are due to having the drill string out of the well for one or more days, without control of the well pressure, hence the possibility of wall collapse or well kick, i.e., an increase in pressure in excess of that exerted by the drilling mud, which, if not controlled, may result in a blow out (Langenkamp, 1994).

Passive listening at the surface, using the noise produced by the bit during drilling as a seismic source (drill-bit VSPs) helps overcome these difficulties and provides geophysicists and drillers with valuable depth/time information. This signal has to be recognized and distinguished from other sources of noise in the well environment. This information is obtained on the well location using reference (pilot) and geophone signals for correlation. The survey can be extended also to distances of several kilometers from the wellhead (offset) by adding extra geophones on the surface seismic line, i.e., the line of surface receivers used for this purpose. Then, at no significant extra cost, multioffset reverse VSP (RVSP) can be recorded.

The drill bit as a source of vibrations has been studied extensively since the 1960's (see Sections 1.4 and 1.5), first to monitor the vibrations produced by the drilling process, and later to obtain seismic-while-drilling measurements. However, reliable while-drilling geophysical results have been obtained mainly in the last ten years. The success of the

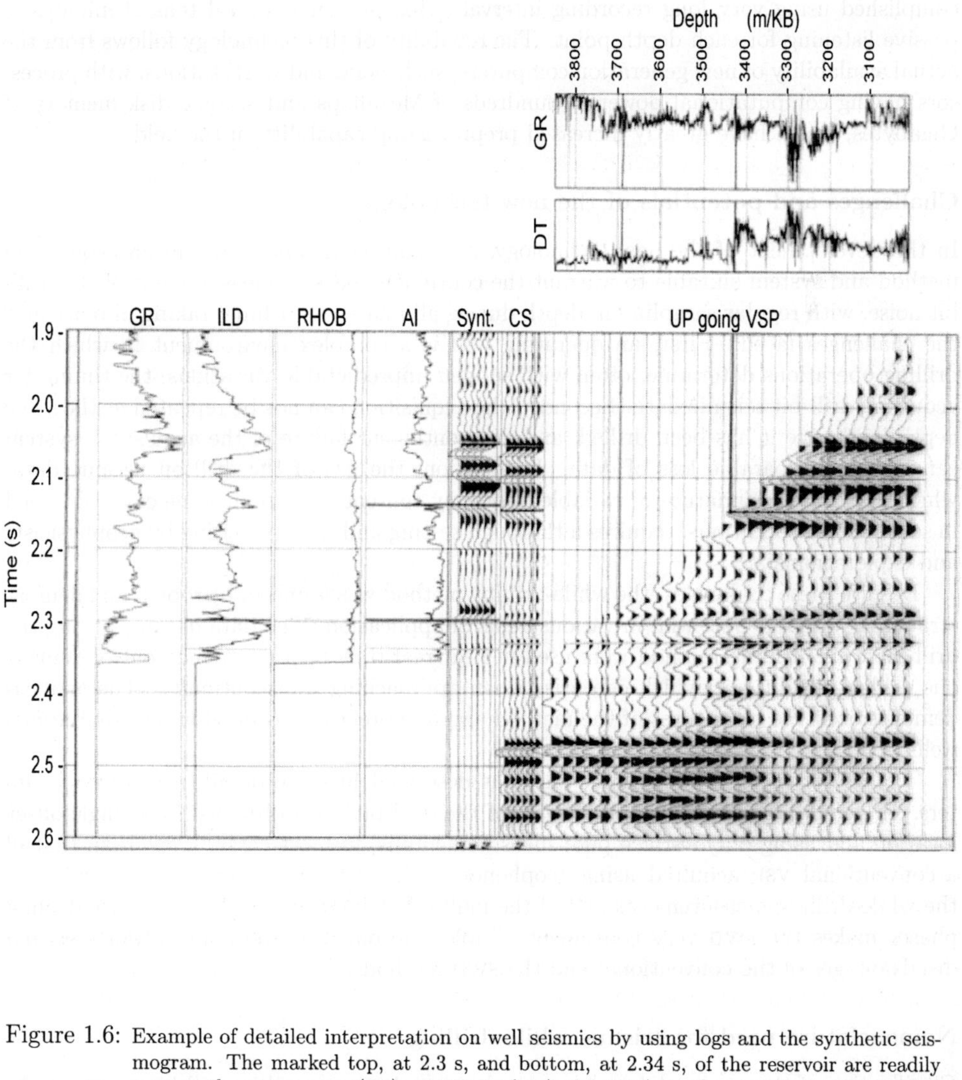

Figure 1.6: Example of detailed interpretation on well seismics by using logs and the synthetic seismogram. The marked top, at 2.3 s, and bottom, at 2.34 s, of the reservoir are readily seen on the gamma ray (GR), resistivity (ILD), density (RHOB), acoustic impedance (AI) logs, synthetic seismogram (Synt), corridor stack (CS), and upgoing wavefield in two-way time (TWT). The sonic (DT) and gamma ray (GR) logs are plotted in depth at the same scale as the VSP. Depths are referred to kelly-bush level (KB).

drill-bit seismic-while-drilling application is a result of an improvement in the conventional seismic method, which uses signals with known signatures and zero start times. The while-drilling technology requires preprocessing of the acquired time-incoherent raw field data to obtain impulsive shaped seismograms with known zero times. This is accomplished using very long recording intervals, that is, using several tens of minutes of passive listening for each depth point. The reliability of this technology follows from the actual availability of new generation computers, such as PC and workstations, with processors having computational power of hundreds of Megaflops and storage disk memory of Gigabytes, which have greatly increased preprocessing capability in the field.

Challenges and potentials of the new technology

In the development of the new technology, one main goal was to set up an acquisition method and system suitable to warrant the continuous passive measurement of the drill-bit noise, with regular sampling in depth during all phases of drilling, taking into account the challenges faced. First, an operating well is a complex environment in which the drilling operations determine, often with sudden unpredictable variations, the timing for acquiring drill-bit seismic data. Secondly, the acquisition can not be repeated at the same depth level once it has been drilled, and any contingent failure of the acquisition system can lead to irreparable loss of data. In addition, the use of the drill-bit seismic data, which may have a signature of variable frequency content and are in some cases obtained in severe noisy conditions, requires a thorough testing and understanding by geophysicists and explorationists.

Despite these challenges, the while-drilling method was found to be robust and demonstrated its valuable potential in various fields of application. The data obtained using the drill-bit while-drilling method were used in near-real time to predict and control dangerous drilling conditions, which can arise when approaching a discontinuity. The analysis demonstrated the robustness of the method and its potentialities in oil and in geothermal wells.

The use of the drill-bit seismic signal is described in more detail in the next chapters. In general, the signal-to-noise ratio of the drill-bit VSPs obtained in a single-offset location and using only surface pilot measurements is, in most cases, lower than that of a conventional VSP acquired using geophones in the well. However, the availability of the while-drilling measurements and of the multioffset information during all the drilling phases makes the SWD very convenient. Table 1.1 compares the main advantages and disadvantages of the conventional and the SWD methods.

Nomenclature used for seismic while drilling

Clarification of the nomenclature for several terms that are used in drill-bit seismic technology is necessary. The more commonly used terms are seismic while drilling (SWD) and reverse vertical seismic profile (RVSP). Reverse VSP means that – unlike for conventional VSP – the source is in the well and the receiver at the surface. Hence, "reverse VSP" has an immediate geometrical meaning, and is commonly used instead of "reciprocal VSP", which refers to the Reciprocity principle (Chapter 3). In other words, reverse and reciprocal VSP are the same.

Table 1.1 – Why seismic while drilling?

Method	Conventional VSP	SWD VSP
Operative risk	Yes (instruments in well, drill string out of the hole)	No (no instruments in well, drill string in the hole)
Global cost	Higher (well stand by and need to pull the drill string out of the hole)	Lower (no well stand by, limited operational assistance)
Data	Few offsets and many levels or many offsets with few levels	Redundancy of WD information (MOVSP virtually at the same cost)
Signal to noise	Higher (in single seismograms)	Lower (with surface pilots) (Presence of noise is reduced by multichannel processing) Comparable (with downhole pilots)
Bandwidth	Larger	Narrower (with surface pilots) Comparable (with downhole pilots)
Data availability	After drilling	While drilling (near real time)

"Seismic while drilling" means that VSP measurements are obtained during the drilling phase. Recently, a tool has been introduced in order to acquire seismic-while-drilling data with the receiver in the drill string and the source at the surface. Unlike the drill-bit seismic while drilling, it uses the normal (non reverse) VSP geometry. This while-drilling technology is called "seismic measurement while drilling" (SMWD[1]) VSP. Consequently, the term seismic while drilling has to be used with some care. However, since SWD is used very often in the literature as a synonym for drill-bit seismic, in the remainder of this book we use the acronym SWD to denote the seismic while drilling by drill-bit technology, if not otherwise specified.

1.4 History of the use of the drill-bit signal

The idea of measuring the drill-bit vibrations during drilling was proposed for a number of reasons. In addition, various approaches have been considered to solve the problem of identifying the drill-bit signal. These approaches, undertaken by several researchers are outlined below.

[1]Mark of Schlumberger

Guy (1961) proposed a new method for the geological control of drilling, which takes into account the "various technical maneuvers and interventions occuring in the course of drilling". This method uses the amplitudes measured by seismometers located at the surface of the ground to record the level of drilling vibrations and obtain information about the nature of the drilled formation. Other studies aimed at understanding the behavior of both the drill-string and rig-structure system and at obtaining information on the rock properties, were based on the theory of impedances (Lutz, Raynaud, Gastalder, Quichaud, Raynald and Muckerloy, 1972). These authors developed a theoretical interpretation of the longitudinal vibrations measured at the top of the drill string to simulate, on the computer, the dynamic phenomena of drilling, and produce an instantaneous "impedance log" of the formation while it is being drilled (SNAP log).

Measuring drill-bit vibrations was also used to analyze the status of the bit. For example, methods for determining the state of wear of a multicone bit were proposed by Stuart (1989), as well as Jardine, Lesage, and McCann (1990). They use cepstral (i.e., log-spectral or, also, log/Fourier) analysis to determine the harmonic frequency spectrum of the bit vibrations and extract the periodicity of the rotating bit and its fundamental frequency, which changes with wear.

More recently, drilling vibrations have been investigated using surface tools and also downhole recording tools inserted in the drill string. These measures are used to adapt the drilling parameters to control the undesired vibrations and phenomena affecting the drill-bit performance, such as stick slip, bouncing and whirling (Hutchinson, Dubinsky and Henneuse, 1995).

Another reason for investigating the drill-string vibrations was to determine the precise bit position during the drilling operations. Several authors studied the use of impulsive sources to obtain results with minimum effort and interruption to the drilling process. Bailey (1974) proposed a (non-passive) continuous bit-positioning system using motion sensors on the drilling apparatus and a plurality of geophones on the surface of the earth to make a measurement from which the position of the bit can be determined. For this purpose, the bit is raised and suddenly released in the borehole to produce an impact and measure the arrival times of the elastic waves in the formation, after they are synchronized with the delay of the signal in the drill pipes. Klaveness (1980) proposed to use a seismic pulse generator in the drill string, near the bit, and an array of seismometers at the surface to determine the drilling direction and the lithology through which the drill is passing. Katz (1984) proposed a method for determining the position of a drill bit by using the coherent acoustic signals generated by the drill bit and recorded at the surface of the earth by a plurality of detectors located in the vicinity of the well. These signals are shifted and their coherency is maximized to focus the precise drill-bit position (see also beam steering in Chapter 6).

Drawing on the advantages of many of these while-drilling methods, Staron, Arens, and Gros (1988) proposed the idea of using the drill-bit vibrations recorded on the top end of the drilling pipes as a sort of reference Vibroseis sweep to obtain while-drilling seismograms. The method consists of "locating on-pipe and ground recordings – corresponding to the same depth levels – grouping these elementary recordings in pairs and correlating and stacking them to produce a correlated signal which is representative of the acoustic energy produced and of the travel times of the waves transmitted in the formation". A particular application of this method is the VSP that is obtained after correcting the correlated seismograms for the travel time of the reference signal recorded on the drill string. This method – of instantaneous acoustic logging within a wellbore – was the basic idea as well as the first effective approach of using the seismic drill-bit technology, notwithstanding the initially low quality of the results.

After Staron, other researchers studied the seismic drill-bit signal. In particular, Rector and Marion (1991) demonstrated – with results of good quality – the use of the drill-bit reverse VSP (RVSP), using roller-cone correlated data. Rector (1989) proposed methods of deconvolving the drill-string multiples by spiking the drill-string pilot autocorrelation. From the promising results obtained by Rector and Marion (1991), the technology of the first seismic-while-drilling commercial system, the TOMEX® system, was developed. Rector and Hardage (1992) studied the bit signal to characterize the radiation pattern of the roller-cone bit wavefields as well as the noisy conditions.

Using a different approach, which does not use pilot recordings at the rig, Widrow (1990) proposed a method to extract the drill-bit signature and signal by means of adaptive filters. These filters are connected to each of the sensors used to acquire a plurality of traces located in different positions in the earth near the surface – in order to measure drill-bit signals with different travelpaths. The drill-bit signature and the reflections are calculated after the convergence of the adaptive filters (Widrow, 1990), obtained by optimizing the crosscorrelation function of the output traces.

In the same period, Rocca, Persoglia, Poletto, and Craglietto (1990) investigated the use of the drill-bit signal for borehole seismic purposes, by using a plurality of pilot sensors in the rig assembly and the noisy drilling yard (Chapters 4 and 5) (Angeleri, Persoglia and Poletto, 1990). They developed a method that enabled them to perform the statistical separation of the pilot signal starting from the noisy traces obtained by the plurality of sensors placed in the different locations, e.g., axial and torque sensors on the rig and pressure sensors in the mud circuit (Angeleri, Persoglia, Poletto and Rocca, 1996; Poletto, Rocca, and Bertelli, 2000). Following these works and experience in explorative wells (Miranda, Aleotti, Abramo, Poletto, Craglietto, Persoglia and Rocca, 1996), ENI E&P Division and OGS developed their own system, SEISBIT®. This system performs automated acquisition, data quality control, and uses a method of selective data processing based on diagnosis of drilling conditions (Miranda, Abramo, Poletto and Comelli, 2000). Furthermore, the investigation extended the SWD to unfavorable conditions – such as downhole-motor and diamond-bit drilling. This was done with the support of downhole tools to acquire improved pilot traces using a local downhole storage memory.

The Institut Français du Pétrole (IFP) developed the TRAFOR system using wired drill pipes and tool joints with electrical connectors to record downhole data while drilling. This system has applications to drill-bit seismic profiling (Naville, Layotte, Pignard, and Guesnon, 1994). Recently, an improved system of electromagnetic communication between bottom and surface (Soulier and Lemaitre, 1993; Soulier, 1997) has been tested to trigger downhole acquisition tools with local storage memory.

A method for VSP by measuring the drill-bit vibrations was also proposed by Ng, DiSiena, and Bseisu (1990). They used an array of seismometers on the earth's surface and an arrangement of sensors mounted in the rotating drill string to measure axial and torsional vibrations. A result is the identification of the relative delay of torsional and axial waves measured by the reference sensors in the drill string. This delay, once the axial and torsional velocity in the drill string are known, allows one to calculate the location of the point in the drill string at which the source signal is produced.

A method for the determination and deconvolution of the drill-bit signature was proposed by Miller, Haldorsen and Kostov (1990). This method uses an array of receivers positioned at the earth's surface to record the seismic signals produced by the drill-bit source. These traces are time-shifted on the basis of analysis of coherence and weighted to estimate the signal (beam forming). A deconvolution filter is computed and applied to the data. A multioffset application is discussed by Haldorsen, Miller and Walsh (1995). Booer and Meehan (1993) used the axial and torsional drill-bit vibrations generated during drilling and measured at the surface at the hookload (i.e., the measurement of the weight of the drill string at hook of the derrick), as well as by torsion sensors to image the drill string considered as an impedance system. Starting from these results, Schlumberger developed a seismic-while-drilling technology, the DBSeis[2] system.

A non-passive drilling approach was proposed by Russian researchers (Schechtman and Schneerson, 1983), who used a method based on the controlled variation of the drilling parameters, which is obtained by changing the velocity of the rotation of the bit. This method produces an unsteady mode for the drill-bit noise with different frequencies within the range limited by the seismic channel. The controlled variation can be also obtained by changing the pressure at the bottom, from zero to the maximum pressure determined by the weight of the drilling bit. Asanuma and Niitsuma (1992) used the polarization analysis of the uncorrelated roller-cone bit vibrations recorded by a network of downhole multicomponent detectors, to map the direct arrivals and reflections in crosshole surveys, showing results with large bandwidth. Among other approaches, we also mention proposals by Neill, Aminzadeh, Sarem, and Quintana (1993) to use the drill-bit signal in adjacent wells to the well being drilled in a cluster of wells of the same site, and by Petersen and Heggernes (1996) to use downhole sensors to measure the pilot drill-bit signal with a wireline connection in the coil tubing.

1.5 Overview of the different approaches to SWD

The seismic-while-drilling tool has been applied worldwide in different contexts using different systems, in some cases based on different approaches for data correlation. TOMEX ®, DBSeis[2] and SEISBIT ® systems have been used extensively, and applications have been also demonstrated by UNOCAL and JNOC, as well as by IFP using the proprietary TRAFOR system. The experiences of these contractors and of oil companies allows us to evaluate the geophysical results obtained. Some applications of the different systems have been published, and are summarized below.

The TOMEX ® system has operated in many surveys offshore and onshore, showing geophysical results with roller-cone bits. This system uses a PC-based technology, with few channels and limited capabilities to manage large seismic data sets. It performs the processing of the pilot signal after subtracting the polarized noise in the signal measured at the head of the drill string (Rector, 1990). The solution adopted to deconvolve the

[2] Mark of Schlumberger

seismic data, with Rector's proposed approach of using an operator derived by the pilot autocorrelation (Rector, 1989), can lead to an incorrect deconvolution of the seismic traces, with distortion of the earth's reflections. This is due to different travelpaths of the drill-string multiples transmitted in the formation with respect to the multiple transmitted to the drill-string head (Poletto, Malusa, and Miranda, 2001).

The DBSeis system has performed well in different applications. This system performs multioffset acquisitions and uses multichannel methods to improve the drill-bit signature, as well as the imaging of the drill-string reverberations. This method, in turn, allows improved calibration of the drill-bit signal delay in the pilot data. This time is used to compensate for correlation delays. The DBSeis system has been tested onshore and offshore, as well as by investigating the possibility of integrating the SWD (as well as while-drilling VSP with receivers in the well) with other technologies used to monitor well drilling and predict dangerous conditions.

The TRAFOR system is based on the transmission of the downhole pilot using a wired (cabled) drill string. The main shortcoming is that it requires the changing of the normal drill pipes to wired drill pipes.

Recently, IFP and jointly ENI E&P Division and OGS have tested the operational solution of using, for SWD purposes, downhole tools with internal memory, wherein data are stored and downloaded after the bit is raised. In this way there is a delay for use of downhole data which are however, if not real-time, available while drilling.

The SEISBIT® system is based on a multi-PC and workstation system able to operate with an intermediate (some tens) or high (some hundreds) number of surface seismic channels (geophone groups). It uses seismic lines with variable settings of onshore recording units. The acquisition is automated, synchronized by the drilling parameters, which are used for continuous automatic quality control, analysis of the acquired data and for optimizing the stack of the correlations of the same depth level. The system performs the preprocessing of the seismic dill-bit data in the field, to obtain a set of seismic traces that can be processed by means of standard VSP packages. As the raw data are available on-line in the field, the preprocessing can be achieved by using different combinations of pilot traces, while the preprocessed results are sent in near real-time to headquarters. This system is synchronized by a satellite global positioning system (GPS) so that it can be promptly interfaced with other systems providing while-drilling data, e.g., downhole tools. This system was used in different operational conditions, such as with crosshole and 3D-RVSP geometry, with roller-cone and polycrystalline diamond compact (PDC) bits, and with downhole motor drilling (DHM).

1.6 Seismic-while-drilling method

In the following chapters we describe the seismic-while-drilling technology taking into account different proposed solutions and systems; the various acquisition systems are extensively described in Chapter 5.

The while-drilling data may be acquired after laying out one or more lines of receivers and several pilot sensors. Pilot receivers are designed to record the on-site noise and the signal produced by the bit without the filtering effect produced by the traversed formations. The geophone traces are located along seismic lines laid in the vicinity of the wells in the direction of the dip of the geological structures of interest, with a sufficient offset from the well. The maximum offset is chosen in relation to the depth of the target, and can be more than 3 km. The signal is sampled in depth every 10 m or 20 m, with regular intervals of penetration, with a continuous control of drilling conditions and signal quality. This is done to avoid degradation or loss of data.

In preprocessing phase the raw field data are transformed into interpretable seismograms. Preprocessing, described in Chapter 6, is essentially based on the identification (or synthesis) of a pilot signal representing the signal emitted by the bit, correlation of this signal with those measured at the geophones, and the subsequent summing of the correlations relating to a given depth level, provided that the drilling depth is essentially unchanged during the acquisition of the depth-level data (Figure 1.7). When the pilot signal is measured at the surface far from the drill-bit source, its delay has to be calculated and added to the correlation times in order to obtain the correct zero time of the signal that travels through the ground.

To obtain seismic signals with sufficiently high signal-to-noise ratio, typically hundreds of megabytes of raw data acquired by using long listening times are processed for each depth level, i.e., correlated with the reference pilot signals.

In this configuration, surface sensors are used at the rig and on the earth's surface to sample the drill-bit signal. These channels are controlled by an acquisition system. Another unit, a workstation, is used for processing, quality control and data management. In the job flow, the raw field data can be either saved on cartridge, or discarded after temporary disk storage. The data are typically collected with a number of seismic channels ranging between fifty and one hundred, but several hundreds of channels are also usable in particular conditions – such as with drill-bit 3D-RVSP that requires an extended areal investigation. In view of the mandatory security rules on oil well sites, to warrant the maximum protection against fires, all the connections in the drilling yard are explosive-proof.

The data contain ambient coherent noise, generated at the rig and in the borehole. This noise masks the signal to some extent. The noise can be removed by means of multichannel filtering techniques. These techniques use the traces measured by noise sensors, i.e., recordings dominated by noise, in order to identify the noise characteristics and remove the noise from the data (Chapter 7; for example, Rector, 1990; Poletto and Craglietto, 1991; and also Angeleri, Persoglia, Poletto, and Rocca, 1996; Poletto, Rocca, and Bertelli, 2000).

The success of the method depends on the possibility of obtaining unambiguous interpretable data during the survey. In actual applications, due to the complexity of the working drill-bit source, a continuous control of the drilling operational conditions is required. The system must be reliable and efficient. In fact, any contingent failure has to be immediately repaired by restarting acquisition because the drill-bit seismic monitoring is non-repeatable at the same position after the drill-bit source is advanced in the borehole.

To reduce crew costs, the acquisition and preprocessing, including pilot correlation and deconvolution, are performed automatically by the system, which is driven by the drilling parameters transmitted from the mudlogging laboratory (Chapters 2 and 5). This cabin operates in the drilling yard by monitoring in real time tens of parameters critical

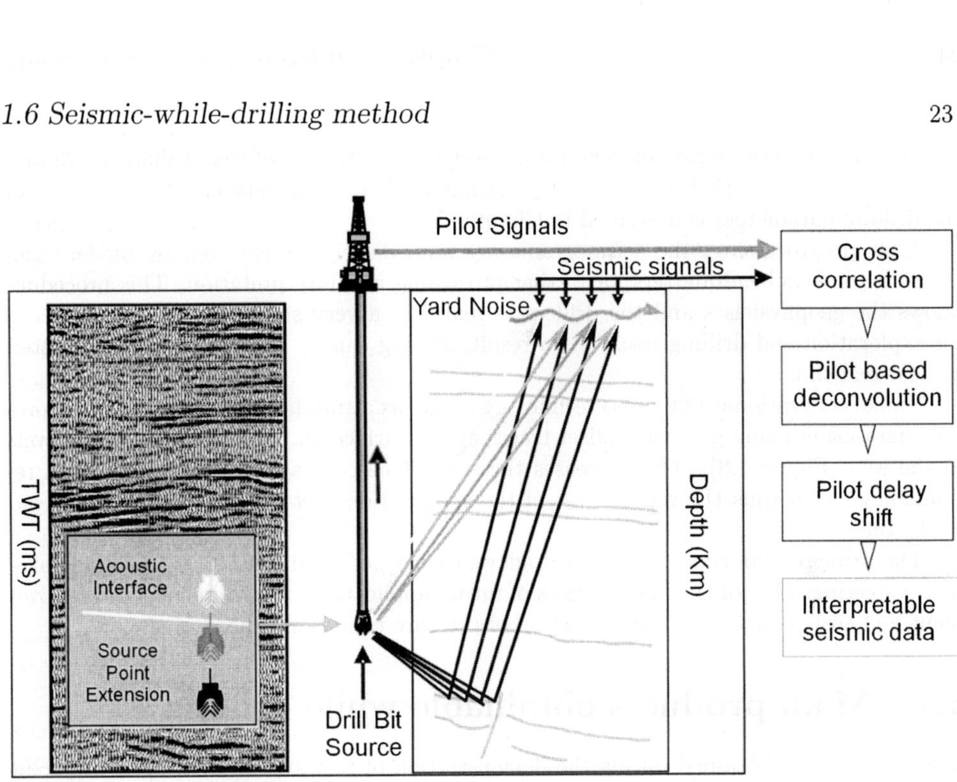

Figure 1.7: The seismic-while-drilling basic concept: the direct and reflected drill-bit signals are used for seismic purposes with the support also of reference (pilot) measurements at the rigsite.

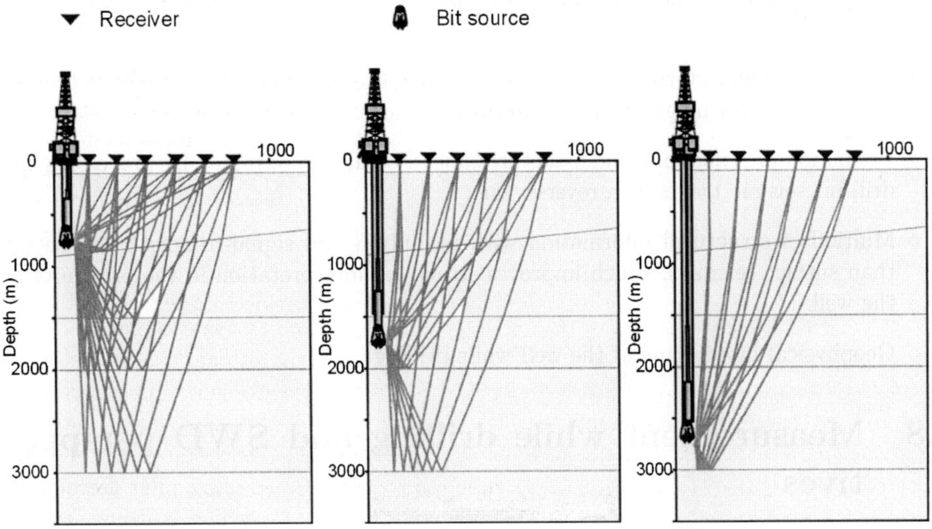

Figure 1.8: The seismic-while-drilling method allows one to acquire data with redundancy, using many depth levels and many offset traces at the surface, in the same well.

for the drilling, such as rate of penetration, weight on bit, level of gas, drilling depth and others. The process used to improve the seismic-while-drilling data using the analysis of the drilling parameters is described in Chapter 7.

After preprocessing, the seismograms are immediately transmitted by modem and telephone line to headquarters for further processing and interpretation. This procedure allows the geophysicists and the drillers to make, in a very short time, useful decisions for exploration and drilling itself. As a result, drilling can be modified and/or corrected in near-real time.

Unlike conventional VSP, SWD technology allows as a standard the acquisition of borehole data using many geophone offset traces at the surface and many source depth levels in the well (Figure 1.8). This increases the subsurface coverage, as well as the data redundancy – and thus the reliability – of the acquired dataset used for prediction ahead of the bit.

The principal information obtained allows casing point calibration and drilling parameter steering. One of the main targets of future applications for drilling safety improvement and cost reduction is the prediction of pressure conditions ahead of the bit.

1.7 Main products obtainable while drilling

The main results obtained during the demonstration of this SWD system under different operating conditions are described in Chapter 8. Five types of products can be identified; these are associated with the applications that can help to assess the drilling results. While the seismic-while-drilling products and applications are discussed in detail in Chapter 8, the main products and applications are, in brief, given below.

- ◇ Checkshot while-drilling, i.e., the identification of the travel times of the direct compressional arrival from the source point (the bit) to a receiver close to the wellhead.
- ◇ Reflectivity characterization, obtained by using the checkshot travel times to process the synthetic seismogram of the intermediate logs.
- ◇ Prediction of reflections and structures ahead of the bit obtained by processing the drill-bit seismic traces as a reverse VSP.
- ◇ Multioffset structural information with resolution and signal-to-noise ratio better than surface seismics, which improves structural interpretation in the proximity of the well.
- ◇ Geophysical monitoring of the well while drilling.

1.8 Measurement while drilling and SWD perspectives

The main development lines of the actual SWD technology concern the extension of the method to complex operational conditions and to the deep-water offshore. A current result is the demonstration of the technology and of the method using the 3D reverse VSP in onshore surveys, with areal dispositions of sensors at the surface and a high number (say, 400) of channels. The SWD technology enables the acquisition of 3D surveys onshore with a significant number of source depth levels (this application is discussed in Chapters 5 and 8).

The more significant perspectives of technical improvement concern the use of downhole tools located in the drill string to provide high quality pilot traces for drill-bit seismic and formation evaluation while drilling. This technology is currently being tested in experimental phases, and allows one to overcome difficulties for effective SWD application related to particular drilling conditions, such as deviated well drilling with downhole motor in the absence of surface rotation, as well as drilling with polycrystalline diamond compact (PDC) and diamond bits.

In this evolution, there is an important interaction between the acoustic monitoring and the drilling technology, which uses signals measured in the drill string for diagnostic purposes and measurement-while-drilling (MWD) telemetry to steer drilling. In particular, the geosteering investigation focuses on the steering of drilling from the geological point of view and recognizing, with a higher definition, the structures around and ahead of the drilled well. Geosteering uses the real-time information provided by the different seismic, acoustic and non-acoustic while-drilling measurements in an integrated intelligent system. The goal is to minimize both the drilling risk and the cost, and to drive with precision the well to the target. The possibility of integrating the drill-bit seismic with while-drilling VSP (WDSP or "reverse-SWD", SWD^{-1}) and acoustic single-well imaging (SWI) technologies is discussed in Chapter 8.

Finally, similar seismic-while-drilling techniques are used also in other fields of drilling. One important application is the tunnel seismic while drilling (TSWD) using the noise produced by tunnel boring machines (TBM). For this, we mention Gangi (1982) as well as Petronio and Poletto (2002) as references.

Another important application of seismic while drilling using the drill-bit signal is for obtaining seismic borehole data in the exploration of geothermal fields, where the SWD method is used in place of the conventional VSP because high temperature in wells makes it impossible to use the electronic instrumentation with the conventional geophones in the borehole. The study of the SWD application in geothermal wells has given important contributions to the development of the method (Naville, Serbutoviez, Throo, Batini, Dini and Cecconi, 2000).

Chapter 2

Principles of drilling

2.1 Introduction

Drill-bit seismic started when geophysicists working with conventional seismics experimented with the idea of measuring the drilling vibrations for reverse VSP purposes (Chapter 1). To achieve this purpose and realize acoustic measurements in the unknown drilling environment – which may operate with variable conditions – the geophysicists had to understand the drilling technology.

Hence, before introducing the drill-bit seismic method, it is useful to give a general overview of the drilling technology, operations and problems. The following justifications are particular notable.

◇ Seismic-while-drilling data are used to support decisions made during critical drilling phases and to forecast geological drilling conditions.

◇ A large part of the drilling plant – and not only the drill bit – is involved in the process of generating drill-bit seismic waves, i.e., the SWD seismic source.

◇ The drilling plant, the well design and the drilling operating procedures determine the logistical conditions and the technical solutions adopted to realize the reference measurements of the drill-bit vibrations. This is true, in particular, with the use of downhole SWD measurements.

In this chapter, we describe the drilling of a well, including the well planning, the main drilling-plant components and the operating procedures, and point out the technical aspects particularly related to seismic while drilling. Similar technologies and methods are used to drill oil, geothermal and scientific-experimental wells.

2.2 Drilling a well

An oil well is drilled to reach a target buried several kilometers under the earth's surface. The maximum depth that can be drilled depends on the "depth-rating" characteristics of the drilling plant, and may reach up to 9 km for current commercial systems with the highest performances. With respect to vertical depth, pressure and temperature conditions, wells are classified as follows:

- ◇ Wells deeper than 4.5 km are defined as "*deep wells*";
- ◇ Wells deeper than 6 km are defined as "*ultradeep wells*";
- ◇ Wells with fluid pressure higher than 690 bar are defined as "*high-pressure wells*";
- ◇ Wells with temperature higher than 150 °C are defined as "*high-temperature wells*".

Recently, experimental vertical wells were drilled in special sites to reach more than 10 000 – 12 000 meters of depth, as extreme (ultradeep) drilling applications (Chur and Oppelt, 1993). In general, the maximum depth target that can be drilled by a given drilling plant depends on many critical factors, such as the complexity of the geological setting, the correct forecast of the drilling conditions and the proper design of the drilling plan. This consideration is also true for deviated and horizontal well drilling, which is a technology currently in use that makes it possible to approach targets located at a distance of several kilometers from the well site.

Modern drilling technology is based on rotary drilling. In fact, after the completion of an initial, conductor, cased borehole of only a few tens of meters using a percussion hammer, the plant starts to perform the deeper drilling of the borehole by acting as a rotating drill machine. For this purpose, the drill facility makes use of a stem of jointed drill pipes (drill string), which is elongatable and has a drill bit at its bottom end to break the rock. The pipes of the drill string are steel (cylindrical) tubes into which flows the drilling fluid (mud), which returns to the surface by circulating in the annulus located between the drill pipes and the borehole. The drilling mud lubricates the borehole and has many other important purposes.

The well is drilled using different hole diameters, and its final shape is that of a telescope oriented with the wider tube in the upper part and the thinner one in the lower part. This is due to the fact that the drilled borehole is progressively protected by inserting and cementing a certain number of steel casings of diameters decreasing with increasing depth. In fact, the length of a free borehole interval that can be drilled without too many risks (collapses and kicks) is limited by depth, by geological condition, and usually can not be the entire well length. The casing points are typically set at the important changes of the formation properties. The correct design of casing is one of the main points of the well design, related to the knowledge (prediction) of the formation properties. Wrong location of the base point at which to set a new casing (casing seat or "shoe") can cause well instability or missing of the target in depth.

2.3 Main well components

In this section, the main drilling-plant components and their functions are introduced.

2.3.1 Drilling site

The well site can be a drilling yard (onshore) or platform (offshore). Near the rig, an area of radius of several tens meters that equals at least the height of the derrick is defined as the "rig-site area", which is protected from other activities. Onshore and offshore site preparation is described by Jahn, Cook, and Graham (1998, p. 42). If no prior drilling activities have been recently carried out in this area, an environmental impact assessment (EIA) will probably be required first by the host country to ensure that drilling activity is acceptable from environmental and legal points of view. EIA may include also the SWD operations in the proximity of the well. Onshore, the size of the rig site will depend on the type of equipment and operational requirements (such as rig type, size of mud pits, amount of required storage space, requirements for personnel accommodation at a safe distance from the rig) and constraints from local conditions (Figure 2.1). Offshore, the size of the site is strictly determined by the layout of the marine drilling rig, which in some cases can be supported by a tender rig, i.e., an auxiliary rig for logistics.

In the rig/drilling yard of wells, danger zones are identified where there can be a presence of either a continuous or occasional explosive atmosphere and risk of fire. The classification of the hazard zones – as recommended by the European IEC classification, corresponding to the North America (USA and Canada) classification – is the following.

- ◇ Zone 0 – An area in which the mixture of explosive gas is constantly present, e.g., the inside of fuel tank.
- ◇ Zone 1 – An area in which the mixture of explosive gas may be constantly present during the normal operation of the plant.
- ◇ Zone 2 – An area in which the mixture of explosive gas is not normally present, but is only present for brief periods of time.
- ◇ Safe area – All other parts of the plant.

Onshore, an area of a 30 meter radius around the derrick and the derrick area above the drilling floor are in hazard zones 1 and 2, so that personnel accommodation, tanks and laboratories are in the safe area. In marine rigs, the accommodation areas are settled in the safe zones, following the rig layout.

2.3.2 Derrick

This is the vertical derrick characterizing the "view" of the well (Figure 2.2). It can be a derrick (conventional) or mast (upsetable). The size (rating) of the derrick is related to the maximum weight load that can be held, which is related to that of the heavier casing pipe, and, hence, to the maximum depth that can be drilled. The total load supported by the derrick structure is distributed on the legs of the derrick (Moore, 1986, p. 5). As an example, consider a casing of 9 5/8 in. diameter, with weight per unit length of 53.5 lb/ft (79.62 kg/m) to be set at 13 500 ft (4115 m). The casing weight in air is 327.6 tons. Because the borehole is filled by mud, the total load in air has to be multiplied by the buoyancy factor

$$C_{\mathrm{B}} = \frac{\rho_{\mathrm{s}} - \rho_{\mathrm{m}}}{\rho_{\mathrm{s}}}, \tag{2.1}$$

where ρ_{s} and ρ_{m} are the density of the steel and of the drilling mud, respectively. The more common values of C_{B} are $0.72 \leq C_{\mathrm{B}} \leq 0.87$ (the density of the drilling mud is

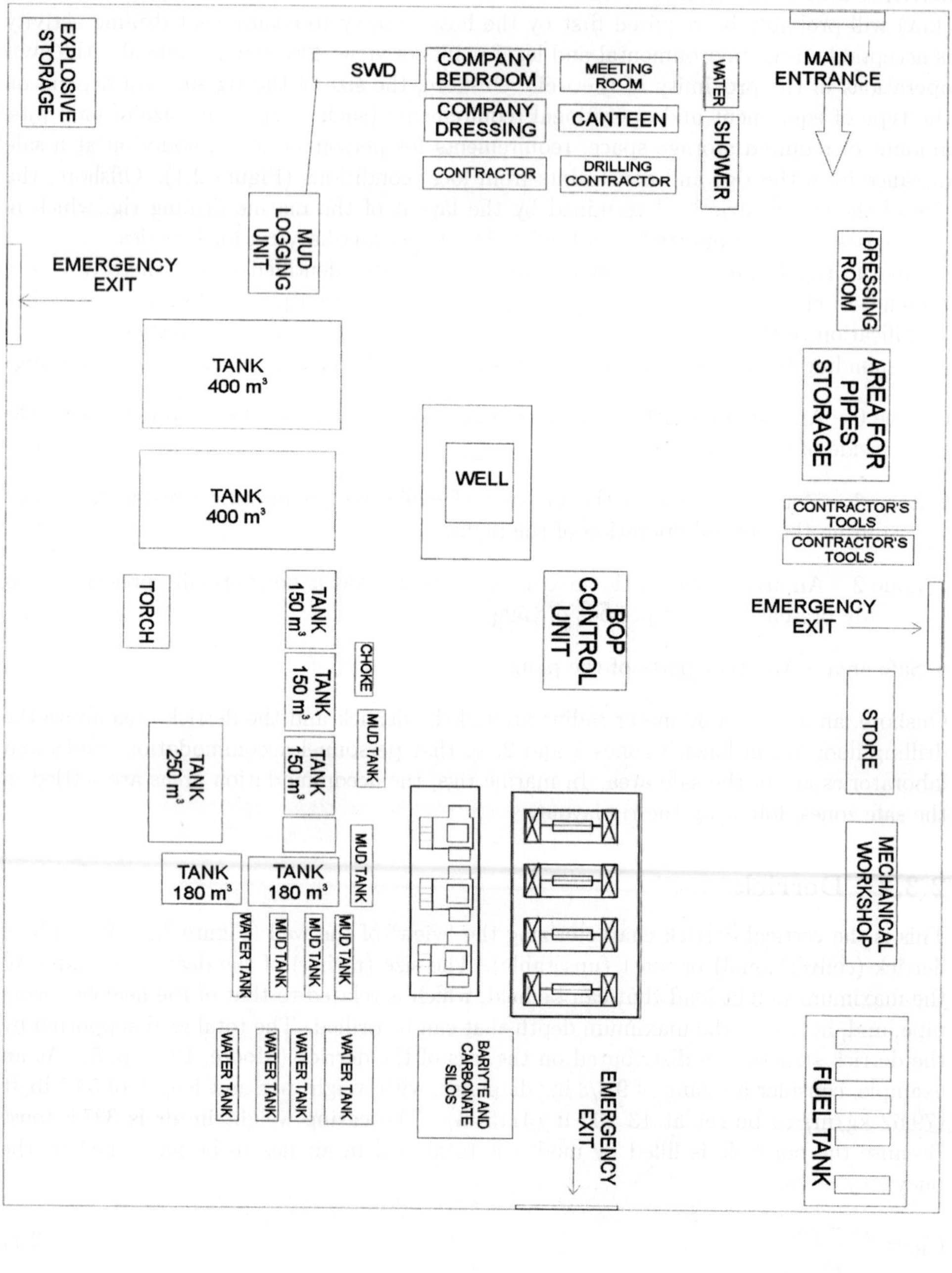

Figure 2.1: Example of onshore rig-site layout map.

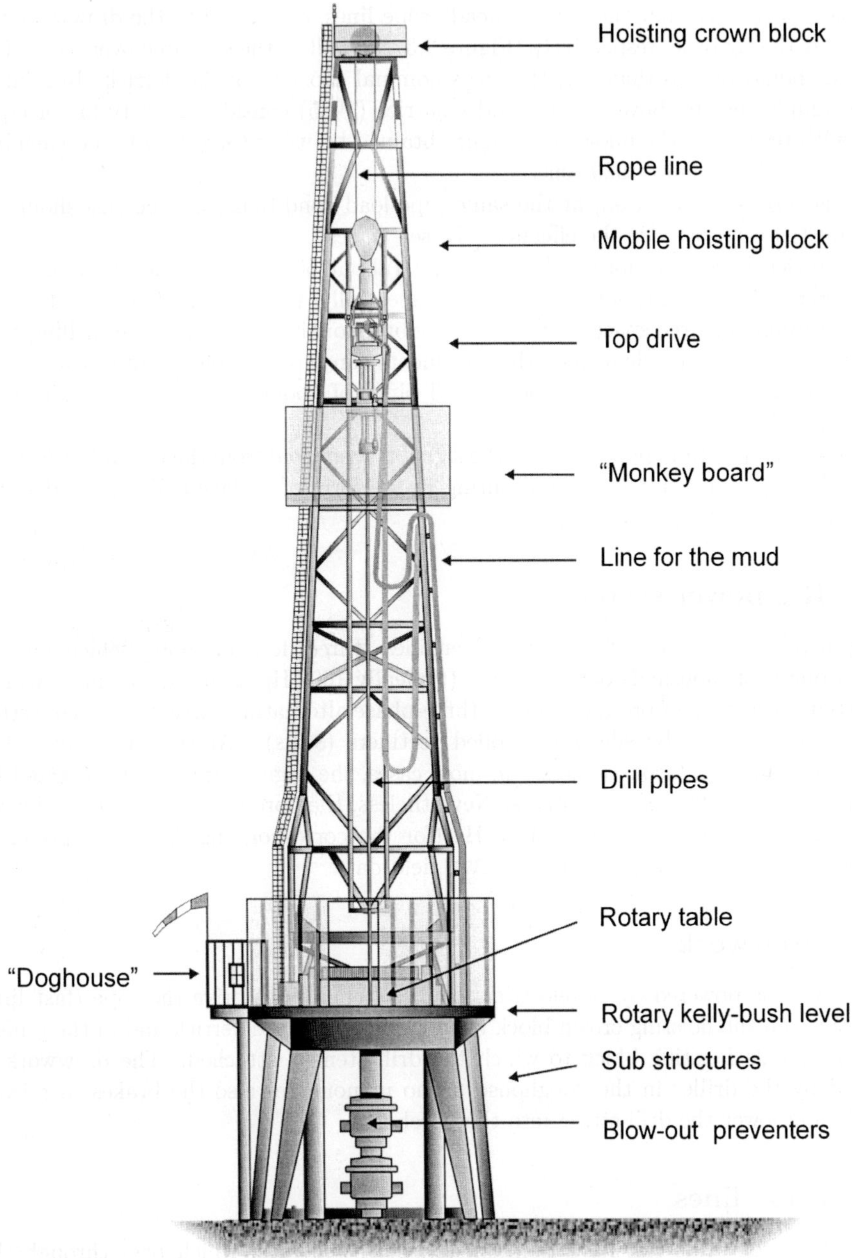

Figure 2.2: Layout of a drilling rig and its main components.

given by the drilling plan and discussed later). The total load is calculated including also the tension of the external "fast" and "dead" rope lines, connected to the drawwork and anchored at the rig base, respectively (Figure 2.3), as well as the dynamic overloads. This total load should be less than half the gross-nominal capacity of the derrick. In relation to the example shown above, Adams and Charrier (1985) consider a safety factor equal to 3.63 with respect to the buoyancy weight obtained by using $C_B = 0.774$, and calculate a total load rating equal to 920 tons.

As a general consideration, at the same pipe-load conditions, marine rigs should be more robust to support also the effects of the sea waves.

The derrick height is more or less constant (about 50 m) and at least 37 m above the base plate (drilling floor or rotary level). Essentially, this eight is determined by the amount of minimum free space between the mobile hoisting block and the drilling floor required to add new pipe elements – these elements can typically be a joint of three pipes of about 27 m of total length, composed of 3 pipes of about 9 m each, or single pipes about 9 m each.

The derrick vibration modes modify the signal transmitted from the bit and drill string through the rig, and result as noise occuring in the form of surface and refracted waves in SWD data.

2.3.3 Rig power system

The rig is powered by a system of diesel engines (three, four or more) which provide electrical power of thousands of horsepower (typically 1000 Hp each). In the most widely used systems, the engines produce 60 Hz – three-phase alternating current (AC), converted to direct current (DC) by silicon-controlled rectifiers (SCRs). At the same time they produce environmental acoustic noise. In most cases, the engines are sound-protected for the environmental safety of the workers. Nevertheless, they produce vibrations in the soil at the typical low frequency peak of 20 Hz (for the corresponding 60 Hz AC power of 3 phases), which is a noise detectable in SWD field data.

2.3.4 Drawwork

The drawwork is powered by DC-electrical or diesel engines to drive the rope (fast line) passing through the hoisting crown block fixed at the top of the derrick and in the pulley, lifting the mobile hoisting block to which the drill stem is attached. The drawwork is controlled by the driller in the "doghouse", who manouevres also the brakes to release the ropes and lower the drill string into the borehole.

2.3.5 Rope lines

The steel-wire-rope line can have from six to twelve reevings, which pass through the hoisting crown block (Figure 2.2) and connect mechanically the derrick to the drill-string system attached to the mobile-hoisting (traveling) block. Two other continuations of the rope line are connected to the drawwork (fast line) and fixed to the derrick at the other end (dead line) (Figure 2.3). The fast line passes through a tool (the "Martin-Decker") which is commonly used to measure the tension and calculate the weight on the bit (WOB). If n_L is the number of lines and F_H the hook load, the tension force of each line (fast, intermediate and dead) in static conditions is

$$t_L = \frac{F_H}{n_L}, \tag{2.2}$$

while the total static load for the derrick is $(n_L + 2)t_L$, as the tension of the fast and dead lines must be added to that of the n_L lines.

In dynamic conditions, a reeving efficiency factor is used to include the friction of the shaves of the hoisting block (Gabolde and Nguyen, 1999; Adams and Charrier, 1985). Note that the cables transmit the axial stress waves and not the torsional stresses, which are exerted by the rotary system. Flexural rope waves are present and are coupled to the vertical vibrations of the mobile-hoisting block, the drill string and derrick. Aarrestad and Kyllingstad (1993) used a coupled damped model to describe the vibrations of the traveling hoisting block and derrick system. The model corresponds to a system fixed at the top and at the bottom ends. They calculate the lateral resonance frequencies of a rope line of length L_c as

$$f_n = \frac{n}{2L_c}\sqrt{\frac{t_L}{m_c}} \sim \frac{n}{2L_c}\sqrt{\frac{F_H}{m_c n_L}}, \tag{2.3}$$

where n is an integer ($n = 1$ for the fundamental mode), m_c is the single-rope-line mass per unit length and the tension force t_L is approximated by equation (2.2). In this analysis, they assume that m_c=4.5 kg/m. (Rope properties specifications are given by API, Std. 9A and RP 9B.)

2.3.6 Mobile-hoisting block

The traveling block, which is coupled by the rope lines to the fixed hoisting crown block is the mobile part of the hoisting system (pulley). It is connected to the hook (or crane block), which is attached to the bail of the swivel or top drive at the top of the drill string. This hanging equipment can be subjected to horizontal and vertical shaking (Aarrestad and Kyllingstad, 1993). Horizontal shaking is less important for SWD with top-drive systems, but it is still present. High levels of vertical shaking produce jerks of the ropes and introduce non-linear noise generated at the surface in SWD data (Poletto, Malusa, and Miranda, 2001).

2.3.7 Rotary systems

The surface rotary systems (rotary drive) are the top drive and the swivel with kelly bush (or rotary table), in general powered by electric engines. There are also bottomhole systems, using downhole motors and turbines inserted in the drill string and powered by the mud flow. Let RPM and TOR be the rotary speed [revolutions/minute] and the torque [kN·m], respectively. The rotary power is given by the product of the rotary speed and torque,

$$W_{\rm rotary} = \frac{2\pi \cdot \mathrm{RPM} \cdot \mathrm{TOR}}{60} \quad [\mathrm{kW}]. \tag{2.4}$$

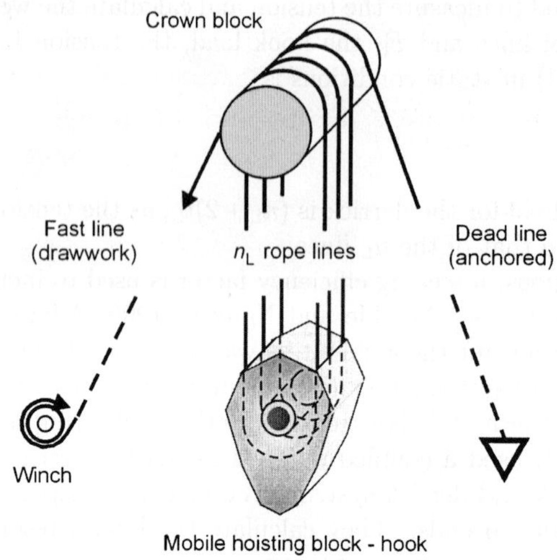

Figure 2.3: The rope-line tension is equally distributed on the $(n_L + 2)$ lines, of which n_L-lines of the mobile hoisting block plus the fast line, which is connected to the drawwork, and the dead line, which is anchored to the rig.

The required torque power depends on the type of system (surface or downhole), hole friction, hole angle and straightness (Adams and Charrier, 1985). For the surface rotary systems, only a part of the power is transmitted to the bit due to friction effects in the borehole (see Chapter 3). Rotary power can be produced in different ways. Note that, assuming that we are looking down, the working drilling rotation is always clockwise (if the observer is above) and in concordance with the screwing verso of the pipe threads.

Top drive

This is a drive motor mounted on the top of the drill string. It is composed of the following parts (Figure 2.4).

- ◇ The head of injection of the mud flow into the pipes.
- ◇ The motor. The top drive can be either an electric motor, or it can be powered by a flow of pressured oil, which is provided through an independent circuit and produced by an external motor. However, often, the top drive is an electrical motor powered by direct current (DC). During drilling, the electric motor is continuously ventilated to prevent any concentration of gas and to avoid risk of fires (hazard zones 1 and 2). The torque (TOR) produced by a direct-current motor is proportional to the current i_a of the induced circuit, and to the current i_f in the coil producing the magnetic field

$$\text{TOR} = k_e i_a i_f. \tag{2.5}$$

therein, k_e is known as the "conversion" constant of the elctromechanic energy. Equation (2.5) is valid also for rotary systems with kelly bush. Possible values of i_a and torque are 700–1000 A and 10 kN·m, respectively. Alternating current (AC) motors are also used. By using AC motors, the probability of producing sparks in the gas-risk area is reduced.

◇ The pipehandler. This is located in the lower part of the top drive (Figure 2.4) and includes the tools used to assemble, connect and disconnect the pipes, such as the rotary head, the pipe hangers and the torque wrench.

The top drive is fixed on a slide, which is lifted by the hook and travels on a vertical rail (guide) mounted on the rig. The rail (consisting of one or two tracks) prevents the rotation of the top-drive slide. However, the top drive can have a limited torsional shaking against the rail, and this movement can introduce noise in SWD data.

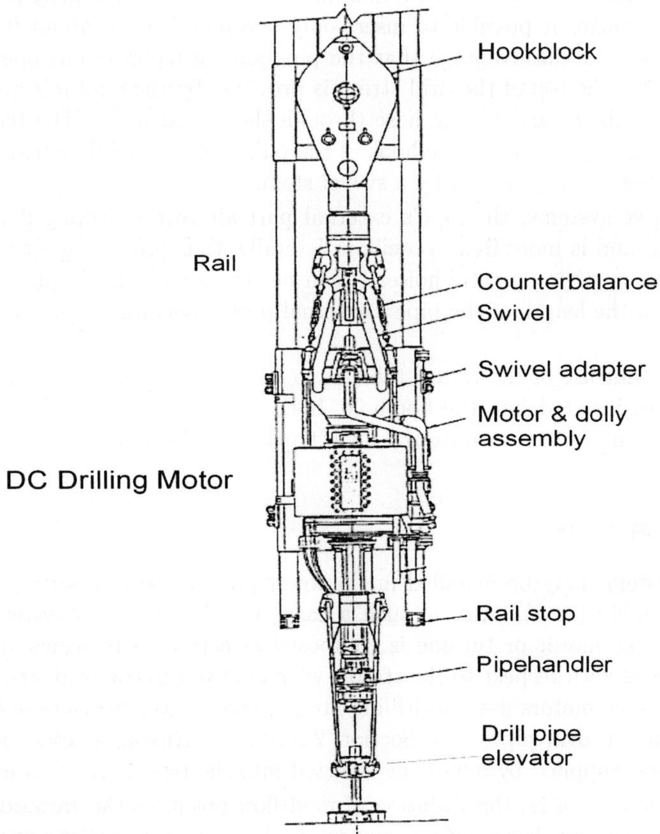

Figure 2.4: Layout of a top drive.

The top-drive system has advantages with respect to the rotary-table system (see below) because it makes it possible to insert a joint of three pipes each time (about 27 m) in the drill string with a simple operation. This system is used worldwide – in particular offshore – because it allows faster drilling due to the reduced loss of operation time. It has the disadvantage that it adds weight to the pipes (a "protection" spring is inserted in it to attenuate occasional load peaks which could damage the drill string during maneuvers), reduces the total load of the plant and adds costs.

Rotary-table or kelly system

This is the other conventional rotary system used at the surface (Figure 2.5). The rotary table is connected by a shaft to the drawwork motor or actioned directly by a motor (for instance, an electric DC motor, for which equation (2.5) can be used). The rotary table imparts rotation to a "kelly" (square or hexagonal) member, which is screwed into the drill string. This method requires the extraction and the insertion of the kelly every time a pipe is joined, and makes it possible to insert only one pipe length (about 9 m) each time. This system has the disadvantage that the pipe-joining (connection) operation is more time consuming. The top of the drill string is provided by the mud injection head, which is connected to the rotary-drilling hose through the "gooseneck". The bail at the top head of the drill string is hung on the hook of the pulley of the mobile-hoisting block, and the head at its bottom is provided by a swivel stem.

Unlike in top-drive systems, the pipe's external part above the drilling floor is not constrained by a rail and is more free to oscillate laterally, thus producing local flexural vibrations that result in environmental noise. It can be shown that the frequency of this noise is dependent on the length of the pipe's external part observable above the drilling floor.

The other important use of the rotary table is to hold in suspension – at the rotary base level – the pipes inserted into the borehole, both with rotary or top-drive system, during the maneuvers performed to raise the bit ("trips") or joint new pipes.

Downhole rotary systems

Downhole-rotary systems may be installed in the lower part of the drill string, denoted as bottom hole assembly (BHA). Also in such a case, a surface-rotation system is still present. The downhole motor or turbine is composed of a rotor – to which the bit is attached – that rotates with respect to the stator, which is fixed to the drill-string pipes. In general, the downhole motors used in drilling are powered by the pressure of the mud flowing internally in the drill pipes (see Section 2.3.13). In Russia, electric downhole motors of large power, supplied by a wireline inserted into the pipes, are often used.

With hydraulic mud motors, the variation of mud flow produces the variation of the rotary speed and torque and, hence, of rotary power. These motors are classified either as "positive-displacement" downhole motors (PDM) or turbines. The efficiency of the hydraulic motors is the ratio of mechanical power (equation (2.4)) and hydraulic power defined as the product of pressure drop (Δp) and flow rate (FLOW).

Downhole motors

Downhole motors use a "positive-displacement" system to produce the relative motion of shaft lobes (rotor) with respect to a sleeve housing (stator). In the inner part of the positive-displacement motor (PDM), stator and rotor are made of hard-rubber cylinders, which are shaped with spiral throats of hypocycloidal vertical cross-sections traversing in the vertical axis of the tool. A hypocycloidal curve is the curve generated by a point of the circumference of a circle rolling without slipping inside a coplanar and stationary circle (Robello and Miska, 1997). This produces an eccentricity of rotation for a shaft of n lobes and a housing of $n+1$ lobes in such a way that, while the shaft lobes maintain contact with the housing lobes (having the same envelope), they are forced to rotate one with respect to the other when the mud flows from chamber to chamber into the motor (Figure 2.6). Motor torque increases and revolution speed decreases as the number of lobes increases. The rotation speed and the torque of a positive-displacement motor are linearly proportional to the mud flow and pressure drop, respectively, and can be expressed as (Gabolde and Nguyen, 1999)

$$\mathrm{RPM} = \left(\frac{\mathrm{RPM}_1}{\mathrm{FLOW}_1}\right) \mathrm{FLOW}, \tag{2.6}$$

$$\mathrm{TOR} = \left(\frac{\mathrm{TOR}_1}{\Delta p_1}\right) \Delta p, \tag{2.7}$$

where Δp is the pressure drop across the motor, FLOW [lit/min] is the mud flow rate, and 1 denotes a reference operating condition given by the manufacturers at the specific mud gravity $\rho = 1.20$ g/cm^3.

Turbines

Turbines have higher rotation speeds than downhole motors, and perform with respect to the rotation speed with a peak of maximum efficiency. The turbine nominal performance is defined by the maximum horsepower output obtained with a nominal input flow rate and mud specific gravity of 1.2 g/cm^3. Variations in nominal specifications of rotation speed and torque are related to the mud flow by a linear and quadratic law, respectively. These nominal values are denoted by $'$ and expressed as

$$\mathrm{RPM}'_{\mathrm{nom}} = \left(\frac{\mathrm{RPM}_{\mathrm{nom}}}{\mathrm{FLOW}_{\mathrm{nom}}}\right) \mathrm{FLOW}, \tag{2.8}$$

$$\mathrm{TOR}'_{\mathrm{nom}} = \left(\frac{\rho}{1.20} \frac{\mathrm{TOR}_{\mathrm{nom}}}{\mathrm{FLOW}^2_{\mathrm{nom}}}\right) \mathrm{FLOW}^2, \tag{2.9}$$

where $\mathrm{RPM}_{\mathrm{nom}}$ and $\mathrm{TOR}_{\mathrm{nom}}$ are the nominal speed and torque specifications at the nominal flow rate $\mathrm{FLOW}_{\mathrm{nom}}$ with mud of specific gravity 1.20 g/cm^3.

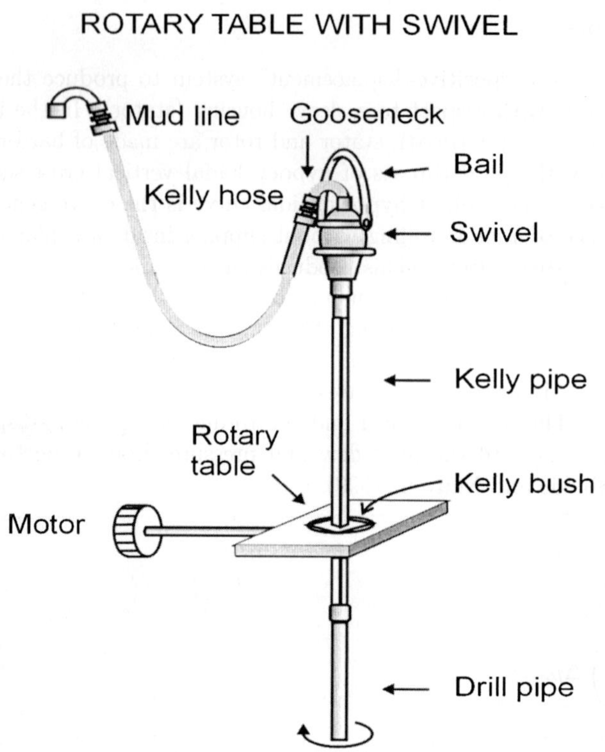

Figure 2.5: Rotary-table system with kelly and swivel.

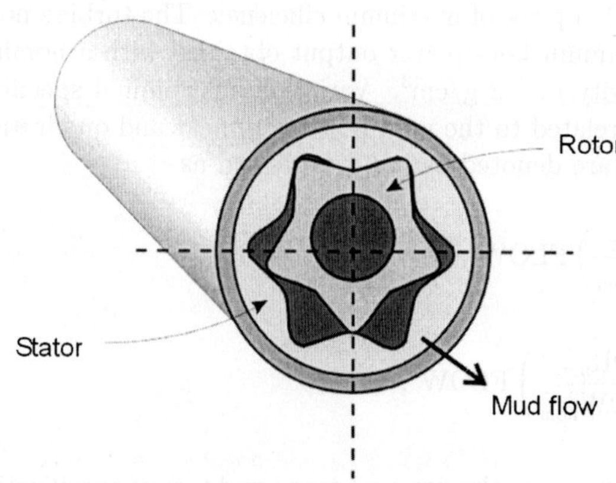

Figure 2.6: Shaft lobes of a 5/6 rotor/stator positive displacement motor (PDM).

Use of positive-displacement motors and turbines

The mechanical powers of positive-displacement motors (PDM) and turbines can be easily calculated by using equation (2.4). The choice of PDM or turbo drilling systems depends on factors such as the maximum allowed rotation speed, the bit type and hydraulic considerations. Typically, the turbines are faster and more heat-resistant than the downhole motors, which more easily control the drilling rotation speed (DeLucia and Herbert, 1984). For performances of PDM and turbo drilling see Gabolde and Nguyen (1999) as well as Bourgoyne Jr., Millheim, Chenevert and Young Jr. (1991).

Downhole motors and turbines are often used together with polycrystalline diamond compact (PDC) and diamond bits (see also Chapter 3), which reach optimum performance and longer life when drilling with regular rotation speeds (Wojtanowicz and Kuru, 1987). Tests of dynamic performance to determine the vibrations of PDC bits with low- and high-speed PDM were made through the application of downhole instrumentation by Lamine, Leseutre, Roberts, Walker, and Jonsson (1998). They concluded that the downhole power dissipated is rarely higher with PDM than with surface rotary, and that downhole-motor lateral vibrations are very sensitive to the load on the bit (downhole WOB), which is difficult to control.

When additional rotation (RPM_{surf}) is imparted from the surface in addition to downhole motor drilling, the total (nominal average) rotation speed at the bit is calculated as

$$\text{RPM}_{\text{tot}} = \text{RPM}_{\text{surf}} + \text{RPM}_{\text{dhm}}. \tag{2.10}$$

The total (nominal) rotation speed can be significantly higher than the RPM_{surf}. It is important to observe that the nominal total speed calculated by equation (2.10) is the speed measured at the surface, which can differ significantly from the true bit rotation speed. This effect is due to the elastic torsional properties of the drill string, which may twist. As a consequence, large downhole oscillations of speed – often related to the bit behavior – may occur when the average rotation speed is measured at the surface (Kriesels, Keultjes, Dumont, Huneidi, Owoeye and Hartmann, 1999). These downhole values can range between zero and more than twice the speed measured at the surface. In such cases, a correct estimate of the drilling rotary power expended at the bit is difficult using only surface data, due to both oscillation of rotary speed and losses of surface power that occur for pipe friction effects (Chapter 3).

2.3.8 Drilling floor or rotary kelly-bush level

The drilling floor is the base "kelly-bush" (KB) level of the well, where the maneuvers are performed, i.e., the drill string is assembled and lowered in the valved borehole through the rotary table with "kelly bush" (Figure 2.5). This point defines the geographic coordinates of the well.

The drilling floor is located in the lower part of the rig, at about 10 m above the ground floor (onshore) or above the main deck (offshore). The maneuvers and the drilling operations are controlled by the driller who operates in a cabin ("doghouse") located on the drilling floor. The driller, following the workplan provided by the well supervisor, changes the weight on bit (WOB) by regulating the drawwork and the brakes, and modifies the rotation speed to optimize the drilling parameters targeted to maximize the rate of penetration and inclination for deviated wells.

2.3.9 Wellhead and blow out preventers

The wellhead is the upper part of the well, located above the casing and under the drilling floor (substructures). The top end of the casing is placed and sealed on it. The wellhead is provided with valves to prevent kicks and blow outs. Offshore, the wellhead can be on the sea floor. The valves (blow out preventers) stacked above the wellhead are of different types. Namely, these are: annular (spherical) preventers; ram preventers, which force two elements to make contact and seal the annulus with or without the pipes (blind rams); and shear-blind ram preventers, which cut the pipe to seal the borehole. Preventer stacks are dimensioned on the maximum expected pressure. Typical classifications consider wellhead equipment designed to operate with up to 5000 psi ($\sim$ 350 bar) and up to 10 000 psi ($\sim$ 700 bar) of working pressure. The wellhead is modular and after the casing is run into the hole at the end of each drill section, each module is added to the wellhead. Onshore, this operation requires normally one day.

2.3.10 Drill string

The drill string connects the bottomhole tools (bit, stabilizers, MWD, LWD, etc.) with the surface, and transmits rotation and weight to the bit. It is composed of steel pipes of different diameters and dimensions, all screwed together. As drilling advances, the top of the drill string lowers and completes its descent when it reaches its lowest operating position above the drilling floor. At this point, the total length of the drill pipes is increased by the addition (jointing or "connection") of other pipes (Figure 2.7). The effect is that the top of the drill string has a variable distance above the drilling floor, while (not considering secondary elongation stretch effects) the total drill-string length remains constant during drilling between two subsequent additions of pipes.

The drill string is composed of two main parts. The upper part – thinner and much longer – consists of the drill pipes (DP) and the lower and more massive part consists of the bottom-hole assembly (BHA) (Figure 2.8).

Drill pipes

These are composed of a sequence of elements with more or less the same dimension and weight (pipes of 5 in. nominal outer diameter are a standard in non ultradeep wells). Drill pipes are composed of a body and tool-joint connectors. Each pipe is connected to the others by threaded tool joints that are thicker and heavier than the body. The drill string can be formed with subsections made of pipes with different tool joints and properties. Table 2.1 shows a typical classification for the drill-pipe components, according to the yield strength. In the SWD analysis of Poletto, Malusa and Miranda (2001), the types of drill pipes most used are S, G and E.

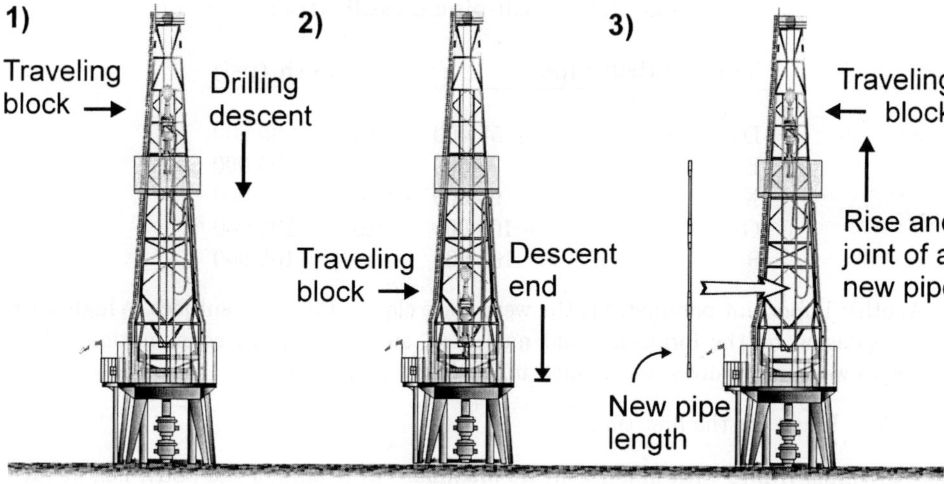

Figure 2.7: Connection of a new pipe length (consisting of three single pipes) in a rig drilling with a top-drive system.

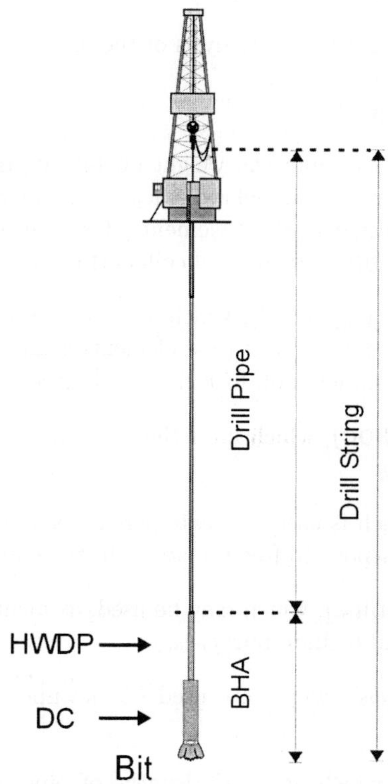

Figure 2.8: Main drill-string components.

Table 2.1 – Drill-pipe classification

Type of drill pipe	Yield strength (psi)		
D	55 000	to	85 000
E	75 000	to	105 000
X	95 000	to	125 000
G	105 000	to	135 000
S	135 000	to	165 000

Another important parameter is the wear-state class. Pipes are subject to high wear, due to abrasion by the formation and mud flow, and to chemical consumption. With respect to wear, pipes are sub-classified in the following manner.

◇ *New.* In use for the first time.

◇ *Premium.* Pipes with uniform wear. Maximum wear 20% of the body, i.e., at least 80% of the pipe transversal section remains.

◇ *Class 2.* Non-uniform wear with remaining wall not less than 70%.

◇ *Class 3.* Any imperfection or damages exceeding Class 2.

Drill-pipe wear is accounted for in the analysis of the drill-string waves in Chapter 3.

Bottom hole assembly (BHA)

The purpose of the BHA is to apply the weight on bit (WOB). The BHA is a complex assembly of different elements listed below and is much more massive per unit length than the drill pipes. The most important elements (for their acoustic properties) are the heavy-weight drill pipes (HWDP) and the drill collars (DC). Some other BHA elements are:

◇ the *stabilizers* (STAB) (Figure 2.9), which serve to center the pipe in the diameter of the well. A stabilizer is a short-pipe element of diameter close to the borehole diameter, i.e., a wall-contact tool, furnished with grooves to circulate the mud;

◇ the *shock absorbers* (SHOC), which limit the amount of high-amplitude axial vibrations in the drill string;

◇ the *jammer* (JAR), which is used to create percussions under high load and traction conditions with the purpose to free the string in the case of sticking of the BHA;

◇ the *downhole motors* (DHM), which may be used, as mentioned before, to have high RPM and for directional drilling purposes;

◇ *instrumented tools (subs)*, which are used for downhole inclination and vibration measurements;

◇ *cross-over subs* (XO), which are short lengths of pipe used between two different pipe sizes or two different thread patterns (Langenkamp, 1994);

- ◇ the *non-magnetic collars* (monel), which are used to separate the downhole instruments, e.g., the magnetic compasses, from the magnetized drill-string steel (the necessary monel length can be calculated by tables, see Devereux, 1998);
- ◇ the *reamers*, which are tools used to enlarge or straighten the borehole. Reamers are furnished with cutting wheels or blades to ream the wall of the well. Using reamers increases friction and torque loss (Chapter 3), thus reducing available rotary energy at the bit. At the same time, additional noise is introduced in SWD data, because a reamer is a secondary source of vibrations. Reamers can be located in the BHA and can be several tens of meters (e.g., 175 m above the bit).

Drill-string loads

The drill string is manufactured to operate under different axial traction and compression conditions (Figure 2.10). During drilling, the hook maintains the hanging drill pipes in tension – the pipes are manufactured to be run in tension and should not be run in compression – and the upper part of the BHA. The effect is that in the BHA there is a "*neutral point*" for axial load in which the resulting axial tension is zero (Adams and Charrier, 1985). Below the neutral load point, the drill collars of the BHA are in compression. The remaining part of the BHA weight not supported by the hook is loaded on the bit. This is the weight on bit (WOB). Buoyancy forces must be included in the calculation.

Assuming a vertical well, the hookload F_{H} can be calculated as the difference between the string weight in mud and the load exerted on the bit (Saleh and Mitchell, 1989; Adams and Charrier, 1985; Devereux, 1998)

$$F_{\mathrm{H}} = C_{\mathrm{B}} \left(\sum_j \Delta l_j \cdot \mathrm{WDS}_j \right) - \mathrm{WOB}, \tag{2.11}$$

where Δl_j and WDS_j are the length and the weight-in-free-air per unit length of the drill-string components, respectively, and C_{B} is the mud buoyancy factor of equation (2.1). The tensile force at any depth z within the drill pipe can be calculated as

$$F_{\mathrm{DS}}(z) = F_{\mathrm{H}} - z \cdot \mathrm{WDP}, \tag{2.12}$$

where WDP is the average weight-in-air per unit length of the drill pipe. The effect of the mud pressure on the pipe-joint areas is not considered in these calculations.

Drill-string dimension and safe rotary speeds

The drill string can be subjected to dangerous vibrations during drilling, and its design is important for the dynamic stability of the well. In addition, the variation of the drilling parameters has to be controlled to avoid values that produce resonances and induce excessive (and dangerous) drill-string vibrations. Resonance occurs when the frequency of the excitation source – such as the rotary speed or the bit-pattern forces – is tuned to a natural frequency of either longitudinal or torsional drill-string modes (Dareing, 1982; Devereux, 1998). If we assume that the drill-pipe model is a uniform rod free to vibrate

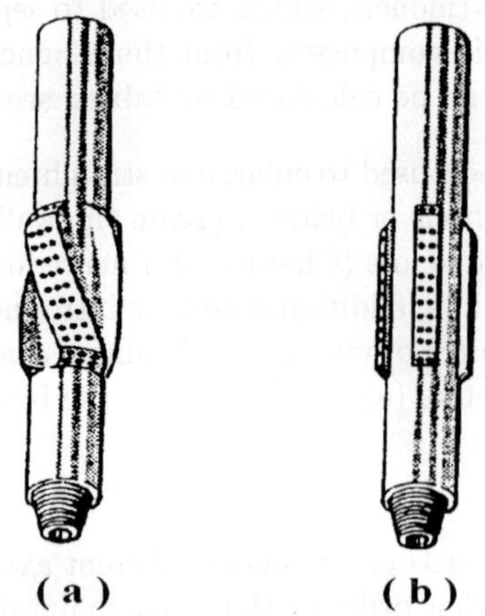

Figure 2.9: Stabilizers of a) spiral and b) straight ribs.

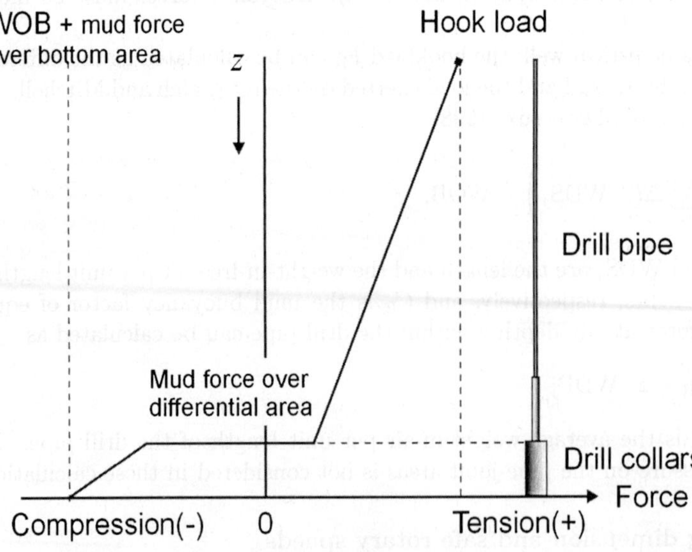

Figure 2.10: Axial drill-string forces in function of depth. The drill pipes are in tension, while the drill collars in the BHA work in compression (modified after Adams and Charrier, 1985). The sudden change of tension force at the top of the BHA is due to the downward force produced by the hydrostatic mud pressure over the increment of area between pipes and drill collars. Another hydrostatic but upward force is exerted at the lower end of the drill-collar section. This force is summed to WOB (Bourgoyne Jr., Millheim, Chenevert and Young Jr., 1991). The overall effect of the hydrostatic forces is accounted for in the buoyancy coefficient of equation (2.11).

at one end (open at the top) and fixed at the other end (fixed at the bottom), we obtain the critical frequency for the fundamental mode of the axial vibrations as

$$f_{cr} = \frac{V_{ax}}{4 \cdot L_{DP}}, \tag{2.13}$$

where L_{DP} and V_{ax} are the pipe length and propagation velocity of the axial waves of pipes, respectively. The fundamental critical rotary speed is calculated from f_{cr} [Hz] as

$$\text{RPM}_{cr} = 60 f_{cr} \qquad [\text{rev/min}]. \tag{2.14}$$

Other critical speeds are odd integer multiples of the critical fundamental speeds RPM_{cr}. So, these resonances result in a "comb" of spikes at the frequencies $(2n+1)f_{cr}$ Hz (n integer). Because the length of the drill string may be several kilometers, the spacing between the comb teeth can be very narrow and the critical frequencies are very close. Hence, careful rotary speed control is required.

We can apply reasoning similar to that used for axial vibrations to fundamental frequencies of the torsional drill-string waves. Note that for a given wave-propagation velocity, the natural frequency depends only on the length of the vibrating rod. In a uniform steel rod, $V_{ax} \approx 5130$ m/s and $V_{tor} \approx 3160$ m/s can be assumed for axial and torsional waves, respectively (Chapter 3). These values do not depend on the rod cross-section (low-frequency approximation). However, these values may differ significantly from the group velocities of axial and torsional waves in real drill strings. In Chapter 3, we will analyze in more detail the drill-string vibrations and the propagation of guided waves in the borehole. This analysis allows us to model the drill-string response, and to determine the group velocity and boundary conditions at the drill-string ends.

In the analysis of the overall vibration response of the drill-string system, Dareing (1982) concludes that the drill-collar assembly, rather than the drill-string total length, is responsible for severe resonances (Dareing, 1982; Wolf, Zacksenhouse, and Arian, 1985). The natural frequency of the fundamental axial vibration of the drill-collar assembly can be calculated as

$$f_{DC} = \frac{V_{DC}}{4 \cdot L_{DC}}, \tag{2.15}$$

where V_{DC} and L_{DC} are the axial velocity and the length of the drill collars, respectively. An analogous expression holds for torsional modes. In the literature, some examples are calculated with the axial uniform-rod velocity of about 5130 m/s. We use a different velocity, namely, the low-frequency approximation of the axial group velocity (Chapter 3), because the mass distribution for the drill collars is not uniform. For example, assuming a drill-collar of 145 m as length and a group velocity of $V_{DC} = 5050$ m/s, we obtain $f_{DC} = 8.7$ Hz.

2.3.11 The bit

The drilling bit, under the weight loaded by the drill collar, transforms most of the rotary energy into the action of breaking the rock and a small percentage into radiated seismic energy (Chapter 3). Different types of drill bits – of different shapes and mechanical action

– can be used to drill a well. The most common are roller-cone bits (also called tricone or rock bits) and bits with fixed cutter inserts, namely the diamond and the polycrystalline diamond compact (PDC) bits. Special types of bits are the core bits, designed to pick up plug-shaped core samples of rock.

The criteria for bit-type selection from hundreds of different drill bits for a given drilling plan is discussed in the remainder of this chapter and is related to factors such as the expected drilling-rate performance, the cost per unit depth and the bit life. The action of the different bit types characterize the vibrations produced while drilling and the signal used for SWD purposes (Chapter 3).

Roller-cone bit

This type of bit usually has three cones with teeth and is designed to break the rock by indention and a gouging action. As the cones roll – at least ideally, as true roll is never achieved for construction and bit performance – across the bottom, the teeth press against the formation with enough pressure to exceed the failure strength of the rock at which rock fracture occurs (Devereux, 1999). The layout of the roller bits is shown in Figure 2.11. The lower part of the bit's body supports the roller cones, usually three (but also one, two or four can be used). Each cone has two or three rows of teeth, which can either be milled from the same block of metal as the bit ("non inserts" bit) – these bits are used in relatively soft formations found at shallow depths (Langenkamp, 1994) – or be of tungsten inserts that are harder and more durable than milled teeth. The external, intermediate and internal rows of a cone each have a different number of teeth (for classification of teeth and inserts by form, and interfit of teeth cones see Adams and Charrier, 1985, pp. 183–185).

Each tooth is like a chisel, and has a maximum height (penetration depth) and a semi-angle (the angle made by its lateral surface with the bit axis, Figure 2.12a). The wedge of a new tooth can either be sharp or flat. Each cone is protected externally by the lobed body of the bit legs. The cones are supported by bearings, which are lubricated and sealed. The bearing axis of the cone forms the cone-journal angle with the horizontal level (or normal to the bit axis, Figure 2.12b). The offset distance is the distance between the cone axis and the the drill-bit center. The offset distance is measured in millimeters. The offset angle is the angle the cone axis would be rotated to make it to pass through the central bit axis. When the three cone axes meet in the bit center, the offset of the bit is zero (Figure 2.12c).

Eronini, Somerton, and Auslander (1982) assumed a predominant shear model for the rock fracturing process, where the roller bit destroys the rock through a teeth indention/fragmentation sequence, with the energy for the indention ("minor" fracturing) coming from the incident axial waves, while the fragmentation ("major" fracturing) is derived from rotary gouging power (Chapter 3).

The teeth of the indentors of the inner rows of a roller-cone bit perform a gouging-scraping action opposite that of the outer row. Sheppard and Lesage (1988) showed measurements of the transverse horizontal-gouging "backward" force that the indentors of the inner row produce, which is, on average, opposite to the transverse force of the outer row. They measured the resulting transverse couple between the inner and outer

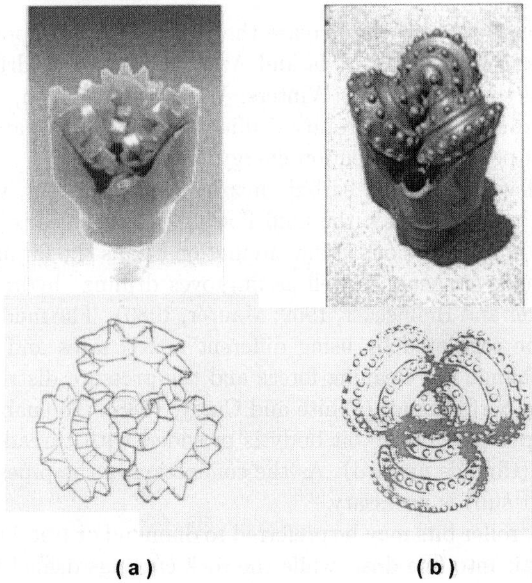

Figure 2.11: Roller-cone bits. a) Milled teeth b) tungsten carbide inserts. These types of roller bits are used in soft shallower and more consolidated deeper formations, respectively.

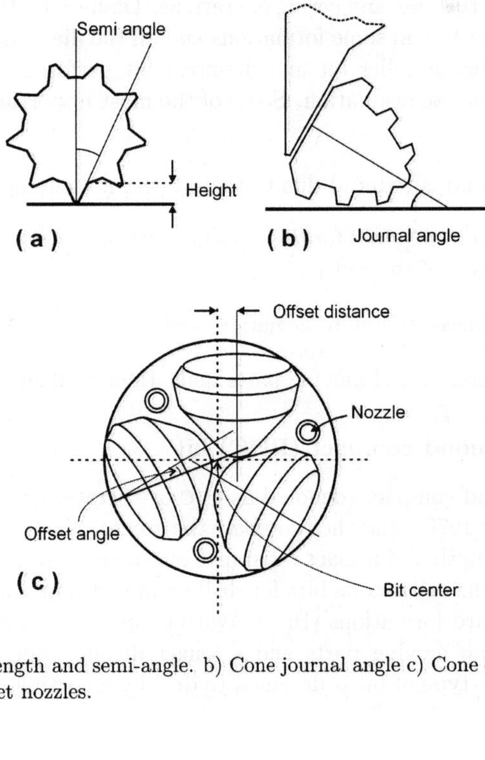

Figure 2.12: a) Tooth length and semi-angle. b) Cone journal angle c) Cone offset distance and angle, and mud jet nozzles.

rows of a single cone. Non-zero offsets increase the gouging and scraping action and the deviation from the true roll of the bit (Ma and Azar, 1989), which drills with a sliding motion proportional to the cone offset (Winters, Warren and Onyia, 1987). indention and gouging energies are calculated in rotary-drilling equations of Chapter 3, in order to estimate the fraction of percussive-indention energy.

The bit is provided with tungsten carbide nozzles (Figure 2.11c), which are located between the cones and through which the mud flowing in the pipes is ejected with high pressure. The mud has many functions. Mud circulation cleans the bit and the hole of the drilled-rock cuttings, removes heat, as well as improves drilling hydraulic performance (see, for instance, Maurer and Heilhecker, 1969; Maurer, 1980). The mud ejection directed against the rock can be regulated by using different nozzle sizes and geometries, thus making it possible to change the cleaning forces and the pressure distribution under the bit and improve drilling performance (White and Curry, 1988). To enable its insertion in the drill string, the upper part of the bit body is provided with threads, so that the bit connection is "pin up" (threads upward). As the connection of the pipes is typically "pin down", a cross-over (bit sub) is necessary.

In exploration wells, roller bits may be preferred to diamond or PDC bits, which destroy the rock by pulverizing it into fine dust, while the rock cuttings drilled by roller bits can be used for formation analysis.

Natural diamond bits

Natural diamond bits are constructed with diamonds embedded into a matrix and are used in conventional rotary, turbine, and coring operations. Diamond bits can provide improved drilling rates than roller bits in some formations and all the diamond bit suppliers provide comparison tables between roller bit and diamond bit performance to aid users in bit selection based on economic evaluation. Some of the most important benefits of diamond bits are:

- ◇ Bit failure potential is reduced due to there being no moving parts.
- ◇ Less drilling effort is required for the shearing cutting action than for to the cracking and grinding action of the roller bit.
- ◇ Bit weight is reduced, therefore deviation control is improved.
- ◇ The low weight and lack of moving parts make them well suited for turbine drilling.

Polycrystalline diamond compact (PDC) bit

Polycrystalline diamond compact (denoted as PDC or "Stratapax" trade name) bits were introduced in the early 1970's and they combine the greater abrasion resistance of the diamond bit with the strength and impact resistance of cemented tungsten carbide. PDC bits have widely replaced the roller-cone bits for drilling in soft and non-abrasive formations, and, in part, also in hard formations (Brett, Warren, and Behr, 1990).

The PDC bit has no moving parts and a longer drilling life; thus, its use allows a reduction in cost. This type of bit is designed to drill by a cutting, shearing and grinding

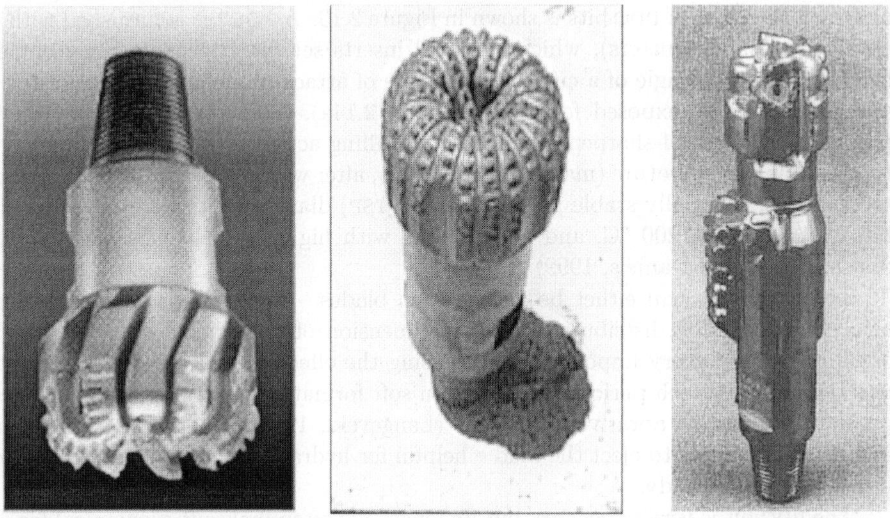

(a) (b) (c)

Figure 2.13: a) Polycrystalline diamond compact (PDC) bit; b) Diamond bit; c) Bicenter PDC bit.

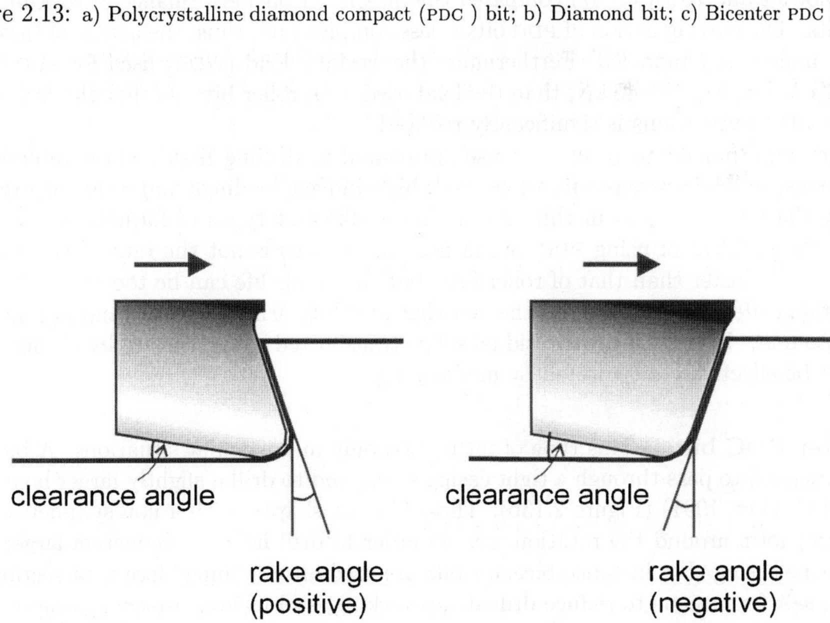

Figure 2.14: Single PDC cutter. Rake and clearance angles.

action. The layout of PDC bits is shown in Figure 2.13. A PDC bit is furnished with chisels (sintered diamond cutters), which are fixed inserts set directly in its body of tungsten carbide. The rake angle of a cutter is the angle of attack made by the front of the cutter and the surface of exposed formation (Figure 2.14a). Some types of cutters have the property of being self-sharpening during the drilling action. Cutters have limitations for their working temperature (maximum ~ 350 °C), after which degradation is very rapid. In recent years, thermally stable polycrystalline (TSP) diamond bits were developed. These bits are stable at 1200 °C, and can be used with highly abrasive formations (Radtke, Smith, Riedel and Daniels, 1999).

The PDC body can either be formed with blades – regularly or irregularly spaced – or a flat body. The distribution and the dimension of the cutters either on the blades or on the body is very important for assessing the effective and stable drilling action of this type of bit, which performs very well in soft formations but may have more stability problems in hard or abrasive formations (Langeveld, 1992a). Like roller-cone bits, PDC bits have jet-nozzles to eject the mud – helpful for hydraulic drilling – and threads on the upper part of the body.

The PDC bit is very expensive, but its longer life often makes it economical for drilling because a lower number of new-bit trips into the well are required. In fact, PDC bit applications result in long footages drilled with the same bit. Only a few bits may be required to drill the whole interval once the right type of PDC bit has been identified. Wojtanowicz and Kuru (1987) compared the maximum bit performance of rock bits and PDC bits. The cutting action of PDC bits is less complex and, thus, the action of these bits can be more easily modeled. Furthermore, the working load (WOB) used for PDC bits is typically lower, e.g., 20–40 kN, than the load used with roller bits, so that the drill-string fatigue due to vibrations is significantly reduced.

The main limitations of PDC-bit use are related to drilling highly abrasive and hard formations, in which severe vibrations with high impact loadings and wear may quickly destroy the bit, as well as in thin alternation of different types of formations. In other words, the problem of using PDC bits in hard formations is not the rate of penetration, which can be faster than that of roller bits, but that their life can be too short. Another undesirable effect for geologists is the fact that PDC bits drill by cutting and grinding the rock into dust. In this case, the residual solids transported to the surface by the mud flow can not be effectively interpreted by mudlogging.

Bicenter PDC bit – This type of bit is used only in particular situations. A bicenter bit is designed to pass through a tight casing string and to drill a slightly larger hole when rotated (Myhre, 1991) (Figure 2.13b). These bits are shaped with a non-symmetric (and eccentric) form around the rotation axis, in order to drill holes of diameters larger than that of the casing. For instance bicenter bits are used after a "liner" (see next section), in moving salt formations to reduce drill-string sticking, and to improve cementing of small and deep casing by increasing the annular space between borehole and casing. In fact, as the bicenter bit maximum diameter is smaller than that of the casing diameter, when the bit rotates, its lower part tends to be centered in the borehole, so that its higher and eccentric blades drill a larger borehole.

Experiences reported in SWD (oral communication at the Schlumberger Cambridge

Conference on SWD, 1996) indicated that bicenter PDC bits produce more usable vibration energy and are more favorable than conventional PDC bits for SWD purposes.

2.3.12 Casing

The casing is used to protect the borehole. It consists of a steel tube, which is inserted in the borehole and cemented to seal its external side. Extreme wear can be expected from the friction produced by the drilling equipment (pipe tool joints) used to drill the next hole section. The selection of casing (steel) grades and weights depends on many factors, including local geology, formation pressures, hole depth, formation temperature, logistics and various mechanical factors. (For current specifications of the performance properties of casing, tubing and drill pipes see American Petroleum Institute, API Bulletin 5C2).

Casing tubulars are placed in a wellbore for the reasons summarized in Table 2.2. The casing installation and cementing are operations repeated several times in a well, and each one takes several days. In these no-drilling periods, borehole logs and wireline measurements are typically acquired. The following sequence of casings is the configuration (casing profile) that can be used for onshore and offshore wells with surface wellhead (Figure 2.15):

1. A *drive* (or structural – conductor) *casing*, which is used to protect the unconsolidated surface formation from the erosion by drilling fluids.

2. *Surface casing*, which is used to prevent the sloughing of unconsolidated shallow formations into the hole, to enable mud circulation, to protect fresh-water sands from contamination from the drilling mud, and to protect against hydrocarbons found at shallow depths. Surface casing is the first casing on which the blow out preventers can be mounted.

3. *Intermediate casing*, which is the string used to ensure adequate blow-out protection and to isolate the formations or hole-profile changes that can cause drilling problems. Intermediate casing is nearly always set in the transition zone, above or below significant overpressures, and in any cap rock below a zone of potential severe loss.

4. *Production casing*, which is the casing string through which the well is completed, produced and controlled.

5. *Intermediate production liner*which is a string of casing that is installed but does not extend all the way to surface. It is hung a short distance above the previous casing base ("shoe") and, usually, cemented over its entire length to ensure sealing within the previous casing.

2.3.13 Pumps

The pumps – at least two for safety reasons – produce the discharge pressure required to force the mud into the drill pipes through the well circulating system. The mud is prepared in the mud tanks (pits) and pumped in the mudline through the flexible mud hose connected to the top of the drill string. The pump hydraulic power is calculated

Table 2.2 – Motivation for casing

Controlling well pressure by containing downhole pressure.
Isolating high-pressure zones from the wellbore.
Isolating permeable zones that are likely to cause differential sticking.
Isolating special trouble zones that may cause hole problems such as lost-circulation zones.
Separating different pressure or fluid regimes.
Providing a stable environment for packers (expanding plugs), liner hangers, etc.
Supporting the weight of the wellhead and blow out preventers (BOP) stack.
Providing a return path for mud to surface when drilling.
Related to production.

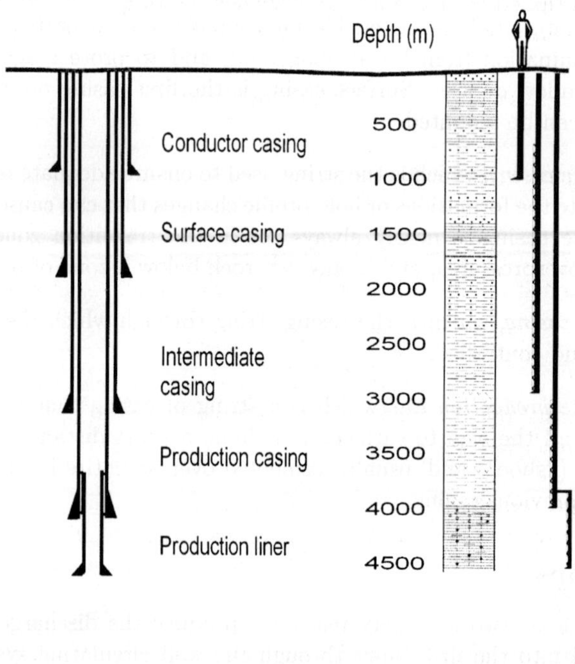

Figure 2.15: a) Casing sequence. b) Example of casing plan.

to compensate for the losses of pressure in the borehole and in the mud circuit, and to provide, through the mud flow, the mechanical power to the downhole motor and turbines used for directional drilling or for vertical drilling. Two types of pumps are used, namely, double-action and single-action pumps (Figure 2.16). The flow rate and the hydraulic power are proportional to the stroke rate (number of pump strokes per minute, SPM). The theoretical flow rate FLOW_t is calculated as (Adams and Charrier, 1985; Gabolde and Nguyen, 1999)

$$\text{FLOW}_t = n_p \pi \left(D_L^2 - 0.5 D_P^2 \right) L_s, \tag{2.16}$$

$$\text{FLOW}_t = \frac{3}{4} n_p \pi D_L^2 L_s, \tag{2.17}$$

for double-acting two-cylinders (duplex) and single-acting three-cylinders (triplex) pumps, respectively, where D_L, D_P, L_s are the liner diameter, the piston-rod diameter, the length of the stroke, respectively, and n_p is the number of strokes per minute. Typical values of flow rate are from 4000 to 1000 lit/min, in function of the borehole diameter. The theoretical hydraulic power is

$$W_h = p \cdot \text{FLOW}_t, \tag{2.18}$$

where p is the discharge pressure. To obtain the true power, previous equations (2.16) and (2.17) are multiplied by a volumetric efficiency factor η_v ($\eta_v \leq 0.98$). Pumps can be provided by dampeners, which smooth their action. However, the acoustic noise pulsation related to the number of strokes per minute is transmitted to the drilling plant and modulates the SWD measured data (Cunningham, 1968).

2.3.14 Drilling mud

Drilling mud is prepared by adding solids to a fluid, which can be water or/and oil. Oil-based muds are used, for instance, to prevent erosion when drilling water-soluble formations such as salt domes. Mud types and their acoustic properties are discussed in more detail in Chapters 3 and 7.

The drilling mud is prepared by adding high- and low-gravity solids to the base fluids, either water or oil (water-based and oil-based muds). This activity is performed by the mud service following the information of the drilling plan. Drilling fluids (mud) have many purposes but their primary functions are the following.

- ◇ Lubricate and reduce the drill-string friction against the wall of the well. Friction produces torque and drag, which together with the material properties of the drill pipe, is the main cause limiting the maximum depth of the borehole (Chur and Oppelt, 1993).
- ◇ Lift formation cuttings to the surface and clean the hole. The cuttings of the drilled rock are transported to the surface as suspensions in the mud. These residual solids of drilling are used to analyze the characteristics for evaluating the drilled formation.

The lag–time required to lift the cuttings from the bit is related to the fluid velocity in the annulus, which is the ratio of the flow rate and area A of the annulus

$$V_{\mathrm{a}} = \frac{\mathrm{FLOW}}{A}. \tag{2.19}$$

The lag–time depends on the cutting's slip velocity which, in turn, depends on the particle properties and whether the flow is laminar or turbulent (Adams and Charrier, 1985). Values of V_{a} from 20 to 50 m/min (from 1.2 to 3 km/hour) are realistic.

◇ Compensate the borehole pressure. An incorrect mud weight can lead to a kick, due to uncompensated pressure conditions in well. This effect can be either produced by using under-weighted mud – hydrostatic pressure lower than formation pressure – or by using over-weighted mud – causing hydrostatic pressure higher than the formation pressure. In the latter case, the mud can invade the walls of the borehole, resulting in a loss of mud in the borehole. If not promptly compensated, mud loss can cause dynamic instability as the level of the mud column lowers in the well and, consequently, the mud pressure lowers also and becomes unable to compensate the borehole-formation pressure. The weight of the mud is a critical parameter decided by the drilling plan in relation to pressure conditions (Section 2.6.3), and for water-based mud may range between 1.0 g/cm^3 and more than 2.2 g/cm^3.

◇ Cool drilling via mud circulation. The downhole temperature can be simply approximated as

$$T(z) = T_0 + z \cdot G_{\mathrm{T}}, \tag{2.20}$$

where G_{T} is the gradient of the formation temperature, T_0 is the mean temperature at the surface, and z is the depth. Because the normal temperature gradient G_{T} is between 2 °C and 3 °C per 100 m depth, the normal bottomhole temperature in a vertical well of several thousand meters can be 100–150 °C. However, medium-depth geothermal fields can have temperatures as high as 450 °C. In addition, the drilling action produces friction, rock breaking and local heat, which is removed by mud circulation. Cooling by mud circulation helps prevent heat failure of downhole drilling tools.

◇ Consolidate the walls of the hole and help to prevent collapse. The formation drilled in an uncased depth interval can be subject to collapse, which can compromise the drilling. Because of rock porosity, the mud penetrates the surrounding formation and forms, in a few hours, a mudcake which helps to consolidate the walls of the well. In very fast drilling conditions, the time of consolidation becomes an important parameter and can limit the drilling rate.

◇ Due to hydraulic action, aid drilling and clean the bit pasted of mud and cuttings. The mud is ejected with high pressure (p = 7800 kPa is a realistic value) through the nozzles of the bits. In roller-cone bits, the ejection direction is oriented in order to help the rotation of the cones. The nozzles of the bits can be changed to optimize the ejection of the mud.

◇ Transmit mechanical power to downhole motors, steerable systems and turbines, via forced fluid flow. These tools are used in deviated drilling and, in many cases, also to steer vertical drilling. The mud is also used to power downhole electric generators. Downhole motor and directional tools need communication with the surface. This is obtained via mud-pulse telemetry. Little mud-turbines inserted in the drill string can produce the electric power required for this downhole instrumentation.

Water-based mud is the most commonly used drilling fluid. Mud weight and downhole pressure influence drilling performance. Oil-based muds often reduce the drilling rates with roller-cone bits, whereas PDC and diamond bits are not affected. Oil-based mud is actually believed to enhance the performance of the PDC bits since it inhibits the hydration of clay (Zijsling, 1987) and the resulting stickiness.

Drilling with air (used in particular situations) requires the use of roller-cone bits because air cannot adequately remove the produced friction heat, which causes bit failure. Cone bits are available with internal porting to keep the roller bearings cool enough. Conversely, although TSP and diamond bits do not have any moving parts, the matrix and the blade structures become weak and break (thermally stable polycrystalline TSP diamonds fail at around 750 °C and natural diamonds at 1200 °C) (Radtke, Smith, Riedel and Daniels, 1999).

2.3.15 Mud circulating line

The circulating mud system (Figure 2.17) includes the rig stand pipe, and the floating mud kelly hose connected to the top of the drill string in order to pump the mud into the drill pipes (inner flow). Then, the mud flows through the nozzles of the bit into the annulus between drill string and borehole (outer flow). To complete its circulation, the mud passes through the mud-flow return line and the solid-removal equipment. The latter removes the fractions of drilling solids (cuttings) transported to the surface. Then, the mud is conveyed in the mud tanks (pits), to which the pump-suction line is connected. The solid-removal equipment includes the mud shale shakers, the dégassers, the désanders, désilters, and centrifuges used to control the solid fraction in the mud composition. The shakers are a source of vibrations and acoustic noise in the drilling environment.

2.3.16 Logistics and laboratories

The following offices and laboratories are usually present in the drilling yard: the company-men, well-assistant and geologist cabin; the driller assistant cabin; the mudlogging cabin; the cabin of the mud-service specialists (mud engineers); the cabin of the deviation-service specialists (if required); the cabin of the specialists of the measurement-while-drilling service (if required); the cabin of the specialists of the SWD service (if required); the cabin of the security for H_2S risk (when necessary); tanks used for driller's and personnel's logistics; and electrical and mechanical workshops.

In particular, the mudlogging laboratory controls day and night safety (presence of H_2S, CO_2, and hydrocarbons) and the drilling conditions. To do so, the drilling parameters are recorded continuously and put on line. Thus, all the measured parameters

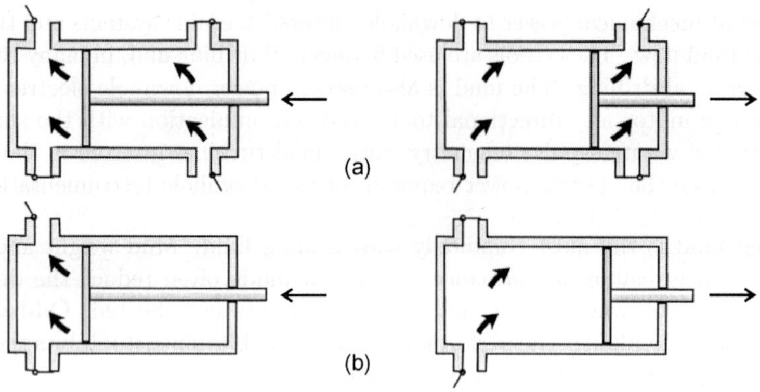

Figure 2.16: a) Double-action and b) single-action pumps.

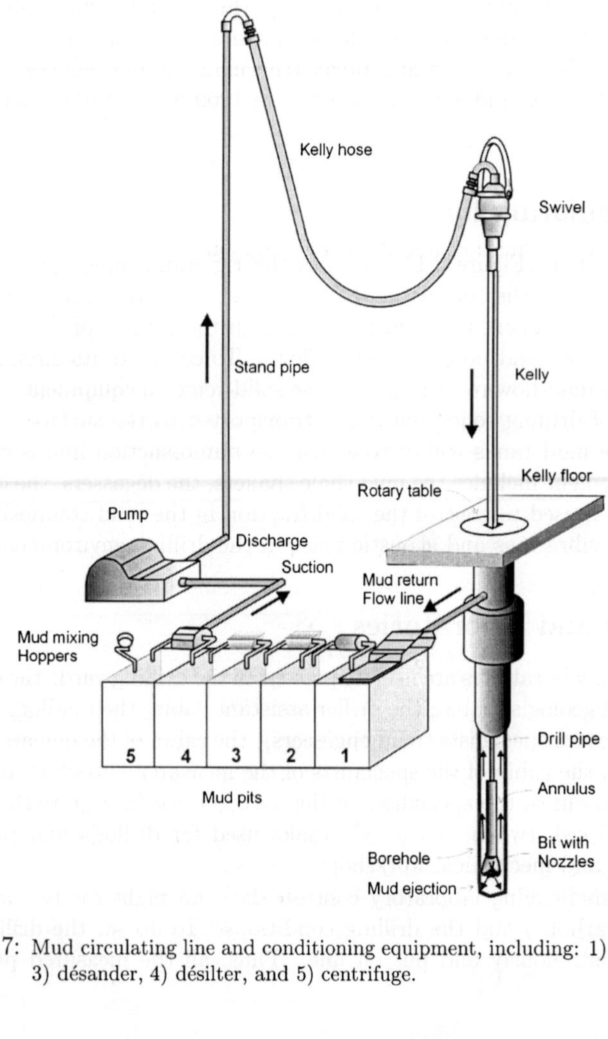

Figure 2.17: Mud circulating line and conditioning equipment, including: 1) shale shaker, 2) dégasser, 3) désander, 4) désilter, and 5) centrifuge.

that characterize drilling become available to all the other laboratories of the yard, and in some cases, via satellite transmission, to the specialists located at remote headquarters.

2.3.17 Drilling parameters and mudlogging

The control of the drilling operations may be performed by a control system located in the cabin of the driller (doghouse) located on the rotary floor, by the measurement-while-drilling (MWD) laboratory if it is present, and typically by the mudlogging unit.

The mudlogging unit controls and reports the surface drilling parameters and provides the necessary information about the drilling status to the drilling crew. These activities are continuously performed 24 h a day during the drilling of the well and include the following:

- ◇ Control of the mud level in the pits and immediate detection of possible mud losses (abnormal lower level) or "kicks" (abnormal higher level).
- ◇ Measuring the presence of gas in the mud to detect dangerous drilling conditions (kick), prevent risk of fires and risk for people (in case of H_2S, CO_2).
- ◇ Analyze the cuttings in the mud to give information about the drilled formation (Formation Evaluation Log).
- ◇ Collect and report information about drill string, bit types, mud, hydraulics and drilling operations.
- ◇ Measure the main mechanical and hydraulic parameters to control the drilling performances, and store these in the data base.

To perform the measurements, dedicated sensors (a typical number is 60) located on the rig, in the mud line and tanks are used. In many cases, these measurements are similar to those used by SWD, but typically sampled at lower rates. In fact, these values are sampled every few seconds and stored more or less every minute after averaging.

All the parameters collected by the mudlogging are coded using the digital wellsite information transfer specification (WITS). Some typical drilling parameters recorded are drilling depth and bit depth, weight on bit (WOB), rate of penetration (ROP), revolutions per minute (RPM), mud inner and outer flows (FLOW-in and FLOW-out). These parameters, and others, are used also by the SWD system, which receives the drilling parameters for automatic acquisition and data quality control.

However, in some cases the precision of the data required by drilling is different than the precision required for seismic investigations. The availability of data with the precision and range necessary for SWD purposes has to be checked carefully (Chapter 5). For example the vertical depth is given by measuring the length of the individual pipe elements of the drill string. This length is shortened by the height of the pipe top end above the kelly floor (kelly position) to obtain the actual bit depth in the well coordinate. Usually this depth can be provided with a precision of 0.25 m, which is acceptable for seismic surveys.

Table 2.3 – MWD components.

Downhole	*Surface*
Sensors	Receiver
Encoder	Decoder
Transmitter	
(with power source)	

2.3.18 Measurement while drilling and mud-pulse telemetry

As *measurement while drilling* (MWD) we define the acquisition of well data during the drilling by using tools assembled on the drilling string. To obtain this information, measurement-while-drilling systems (MWD) have been developed. The downhole measurements are used together with the surface parameters and provide valuable information (about downhole WOB, torque, temperature, inclination,...) to perform a more effective analysis of the true drilling conditions at the bit.

These measurements-while-drilling systems are based on the components of Table 2.3.

Two types of MWD systems are currently used:

Mud pulse

This system uses pressure waves in the mud; three different ways are used to produce the signals.

- ◇ An active (positive-mud-pulse) system produces a restriction of the annulus by a valve and an increase in the mud pressure, which travels in the form of acoustic guided wave in the borehole (Chapter 3).
- ◇ A negative-mud-pulse system works by opening a valve in the tool and thus discharging the mud and reducing the pressure.
- ◇ Continuous modulation produced by a internal rotating disk can continuously variate the transit section and the mud pressure. This system (mud siren) has better performance and rate of transmission of 10 bits per second.

The mud-pulse MWD can transmit signals at great distances because the attenuation is low. However the rate of transmission is limited and affected by the presence of solids in the mud.

Electromagnetic

This system uses electromagnetic guided waves in the formation and casing to transmit data to the surface (with rates of 100 bits per second). It offers the advantage of obtaining downhole-surface synchronization with good precision. Currently, the high attenuation limits its performance to shallow wells.

2.3.19 Logging while drilling

The well data acquired while drilling can be classified into different categories:

◇ Directional data: making it possible to determine the well path and BHA orientation.

◇ Drilling mechanical data: giving information about torsion and weight at the bit.

◇ Correlation data: correlation logs (e.g., short normal resistivity, GR) for well positioning in a given geological model.

◇ Formation evaluation data: set of logs suitable for petrophysical formation evaluation (e.g., porosity, hydrocarbon saturation, etc...).

The main advantage of this technology is that it provides good-quality log data while drilling. These data can be used to make immediate decisions and can be integrated with information used to steer the well while drilling if the LWD tool is used close to the drill bit. For a general description of borehole logging and formation evaluation principles see, for instance, Labo (1987).

The acquisition of petrophysical data during drilling operations is called *logging while drilling* (LWD) or *formation evaluation while drilling* (FEWD), while the directional and mechanical data are classified as MWD measurements. The LWD data can be acquired using two methods: "real-time" acquisition, by transmitting data to the surface in real time using the mud-pulse system; and "downhole-memory" acquisition, by storing data in the LWD tool memory and retrieving it from the memory each time the tool is run out of the borehole, e.g., at a bit change. The first data acquisition system is used when it is necessary to get data in real time to detect casing points, coring points, overpressured levels, hydrocarbon zones or in order to "geosteer" the well (Section 8.12.5) or obtain real-time information about the petrophysical characteristics of the drilled formation. The second data acquisition system is usually combined with the first one, allowing data acquisition with a higher sampling rate with respect to the transmitted data, for a more accurate formation evaluation. It is recommended to always transmit data to the surface in order to verify that the downhole tools work properly. MWD and LWD data are of very high value. On-site-validated data need to be of appropriate quality to be used in conventional log interpretation, as it is required for wireline log data. Fundamental aspects of this data quality relate to the proper use of the tools, the adherence to the procedures, the definition of a measure of uncertainty, the proper recording of data and integrity while storing and archiving.

LWD data are generally acquired for two reasons: real-time correlation and formation evaluation. In both cases, a number of factors (costs, benefits, risks, etc.) must be considered in order to achieve any real financial benefit to economically justify the use of LWD. The majority of cost savings are due to rig time reduction associated with wireline operations, conventional slick-line directional surveys and setup charges (particularly for offshore wells). Further cost savings can be derived from improved rates of penetration (ROP). In fact, by eliminating undesired drilling problems, better survey accuracy and real time toolface data can be obtained, resulting in smoother wellbores, faster and more accurate penetration of the target and a reduced risk of a lost well or BHA due to borehole instability, fishing and sidetracks. If only a single wireline logging service has to be run, much of the potential cost savings may be lost. So a well requiring auxiliary wireline information, e.g., dipmeter, sidewall core or formation testing, acoustic images, is less likely to be a good candidate for LWD/wireline replacement, in terms of financial reasons alone. The best wireline replacement candidates are usually limited to wells where time consuming pipe-conveyed logging (TLC) operations are required or where good geology databases exist. LWD may also be appropriate in areas of deep invasion, e.g., low pressure in depleted reservoirs, or when acquisition of wireline data is problematic due to washouts, ledges or doglegs. In some wells, LWD information is critical for either the drilling mechanics or well evaluation. Examples of these wells are:

High-profile exploratory wells, that is, wells in which the geological condition may be problematic for drilling. In this case, LWD is used for correlation, pressure detection, to pick casing points, to identify potential production intervals for early evaluation, or for "insurance logging" in the event that a wellbore may be lost. A typical example is the deep-water wildcat.

Highly deviated and horizontal wells, where obtaining pipe-conveyed (TLC) or conventional wireline logs is extremely difficult and/or risky, or where acquiring real-time information on the drilled formations is fundamental for the optimum achievement of the well objectives (geosteering).

Production wells, requiring a casing point above a greatly depleted reservoir with a high risk of lost circulation or pipe sticking.

During the last few years, there has been a fast growth in the LWD technology with reduction of tool-diameter sizes and improvement of tool types. The new generation of resistivity tools provides, for example, curves with different depth of investigation allowing log analysts to determine the invasion profile, while the azimuthal measurements help to measure formation heterogeneity. According to the purposes and the operational constraints, the tool type and the data sampling should be defined prior to logging. The LWD data can be effectively used for quantitative formation evaluation (FE), provided the dynamic environment is taken into account and the needed corrections applied. A big effort of integration between LWD service companies and the FE software providers is then necessary in order to allow the effective use of the petrophysical software chains for LWD data analysis. The LWD technique is, moreover, recommended in high-profile exploratory wells for correlation, pressure detection, insurance logging or early evaluation of potential pay intervals or in highly deviated/horizontal wells where obtaining pipe-conveyed or conventional wireline logs could be extremely risky or expensive.

2.3.20 Wellsite communication systems

As modern drilling integrates more and more sophisticated technologies, communication becomes more and more important. The well communication systems are described below.

Internal communication in the rigsite – The digital drilling parameters are transmitted by mudlogging via a local network – by using the wellsite-information-transfer-specification (WITS) protocol – to all the other cabins, where they are displayed on screen. Intercom connections are also avaliable for oral communication. Networks can be also used for local communication by TCP/IP protocol in current rig sites.

External communication between rig and remote quarters – The drilling data and conditions are communicated to the headquarters via a dedicated radio link. This may include satellite and standard telephone connections. In this way, drilling decisions are continuously shared among and supported by the operating districts. Communication via the internet may be also used.

Communication between surface and downhole and vice versa – The communication between surface and the bit location is used to receive information from and transmit commands to downhole systems that perform processes, which need to be controlled in real time. Typically, this communication is obtained by measurements while drilling (MWD) mud-pulse telemetry (Section 2.3.18).

2.4 Drilling offshore

Drilling in deep and ultra-deep water is one of the main goals for the development of oil fields in new exploration areas, and while-drilling monitoring of these applications is one important goal of the drill-bit SWD. Drilling offshore is expensive, particularly with the technologies needed for facing the severe environmental sea problems. Cost increases drastically with the water depth. In Table 2.4, the wells are classified with respect to the water depth.

Drilling offshore is performed by using marine rig systems, described below. Supply vessels continuously support the rig, provide the drilling materials, and generate also environmental noise for SWD. Marine rigs have sleeping quarters for the crew and galley facilities, standby boats for safety purposes, a heliport, and are organized in self-contained or tendered configurations. Life and operations must follow strictly the safety rules on board. The marine rigs (Figure 2.18) may be either bottom supported systems or floating systems.

Bottom supported systems

These systems stand on the sea floor and include the following types (Adams and Charrier, 1985):

Barge and submersible rig (Figure 2.18a) – These are mobile rigs that are sunk to the bottom by flooding the vessel compartments. Barges are used typically in 2 to 6 m of water. In the presence of poor seafloor support, shell mat padding is required. Submersible rigs are provided with stabilizing columns of large diameter that are placed on a submersible hull, which is put down on the sea floor. The maximum water depth for these rigs is 100 m.

Jackup rig (Figure 2.18c) – These rigs have three to five legs which can be 60 to 120 m long and can be lowered and fixed by adapting their length to the sea-floor conditions at the site. After drilling the well, the legs of jackups are raised and the rigs become mobile rigs that can be towed by a vessel. In areas with soft soil, mat bases are used under the rig legs.

Platform rig (Figure 2.18e) – These rigs are installed on the production platform in production areas and are used to drill a number of directional wells (a cluster) from the same platform, thus reducing costs. Platforms rigs can be self-contained or provided with a tender to support personnel and auxiliary drilling equipment.

Floating systems

This category includes the following types:

Semi-submersible rig (Figure 2.18d) – These are floating rigs that do not rest on the sea floor during drilling, and have no restrictions on the length of the legs. Their position can be maintained either by active propulsion (dynamic positioning system), which continuously compensates the changes of position controlled by a global position satellite (GPS), or by anchored chains. These rigs are characterized by their large size (rating weights from 8 000 to 20 000 tons), and have high stability with a proper oscillation motion with the period outside the range of the period of sea waves. These rigs can operate in medium-deep water.

Drillship rig (Figure 2.18f) – These rigs are self propelled and are provided with dynamic positioning or can by anchored to maintain position, and can operate in medium-deep and deep water.

It has been shown (oral communication at the Schlumberger Conference on SWD , Cambridge, 1996) that, for floating rigs, the floating effects can influence the SWD signal transmitted in the drill string. In fact, floating may modify the reflection of drill-bit waves at the top of the drill string (Chapter 4). In Chapter 5, we will analyze the impact of the different rig technologies on seismic-while-drilling measurements.

The wellhead can be on the sea floor or on the rig. With floating rigs, it is necessary to connect the sea-bottom wellhead to the surface rig. The rig is connected to the wellhead through a riser. This is a special tube that protects the drill string and contains the mud output line. It has to be flexible and elongatable as a telescope in order to support limited deviations of the rig; it must be light so that it does not compress its lower part, and robust enough to support high pressure differences between mud and sea water. The riser is fixed by two connectors (one at the rig and the other at the sea floor). Rigs have a rapid release system for the riser, for emergency. This system makes it possible to free and move the rig from its position when necessary due to bad sea conditions.

2.5 Directional and deviated wells

Because drilling a vertical well, in many cases, is either not desirable or possible, directional drilling of a deviated or horizontal well may be preferable or necessary. The definitions of vertical and deviated wells are the following:

Table 2.4 – Depth classification of offshore wells

Name	Water depth [m]
Shallow water	less than 100
Medium-shallow water	from 100 to 400
Medium water	from 400 to 1500
Deep water	from 1500 to 2000
Ultradeep water	after 2000

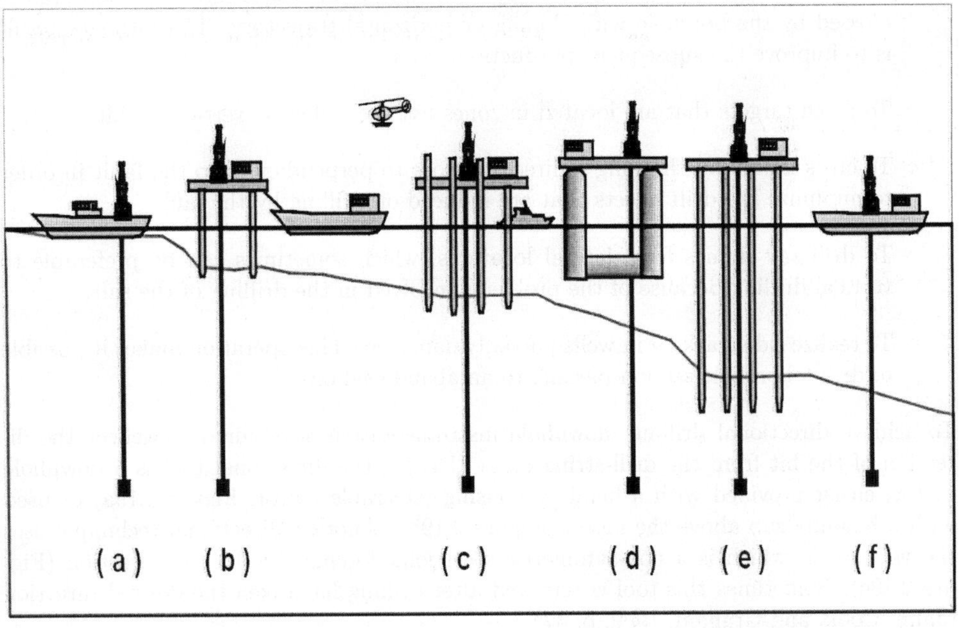

Figure 2.18: Marine rigs. a) Swamp barge, b) jacket with tender, c) jack-up, d) semi-submersible, e) fixed platform, f) drillship. Modified after Jahn, Cook and Graham (1998).

Vertical well – This is a well in which the initial horizontal coordinates (x northing and y easting, say) and the horizontal geographic coordinates of the well target are almost equal.

Deviated well – This is a well in which the initial horizontal coordinates and those of the well target are different.

Wells can be deviated for two reasons. The deviation can be planned, and in this case a directional or "horizontal" well is drilled. In other cases, spontaneous deviation can occur due to drilling problems, which have to be controlled by adapting the drilling parameters (see Section 2.6.4), and may require corrective actions. Acceptable values of spontaneous deviation angles are less than 10°.

2.5.1 Directional drilling

Directional drilling is undertaken for the following reasons:

◇ To optimize drilling, as several wells and side tracks (i.e., another well drilled using the upper part of the first well) can be drilled from the same onshore site or marine platform.

◇ To optimize the approach to the target, so that an increased interval of reservoir is crossed by the borehole with sloping or horizontal trajectory. The ultimate result is to improve the subsequent production.

◇ To reach targets that are located in zones not accessible by vertical drilling.

◇ To cross faults, by choosing a direction close to perpendicular to the fault in order to minimize the drift effects that are induced on drilling by the fault.

◇ To drill salt domes from lateral locations, which sometimes can be preferable to vertical drilling because of the problems involved in the drilling of the salt.

◇ To realize side tracking in wells partially damaged. This operation makes it possible to drill a borehole section parallel to an abandoned one.

To achieve directional drilling, downhole instrumentation is required to deflect the direction of the bit from the drill-string axis. Usually, the directional tool is a downhole motor, either provided with a bending housing (steerable motor; Figure 2.19a) or used with a bending sub above the motor (Figure 2.19b). Another directional technique uses the whipstock, which is a non-symmetric steel joint forcing the drilling direction (Figure 2.19c). Sometimes this tool is removed after drilling has taken the desired direction (Jahn, Cook, and Graham, 1999, p. 47).

The following main phases are involved in drilling deviated wells:

1. An initial vertical section is drilled by using a standard drill string, up to the point (kick-off point) at which the planned deviation starts (Figure 2.21).

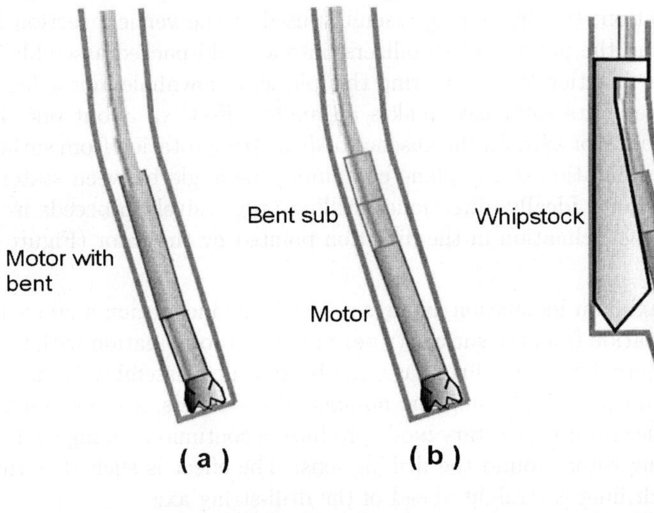

Figure 2.19: Deviation tools. a) Steerable downhole motor provided with bent housing; b) Downhole motor with bent sub above; c) Whipstock (this tool is inserted for deviation and sometimes is removed after that deviation has taken the desired direction).

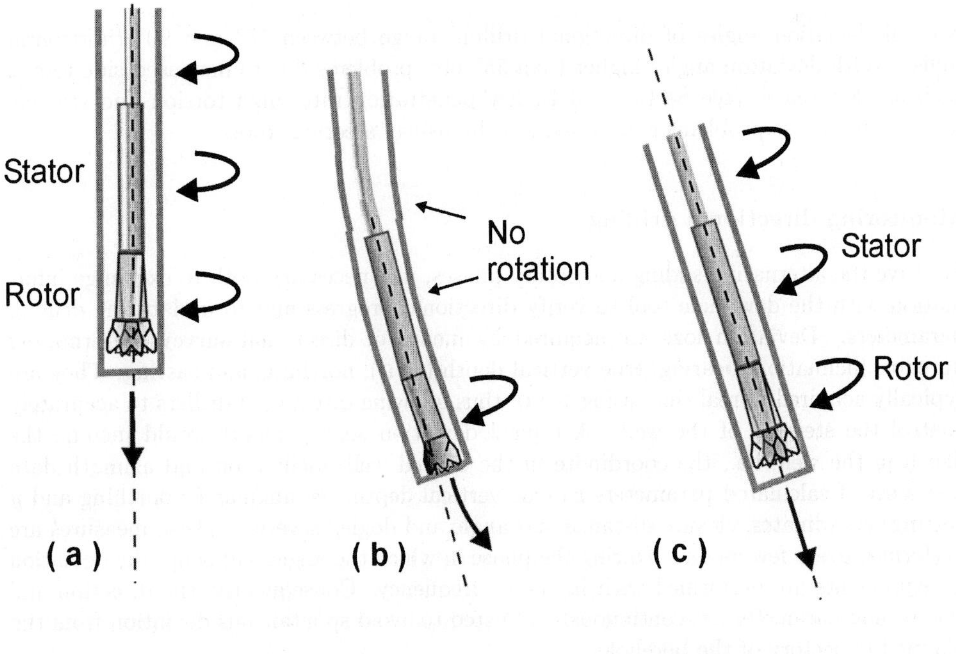

Figure 2.20: Sequence of a) "rotary", b) "sliding" and c) again "rotary" drilling modes.

2. After the kick-off point, the drilling angle increases to its maximum inclination. During deviation, the drill-string assembly used in the vertical section is changed, by diminishing the number of stabilizers, into a "mild-packed assembly" with only one stabilizer (Section 2.6.4). During this phase, a downhole-motor for directional drilling is used. Its rotor axis makes an angle (offset) of about one degree with respect to the stator axis. In the absence of drill-string rotation from surface (sliding mode), the orientation of the plane containing the angle between stator and rotor remains constant. Ideally, directional drilling progressively proceeds in this plane, with increasing inclination in the direction pointed by the rotor (Figure 2.20b).

3. Once the maximum inclination angle is reached, the inclination angle is maintained by adding rotation from the surface (rotary mode) in combination with the downhole rotation (Figure 2.20c). In this phase, the bottom hole assembly is more rigid (this effect is obtained by increasing the number of stabilizers, see Section 2.6.4). The rotation of the stator with rotary mode produces a continuous change of the direction of the bending rotor around the drilling axis. The effect is such that the resulting direction of drilling is straight ahead of the drill-string axis.

4. This "tangential" drilling condition may either continue straight ahead, or change again when another point, the drop-off point (Figure 2.21), is reached. Here opposite deviation starts to return to the vertical inclination with a procedure similar to that already used in the kick-off phase, with a mild drill-string of type "pendulum assembly" (Section 2.6.4).

Normal deviation angles of directional drilling range between 15° and 90° (horizontal wells). With deviation angles higher than 55°–60°, problems due to increased side forces, such as "key-seats" (see Section 2.6.4), low penetration rate, high torsion and sticking may occur. These problems can be reduced by using "steering" tools.

Monitoring directional drilling

To drive the alternating sliding and rotary phases, it is necessary both to exchange information with the downhole tool to verify directional progress and to update the drilling parameters. Deviation logs are acquired by means of directional surveys determining borehole inclination, bearing, true vertical depth (TVD), northing, and easting. They are typically acquired in real time using MWD, thus allowing directional drillers to accurately control the steering of the well. A typical deviation survey report would include the depth in the well, i.e., the coordinate in the curved well, inclination and azimuth data and a set of calculated parameters as true vertical depth, rectangular (x northing and y easting) coordinates, closure distance and angle and dogleg severity. These measures are performed every few meters. During the phase in which the angle builds up, the deviation measurements are performed with increased frequency. Consequently, the direction and the drilling parameter are continuously adjusted to avoid spontaneous deviation from the planned trajectory of the borehole.

Axial loads in directional wells

The equation used to calculate the tension transmitted through the pipe and bit load (equation (2.11)) has to be modified in directional wells, by taking into account the reaction force of the wall against the inclined drill string. This reaction force is opposite to the weight component normal to the wall of the inclined well. Let α_j be the inclination (Figure 2.21) of the j-th string section of weight in air WDS$_j$, Δl_j be its length, and C_B be the mud buoyancy factor. The hookload can be calculated as

$$F_H = C_B \sum_j \cos \alpha_j \Delta l_j \cdot \mathrm{WDS}_j - \mathrm{WOB}, \tag{2.21}$$

where WOB is the weight load on bit. Drag friction forces are not included in equation (2.21). However, drag-friction forces are relevant – often critical – in directional wells.

Drill-string rotation and axial forces

In general, the transmission of WOB is facilitated by the rotation of the drill pipes. Contrary, in the sliding mode larger drag static-friction occurs, and may cause a reduction of WOB transmission. For example, in a 50° inclined well, less than 10 % WOB losses due to drag friction in the rotary mode have been measured, while 20–40 % WOB losses have been measured in sliding mode (Lesage, Falconer, and Wick, 1988). Hence, the value of the true weight on bit can significantly differ from the value calculated without drag friction. In the absence of drill-string rotation from the surface, drag friction effects increase several times (Lesage, Falconer, and Wick, 1988). In addition, string buckling may occur in pipes subject to higher drag forces (He, Hasley, and Kyllingstad, 1995). In general, sliding and rotary drilling modes – in directional as well as in vertical wells – are important conditions for SWD. With sliding, frequently there is loss of drill-bit signal transmission through the drill string (Chapter 3).

2.5.2 Horizontal and extended-reach drilling

Horizontal drilling was introduced to improve the productivity of wells and enhance oil recovery because of the increased draining surface in the horizontal dimension of the reservoir. This technology is much more expensive than vertical drilling and requires accurate directional surveying and gathering of MWD information. Nevertheless, it makes it possible to drill the reservoir along its larger dimension and to improve recovery by a factor of six.

Horizontal well are classified by their radius of curvature (or the equivalent build up rate) (Jahn, Cook, and Graham, 1999):

◇ Short radius (from 50 to 20 deg/30-m).

◇ Medium radius (from 8 to 20 deg/30-m).

◇ Long radius (from 0.1 to 8 deg/30-m).

Extended reach drilling (ERD) is defined as having the horizontal displacement at least twice the vertical well depth. These wells are used when platforms exist in marginal areas, and to reduce the number of required platforms.

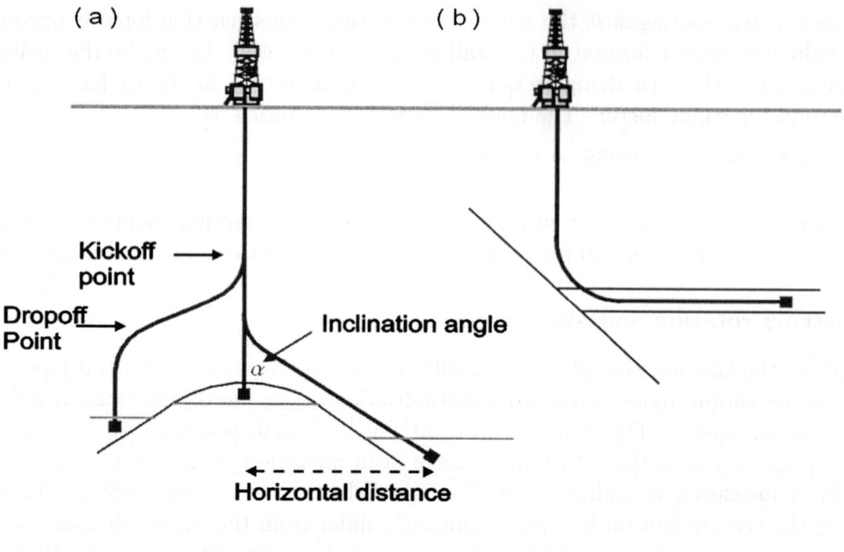

Figure 2.21: Layout of directional wells. a) Deviated and tangent, b) horizontal wells.

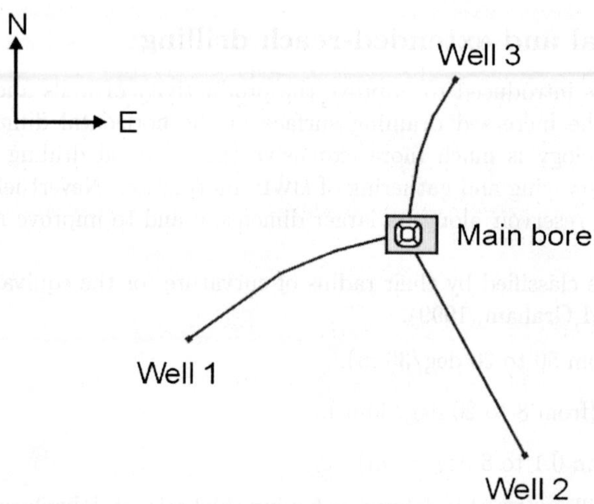

Figure 2.22: Multilateral wells drilled from the same main bore (top view).

2.5.3 Multi-lateral wells

This technology makes it possible to drill several lateral wells and branches from a principal well (main bore, Figure 2.22). This technology has higher risks for drilling, in particular for the lower well sections. However, it gives advantages in production and makes it possible to produce from reservoirs whose conditions are unfavorable for conventional techniques.

2.5.4 Steering of drilling

Steerability is defined as "the ease with which the course or direction of a well can be changed". New bit design features have been applied to allow and enhance steerability. Fixed PDC cutters placed along the gage, shorter cutters, gages and overall bit lengths, as well as flattened profiles to increase sidecutting aggressiveness, are all factors that enhance steerability.

Furthermore, advanced downhole drilling tools are used to steer drilling and improve well verticality (Oppelt, Chur, Feld and Juergens, 1991; Chur and Oppelt, 1993). The recent developments of MWD systems allow the exchange of information between tool and surface without drilling interruptions (Ligrone, Oppelt and Calderoni, 1996; Poli, Donati, Donati and Ragnitz, 1996).

To solve the problem of maintaining the drilling direction, automatic devices – such as the rotary closing loop system (RCLS) – were developed from this concept (Figure 2.23). The steering tool consists of a downhole motor integrated with other downhole tools. These are: the measurement-while-drilling (MWD) device, the mud pulser, a variable stabilizer with retractile arms (ribs) that make it possible to control the inclination and the direction while drilling, an AC local power generator, inclination sensors and the electronics used to drive the tool. When a spontaneous deviation from the theoretical trajectory occurs, the sensors detect it and the tool compensates for it by retracting or extending the ribs of the steerable stabilizer. In this way, the RCLS generates the lateral force needed to return to the right direction. The system transmits the information to the surface operator, who analyzes it and changes the tool operating parameters.

It was shown that using such a system produces wells with improved stability, improves rate of penetration (ROP), reduces power dissipation for friction effects, and produces better well profiles. Steering performed by using geological and geophysical while-drilling information (geosteering) is described in Chapter 8.

2.5.5 Slim holes and coil tubing

A slim hole is a well in which the first diameter is small, i.e., 9 in. or 7 in. or less, and the diameter of the production casing is 2–3 in. Slim holes are drilled for several reasons using a small-size rig.

Coiled tubing drilling uses – unlike conventional pipe addition system – a continuous pipe without interruption and with downhole motor. The tube is unrolled down into the well from a drum. This technology has many advantages, one of which is the possibility to insert wires for SWD measurement and MWD communication.

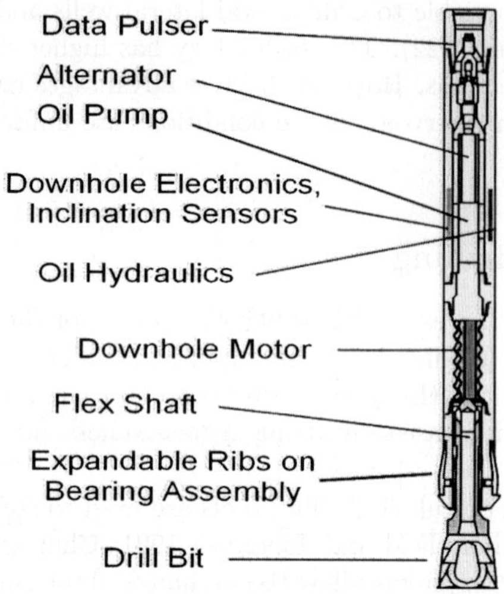

Figure 2.23: Layout of the straight-hole drilling device (SDD) steering tool. (Courtesy of ENI E&P Division).

Coil tubing is used only in slim holes in the last section of the well, when drilling the production phase (reservoir section).

2.6 Designing a well

The well design is defined as the desired final status of the well (Devereux, 1998). It involves many aspects of which we consider only the following points more related to seismic while drilling.

- ◇ Evaluation of the borehole pressure gradient.
- ◇ Selection of the casing points (seats).
- ◇ Design of the mud plan.
- ◇ Design of the BHA and drill-string stabilization.
- ◇ Bit planning.

The design of the daily drilling program is built up on the basis of general drilling considerations and, if available, previous well information. The daily program presents the expected depth of the well with respect to days. This information is collected and used in the SWD feasibility study (Figure 2.24).

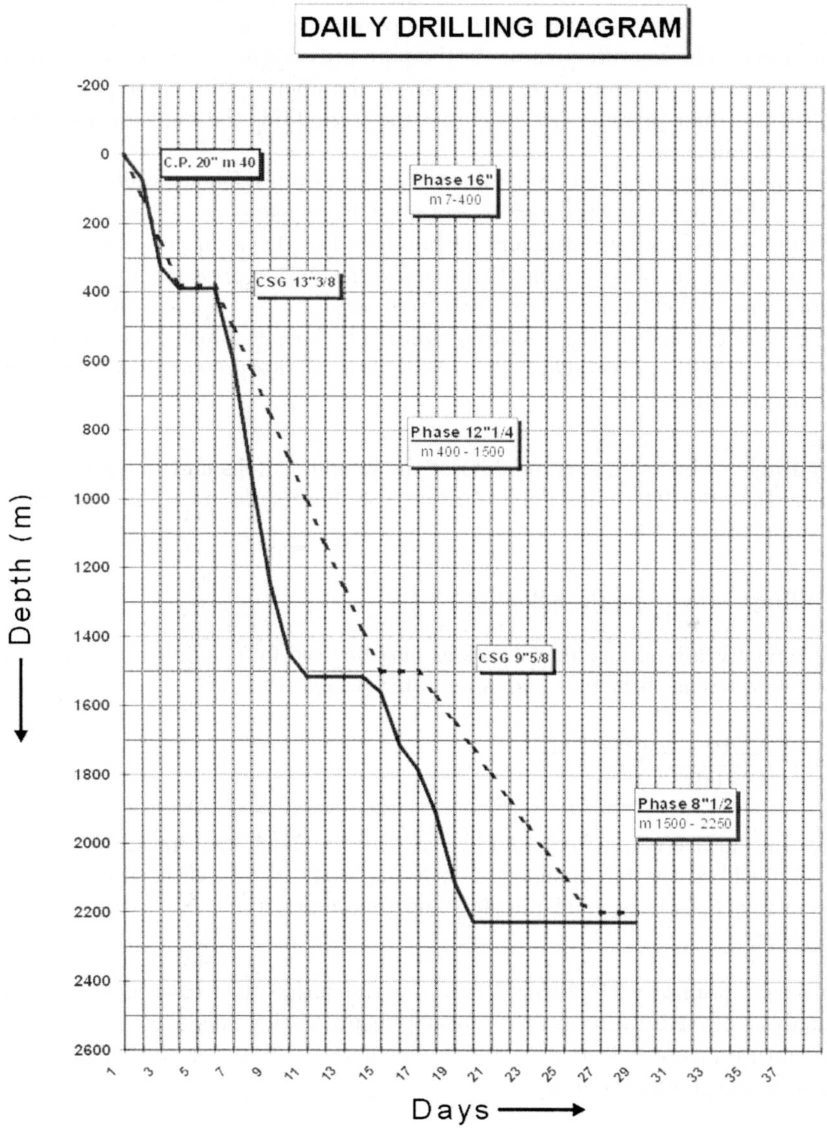

Figure 2.24: Expected (dashed line) and measured (continuous line) daily drilling program (drilling depth versus days). Interruptions of drilling for operations between two different casing phases can be observed near depths of 400 and 1500 m.

2.6.1 Evaluation of the borehole pressure

To drill safely, a well-drilling program must contain the analysis of the different pressure types – overburden, pore, and fracture – and the temperature profiles. The overburden pressure is the pressure of the rock matrix. The pore pressure is the pressure of the fluid in the rock pores. The fracture pressure is the pore pressure at which the pressure itself starts to fracture the formation. In practice, this is the upper limit for the pore pressure, because after the fluid pressure has fractured the formation, the fluid escapes and the pore pressure diminishes.

The overburden pressure can be obtained from logs from reference wells or from seismic analysis. The pore pressure can be obtained from reference-well data (formation and production tests, electric logs, and calculations based on surface drilling parameters) or from seismic analysis. Normal (hydrostatic) formation pressure gradient is about 1.03 kg/cm^2/10m ($\sim$ 0.465 psi/ft), but varies from region to region. Once the overburden and pore pressure are known, the fracture gradient can be calculated, for example by using the Terzaghi equation (Terzaghi, 1943). Assessment of overpressure can be performed before, while, and after drilling.

Before drilling – by using existing data of the reference wells and/or seismic surveys (for instance, Bowers, 1995; Bilgeri and Carlini, 1988; Sayers, Johnson and Denyer, 2000; Kan and Swan,2001).

While drilling – by using the following:

◇ Real-time drilling-parameter indications may be used for example by using the empiric "*d*-exponent" formula (Jorden and Shirley, 1978)

$$d = \frac{\log(\mathrm{ROP}) - \log(60\mathrm{RPM})}{\log(12\mathrm{WOB}) - \log(10^6\,D)}, \tag{2.22}$$

where the surface parameters are given as ROP [ft/hour], RPM [rev/min], WOB [lbs] and D [inch] is the bit diameter. Bottomhole MWD data such as torque, WOB, gamma ray, resistivity, mud pressure and temperature are also used as real-time indicators.

◇ Indicators obtained with the mud raised at the surface (lag-time indicators), such as analysis of cuttings to evaluate formation properties, mud-gas monitoring and temperature measurements analyzed to calculate the heat transfer in the borehole (Kårstad and Aadnøy, 1997).

◇ Seismic while drilling to correct the initial geophysical information and predict pressure conditions ahead of the bit. A more detailed discussion on the potential of SWD for prediction and overpressure evaluation is contained in Chapter 8.

After drilling – which includes the analysis of density, sonic, resistivity (Eaton, 1975), neutron, induction and temperature well logs.

2.6.2 Selection of the casing depths (seats)

One of the most critical aspects of the well design is the selection of the depths at which to set the casing. For this purpose, the following parameters are taken into account:

◇ The total depth of the well.

◇ The pore- and fracture-pressure gradients.

◇ The probability of encountering shallow gas pockets and problem zones.

◇ The depths of the targets.

◇ The time limits on open-hole drilling.

◇ The compatibility of the casing program with the wellhead systems.

◇ For production wells, the compatibility of the casing program with the completion program.

◇ Casing availability – size, grade and weight – and economics.

The selection is performed by evaluating the existing seismic and geological documentation and the drilling data of wells in the area.

The key factor determining the depth locations of the casing seats is the assessment of pore-pressure and fracture-pressure profiles in the well. As the pore pressure in a formation being drilled approaches the fracture pressure at the last casing seat, the installation of an additional string of casing is necessary. This is done to prevent the weight of the mud (used to compensate the pressure of the formation being drilled) from fracturing the formation at the last casing seat. Figure 2.25 is an example of idealized casing-seat selection. It does not include any safety or drill-pipe trip margins, which would, in practice, be taken into account.

2.6.3 Design of the mud plan and subsurface well control

The mud selection is based on many factors, including the mud's chemical properties and viscosity. The design of the mud-weight plan is important. From a general point of view, it would be desirable to use the lowest possible weight to achieve the maximum drilling rate and minimize the problems of lost circulation. However, the pressure applied by the mud must be greater than the highest formation pressure to effect borehole pressure control.

To determine the required mud weight, it is necessary to predict both the formation pore pressures and the fracture pressure. The selected mud weight must exceed the formation pore pressures and, at the same time, not exceed the fracture pressure in each section. For these reasons, a balancing between satisfying well-pressure control and not exceeding the rock strength in weak zones is required.

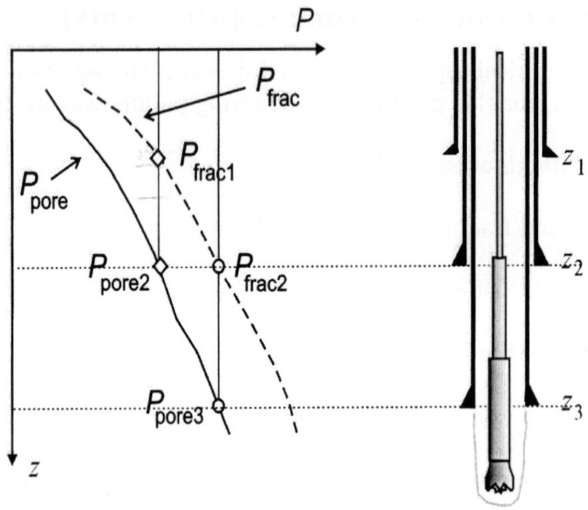

Figure 2.25: The casing is set at depth z_1, where the pore pressure is P_{pore1} and the fracture pressure is P_{frac1}. Drilling continues to depth z_2, where the pore pressure P_{pore2} has risen to almost equal the fracture pressure (P_{frac1}) at the first casing seat. Another casing string is therefore set at this depth, with fracture pressure (P_{frac2}). Drilling can thus continue to depth z_3, where pore pressure P_{pore3} is almost equal to the fracture pressure (P_{frac2}) at the previous casing seat. Modified after Magarini and Monaci (1999).

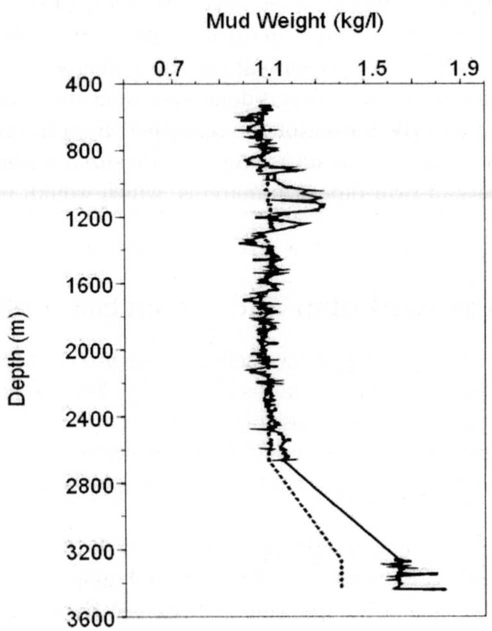

Figure 2.26: Mud weight, planned before (dotted line) and used (continuous line) while drilling.

It is important that overpressure conditions are predicted and monitored while drilling. Once the formation pressures for a well section are known, a safety margin is added and the mud density is calculated as

$$w_{\rm m} = \frac{p_{\rm f} + p_{\rm SM}}{g \cdot z_{\rm true}}, \tag{2.23}$$

where g is the acceleration of gravity, $p_{\rm f}$ and $p_{\rm SM}$ are the pore pressure and the security margin, respectively, and $z_{\rm true}$ is the true vertical depth (TVD). For example, if a formation has a pressure of 27.7 MPa at 2600 m and a safety margin of 4.1 MPa is desired, the required mud density is $w_{\rm m} = 1.25$ g/cm^3. Safety mud-density margins are usually around 0.025 g/cm^3 but may vary according to conditions. The mud density is increased by the addition of heavy solids, such as barite (the properties of the different mud compositions are analyzed more in detail in Chapter 4 for acoustic purposes).

The mud pressure for a given depth and mud density is given by (Figure 2.26) and, consequently, we can write

$$p_{\rm h} = g \cdot w_{\rm m} \cdot z_{\rm true}, \tag{2.24}$$

assuming that the mud density is constant in the borehole at all depths. The pressure gradient in the mud column is

$$\frac{\partial p_{\rm h}}{\partial z_{\rm true}} = g \cdot w_{\rm m}, \tag{2.25}$$

that is the pressure gradient equals the mud weight per unit volume (mud weight). For this reason, drillers often prefer to use the concepts of "pressure gradient" and "mud weight" to calculate the pressures.

2.6.4 Design of the bottom-hole assembly

Understanding drill-string and bottom-hole-assembly (BHA) behavior is important for seismic-while-drilling purposes because it determines many operational and noise conditions and acoustic signal transmission from bottomhole (see also Chapter 3). Hence, the analysis of the BHA design and drill-string stabilization is important for SWD also.

A requirement for the BHA design is to maintain the vertical drilling condition if deviation is not planned. For instance, a tendency to deviate from the straight-vertical path may be expected for a vertical borehole to be drilled in dipping formations. Therefore, the selection of a well-designed bottom-hole assembly is done to control the veering off-line, which has to be maintained within acceptable pre-planned limits (Magarini and Monaci, 1999).

The following problems are related to non-vertical drilling and deviation.

Dogleg severity, drill-pipe fatigue and key-seat problems

Doglegs refer to large changes of direction occurring over short well paths (Figure 2.27a). Doglegs have a short curvature radius and can produce high bending stress on the pipe (see Section 2.6.6) with risk of pipe failure (Adams and Charrier, 1985). In addition, "key seats" (Figure 2.27b) can also be excavated by the rotating pipe. This is due to the high friction against the wall produced by the lateral force of the pipe under tensile load and constrained to bend at the dogleg position. The rate of change of the angle is measured in degrees per 30 m of drilled interval. A method for calculating the dogleg severity is given by the API formula (Gabolde and Nguyen, 1999), which takes into account the inclination and the azimuth direction of the well path at the initial and final points of the analyzed interval.

Figure 2.27a shows a severe dogleg condition which produces fatigue failure in drill pipes. The tensile stress at point B is greater than at point A. As the pipe is rotated, point A moves from the inside to the outside of the bend, so that the pipe is subjected to both minimum and maximum tension every rotation. These cyclic stress reversals cause fatigue failures in the drill pipes, which occur usually within the first two feet (0.6 m) of the body adjacent to the tool joint (where the bending stress concentrates because of the abrupt change of the section).

To avoid rapid fatigue failure of pipe, the rate of change of the hole angle must be controlled. Suggested limits can range from 4° to 6° per 30 m, depending on the grade of the pipe and on the tensile load on the pipe. Considerations about dogleg severity and key seats as causes of additional noise for SWD with directional-well drilling are discussed in Chapter 3.

Sticking of pipes

Sticking can occur when the borehole walls heave or collapse and also when extra large outer-diameter drill collars contact a key seat during the trip out of the drill string (Magarini and Monaci, 1999). Logging tools and wire line can become stuck in key seats, and the wall of the hole can also be damaged, causing future hole problems. Furthermore, running the casing pipe through a high-severity dogleg can cause serious problems. If the casing becomes stuck in a dogleg, it will not extend through the productive zone. This would make it necessary to drill out the casing base (shoe) and set a smaller size casing through the productive interval. Even if this corrective action is successful, the casing could be severely damaged and compromise future production operations.

Other problems related to doglegs

Doglegs force the casing against the wall of the hole, preventing the circulation of the cement between the wall-hole and the casing and preventing a good bond at this point. Furthermore, the lateral force of the drill pipe rotating against the casing in the dogleg or dragged through it in a pipe trip, can cause high wear of the casing. This could cause drilling problems and/or a possible serious kick. In addition, high casing wear conditions while drilling – due to pipe-tool-joint contacts – may cause noise for SWD data.

2.6.5 BHA rigidity and drill-string stabilization

A main factor to be considered in the design of the bottom-hole assembly is the bending characteristics of the drill stem. In an inclined drill stem with no weight on the bit (WOB), the component of the weight of the string portion between the bit and the point of contact

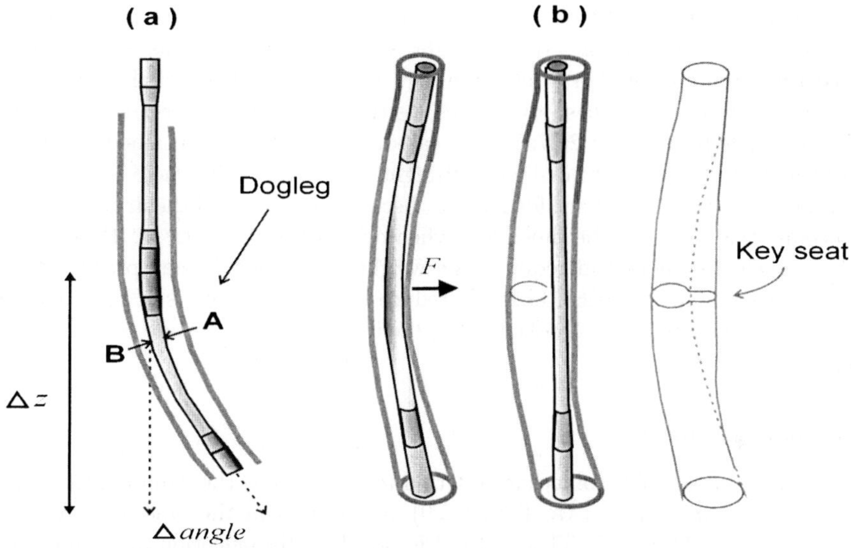

Figure 2.27: a) Dogleg severity measured by angle over depth increment. The rotating pipe is in maximum tension at point B, and it is in minimum tension at point A. Alternation of stress causes failure. b) Rotating pipe with lateral force may cause wear and "keyseat" shaping of the borehole. This causes pipe sticking.

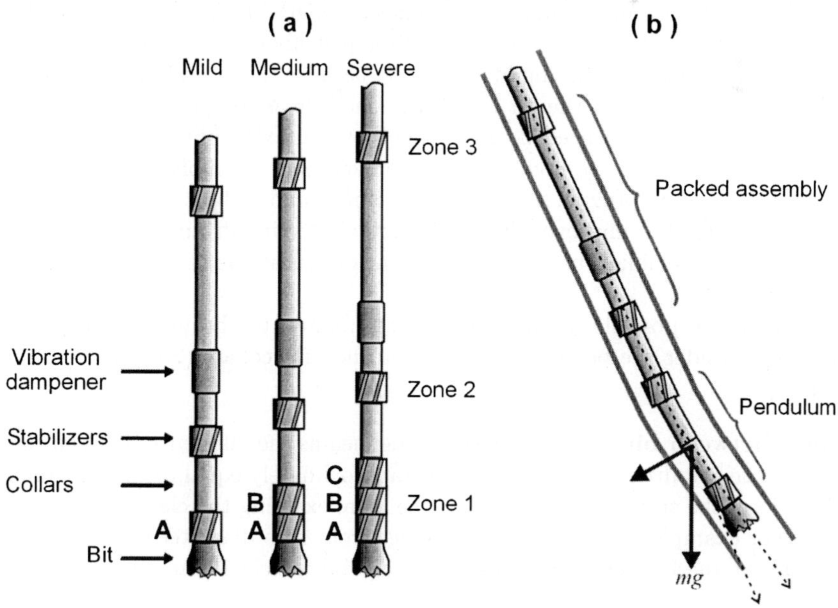

Figure 2.28: a) Mild, medium and severe packed-hole assemblies. b) Pendulum assembly. Modified after Adams and Charrier (1985).

with the borehole wall is the only force acting on the bit. This force tends to bring the hole back towards the vertical. When weight is applied to the bit, as is required to drill the formation, there is another force on the bit, which tends to direct the hole away from the vertical. The result of these two forces, vertical and axial, respectively, may be that the drilling angle (see Figure 2.27a) is increased, decreased or maintained constant.

In general, the problem of drilling a straight or nearly vertical hole is much easier in soft formations than in very hard formations, because the effects of the drill-string bending and encountering dips may be much less when drilling soft formations. Hard formations that have high dip angles and require high weight on bit (WOB) may present difficulties for drilling a straight or vertical hole. To reduce the possible causes of bit deviation and the problems associated with crooked holes, the following solutions are considered in BHA design.

Packed-hole assembly

The packed-hole assembly (sometimes referred to in jargon as the "gun barrel" approach) has a series of stabilizers (STAB, Figure 2.9) and is used in the hole already drilled to guide the bit straight ahead. This method is based on the selection of a BHA with drill collar with the necessary flexural rigidity (stiffness) (see Section 2.6.6) and wall-contact tools (stabilizers) to force the bit to drill in the direction of the hole already drilled. If the proper selection of the drill collars and bottomhole tools is made, only gradual changes in hole angle will develop during drilling (this should create a useful hole for further drilling and completion).

The wall-contact assemblies must have sufficient length of contact to ensure alignment with the hole already drilled. A single stabilizer just above the bit generally acts as a fulcrum or pivot point and will "build" the angle because the lateral force of the unstabilized collars above will cause the bit to push to one side as weight is applied. Therefore, other stabilizing points are needed, and three (soft-packed), four (mild-packed) or five (severe-packed) stabilizers are used to form a packed-hole assembly. Figure 2.28a shows basic assemblies used to provide the necessary stiffness and stabilization for a packed hole assembly. However, the improvement of the actual steering-drilling technology reduces the need for such assemblies, based on these BHA configurations of variable "rigidity", for directional control.

To evaluate the possible mechanical impacts related to the use of the packed-hole assembly, we consider the bending properties of the drill collars (Section 2.6.6) and the following aspects:

Clearance between bit and stabilizers – defined as the difference between the diameter of the stabilizer and that of the bit (approximately equal to the hole diameter). The closer the stabilizer is to the bit, the more exacting the clearance requirements are. For example, a 1/16 in. under gauge from hole diameter can be satisfactory just above the bit, and 1/8 in. clearance can be critical 100 m above the bit.

Wall support and length of the stabilizers – The BHA must adequately contact the wall of the hole to stabilize the bit and centralize the drill collars. The length of contact needed between the tool and the wall of the hole will be determined by the formation. The contact surface must be sufficient to prevent the stabilizer from digging into the hole wall. If the formation is hard and uniform, a short narrow contact surface is adequate and will ensure proper stabilization. If the formation is soft and unconsolidated, a long stabilizer may be required. Hole enlargement in formations that erode quickly tends to reduce the effective alignment of the BHA.

The proper design of a packed bottom-hole assembly requires the knowledge of the crooked-hole tendencies (classified as mild, medium and severe) and the drillability of the formations to be drilled (hard to medium hard, abrasive, non abrasive and medium hard to soft formation).

Pendulum assembly

The "pendulum" method is based on the principle that – when no downhole-motor steering is used – the only force available to straighten a deviated hole back towards the vertical is the weight of the drill collars between the point of contact (stabilizer) and the bit. In the packed-pendulum technique, the collars are slung below the regular packed-hole assembly (Figure 2.28b). The pendulum method is used only as a corrective measure to reduce the angle when the maximum permissible deviation has been reached. The forces acting upon the bit are:

◇ The axial load supplied by the weight of the drill collars.

◇ The lateral force, the weight of the drill collar between the bit and the nearest point of contact with the wall of the hole by the drill collar, i.e., the tendency of the unsupported length of the drill collar to swing over against the low side of the hole due to gravity.

◇ The reaction of the formation to these loads may be resolved into two forces, one parallel to the axis of the hole and one perpendicular to the axis of the hole.

When the hole deviation has dropped to an acceptable limit, the pendulum collars are removed and the packed-hole assembly again is used.

Table 2.5 – General guidelines for BHA design

Packed-hole assemblies should be used unless otherwise dictated by hole conditions.

Standard packed-hole assembly
Bit
Near-bit stabilizer
Short drill collar (7ft = 2.1 m)
String stabilizer
K monel drill collar
String stabilizer
2 drill collars
String stabilizer

Pendulum forces and bending angles, considering BHA dimensions, stabilizer spacing, formation dip angle and anisotropy in the bit/rock interaction have been calculated, for instance, by Williamson and Lubinski (1987). Figure 2.29 shows a BHA record, typically available in the field, which can be used to calculate the vibration properties of pipes (Chapters 3 and 6).

Reducing weight on bit

A reduction of the bit weight is usually required when changing from a packed-hole assembly to a pendulum-drilling operation. In general, by reducing the weight on the bit, the bending tendency of the drill string is reduced and the hole will be straighter. One of the earliest techniques for straightening the hole was to reduce the weight on the bit and speed up the rotary table. In recent years, it was found that this is not always the best procedure because it sacrifices considerable penetration rate. Worse than that, reducing WOB frequently causes doglegs. The straightening of a hole by reducing bit weight should be done very gradually so that the hole will tend to return to the vertical without sharp bends (Magarini and Monaci, 1999).

As a "rule of thumb", it is recommended a weight-on-bit value of 1.2 to 2.0 tons per inch of hole diameter. In any case, the WOB is a function of the type of rock to be drilled, the type of drilling operation (such as mud-motor drilling, vertical or deviated drilling or rotary drilling), the bit type, the capacity of the cutting treatment system in the mud circuit, etc.

2.6.6 Stiffness of the drill collars

Axial forces applied at either end of a bar are equivalent to a couple of moment proportional to the pipe Young modulus, to "flexural rigidity", and are inversely proportional to radius of curvature (Love, 1952, pp. 129–130). Assume that the pipe tube bends in the x direction, and that R is the radius of the circle which approximates the bending curvature in the plane xz (Figure 2.30a). If we take the stress in the vertical z direction to be equal to $-YR^{-1}x$, where the curvature radius R is a constant and Y is the Young modulus of the tube, then the bending couple about the y axis (this axis is perpendicular to the xz plane of the couple of bending, Figure 2.30a) can be expressed as

$$M = -\frac{YI}{R}, \tag{2.26}$$

where, I is the transversal moment of inertia per unit mass and length of the bar, calculated with respect to the principal axis that passes through the centroid of the tube and parallel to the y axis. The radius of curvature R may be expressed with sufficient approximation by

$$R = \left(\frac{\partial^2 x}{\partial z^2}\right)^{-1}. \tag{2.27}$$

The quantity

$$YI \tag{2.28}$$

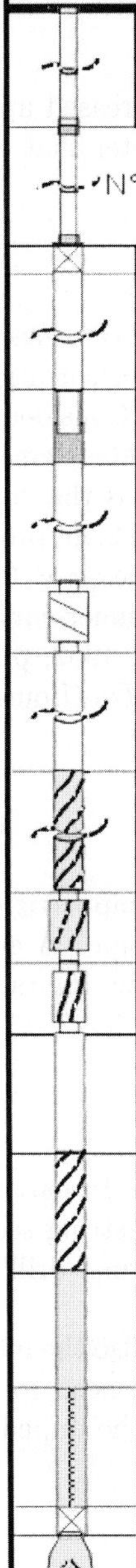

COMPONENT	PARTIAL LENGTH (m)	CUMULATIVE LENGTH(m)	O.D. (inch)	I.D. (inch)	PARTIAL WEIGHT IN AIR (t)	PARTIAL WEIGHT IN MUD (t)
DP 5" S135	5258,84	5591,51	5"	4"9/32	213,4	184,6
N° 15 HWDP	137,99	332,67	5"	3"	39,9	34,5
XO	0,47	194,68	6"1/2	2"13/16	29,7	25,7
N° 3 DC	27,58	194,21	6"1/2	2"7/8	29,1	25,2
JAR	10,33	166,63	6"1/4	2"1/2	25,2	21,8
N° 8 DC	75,40	156,30	6"1/2	2"7/8	23,8	20,6
STAB	1,89	80,90	6"1/4	2"13/16	13,5	11,7
N° 2 DC (sic)	18,65	79,01	6"1/2	2"7/8	12,6	10,9
N° 2 DC	18,65	60,37	6"1/2	2"7/8	10,3	8,9
STAB	1,48	41,71	6"3/4	2"13/16	7,7	6,7
KEY SIT WIPER	1,48	40,23	6"3/4	2"13/16	7,1	6,1
N° 1 DC (sic)	9,46	38,75	6"1/2	2"7/8	6,5	5,6
N° 1 DC (sp)	9,46	29,29	6"1/2	2"7/8	5,2	4,5
MONEL	9,46	19,83	6"1/2	2"7/8	3,9	3,4
MWD	9,46	10,37	6"1/2	2"7/8	2,6	2,3
XO	0,66	0,91	6"1/4	2"13/16	1,3	1,1
BIT	0,25	0,25	8"1/2		0,5	0,4

Figure 2.29: Example of BHA record, with cumulative drill-pipe components, lengths, outer (O.D.) and inner (I.D.) diameters and weights.

is also called "flexural rigidity" (Love, 1952, p. 369). For a pipe of cylindrical section of outer and inner diameters D_O and D_I, we have

$$I = \frac{\pi}{64}\left(D_\mathrm{O}^4 - D_\mathrm{I}^4\right). \tag{2.29}$$

So, if the diameter of a cylindrical rod (bar) is doubled, its stiffness is increased up to 16 fold. Hence, it is important to select the drill collars of maximum diameter that can be safely run.

Design of the cross sections and bending stresses

The normal aim in drill-string design is to avoid abrupt changes of the cross-sectional areas of its components (Figure 2.30b). Abrupt changes can lead to concentrations of bending stresses, which can lead to a twist off if rotary is not stopped. Consider again the geometry used above to calculate the flexural rigidity of the collars. The elements of the pipe (beam) material that in the unstressed state are at $x = 0$ (this is the "neutral plane"), undergo no extension or contraction when a couple M (of equation 2.26) produces the curvature $R = M/YI$. For material points outside the neutral plane ($x \neq 0$), the amount of longitudinal tension or compression – for outer ($x < 0$) and inner-curvature ($x > 0$) side points, respectively – is the absolute value of Mx/I (Love, 1952, p. 130). Flexural rigidity YI is often plotted versus the outer diameter of the pipe D_O (Bourgoyne Jr., Millheim, Chenevert and Young Jr., 1991). The ratio

$$\mathcal{R} = \frac{I}{D_\mathrm{O}} \tag{2.30}$$

is directly related to the longitudinal resistance of pipe to bending (or pipe "rigidity", Gabolde and Nguyen, 1999). $\mathcal{R}$ is defined as the polar modulus of the pipe. A general recommendation is that at a crossover from one tubular size to another size, the ratios of the polar moduli should satisfy the condition

$$\frac{\mathcal{R}_\mathrm{thick}}{\mathcal{R}_\mathrm{thin}} \leq \alpha_\mathrm{cross}, \tag{2.31}$$

where $\alpha_\mathrm{cross} \leq 5.5$ for soft formations and $\alpha_\mathrm{cross} \leq 3.5$ for hard formations. In particular, the need to limit large abrupt variations for the dimensions of close drill-string sections leads to using sections of heavy-weight drill pipes (HWDP) instead of normal drill pipes (DP) directly above the drill collars (DC).

The recommended constraints for the bending of equation (2.31) limit also the internal drill-string reflectivity to axial and torsional waves. In other words, the need to avoid abrupt changes between drill-string elements also limits the variation of the impedances for signal propagation (Chapter 3).

Bottom-hole-assembly buckling

When small weight and torque is applied on the bit, the string remains straight if the hole is straight. As WOB is increased, a critical weight W_CR1 is reached and the drill string buckles and contacts the wall of the hole. If the weight on the bit is further

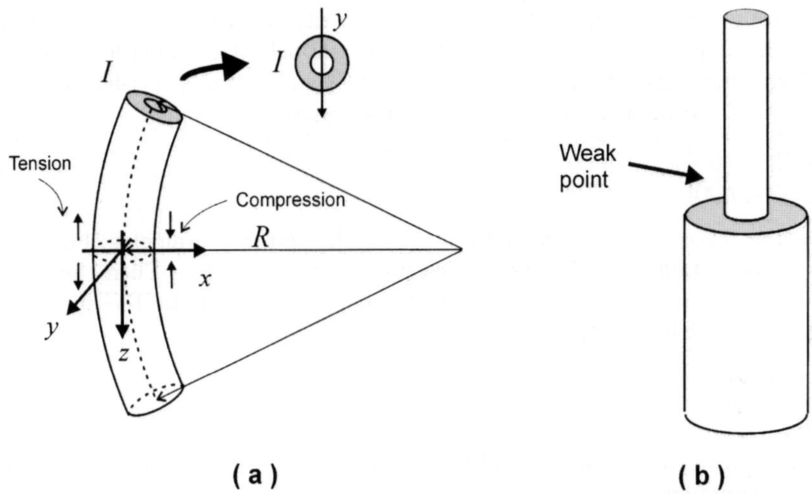

Figure 2.30: a) Bending couple and curvature may be caused by vertical (axial) stress. The moment of inertia I is calculated with respect to the axis that passes through the centroid of the annulus and is parallel to the axis y perpendicular to the xz plane in which the tube bends with radius of curvature R. b) Weak point, where bending stresses may concentrate.

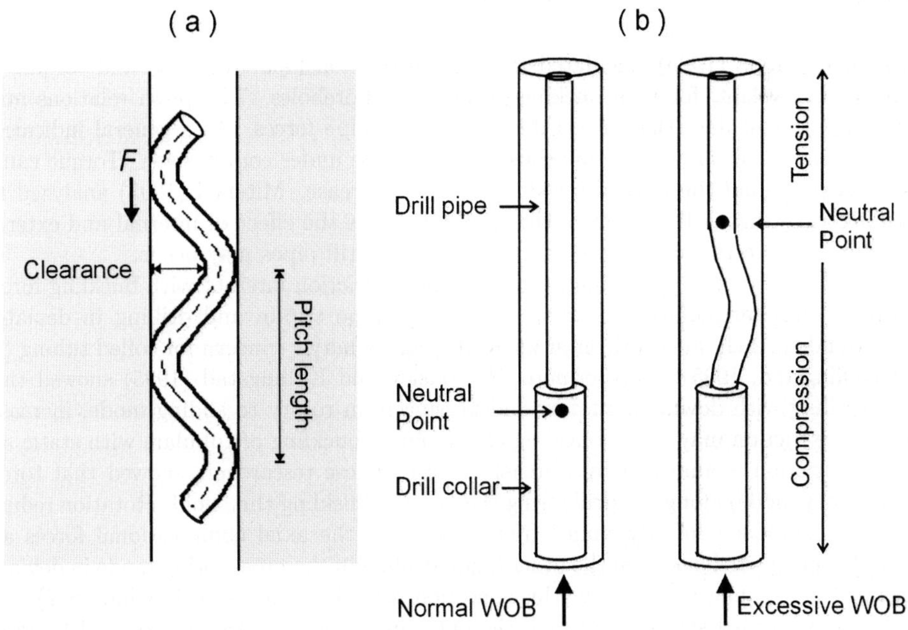

Figure 2.31: a) Helical buckling with axial force F, pitch length and clearance. b) Buckling and neutral-point position. This shift effect is more frequent in directional drilling.

increased, a new critical value W_{CR2} is reached at which the drill string buckles a second time ("buckling of the second order") and so on. Critical values are calculated also for torque. Miska and Cunha (1995) analyzed the critical helical buckling for pipe tubulars subjected simultaneously to an axial load force (F) and a torque (TOR) in frictionless systems. Helical buckling can be described by the equations

$$x = r\,cos\left[2\pi z/p(z)\right],\quad y = r\,sin\left[2\pi z/p(z)\right], \tag{2.32}$$

where r is the "clearance" radius, and $p = p(z)$ is the "pitch" parameter (Figure 2.31a). When $p(z)$ is constant, equation (2.32) describes a circular helix. Then the parameter p, assumed $p^2 \gg (2\pi r)^2$, can be related to the curvature radius by the approximation (Cheatham and Pattillo, 1984)

$$R^{-1} = \frac{(2\pi)^2 r}{p^2}. \tag{2.33}$$

Let us assume a weightless string. With only axial force, the force-pitch relationship is (see bending equation (2.26); and Miska and Cunha, 1995)

$$F = \frac{4\pi^2 YI}{p^2}. \tag{2.34}$$

With only torque, the torque-pitch relationship is

$$\mathrm{TOR} = \frac{2\pi YI}{p}. \tag{2.35}$$

Miska and Cunha (1995) calculated the general relationships with load and torque, including pipe weight, for pipe buckling in deviated boreholes. The pitch relations make it possible to calculate the critical dimensions and pipe forces. As a general indication, axial loading forces may cause buckling of drill collars under compression. Torque causes the helix pitch and the critical buckling force to decrease. Mitchell (1995) analyzed the buckling force that is the result of the axial force plus the effect of internal and external pressures. Due to this composite effect, buckling of drill pipes may occur.

In real conditions, pipe weight, drag and torque friction interact with buckling forces, which become particularly important for casing, pipe trip-in and drilling in deviated, long extended reach, and horizontal wells. It is also a major concern for coiled tubing (He and Kyllingstad, 1995). Furthermore, He, Hasley and Kyllingstad (1995) showed that, when drilling with downhole motors and passing from rotary to sliding mode, increased static drag friction may cause buckling of the pipes (buckling of tubulars with static and dynamic friction is analyzed by Mitchell, 1995). These researchers showed that torque (with rotary mode) along the drill string changes the buckling threshold. Rotation reduces the axial component of "dynamic" friction, reduces the axial compressional forces and makes buckling less likely. On the other hand, drilling in the pipe-sliding mode in deviated wells makes buckling more likely (this condition is analyzed for SWD in Chapter 7).

When a buckled string is rotated, the outer fibers of the tube are stressed by forces against the borehole, leading to increased torque and drag friction, and additional noise for SWD can be expected from the wall contact. These stresses increase with the diameter of the hole (i.e., higher clearance and curvature of equation (2.33)) and can result in fatigue failure of the string. Again, when a drill string buckles in a straight hole, the bit is no longer vertical and a perfectly vertical hole can not be maintained. Therefore, in the design of the BHA, it is important to determine the critical values of WOB at which buckling occurs. Relations for calculating sinusoidal buckling are given by Gabolde and Nguyen (1999). Under normal conditions, some buckling of the drill string is inevitable; therefore, stiffer collars and stabilizers should be used for control of the hole angle. The drill string is designed in order to drill with the neutral point in the drill collars. If too large WOB is loaded on the bit, the neutral point moves above the drill collars (Figure 2.31b), into the slender drill pipes which buckle due to compression (Bourgoyne Jr., Millheim, Chenevert and Young Jr., 1991).

2.6.7 Bit planning

The selection of the bits for a well program is an important decision that has a big impact on the well costs. The factors considered in selecting the type of bit and optimizing the drilling operations are the formation type and properties, the method of drilling (turbine, rotary, air), the mud system and rig costs.

In Chapter 3, we will see that the selection and the wear status of the bit also impacts and influences the quality of the data measured by the SWD passive-listening method both in terms of different drilling rates and methods and of produced signals.

In the following discussion, we consider the drilling parameters used to select the drill bits, which are:

Formation hardness/abrasiveness

Formation hardness and/or abrasiveness determines the major classification of the bits. As we will see in Section 2.7, the International Association of Drilling Contractors' (IADC) classification generally categorizes the bits by the hardness of the formation they can drill.

With respect to the formation hardness/abrasiveness, the following recommendations are given (Magarini and Monaci, 1999):

◇ Choice of bit. Where a selection of different bits can be used – say to drill a soft formation – the choice will depend on other considerations such as mud type, hole size, and drilling costs. For example, compared to roller cones, PDC bits have a longer expected life. Thus, by using PDC bits, the cost of rig standby time can be significantly reduced because fewer trips of the bits in the well are needed.

◇ Performance of roller bits. McGehee, Dahlem, Gieck, Kost, Lafuze, Reinsvold, and Steinke (1992a) gave the following indications regarding the performance of the roller bits. Soft formations are generally drilled most effectively by a combination of deep tooth penetration and a gouging-scraping action (i.e., long teeth and high offset). "Softer bits" are typically applied with lower WOB and higher RPM. In contrast, hard formations are drilled most effectively by a chipping and crushing action (harder bits with shorter and more closely-spaced teeth, low offsets). Harder bits are typically applied with high WOB and lower RPM.

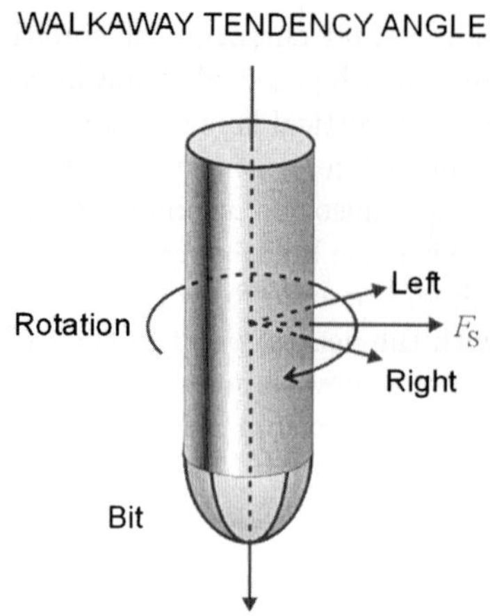

Figure 2.32: Concept of "walk tendency angle". A side force (F_s) is applied to the bit. The walk angle is the angle in the plane perpendicular to the bit axis between the direction of the side force and the lateral displacement. When the lateral displacement is on the left, on the right, or has the same direction as the side force, the bit has a left, right or neutral walk tendency, respectively.

- ◇ Performance of PDC bits. PDC bits perform successfully in soft and non-abrasive formations. In general, the bit breaking action in hard and abrasive formations is by failure in the compressive mode; bits which use shear action are not very successful. In this case, roller bits in the IADC code range 6-1-7 (see Section 2.7) or higher are usually more successful as they have been designed for abrasive wear. Abrasive wear may be very damaging to bits which drill by shear-failure action.

- ◇ In general, PDC bits drill faster than mill-tooth or diamond bits in soft to medium-soft rocks unless they are "sticky".

- ◇ In sticky formations (e.g., hydratable clay rocks), the cuttings stick to the teeth or to the bit structure and lower drilling efficiency. Bits designed for sticky formations have a high degree of teeth inter space (inter fit) and hydraulic capabilities (such as center jetting). In this case, PDC, diamond and short tooth roller-cone bits are particularly unsuccessful (unless PDC bits are used with oil-based mud). Sticky drilling conditions may be a cause of attenuation for the signal produced by the drill-bit source.

Influence of mud type on bit performance is discussed in Section 2.3.14.

Directional control and bit type

Directional control is affected by the method of drilling, the BHA design and the type of bit. In particular, the directional behavior of the bit depends on the offset of the roller-cone bits (Figure 2.12c), the cutting structure on the cone, and the weight on bit. The bit "walk tendency" is measured by the "walk angle" in the plane perpendicular to the bit axis (Menand, Sellami, Simon, Besson and Da Silva, 2002). The concept is represented in Figure 2.32.

The following considerations affect directional control.

Rotary drilling operations are inclined to give "right-hand" walk (which may produce a "right-hand" helicoidal spiral shaped hole). With roller bits, this tendency is increased as the offset of the cone from the bit center increases – due to the fact that the bit scraping action increases with cone offset (McGehee, Dahlem, Gieck, Kost, Lafuze, Reinsvold and Steinke, 1992a). At the same time, the drilling rate is increased by using roller cones with higher offsets. This advantage has to be balanced with the increased difficulty of maintaining the directional control.

Turbine drilling may have the tendency to "left-hand" walk. This effect can be controlled by the turbine used, bit gauge length, and BHA stabilization. In addition, some bit manufacturers developed two- and four-coned roller bits purely for directional purposes.

Dipping formation normally induce movement towards the dip. Roller bits are also available with a special cutting structure to drill with formation dips. The feature of these special manufactured bits consists of outside teeth that dig into the dipping formation to prevent the movement in the dip direction.

High weight on bit tends to increase directional control problems. Low weight on bit help maintain straight holes with the penalty of reduced drilling rate. Due to this fact, PDC bits, which use relatively lower bit weights and no cones (which may give problems of directional control because of the cone offset), are favored in the bit selection for improved directional control.

Bit selection and drilling method

There are no restrictions in the selection of the bit associated with the type of drilling method, which can use surface or downhole rotary systems. A general consideration is that when surface rotary systems are used to drill deep wells, drilling with PDC bits is preferred because it reduces the weight on bit. Consequently, the torque due to the high friction against the wall of the well is also reduced because there is less pipe bending.

With downhole turbine drilling, the efficiency of the turbine tools makes it convenient to use bits with long-life expectancies, such as PDC, diamond and journal-bearing insert bits, but not the roller-cone bits. The manufacturers of roller-cone bits specify the maximum warranted life of the bearings in terms of maximum number of rotations without risk of failure. Failure can cause the loss of a cone in the borehole, thus causing drilling interruption and severe problems of fishing.

Coring bits

Roller-cone, PDC and diamond bits (coring head) are used for coring operations. The bits used for coring are designed to minimize the flushing of the formation fluids from the core by the mud (Magarini and Monaci, 1999).

Bit size

Large-size bits are only roller cone, which are available off-the-shelf in almost all sizes in the range from 6 in. to 26 in., in almost every type, cutting structure and jetting system. PDC and diamond bits are not available off-the-shelf in the sizes over 16 in. In deep wells with small holes, i.e., 4 1/2 in. to 8 1/2 in., PDC bits have much better performance because they have no moving parts. Rotary bits have high failure rates because their bearing areas are small (Magarini and Monaci, 1999).

Bit and rotary speed

The rotary drilling speeds (measured in RPM [rev/min]) has less impact on the rate of penetration (ROP [m/h]) of toothed rotary bits than the weight on bit, and that impact is difficult to evaluate. Sometimes the apparent inconsistencies of the analyzed data are due to vibrations originating at the bit at critical rotary speeds (for instance, see the lobed-bottom-pattern analysis of Chapter 3). These vibrations may contribute to the rock failure and to aid the drilling process. On the other hand, the vibrations are an undesirable phenomenon because they may cause the failure of the drill bit (bearings and bit teeth), of the drill-string collars and drill pipe.

In Chapter 3, the drilling energy and rotary-drilling equations are analyzed in more detail. Ideally, the drilling rate should be linearly proportional to the rotary speed because the contacts between the bit teeth and the rock formations are proportional to the rotary speed. This approximation is true only if the assumption that the contact is "equally effective" at both low and high rotational speeds is true. However, this ideal linear relation is not substantiated by the experimental data, and, in fact, the penetration rates are less than proportional to the rotary speed.

In practice, one should start with low rotary speed, and should increase it until an optimum penetration rate is achieved without vibration. A general recommendation is that, as WOB is increased, the RPM should be decreased and vice versa. Most bit suppliers provide data on optimum bit weight and rotary drilling speeds for specific conditions of operation.

Evaluation of bit wear

The evaluation of bit-wear condition after the bit is recovered at the surface is an important step in bit design and may be used to optimize the next bit plan and improve existing technology. For these reasons, the wear state of a bit is reported in the bit record by using the code provided by IADC to describe why a bit was pulled or a run was terminated (see McGehee, Dahlem, Gieck, Kost, Lafuze, Reinsvold and Steinke, 1992b, for roller cone bits; and IADC Fixed Cutter Drill Bit Committee, 1992a, for PDC bits).

The tooth-bit-wear classification and the impact in the production of downhole vibrations, either when using roller-cone or PDC bits, are analyzed in Chapter 3 in order to characterize the while-drilling drill-bit signal.

2.7 Classification of drill-bit types

2.7.1 Roller-bit classification according to IADC

The 1992 IADC (International Association of Drilling Contractors) roller bit classification is a 4-character design-based code. The first three characters are numeric and the fourth character is alphabetic. The sequence of the numeric characters is: "series", "type", and "bearing/gage". The alphabetic fourth character defines "features available" (McGehee, Dahlem, Gieck, Kost, Lafuze, Reinsvold and Steinke, 1992a). The system uses a three digit code for classification constructed as follows: A, B, C where:

◇ A: Is a number from 1–8, which is the major class or "series".

◇ B: Is a number from 1–4, which is the subgroup or "type".

◇ C: Is a number from 1–7, which is the specialty feature "bearing/gage".

A Major-group (or "series") classification

The major classification number denotes the formation types in which the roller bits should be used (Table 2.6).

B Sub-group (or "type") classification

Each major group (series) is sub-divided into four sub-groups or "types" representing degree of hardness. Type 1 refers to bits designed for the softest formation in a particular major group. For example, a 1-2 bit is a mill-tooth bit designed to drill formations of a slightly greater compressive strength than required for a 1-1 bit, etc. (Table 2.7)

C Specialty feature

The code numbers corresponding to the specialty features are shown in Table 2.8.

Bit cones

The bits listed in the major classification have two types of cones. Originally, cutter bits had cone teeth machined out of the cone material by a mill; hence, they were termed "mill-tooth" bits. These bits wear quickly when hard abrasive rocks are encountered. In this case, drilling can be performed by cones which have teeth inserted into the cone made of more wear resistant materials such as tungsten carbide. The inserts are of varying shapes to obtain the best penetration in a particular rock. The cone teeth of a mill-tooth bit can be heat-treated to provide better wear resistance, but are only good up to classification 3. Insert bits are used for ranges 4 through 8 (Table 2.7).

Cone offset also has a significant effect on the penetration rate due to the shear mechanism that best suits the formation type.

Table 2.6 – Roller-bit major group classification

Group number	*Formation*
Mill-tooth bits	
1	Soft formations of low compressive strength and high drillability
2	Medium to medium-hard formations with high compressive strength
3	Hard semi-abrasive or abrasive formations
Insert bits	
4	Soft formations of low compressive strength and high drillability
5	Soft to medium formations with low compressive strength
6	Medium-hard formations with high compressive strength
7	Hard semi-abrasive or abrasive formations
8	Extremely hard and abrasive formations

Table 2.7 – Roller-bit type and classification

Bit type	*Class*	*Formation type*	*Tooth description*	*Offset angle* (°)
	1-1, 1-2,	Very soft	Hard-faced tip	3–4
	1-3, 1-4	Soft	Hard-faced side	2–3
Mill tooth	2-1, 2-2	Medium	Hard-faced side	1–2
	2-3	Medium hard	Case hardened	1–2
	3	Hard	Case hardened	0
	4	Very soft		
	5-2	Soft	Long blunt chisel	2–3
	5-3	Medium-soft	Long sharp chisel	2–3
Insert	6-1	Medium shales	Medium chisel	1–2
	6-2	Medium limes	Medium projectile	1–2
	7-1	Medium hard	Short chisel	0
	7-2	Medium	Short projectile	0
	8	Hard chert	Conical or hemispherical	0

Table 2.8 – Bearing and special feature codes

Code number	*Feature*
1	Standard roller bearing
2	Roller bearing air cooled
3	Roller bearing gage protected
4	Sealed roller bearing
5	Sealed roller bearing and gage protected
6	Sealed friction bearing
7	Sealed friction bearing and gage protection

2.7.2 Diamond bit classification

Two types of diamond bits are used for special applications where their cutting action is most efficient. These are natural diamond (Section 2.3.11) and the PDC (polycrystalline diamond compact) bits (Section 2.3.11).

PDC bits

The advancement in technology in PDC design and performance in recent years has been significant and there are now a wide range of bits from many manufacturers available. Due to the diversity of bits and bit features, there is no IADC classification system similar to roller bits, but simply a code to provide a means of characterizing the general physical properties of fixed cutter drill bits (IADC Fixed Cutter Drill Bit Committee, 1992b).

Fixed cutter classification according to IADC

To catalog the wide range of fixed cutter bits including natural diamond and PDC, IADC introduced the following classification system. The classification system consists of a four-character code (Table 2.9). The Code-1 subgroup classification is simply a five letter designation categorizing the type of cutter and body material. The Code-2 numbers (1–9) categorizes the bit profile by shape. The Code-3 numbers (1–9) describe the hydraulic features. The Code-4 numbers (1–9) categorizes the cutter size and cutter material.

An example bit code would be M442. It equates to a PDC bit with matrix body, medium taper-deep shape, changeable jets-ribbed design with large size cutter of medium density.

Table 2.9 – IADC fixed cutter classification code

Code 1	Cutter type and body material (D, M, T, S, O)
Code 2	Bit profile (1-9)
Code 3	Hydraulic design (1-9)
Code 4	Cutter size and density (1-9)

Table 2.10 – Code 1 - Cutter type and body material

Group letter	*Cutter type and body material*
D	Natural diamond matrix body
M	PDC matrix body
T	TSP matrix body
S	PDC steel body
O	Other

Table 2.11 – Code 2 - Bit profile

Code 2	*Bit profile*
1	Long taper deep shape
2	Long taper medium shape
3	Long taper shallow shape (parabolic)
4	Medium taper deep shape
5	Medium taper medium shape
6	Medium taper shallow shape (rounded)
7	Short taper deep shape (inverted)
8	Short taper medium shape
9	Short taper shallow shape (flat face)

Table 2.12 – Code 3 - Hydraulic design

	Type of dusage		
	Changeable sets	*Fixed ports*	*Open throat*
Bladed	1	2	3
Ribbed	4	5	6
Open Faced	7	8	9

Table 2.13 – Code 4 - Cutter size and density

	Density of cutters		
	Light	*Medium*	*Heavy*
Large	1	2	3
Medium	4	5	6
Small	7	8	9

Chapter 3

General theory: drill-bit seismic waves

3.1 Introduction

This chapter describes the theory related to the drill-bit signal production. In particular, we develop the analysis with the energy balance of the drilling process, calculate the vibrations produced by drilling bits and the wavefields radiated by the bit source.

In Chapter 1, we introduced the concept that seismic while drilling can be performed either by using reciprocal or normal VSPs (Figure 3.1). These give equivalent seismic products in the sense of the "reciprocity principle".

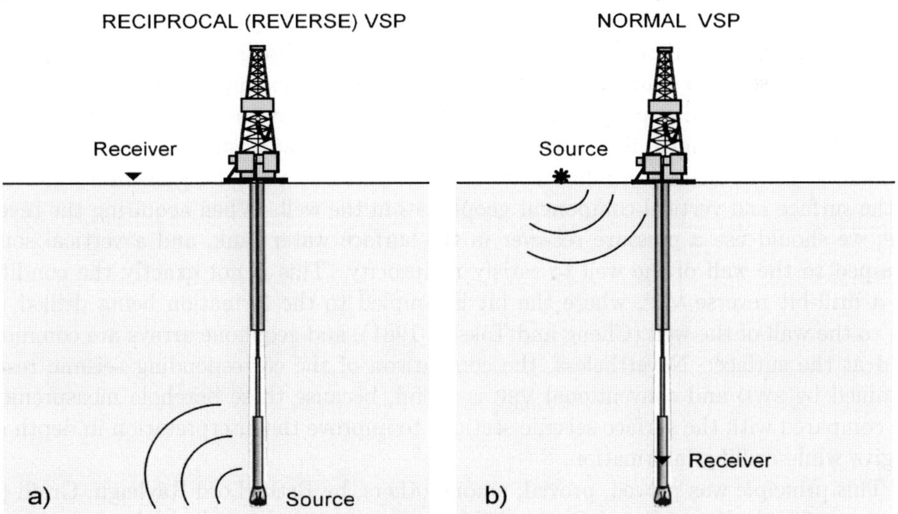

Figure 3.1: a) Reciprocal (or reverse) and b) normal VSP.

3.2 Reciprocity principle

The reciprocity principle applied to seismology states that *"the seismic trace from a source at A to a geophone at B is the same as from a source at B to a geophone at A if sources and receivers are properly oriented, i.e., similarly coupled to the Earth"* (Sheriff, 1999). This principle has important applications in seismology. For example, it is used to measure data of surface refraction profiles and to study noise fields. The reciprocity principle is used also to show the equivalence between normal and reverse VSP data. It states that the geophysical information provided by the drill-bit source and surface receivers corresponds to that obtained by using receivers in the well and surface seismic sources.

In real reverse VSP applications, the reciprocity conditions are, in general, not exactly satisfied. To satisfy the reciprocity conditions, it is necessary that the source and the receiver interchange position, by maintaining the same orientation in the same location. In other words, the directivity properties have to be regarded as attached to the acquisition point fixed to the earth, and not to the type of tool (source or receiver).

Claerbout (1976) summarizes the principle's concept as follows. Assume two experiments are obtained by interchanging the position of source at the point A and receiver at the point B. The reciprocity principle's basic equation says that the inner product of the source and receiver vectors at A equals the inner product of the receiver and source vectors at B. We observe that these inner products are between vectors of different experiments. Let $\mathbf{E} = (V, F)$ be the excitation vector of external relative volume injection and force, and let $\mathbf{R} = (p, u)$ be the receiver-response vector of pressure and displacement. We have

$$\mathbf{E}_{\mathrm{A}}^{\mathrm{T}}\mathbf{R}_{\mathrm{A}} = \mathbf{E}_{\mathrm{B}}^{\mathrm{T}}\mathbf{R}_{\mathrm{B}}, \tag{3.1}$$

where superscript ^{T} denotes transpose (for a more general discussion of integral equations of reciprocity see also Carcione, 2001, pp. 157-165). For example, assume vertical displacement is measured at position A with a pressure source at position B. Reciprocity says we obtain the same signal if we use a vertical-force source at position A and measure the pressure at B. An onshore wireline VSP can be acquired by using air guns in a water tank at the surface and vertical-component geophones in the well. When acquiring the reverse VSP, we should use a pressure receiver in the surface water tank, and a vertical source clamped to the wall of the well to satisfy reciprocity. This is not exactly the condition for a drill-bit reverse VSP, where the bit is coupled to the formation being drilled and not to the wall of the well (Cheng and Toksöz, 1981), and geophone arrays are commonly used at the surface. Nevertheless, the comparison of the corresponding seismic results obtained by SWD and conventional VSP is useful, because these borehole measurements are compared with the surface seismic sections to improve the interpretation in depth and to give while-drilling information.

This principle was proved, proved, among others, by Betti, Lord Rayleigh, Graffi (see Carcione, 2001). Knopoff and Gangi (1959) demonstrated the reciprocal relationship of volume integrals of force sources and particle displacements in heterogeneous, anisotropic, elastic media. White (1960) used the reciprocity theorem to compute the low-frequency radiation pattern of vertical and horizontal force sources – as well as the sensitivity pattern of vertical and horizontal geophones – and experimented with the reciprocity principle in a combined electromagnetic, solid, liquid and gaseous system. More recently, Arntsen and Carcione (2000) provided examples of reciprocity relationships for distributed sources and receivers in inhomogeneous, anisotropic, viscoelastic solids.

3.3 Normal while-drilling VSP

Normal while-drilling VSP is an alternative approach to the drill-bit passive SWD described in this book for obtaining while-drilling seismic information. While-drilling normal VSP (NVSP) is an innovative technology acquired with normal geometry (Chapter 1), similar to that of the conventional wireline one. In these measurements, downhole sensors and surface sources are used. In some applications, the sensors are connected by wire lines and lowered into the drill string normally used to drill the well, thus avoiding the extraction of the pipes during acquisition.

This novel while-drilling method, of acquiring data with the drill string in the well (Figure 3.1b), is described below. It allows geophysicists to use known sources at the surface and to obtain reliable datasets for time-depth (checkshot) analysis and also imaging with single-offset datasets (Haldorsen, Esmersoy, Hawthorn, Krasovec, Raikes, Harrold, Day and Clippard, 2003). However, in comparison with the reverse drill-bit seismic method, this technology is more expensive and less convenient for multioffset 2D and 3D investigations due to the necessity of having to repeat a high number of source points.

3.3.1 Seismic measurement while drilling SMWD

An alternative while-drilling mode of acquisition of borehole seismic data is provided by the "Seismic MWD" system of Schlumberger (SMWD[1]). This system consists of a logging-while-drilling (LWD, see Chapter 2) tool with seismic sensors in the bottom-hole assembly, a seismic source with control equipment at the surface, and a measurement-while-drilling (MWD, see Chapter 2) system for real-time telemetry. The seismic source is fired during pipe connections (Moufarrej and Thompson, 2002). Seismic data are processed downhole, and checkshot information is sent to surface via mud-pulse telemetry. All seismic data are recorded in downhole memory in case further processing is required. The initial focus of the development of this tool is the checkshot measurement to obtain the time-depth relation for tying the seismic images to the bit position.

Seismic MWD system

The seismic energy is produced from a conventional source such as an air gun (Figure 3.2). Offshore, the source can be deployed from the rig or a boat for offset surveys. Seismic energy is received by the SMWD tool, containing sensitive receivers, a processor and memory. The SMWD tool is instrumented with a three-component geophone and hydrophone (4C). The seismic signals are acquired, locally stored and processed to provide key information for transmission uphole. This technique allows checkshot data (Chapter 8) to be sent to the surface in real time by MWD transmission. The time-depth data are used to display the well position on the seismic map at the wellsite or in the office. Waveforms

[1] Mark of Schlumberger

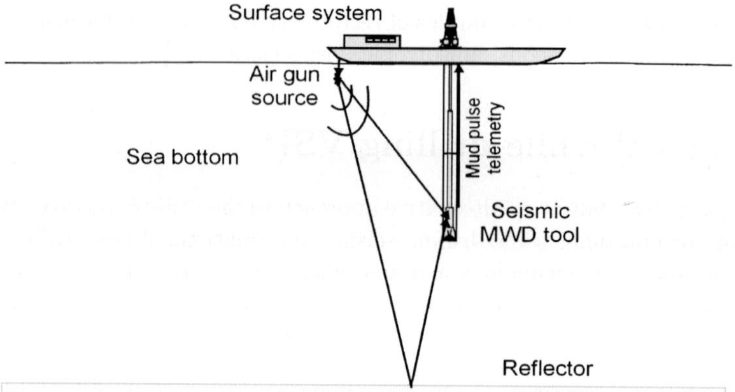

Figure 3.2: SeismicMWD[2] system: surface source, downhole receivers, waveforms recorded in downhole memory and downhole processing.

are recorded in the tool memory for later VSP processing after a bit trip. Although all borehole seismic surveys can be sensitive to noise, a patented technique enables source activation and data acquisition during drilling pauses when the downhole environment is quiet. Suitable times to acquire data are during pipe connections while drilling and bit tripping. Typically, source activation is completed with no interference to drilling operations.

SMWD operational conditions and features

It was reported that the SeismicMWD[2] has been proven in a wide range of environments, including all well geometries with vertical depth greater than 7500 m, open hole and cased hole environments, hard and soft formations, moored and dynamically positioned rigs, unlimited water depth, zero-offset and vertical-incidence VSP recorded in a deviated well.

The tools are designed to permit operation with flow rates up to 9000 lit/min and pressures up to 170 MPa.

Table 3.1 – Main features
Real-time checkshot
Recorded waveforms for VSP processing
Rugged LWD technology
Hole sizes 8.5 in. to 17.5 in.
Multicomponent measurements
High flow rate and pressure ratings
Combinable with all Schlumberger's LWD tools

[2]Mark of Schlumberger

3.4 Drill-bit seismic source

The basic idea of the passive drill-bit seismic method is to use the drill-bit noise as a seismic source. The motivation and the history of this idea are presented in Chapter 1. In the following sections, we consider the main requirements to use the drill-bit vibrations as an effective seismic source.

A main issue in seismic-while-drilling (SWD) technology is to calculate the dynamic and radiation properties of the drill-bit source under different operational conditions. The rotary-drilling energy requirements, power losses, local crack effects, radiation and near-field effects are analyzed in detail in order to quantify the waves produced by a SWD vertical force acting in a bounded and unbounded medium. Results can be expressed in terms of complex integrated impedances and rotary-drilling equations. The calculations – extended to the waves in the drill string – are used to quantify the performance of actual drill-bit sources.

3.4.1 Drill-bit signal characterization

Recently, interest has grown in the performance of the drill-bit source in order to improve seismic-while-drilling VSP acquisition under poor drilling conditions, such as with diamond bits (Poletto and Savini, 1997), and to use the drill-bit signal for downhole geosteering purposes (Lesso and Kashikar, 1996; Hokstad, Sollie and Carlsen, 2001) (Chapter 8).

A main issue is the repeatability and controllability of the drill bit as a downhole acoustic source. In fact, drilling is performed with many different configurations of mechanical drilling equipment, with various operational parameters and in different geophysical settings that may change the properties of the radiated signal. For these reasons, the drill-bit method requires control of the acquisition conditions needed for effective applicability of SWD. Methods based on signal-to-noise analysis in data beam forming (Haldorsen, Miller and Walsh, 1995) and on evaluation of statistical properties of the signal (Poletto, Rocca and Bertelli, 2000) can help the while-drilling quality control. However, it is common practice in the SWD surveys to choose the acquisition parameters on the basis of experience, rather than on specific criteria accounting for the dynamic conditions at which the drill-bit source is used. Hence, a more analytic approach is necessary to describe the behavior and calculate the performance of the bit source. This is based on knowing the rotary-drilling process and the properties of the radiated fields.

Rotary drilling is the conventional technique used to excavate oil wells. This technology is extensively described in Chapter 2. The rotary energy imparted to the bit is expended in the rock-breaking action, with performances that are monitored in terms of penetration rate for drilling optimization. Part of this energy is radiated in the form of vibrations in the surrounding rock. Below, we consider the main energy requirements that must be met in order to use the drill-bit vibrations as an effective seismic source. The possible drilling conditions are described together with the dynamic characteristics of the bit sources and their frequency content.

The calculations show that the drill-bit seismic source produces sufficient radiated energy, and that the frequency bandwidth of the signal is broad, so that it can be used for correlation in SWD applications.

3.5 Total drilling power

The performance of the drill-bit source is related to the drilling power expended during data acquisition. The total drilling power at the bit can be calculated by adding the vertical work per unit time of the axial force (weight on bit) to the torque power. Let T, ω, F and z be torque, angular velocity, vertical (axial) force and drilling depth, respectively. Thus, we have

$$W_{\text{drill}} = T\omega + F\frac{dz}{dt}. \tag{3.2}$$

In terms of typical drilling parameters, equation (3.2) becomes

$$W_{\text{drill}} = \frac{2\pi\,\text{TOB} \cdot \text{RPM}}{60} + \frac{\text{WOB} \cdot \text{ROP}}{3.6 \cdot 10^3} \qquad [\text{kW}], \tag{3.3}$$

where ROP [m/h], WOB [kN], TOB [kN m], and RPM [rev/min] denote rate of penetration, weight on bit, torque on bit and rotary speed, respectively. Table 3.2 shows typical values of these drilling parameters. If we assume ROP = 10 m/h, WOB = 98 kN (10 ton), TOB = 6 kN m, and RPM = 120 rev/min, we obtain the total drilling power

$$W_{\text{drill}} = 75.4 + 0.272 \cong 75.7 \qquad [\text{kW}]. \tag{3.4}$$

The total energy expended during one minute of drilling is $\approx$ 4.5 MJ and the total energy expended during one hour of drilling (also, in 10 m of bit descent) is $\approx$ 272 MJ. Note that the only term that substantially contributes to the drilling energy in equation (3.3) is the rotary power (first term on the right), which is, ultimately, transformed into the drilling action of the bit pushed against the rock by the weight on bit i.e., the vertical work expended by potential energy per unit time $\sim$ 0.3 % of torque power.

When calculating the drill-bit power by equation (3.3), it is important to determine the downhole drilling parameters correctly. In general, the surface measurements differ significantly from the measurements made at the bit location (Falconer, Belaskie and Varlava, 1989). The magnitude of the surface torque can be several times the value of the downhole torque applied on the bit (TOB), and the downhole WOB may differ from the WOB calculated at the surface. The torque loss is due to the friction between the rotating drill string and the wall of the borehole, as well as to mud viscosity. A great decay of torque in the rotating drill string makes the surface and downhole-drilling-energy values differ significantly. Where while-drilling downhole measurements (MWD) are not available, only a rough value of the downhole average energy can be estimated by calculating the torque losses or by assuming approximate trends based on experimental values for torque friction.

Table 3.2 – Realistic ranges for drilling parameters

Parameter	Unit	Minimum	Typical-average	Maximum
TOB	kN m	1000	3000–6000	15 000
RPM	rev/min	60–80	120–180	500
WOB	kN	0–1	100–200	300
ROP	m/h	0–1	10	300

3.5.1 Energy losses for drill-string torque friction

In rotary drilling, the torque friction of the pipes rotating in the borehole is a main cause of surface-power loss, and the surface measurements are not representative of the power at the bit. In deviated wells, the torque loss may be determined using the Coulomb torque-friction coefficient η_{rot}. By definition, the Coulomb friction coefficient is the ratio of friction force to the normal force, i.e.,

$$\eta_{\mathrm{rot}} = \frac{\mathrm{TOR}_{\mathrm{loss}}}{F_{\mathrm{s}} \cdot r_{\mathrm{DS}}}, \tag{3.5}$$

where $\mathrm{TOR}_{\mathrm{loss}}$ is the torque loss over a drill-string section of radius r_{DS}, and F_{s} is the side force pushing the string against the wall of the borehole (such as the side component of the string's weight). Generally accepted values of the friction coefficient lie between 0.25 and 0.4 – which indicatively correspond to cased- and open-hole conditions, respectively (Lesage, Falconer and Wick, 1988).

In vertical wells, where the side forces are assumed to be negligible, equation (3.5) is invalid for calculating the torque losses due to drill string and bottom-hole-assembly (BHA) contacts with the wall of the borehole. In this case, torque losses may be estimated by relating surface (at rotary motor) and downhole (at bit) reference measurements (Cunningham, 1968). Falconer, Belaskie and Varlava (1989) showed an example in which the surface torque was approximately twice the bit torque when the bit and a stabilizer were in different formations, shale and sandstone, respectively. Furthermore, in deep vertical wells, the rotary friction caused by drilling-mud viscous damping is also important. Eronini, Somerton and Auslander (1982) calculated torque losses of 75 % for typical wells.

3.5.2 Effects of drag friction

A Coulomb drag-friction coefficient is defined as (Falconer, Belaskie and Varlava, 1989)

$$\eta_{\mathrm{drag}} = \frac{\mathrm{WOB}_{\mathrm{loss}}}{F_{\mathrm{s}}}, \tag{3.6}$$

where $\mathrm{WOB}_{\mathrm{loss}}$ is the weight loss for drag-friction over the drill-string section subject to side force F_{s}. Weight losses for drag-friction cause decreases in the effective weight on bit, measured downhole, relative to the weight on bit measured at the surface. This effect is important for deviated and horizontal wells. Falconer, Belaskie and Varlava (1989) measured while drilling the drag (also, torque) loss over the drill string and calculated the average drag-friction coefficient from the MWD tool to the rotary table. This effect is generally less than 10% in the rotary mode, i.e., with drill-pipe rotation. However, static drag friction in the sliding mode, i.e., without drill-pipe rotation, is higher. Falconer, Belaskie and Varlava (1989) measured a variation of the drag-friction coefficient from 2 % to 16 % when passing from rotary to sliding mode in a "S"-shaped well. Lesage, Falconer and Wick (1988) measured the difference of surface and downhole WOB and calculated the sliding friction factor in field examples. In the static condition (sliding mode), drag friction may increase from less than 10% to 20–40% when a turbine or downhole motor is used without pipe rotation. This affects the calculation of the effective WOB and, also, can lead to a loss of SWD signal transmission through the drill string in vertical wells

(Poletto and Miranda, 1998). Because the drilling energy expended by vertical-axial force is low when compared to the rotary energy (equation (3.3)), the drag-friction effect can be considered negligible in the computation of the total energy expended for drilling. Nevertheless, this effect cannot be considered negligible when the axial force actually exerted on the bit (WOB) is used to calculate the radiated energy (Sections 3.8 and 3.12).

3.5.3 Downhole motor drilling

When downhole positive-displacement motors (PDM) or turbines are used alone, or in addition to surface rotary power, the downhole RPM, hydraulic power and efficiency to produce mechanical power are obtained from their performance specifications (Section 2.3.7). These are expressed as functions of mud flow and specific mud gravity (Gabolde and Nguyen, 1999). Also in this case, the rotary power can be expressed in terms of torque and total RPM at the bit (equation (3.3)).

3.6 Energy analysis in terms of drilling parameters

We will show in Section 3.7.3 that the radiated energy is only a small percentage of the rotary drilling energy (equation (3.3)). For this reason, radiation effects are usually not considered by drillers in the analysis of rotary energy with drilling parameters. At the same time, simplification is made also in the analysis of radiated power and near-field effects discussed in the previous sections. In the first case (rotary drilling), the energy is essentially calculated by rotary-torque power. In the second case (radiated waves), the energy is the work expended by an axial force applied to the bit. The radiation model is calculated by assuming zero drilling rate, i.e., by assuming that the force is applied at the bit without rock failure and bit penetration. In real drilling conditions the bit-rock interaction is more complex (Chin, 1994). For instance, a roller-cone bit performs as a displacement source. The instantaneous rate of penetration, the axial bit displacement and the local strain are not independent parameters. Chin (1994, pp. 148–150) discussed the general conditions that relate these variables in order to model both the bit kinematics and drilling ahead. Assuming that the interacting process between a bit and rock during drilling is a stable random process (Ma, Zhou and Deng, 1995), we can obtain information about the expended energy by relating the average drilling action and the drilling parameters (Poletto, 2002a) (see also Chapters 2 and 5).

In the following sections, we have disregarded the energy losses due to drill-string friction, as well as the damping due to the presence of drilling mud, and have considered the energy required by the bit to drill the rock.

3.6.1 Specific energy – required to drill a unit volume of rock

The energy expended in drilling can be expressed in terms of specific energy, which is used in the drilling literature for bit selection (Rabia, 1985) and evaluation of the drilling conditions. The specific energy can be defined in relation to the total drilling power, given by equation (3.3), as

$$\mathrm{E_S} = \frac{3600 \cdot W_{\mathrm{drill}}}{A_{\mathrm{B}} \cdot \mathrm{ROP}}, \tag{3.7}$$

where A_{B} is the borehole area and ROP is the rate of penetration (see Table 3.2).

$\mathrm{E_S}$ represents the energy required at the bit to drill a unit rock volume – note that here *specific* has the meaning of *energy density* instead of *energy per unit mass.* The specific energy has the dimension of a stress. Pessier and Fear (1992) pointed out that the minimum (optimal) specific energy is reached when $\mathrm{E_S}$ is roughly equal to or approaches the compressive strength σ of the rock being drilled, which is defined as the differential stress at which rock failure – as determined by laboratory triaxial loading tests – occurs. In other words, specific energy higher than σ should be related to energy dissipation and to a non-optimal choice of the drill bit for a given type of rock, or to the use of damaged and worn bits that produce low rates of penetration (ROP).

Simon (1963) analyzed the energy balance in percussive rock drilling, and found that the expenditure of energy required to drill a unit volume of rock (i.e., $\mathrm{E_S}$) varies from approximately 35 J/cm^3 (35 MPa) to approximately 700 J/cm^3 (700 MPa). In his analysis, he assumes that this energy density is roughly twice the compressive strength of the rock – as measured by a uniaxial loading test – and indicates also that it may range from roughly one to several times the rock's compressive strength σ.

3.6.2 "*Perfect-cleaning*" theory of drilling

In a very interesting work, Maurer (1962) analyzes the rotary drilling with a roller-cone bit. This process consists of two fundamental operations: the formation of craters under the bit teeth and the removal of the broken rock from the craters. Maurer (1962) analyzes the energy expended in the formation of the individual crater, i.e., energy versus volume, and calculates the relations between the drilling parameters in the "perfect cleaning" theory of rotary drilling. This state is assumed to exist when all the broken rock is removed from the craters between impacts and no regrinding occurs under the bit. He calculates the following results under perfect drilling conditions. First, the drilling rate is directly proportional to rotary speed, to weight on bit squared and inversely proportional to rock strength squared. Secondly, the torque is proportional to weight on bit squared. We have

$$\mathrm{ROP} \propto \frac{\mathrm{RPM} \cdot \mathrm{WOB}^2}{D^2 \sigma^2}, \tag{3.8}$$

$$\mathrm{TOB} \propto \mathrm{WOB}^2, \tag{3.9}$$

where σ is here defined as the drillability strength of the rock (inversely proportional to chisel penetration) and $D = 2r_{\mathrm{b}}$ is the diameter of the bit and of the borehole. Note that under perfect-cleaning conditions, the penetration rate is proportional to rotary speed and that the torque on a roller bit (TOB) is independent of rotary speed (RPM). Maurer (1962) analyzes also deviation from equations (3.8) and (3.9) under real drilling conditions. A general analysis of drilling mechanics data with different types of bits is discussed in Sections 3.6.3 and 3.6.4.

3.6.3 Dimensionless drilling parameters

Falconer, Burgess and Sheppard (1988) introduced dimensionless drilling parameters as diagnostic indicators to separate the lithology effects from drilling mechanics data. These are the dimensionless downhole torque and rate of penetration, defined below.

Dimensionless torque – Rotary drilling converts the load on the bit into indention and gouging forces that produce a moment and a torque resistance to rotation proportional to the bit diameter, D. Hence, the dimensionless torque is defined as

$$T_D = \frac{\mathrm{TOR}}{\mathrm{WOB} \cdot D}. \tag{3.10}$$

Dimensionless rate of penetration – The rate of penetration is ideally proportional to the bit rotary speed RPM, and the dimensionless rate of penetration is defined as

$$R_D = \frac{\mathrm{ROP}}{\mathrm{RPM} \cdot D}. \tag{3.11}$$

3.6.4 Rotary-drilling model (dimensionless parameters)

Falconer, Burgess and Sheppard (1988) analyzed the torque drilling model by adopting an energy-balance approach and calculated the relation between the work per revolution produced by the downhole torque and the vertical (axial) distance x_{p} penetrated and gouged by the bit. They assumed that the friction of the borehole walls against the rotating bit is negligible. To evaluate the relation between vertical force and rotary drilling, they assume that the force F per unit width needed to push a sharp tooth into the rock is proportional to the maximum cross-sectional area embedded in the rock; hence,

$$F(x_{\mathrm{p}}) = \sigma_{\mathrm{p}} x_{\mathrm{p}} \cdot \tan \Theta\,, \tag{3.12}$$

where x_{p} is the depth of penetration and Θ is the tooth semi-angle for roller-cone bits (Figure 2.12), or the rake angle for polycrystalline diamond compact (PDC) bits (Figure 2.14). The constant σ_{p} has the dimension of a stress that represents the in-situ rock hardness (or penetration strength, not to be confused with rock strength as calculated by uniaxial load tests; however, the authors have pointed out that in brittle fracturing it can have an equivalent meaning). The work done by the indention force of equation (3.12) is calculated by integrating $F(x_{\mathrm{p}})$ from 0 to x_{p} and is proportional to x_{p}^2. Falconer, Burgess and Sheppard (1988) calculated also the work of the transverse gouging action, which is assumed to be proportional to the depth of penetration x_{p}. The total torque power per unit revolution frequency (i.e., $\omega_{\mathrm{rev}} = 2\pi$) can be expressed by the sum of the gouging and indention terms as

$$2\pi\,\mathrm{TOB} = \mathrm{b}_4 \cdot \tau\, x_{\mathrm{p}}\, D^2 + \mathrm{b}_5 \cdot \sigma_{\mathrm{p}}\, x_{\mathrm{p}}^2\, D, \tag{3.13}$$

where b_4 and b_5 are dimensionless constants used by Falconer, Burgess and Sheppard (1988), which depend on the bit properties and number of impacts per revolution. The constant τ has the dimension of a stress and represents the intrinsic rock shear strength. The first term, linear in x_{p}, on the right side of equation (3.13) corresponds to the work expended in gouging. The second term, quadratic in x_p, corresponds to the work expended in pushing the bit teeth into the formation in a single revolution.

The torque-power equation (3.13) can be rewritten in implicit form with respect to x_p and in terms of the dimensionless drilling parameters given by equations (3.10) and (3.11), as

$$T_D = b_6 \cdot \tau/\sigma_p + b_8 \cdot R_D^{1/n}, \tag{3.14}$$

where b_6 and b_8 are dimensionless constants of proportionality used by Falconer, Burgess and Sheppard (1988). The constants b_6 and b_8 account for the gouging and indenting bit action, respectively. The integer n relates the dimensionless rate of penetration to x_p, being $T_D \propto (x_p/D)^n$. Usually, n is equal to 1 or 2, depending on the average crater geometry. In other words, it depends on possible superposition of craters created by different teeth.

Equation (3.14) is assumed valid both for roller-cone and PDC bits. The constants b_6 and b_8 have been discussed in detail by Falconer, Burgess and Sheppard (1988). For instance, the indention coefficient can be assumed to be $b_8 = 0.08$ for a milled-tooth roller cone. Because dimensionless torque and rate of penetration can be measured, it follows from equation (3.14) that it is possible to calculate the vertical (or axial) indention energy relative to the total rotary energy (T_D). In general, the second (indention) term is small compared to the first (gouging) term of the torque equation. It depends on n and is of the order of small percentages of the total torque. Once the penetration energy involved in the rock fracturing has been calculated, we have to evaluate which part of the work done by the vertical (axial) force corresponds to elastic radiated energy.

3.7 Energy balance in rock fracture

The drilling energy is expended in different ways during the drilling process, and several effects are produced by the working bit. The "main" and desired effect is the breaking of the rock. Another effect is the emission of vibrations propagating in the surrounding formation, in the drill string and in the borehole. Other effects are related to the dissipation of energy in the form of heat due to the action of the drill bit, and the friction of the pipes rotating in the borehole. The latter can be assumed to be negligible when considering only the friction of the rotating bit against the wall of the borehole (Falconer, Burgess and Sheppard, 1988).

Let W_{drill} be the total drilling power of equation (3.3). It can be represented as

$$W_{\text{drill}} = W_{\text{break}} + W_{\text{heat}} + W_{\text{vibr}}, \tag{3.15}$$

where W_{break}, W_{heat} and W_{vibr} are the power used to generate the rock fragments, the heat-dissipated and the vibration power, respectively. The balance of the energies involved in rock fracturing is discussed in detail by Simon (1963). In his basic analysis, Simon considers the following main energy requirements for fracturing rock.

3.7.1 New-surface energy

This is defined as the energy expended when a complementary pair of new free surfaces is formed. This energy corresponds to the work done against the cohesive forces that hold the material together. In our formulation, this effect is represented by W_{break}. Simon (1963) uses the Griffith criterion[3] (Jaeger, 1969) for brittle failure to calculate the new-surface energy involved in crack propagation under different tensile and compressive loading conditions. He estimates that this energy represents less than 1 % of the total energy required to drill the rock. However, other authors obtain higher fractions. For instance, Bernabe and Brace (1990) estimate it to be from 1 % to 10 %, while Olgaard and Brace (1983) estimate that the new-surface energy is no more than 20 % of the total energy (Note: this discussion was received, in a personal communication, by Anthony Gangi who still believes that these values are underestimated).

3.7.2 "*Elastic-strain*" (heat) energy

Because it is impossible to impart only the new-surface separation energy, "*the best that can be done in practice is to apply repeated concentrated loadings on successive portions of the exposed rock at the bottom of the hole*" (Simon, 1963). This action results in a system of distributed stress that would be favorable for breaking the rock fragments. Somerton, Timur and Gray (1961) measured the permanent strain due to cyclic loading in laboratory tests on rock specimens drilled with roller bits. These researchers showed that the stress-strain relations are non-linear, dependent on the past stress history and loading rate, and consequently the dynamic Young's moduli (of the rocks altered by the cyclic loading) are substantially greater than Young's moduli determined by static tests.

By analyzing the different modes of tensile and compressive loading, Simon (1963) calculated that, after failure has occurred, most of the elastic energy stored up to the point of failure is dissipated in "stress-wave initiated vibrations". He assumed that the rapid release of the loads at the faces of the propagating crack following failure produces stress waves that are "*reflected back and forth between the fragments of the medium and soon dissipate into heat energy*". So, despite the name "elastic strain" given by Simon (1963) to this energy component, in our analysis, this effect should contribute greatly to W_{heat}. It could be of interest for high-resolution SWD to know which part of this energy is transformed into heat due to friction and which part consists of measurable high-frequency waves produced by drilling cracks. Maurer (1965) analyzed friction in shear failure of rock under compression. Shear failure predominates in drilling because it is difficult to produce tensile stresses under the high compressive stresses of the earth. Friction in rock fractures produces heat – temperatures close to the melting temperature were measured at the rock fracture surface. Maurer (1965) discussed brittle failure and friction along the fracture surface, as well as the transition from brittle towards ductile fracture, which occurs when the friction along the fracture surface exceeds the shear strength of the rock. He showed that this friction, i.e., resistance to sliding of the fracture surfaces with asperities and subject to molecular attractions, depends also on confining pressure.

[3]In the theory of brittle strength with failure, Griffith analyzed crystal failure and calculated the tensile strengths of crystals assuming the presence of a large number of minute cracks, the *Griffith cracks* (Jaeger, 1969), p. 85.

It has to be emphasized that the mechanical process of drilling requires the expenditure of the heat-dissipated energy, which is "*as necessary as ... the new-surface energy*", as Simon (1963) pointed out. Excess heat energy – affected by bit selection, hydraulics, pressure and mechanical and drilling conditions – can be evaluated by using specific energy. Pessier and Fear (1992) suggested that the excess heat generated per unit volume of removed rock is proportional to the difference in the specific energy E_S (equation (3.7)), of two bits drilling at different E_S levels. The difference in heat energy per unit volume of drilled rock is

$$H = C\,(E_{S2} - E_{S1})\,. \tag{3.16}$$

If H and E_S are measured in cal/m^3 and Pascal, respectively, we have $C = 0.239$ cal/J. Pessier and Fear (1992) used the specific energies $E_{S1} = 280$ MPa and $E_{S2} = 560$ MPa for two bits with diameters of 12 1/4 in. (31.1 cm) both drilling at a rate of penetration of 6 m/h. We obtain a difference in the drilling power of about 35 kW, which corresponds to a difference in heat drilling power – given by H·(ROP/3600)·(bore-hole area) – of about 8370 cal/s. This amount of power – as pointed out by Pessier and Fear – is certainly not negligible when compared to the total drilling power (see the example of equation (3.4) in which the total power is 75 kW).

3.7.3 Stress waves produced in loading/unloading

In the title of this subsection we have substituted "*produced*" for the original term "*dissipated*" used by Simon (1963) because it defines the energy radiated out as stress waves in the far field – that is, the drill-bit signal energy. It does not include the elastic energy temporarily stored in the near field nor the energy of inelastic deformation in the crushing and yielding of the rock close to the bit. Simon (1963) (see also Hunter, 1957) calculated the energy of the waves radiated for a concentrated loading (vertical source) on the surface of a semi-infinite medium (horizontal half space) in order to simulate the axial action of a bit chisel struck by a weight. After Miller and Pursey (1954), Hunter (1957) calculated the radiation of energy in the form of elastic waves generated by a transient localized force acting normally to the free surface of a semi-infinite solid. The radiation analysis was applied to the collision of a small body with the plane surface of a massive specimen, in the Hertzian approximation, which takes no account of the dissipation of energy (see Love, 1952). Starting from Hunter's analysis, Simon (1963) calculated the total energy radiated during loading by integrating the power from zero up to the time t_0 at which the maximum loading force occurs (bit forces are discussed in Sections 3.12 and 3.14). This gives

$$E_R = f(\nu)\frac{1}{Yc}\int_0^{t_0}\left(\frac{dF}{dt}\right)^2 dt, \tag{3.17}$$

where $c = (Y/\rho)^{1/2}$ and Y are the theoretical thin-rod (bar) velocity and Young's modulus, respectively, of the rock of density ρ. In previous equation, ν is the Poisson ratio of the medium, and $f(\nu)$ is a function slowly varying with respect to ν, which can be expressed as

$$f(\nu) = \frac{2\beta(\nu)(1+\nu)}{\pi}\sqrt{\frac{1-\nu^2}{1-2\nu}}, \tag{3.18}$$

where $f(\nu)$ approximately equals 0.6 in a Poisson medium, i.e., where $\nu = 0.25$ and $\beta(0.25) = 0.5374$ (see Hunter, 1957, equations from (17) to (20)). This theoretical calculation is performed by assuming that the dimensions of the region deformed inelastically is small compared with the dominant wavelength of the produced wave. By assuming the model of a dropping weight to simulate the indention impact of a cone tooth (chisel), Simon showed that the ratio between radiated energy, E_R, and the impact energy, E_0, during loading can be expressed as

$$\frac{E_R}{E_0} = f(\nu)\left(\frac{\pi^2}{2}\right)\left(\frac{K}{Y}\right)\left(\frac{l}{ct_0}\right), \tag{3.19}$$

where K is the constant of proportionality between force per unit chisel length and displacement – it has the dimension of a stress – and l is the length of the chisel edge. In his calculations, Simon (1963) assumed $K/Y \approx 10^{-2}$, $l = 1$ in. (2.54 cm), $t_0 = 2$ ms, and obtained values of the order of 10^{-4} for the ratio E_R/E_0.

Moreover, he estimates that the stress-wave energy produced on the unloading of a roller-cone tooth would be about three times higher than that resulting from the loading process. This difference is due to the fact that the tooth moves out of engagement with the rock faster after a rock chip has been broken out as a consequence of the loading process, because it has a shorter distance to travel when disengaging (see Section 3.12.1). In his calculations, Simon (1963) assumes that the load force would be released more rapidly, $(dF/dt)^2$ would be nine times greater, t_0 three times smaller and E_R should result in about three times the energy radiated during loading. Similar values may be estimated when considering the simultaneous action of more teeth indenting the rock. For example, in a working roller bit, the number of the teeth N_T that bear the load at any one time is typically of the order of 1 to 4 (Burgess and Lesso 1985), and the total chisel length l consequently increases.

The basic concept of equation (3.17) is that the power radiated in the medium is proportional to the square of the instantaneous variation of the loading force F, i.e., $(dF/dt)^2$. This is a key point in understanding the behavior of the drill-bit source, and in general of vibrating sources. From this concept, it follows that another force having equal maximum magnitude of F and acting with twice the rate in a time interval of half the duration produces twice the radiated energy. In other words, for equal maximum magnitude of the bit force, the high frequency, abrupt and irregular discontinuous changes are more important than the regular and slow periodic action of the bit force.

The structure of the roller-cone bits and of polycrystalline-diamond-compact (PDC) bits is described in Chapter 2. Their drilling action is described in detail below (Poletto, 2002b), in Hardage (1992) for SWD purposes (with roller-cone bits), and in the drilling literature (for instance, Bourgoyne Jr., Millheim, Chenevert and Young Jr., 1991; Devereux, 1999). The analysis of the theoretical forces for an idealized roller bit shows that the chipping action of the teeth wedges is a primary cause of discontinuities of vertical, horizontal and torque forces when a tooth moves out of contact with the rock (Biggs, Cheatham and Rice 1969). Examples of unloading rates, which are faster than loading rates (corresponding to about five times the loading rates) are shown by Sheppard and Lesage (1988), who measured the forces at the teeth of a drilling roller-cone bit. For these conditions, higher radiated-to-unloading energy ratios, of the order of 10^{-3}, could be expected. After Simon, several other authors calculated and analyzed the kinematic and dynamic behavior of drill bits using laboratory and field measurements. These results are summarized in Section 3.12, which is dedicated to describing the performance of different types of drill bits.

Finally, Simon calculates that during the sporadic propagation of rock cracks, E_R/E_0 can be 0.1 or closer to 1.0, and that high-frequency energy is involved during these "catastrophic" events. This high-frequency energy accounts for the local heat dissipation of elastic strain energy. He concludes that the energy of the radiated waves is a negligible fraction of the total energy and does not substantially affect the drilling efficiency. Table 3.3 and Figure 3.3a summarize Simon's result. However, these results may be underestimated in the evaluation of the new-surface energy, which may be, if we assume the opinion of other authors, 10 % or more than 10 % the total drilling energy expended by the bit.

For SWD purposes, we have to determine the contribution of the total energy radiated during long drilling intervals – such as those used for SWD acquisition. This energy can be very high (Figure 3.3b). In other words, while the fraction of the power that goes into radiated waves is small, the actual amount of energy in the long-time periods used in SWD acquisition is comparable to that of conventional seismic sources.

In our analysis, the power of the waves radiated into the rock is part of W_{vibr}. However, we have to take into account that the Simon's results are calculated for a source acting on a free surface, and not for a source "buried" in an unbounded infinite medium (see also Gangi, 1987). Furthermore, they do not include the significant part of the energy transmitted into the drill string and borehole. Simon's results in equation (3.17) and (3.19) include only body and surface waves radiated into the rock. They are in agreement with the analysis of the impedances, for which – with equal excitation force F – the energy radiated into the rock decreases in the presence of hard rocks, i.e., with high values of Yc in equation (3.17). We will see in Section 3.10 that the partitioning of the radiated energy produced by the drill-bit displacement source as waves radiated into the rock and drill string is related to the reflection coefficient at the bit/rock interface. This partitioning is important for characterizing the SWD signals, as the energy level of the SWD data is commonly obtained by calculating the coherent cross-energy of the signals measured in the rock and drill string.

3.8 Radiation of energy from the bit (far-field effects)

To quantify the downhole bit-seismic-source energy, Poletto (2002a) analyzed the downhole vibrations in the rock and in the drill string as produced by a downhole force. The purpose of this analysis was to compare the radiated field to the vibration levels measured in the drill string.

Table 3.3 – Simon's drilling-energy balance (source at the surface)

Type of energy expenditure	*Fraction (%)*
New surface	~ 1
Elastic strain into local ("necessary") heat	~ 99
Radiation (far field)	$\sim 10^{-4}$
Total	100

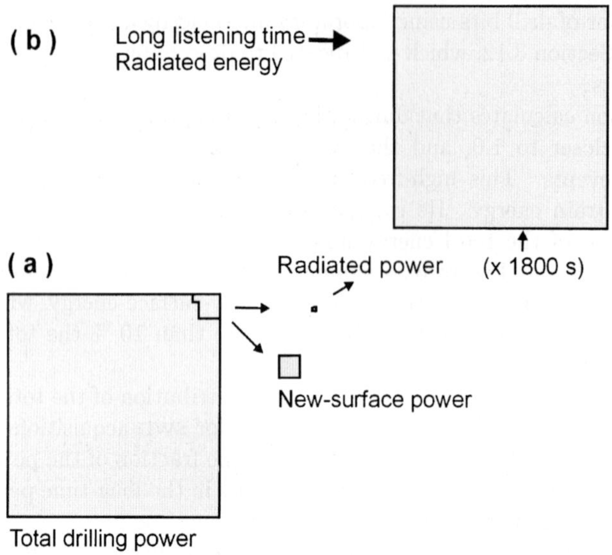

Figure 3.3: Power and energy are represented as proportional to area. a) The requirements of drilling energy as calculated by Simon (1963). The radiated power is only a small amount of the total drilling power at the bit. In Simon's analysis the new-surface power is about 1 %. However, this value seems to be underestimated and may be higher (10 % or more). b) Radiation energy measured in long listening times (e.g., 1800 s) can be comparable to that from conventional seismic sources (also, to drilling energy expended in 1 s).

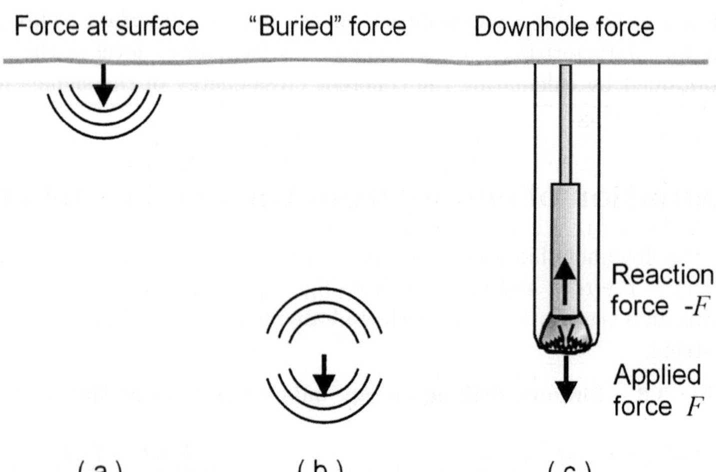

Figure 3.4: Vertical force as drill-bit source model. a) Applied at the surface, b) in a point of an infinite medium and c) downhole.

The drilling literature provides several examples of theoretical calculations and experiments performed to measure the drill-bit forces. To take into account the dynamic behavior of the bit, Poletto (2002a) reformulated the expression for the radiated energy into a half-space (equation (3.17)) on the basis of the drill-bit radiation pattern in an unbounded infinite, homogeneous and isotropic medium (Gangi, 1987). Assume, without loss of generality, we have vertical drilling and let, for simplicity sake, the force exerted by the bit be represented by the harmonic vertical component

$$F = F_0 \sin \omega t, \tag{3.20}$$

where $\omega = 2\pi/T$ is the angular frequency. In this approximation, we have not accounted for the gouging torsion – i.e., the SH component. This bit force is measured, for instance, by Sheppard and Lesage (1988), by Lesage, Falconer and Wick (1988), included in the calculation of the roller-bit radiation pattern by Rector and Hardage (1992) and, in finely stratified media, by Carcione and Carrion (1992). Moreover, we have not accounted for the lateral-force components (Langeveld, 1992a) of the point source (Malusa, Poletto and Miranda, 2003).

3.8.1 Radiation from a surface harmonic force

We have rewritten the surface-loading results of Simon below. Using equations (3.17) and (3.20), the loading energy radiated in the time interval of length $t_0 = T/4$ becomes, after some calculations,

$$E_{\rm R} = f(\nu) \frac{\pi}{4} \frac{\omega F_0^2}{Yc}. \tag{3.21}$$

Hence, Simon's average power radiated during loading – as well as during unloading with constant ω – can be rewritten as

$$W_{\rm R}^{\rm surf} = \frac{E_R}{(T/4)} = \frac{f(\nu)}{2} \frac{\omega^2 F_0^2}{\rho c^3}, \tag{3.22}$$

where we have used the relation $c = (Y/\rho)^{1/2}$ for the thin-rod (bar) velocity. This is the energy radiated by the source F pulsating at the surface of the halfspace (Figure 3.3a). In a Poisson medium ($\nu = 0.25$), we have $f(0.25) = 0.586$ (see equation (3.18) and Hunter, 1957), $c = 0.913 \cdot \alpha$, where α is the compressional P-wave velocity, related to c by

$$c = \sqrt{\frac{(1+\nu)(1-2\nu)}{(1-\nu)}}\, \alpha. \tag{3.23}$$

Hence, equation (3.22) can be rewritten as

$$W_{\rm R}^{\rm surf} = 4.836 \frac{\omega^2 F_0^2}{4\pi\rho\alpha^3}. \tag{3.24}$$

Here we obtain the same total power as calculated by Miller and Pursey (1954) for the radiation in a Poisson semi-space from a pulsating vibrator plate (Cassand and Lavergne, 1971, p. 206).

3.8.2 Radiation from a downhole harmonic force

Now consider an unbounded, homogeneous and isotropic medium in which a "buried" vertical force F given by equation (3.20) acts. The vertical-force model was proposed by Rector and Hardage (1992) to calculate the radiation pattern of a working roller-cone bit. This is a simplified source model; Poletto (2002a) used the same approach to calculate the far-field power radiated in the formation (Figure 3.4b). Moreover, we have assumed a vertical-reaction-force source to calculate the energy transmitted into the drill string in the form of one-dimensional axial waves, and to calculate the partition of these energies (Figure 3.4c) (for a discussion about the excitation process of a roller-cone bit acting as an axial-displacement source see also Chin, 1994, pp. 146–155).

Let us consider a wave $f = f(\omega t - kr)$, radiated in the far field with velocity $c = \omega/k$, where ω is the radiation frequency, $k = 2\pi/\lambda$ is the wavenumber, λ is the wavelength and r the radial distance from the point source. The far-field and near-field concepts are explained, for instance, by Aki and Richards (1980). The far-field approximation holds when

$$r \gg \frac{c}{\omega} = \frac{\lambda}{2\pi}, \qquad (\text{or } kr \gg 1), \tag{3.25}$$

in which case the near-field and intermediate-field terms – which decrease as $1/r^3$ and $1/r^2$, respectively – become negligible. Rector and Hardage (1992) use the expressions given in White (1965) (see also, Love (1952)) to represent, at a point located in the far field, the displacement produced by a vertical force $F(t) = F_0 f(t)$, where F_0 is a constant magnitude. The displacement components are given in spherical coordinates as

$$u_r = \frac{F_0 \cos\phi}{4\pi\rho\alpha^2 r} f\left(t - \frac{r}{\alpha}\right), \tag{3.26}$$

$$u_\theta = 0, \tag{3.27}$$

$$u_\phi = -\frac{F_0 \sin\phi}{4\pi\rho\beta^2 r} f\left(t - \frac{r}{\beta}\right), \tag{3.28}$$

where ϕ is the angle between the radius and the vertical axis, θ is the azimuth angle (Figure 3.5), and α and β are the velocities of compressional and shear waves in the formation, respectively. The displacements u_r and u_ϕ correspond to the P and SV waves, respectively. Using the time-harmonic force of equation (3.20), we obtain

$$u_r = \frac{F_0 \cos\phi}{4\pi\rho\alpha^2 r} \sin\omega\left(t - \frac{r}{\alpha}\right), \tag{3.29}$$

$$u_\phi = -\frac{F_0 \sin\phi}{4\pi\rho\beta^2 r} \sin\omega\left(t - \frac{r}{\beta}\right). \tag{3.30}$$

Let the particle velocity be $v = \partial u/\partial t$. The amount of energy transmitted per unit time across unit area normal to the wave direction of propagation, i.e., the energy flux, of a wavefront radiated with propagation velocity $c_{\rm wav}$ is given by $P_{\rm wav} = \rho c_{\rm wav} v^2$ (Aki and

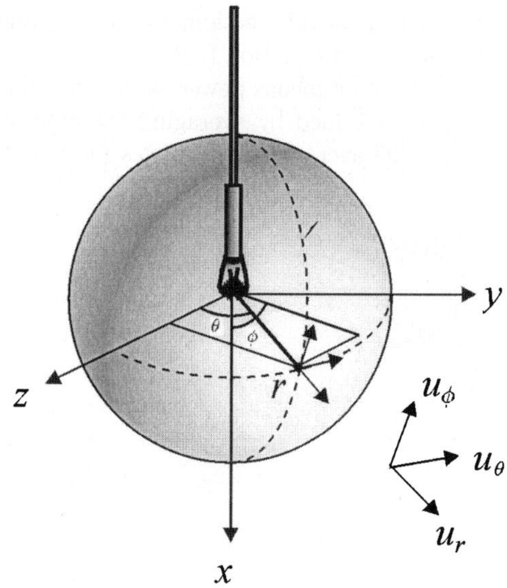

Figure 3.5: Displacement vector in spherical coordinates.

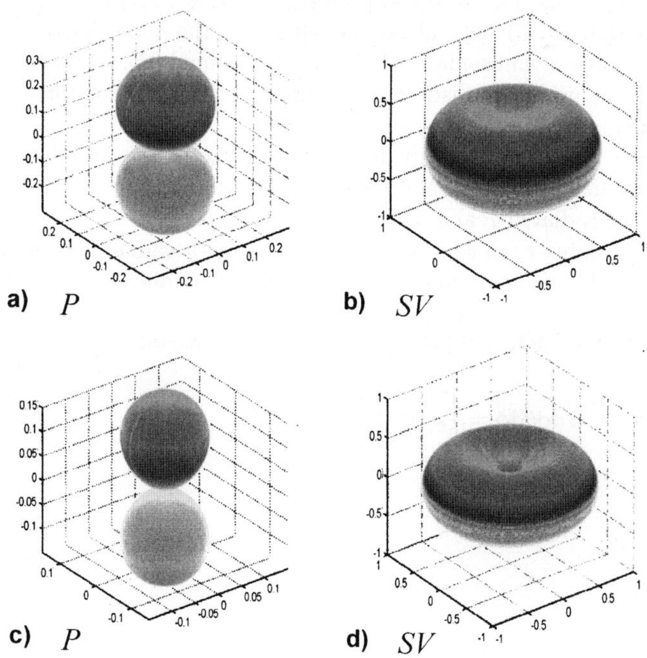

Figure 3.6: Vertical force model of the source. a) Compressional and b) shear radiation lobes of displacement. c) Compressional and d) shear radiation lobes of power.

Richards, 1980, p. 127). In analysis of this section, we use harmonic waves propagating spherically in the far-field condition of equation (3.25).

We are interested not in the instantaneous power, which depends on the instantaneous sine phase, but in the intensity obtained by averaging the energy flux per unit surface over a period T ($T = 2\pi/\omega$). The average energy fluxes for radial P and tangential SV components are given by

$$\bar{P}_{\mathrm{r}} = \frac{1}{T}\int_0^T \rho\alpha v_{\mathrm{r}}^2(t)\mathrm{d}t \ = \ \frac{1}{2}\rho\alpha V_{\mathrm{r}}^2, \tag{3.31}$$

$$\bar{P}_{\phi} = \frac{1}{T}\int_0^T \rho\beta v_{\phi}^2(t)\mathrm{d}t \ = \ \frac{1}{2}\rho\beta V_{\phi}^2, \tag{3.32}$$

where

$$V_{\mathrm{r}}(\phi) = \frac{\omega F_0 \cos\phi}{4\pi\rho\alpha^2 r}, \tag{3.33}$$

$$V_{\phi}(\phi) = \frac{\omega F_0 \sin\phi}{4\pi\rho\beta^2 r}, \tag{3.34}$$

and where we have used the relations

$$\frac{1}{T}\int_0^T \cos^2\omega\,(t - r/\alpha)\;dt \ = \ \frac{1}{T}\int_0^T \cos^2\omega\,(t - r/\beta)\;dt \ = \frac{1}{2}. \tag{3.35}$$

Displacement and energy-flux radiation patterns are shown in Figure 3.6. Note that the integrals of equation (3.35) depend on the square of the time derivative of the force. In this case, the force is sinusoidal and the integrals equal 0.5. However, we will see in Section 3.12 that, by using a realistic time variation of the indention force of a cone tooth, the integral can equal 1.5, i.e., three times higher.

The total value of the mean energy flux during a period T, i.e., the total average power radiated in the far field (Kaufman and Levshin, 2000, p. 179), is calculated by integrating the intensity of the radiated P and SV waves over the spherical surface, $\mathcal{S}$, of a wavefront with an arbitrary radius r, so that we can take as large a value for r as desired in order to satisfy the far-field condition of equation (3.25). We obtain

$$W_r = \int_S \bar{P}_{\mathrm{r}}(\theta,\phi) r^2 \sin\phi \mathrm{d}\theta \mathrm{d}\phi = \frac{1}{2}\rho\alpha r^2 \int_0^{2\pi} d\theta \int_0^{\pi} V_{\mathrm{r}}^2(\phi)\sin\phi\, d\phi, \tag{3.36}$$

$$W_\phi = \int_S \bar{P}_{\phi}(\theta,\phi) r^2 \sin\phi \mathrm{d}\theta \mathrm{d}\phi = \frac{1}{2}\rho\beta r^2 \int_0^{2\pi} d\theta \int_0^{\pi} V_{\phi}^2(\phi)\sin\phi\, d\phi, \tag{3.37}$$

which gives

$$W_r = \frac{1}{6}\,\frac{\omega^2 F_0^2}{4\pi\,\rho\alpha^3}, \tag{3.38}$$

$$W_\phi = \frac{1}{3}\,\frac{\omega^2 F_0^2}{4\pi\,\rho\beta^3}, \tag{3.39}$$

for compressional P and shear SV waves, respectively. These radiated powers are part of W_{vibr} in equation (3.15).

3.8.3 Integrated downhole-radiation impedance

The impedance that a medium presents to a given motion is a measure of the amount of resistance to particle motion (Aki and Richards, 1980, p. 137). By definition, the radiation impedance in elasticity is the ratio of stress to particle velocity – in the far field it equals the acoustic impedance, $\rho\alpha$ and $\rho\beta$ for P and SV waves, respectively. If complex, the real and imaginary parts of the radiation impedance are also called specific acoustic resistance and reactance, respectively.

Because we are interested in the total powers radiated both in the formation and in the drill string, we introduce integrated impedances. Here, the term "total power" has the meaning of integrated over the whole surface of the spherical surface in the formation or of the cross-section in the drill string. To express the radiated power in terms of integrated "resistance" of the formation to radiation at a given harmonic frequency, Poletto (2002a) re-formulated equations (3.38) and (3.39) by introducing an equivalent, "effective", radiation radius. The effective radii R_α and R_β for compressional P and shear SV waves are defined by the equations

$$R_\alpha^2 = \frac{1}{I_\alpha k_\alpha^2}, \tag{3.40}$$

$$R_\beta^2 = \frac{1}{I_\beta k_\beta^2}, \tag{3.41}$$

where $k_\alpha = \omega/\alpha$ and $k_\beta = \omega/\beta$. I_α and I_β are the normalized surface integrals of the unit energy-flux patterns $\bar{P}_r = \cos^2\phi/r^2$ and $\bar{P}_\phi = \sin^2\phi/r^2$, respectively, calculated as

$$I_\alpha = \frac{1}{4\pi}\int_0^{2\pi} d\theta \int_0^{\pi} \cos^2\phi \;\sin\phi \, d\phi = \frac{1}{3}, \tag{3.42}$$

$$I_\beta = \frac{1}{4\pi}\int_0^{2\pi} d\theta \int_0^{\pi} \sin^2\phi \;\sin\phi \, d\phi = \frac{2}{3}. \tag{3.43}$$

It can be observed that the effective radii R_α and R_β are of the same order of magnitude of the dominant wavelength as the reference radii used to define the near-field conditions (3.25).

By using R_α and R_β, the integrated impedance can be expressed by the product of the radiation impedance (or far-field impedance) and the "effective" area as

$$A_\alpha = 4\pi R_\alpha^2 = \frac{4\pi}{I_\alpha k_\alpha^2}, \tag{3.44}$$

$$A_\beta = 4\pi R_\beta^2 = \frac{4\pi}{I_\beta k_\beta^2}. \tag{3.45}$$

We have

$$\mathcal{Z}_\alpha = A_\alpha \rho\alpha = \frac{12\pi\alpha^3\rho}{\omega^2}, \tag{3.46}$$

$$\mathcal{Z}_\beta = A_\beta \rho\beta = \frac{6\pi\beta^3\rho}{\omega^2}, \tag{3.47}$$

for the downhole P and SV radiated signals, respectively. The integrated impedances of equations (3.46) and (3.47) have the dimension of force over particle velocity (while the radiation impedance has the dimension of stress over particle velocity). The variation of the integrated radiation impedances with frequency ($\mathcal{Z}_\alpha, \mathcal{Z}_\beta \propto \omega^{-2}$) is in agreement with the average radiation impedance curves as calculated for a surface Vibroseis source by Baeten and Ziolkowski (1990). Hence, average radiated powers (equations (3.38) and (3.39)) can be expressed in terms of force and integrated impedance as

$$W_r = \frac{F_0^2}{2} \frac{1}{\mathcal{Z}_\alpha}, \tag{3.48}$$

$$W_\phi = \frac{F_0^2}{2} \frac{1}{\mathcal{Z}_\beta}. \tag{3.49}$$

The form of equations (3.48) and (3.49) is preferable for comparing the impedances of the radiated and drill-string energy – because the "characteristic impedance" (or mechanical impedance) in one-dimensional drill strings is defined as the ratio of force over particle velocity (see Section 4.2.1), rather than stress over particle velocity – while equations (3.38) and (3.39) are preferable for calculating the power radiated by a known periodic force.

3.8.4 Total P+SV power radiated in the formation

The total power radiated in the formation is obtained by the sum of the radiated power of compressional and shear waves as

$$W_\mathrm{R} = W_r + W_\phi = \frac{1}{6} \frac{\omega^2 F_0^2}{4\pi\,\rho} \left(\frac{1}{\alpha^3} + \frac{2}{\beta^3} \right). \tag{3.50}$$

In terms of total radiation impedance, we have

$$W_R = \frac{F_0^2}{2} \frac{1}{\mathcal{Z}_\mathrm{R}}, \tag{3.51}$$

where the total impedance is given by

$$\frac{1}{\mathcal{Z}_\mathrm{R}} = \frac{1}{\mathcal{Z}_\alpha} + \frac{1}{\mathcal{Z}_\beta}. \tag{3.52}$$

Assuming a Poisson medium, denoted by superscript $^{(\mathrm{P})}$, we have $(\alpha/\beta) = \sqrt{3}$, and

$$W_\mathrm{R}^{(\mathrm{P})} = \frac{1}{6} \frac{\omega^2 F_0^2}{4\pi\,\rho\alpha^3} \left(1 + 6\sqrt{3}\right). \tag{3.53}$$

In this case, the total $P + SV$ radiated power produced by a downhole sinusoidal force can be written as

$$W_\mathrm{R}^{(\mathrm{P})} = \frac{F_0^2}{2} \frac{1}{\mathcal{Z}_\mathrm{R}^{(\mathrm{P})}} \approx 0.151 \, \frac{\omega^2 F_0^2}{\rho\alpha^3}, \tag{3.54}$$

where

$$\mathcal{Z}_{\rm R}^{(\rm P)} = \frac{12\pi\rho\alpha^3}{\omega^2(1+6\sqrt{3})} \approx \frac{1}{0.302}\frac{\rho\alpha^3}{\omega^2}, \tag{3.55}$$

is the radiation impedance calculated for the given Poisson medium and frequency. Compared with equation (3.22) calculated for radiation from a surface force, equation (3.54) could be roughly interpreted in the following way: a given excitation force energizes the half-infinite medium with approximately twice the radiated power than the infinite one.

From equation (3.53) it follows that the average compressional P-power is about 10 % of the total radiated power. However, when averaging the compressional power of equation (3.38) over a spherical wavefront of area $4\pi r^2$ – in order to compare it with the power of a pressure source with a spherical radiation pattern – remember that the radiation power of the P component produced by the force in the directions of the maximum radiation (i.e., $\phi = 0,\ \pi$) is three times its average power density. In the directions $\phi = 0,\ \pi$ of maximum compressional radiation, the compressional power density at radius r is

$$P_r^{(\rm max)} = \frac{1}{32\pi^2 r^2}\frac{\omega^2 F_0^2}{\rho\alpha^3}. \tag{3.56}$$

3.8.5 Radiation from a non-harmonic force.

In general, when assuming a force of arbitrary time dependence $F(t)$ acting in the time interval $(0,\ T)$, we calculate the integrals

$$I_f = \frac{1}{T}\int_0^T \left(\frac{\partial F}{\partial t}\right)^2 dt, \tag{3.57}$$

and multiply the previous power results by the scaling factor

$$\frac{2I_f}{(F_0\omega)^2}. \tag{3.58}$$

3.9 Near-field effects

Another approach to bit-source analysis is based on energy balance in terms of rotary-drilling equations expressed in terms of the drilling parameters (Section 3.6.3). To relate the rotary-drilling equations to the vibrations in the formation close to the bit, we need to account for the near-field effects. In fact, while the far-field terms dominate in the far field, the near-field terms are important in the near field (Aki and Richards, 1980, p. 119) (Figure 3.7), i.e., to calculate the bit vibrations and exchanges of energy with the drill string. This analysis assumes that the bit front is a flat rigid plate, and does not consider the presence of the cylindrical hole (Heelan, 1953; Meredith, 1990; Meredith, Cheng, Toksöz and Bouchon, 1990).

The general expression of the total particle displacements (near- and far-field) produced by a single vertical force $F_0 f(t)$ in an unbounded homogeneous, isotropic, elastic solid is given as (White, 1965, p. 214)

$$u_r = \frac{F_0\cos\phi}{4\pi\rho r\alpha^2}\left[\frac{2\alpha^2}{r^2}\int_{r/\alpha}^{r/\beta} t' f(t-t')dt' + f\left(t-\frac{r}{\alpha}\right)\right], \tag{3.59}$$

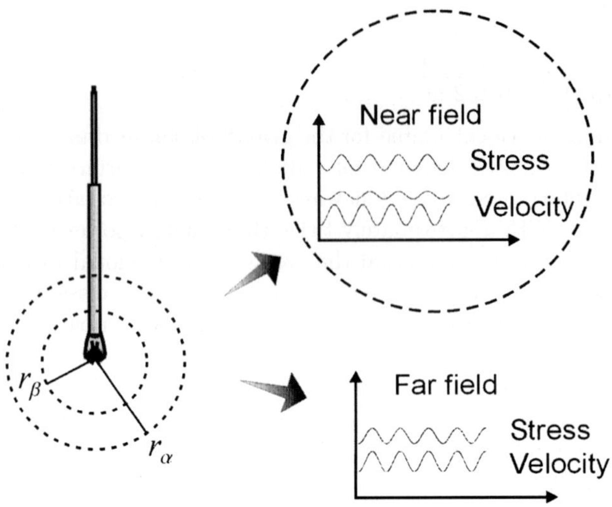

Figure 3.7: Near- and far-field radii $r_\alpha = 1/k_\alpha$ and $r_\beta = 1/k_\beta$, respectively, for drill-bit spherical waves. Assuming frequency 30 Hz, $\alpha = 4200$ m/s, $\beta = 2425$ m/s, we obtain $r_\alpha \sim 22$ m, and $r_\beta \sim 13$ m. In the near-field, particle velocity is the sum of components which are in- and out-of-phase with stress. In the far-field, stress and particle velocity are in phase.

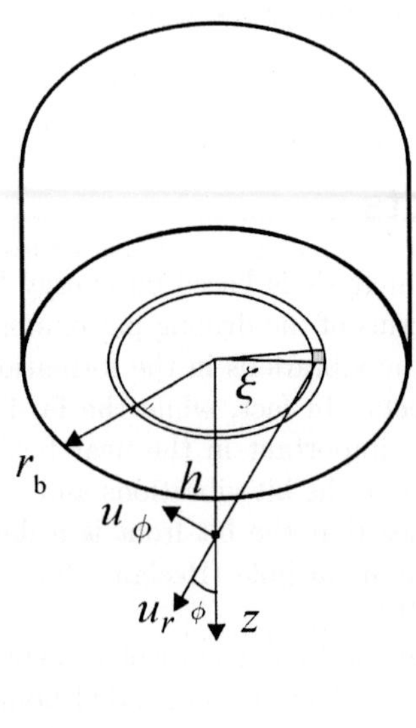

Figure 3.8: System of coordinates used to calculate displacement in front of an ideal flat bit: x is the radial distance from bit center and h is the distance of the point ahead of the bit in which the axial contribution from radial and tangential components is summed.

$$u_\phi = \frac{F_0 \sin\phi}{4\pi\rho r \beta^2}\left[\frac{\beta^2}{r^2}\int_{r/\alpha}^{r/\beta} t' f(t-t')dt' - f\left(t - \frac{r}{\beta}\right)\right], \tag{3.60}$$

for radial P and tangential SV displacement waves, respectively. Integrating these equations (Love, 1952, p. 306) and assuming a time-harmonic force $f(t) = \sin\omega t$, we have (see also Pilant, 1979, p. 79)

$$u_r = R_0\left[R_1 \sin\omega\left(t-\frac{r}{\alpha}\right) + R_2\cos\omega\left(t-\frac{r}{\alpha}\right) + R_3\cos\omega\left(t-\frac{r}{\beta}\right) + R_4\sin\omega\left(t-\frac{r}{\beta}\right)\right], \tag{3.61}$$

$$u_\phi = S_0\left[S_1 \sin\omega\left(t-\frac{r}{\beta}\right) + S_2\cos\omega\left(t-\frac{r}{\beta}\right) + S_3\cos\omega\left(t-\frac{r}{\alpha}\right) + S_4\sin\omega\left(t-\frac{r}{\alpha}\right)\right], \tag{3.62}$$

where

$$\begin{aligned}
R_0 &= \frac{F_0\cos\phi}{4\pi\rho r\alpha^2} & S_0 &= \frac{F_0\sin\phi}{4\pi\rho r\beta^2}\\
R_1 &= 1 - \frac{2\alpha^2}{r^2\omega^2}, & S_1 &= -1 + \frac{\beta^2}{r^2\omega^2},\\
R_2 &= -\frac{2\alpha}{r\omega}, & S_2 &= \frac{\beta}{r\omega},\\
R_3 &= \frac{2\alpha^2}{\beta r\omega}, & S_3 &= -\frac{\beta^2}{\alpha r\omega},\\
R_4 &= \frac{2\alpha^2}{r^2\omega^2}, & S_4 &= -\frac{\beta^2}{r^2\omega^2}.
\end{aligned} \tag{3.63}$$

We observe that only R_0 and S_0 depend on the angle ϕ and do not depend on the frequency ω. Moreover, in the far field $R_1 = 1$, $S_1 = -1$ and the other terms with higher index are zero.

3.9.1 Phase of the harmonic wave components

Equations (3.61) and (3.62) can be rewritten in terms of orthogonal (sine and cosine) harmonics, as

$$u_r = R_0\Re\left[Q_r^* e^{\imath\omega t}\right] = R_0|Q_r|\Re\left[e^{\imath(\omega t-\varphi_r)}\right] = R_0\left[Q_1\sin\omega t + Q_2\cos\omega t\right], \tag{3.64}$$

where $\Re$ denotes the real part, '$*$' denotes complex conjugate and

$$Q_1 = R_1\cos k_\alpha r + R_2\sin k_\alpha r + R_3\sin k_\beta r + R_4\cos k_\beta r, \tag{3.65}$$

$$Q_2 = -R_1\sin k_\alpha r + R_2\cos k_\alpha r + R_3\cos k_\beta r - R_4\sin k_\beta r, \tag{3.66}$$

$$Q_r = Q_2 + \imath Q_1 = |Q_r|e^{\imath\varphi_r}, \tag{3.67}$$

with $k_\alpha = \omega/\alpha$ and $k_\beta = \omega/\beta$. Similar relations, obtained for the transverse displacement component by substituting S_i in place of R_i and interchanging k_α and k_β, are as follows:

$$u_\phi = S_0\Re\left[Q_\phi^* e^{\imath\omega t}\right] = S_0|Q_\phi|\Re\left[e^{\imath(\omega t-\varphi_\phi)}\right] = S_0\left[Q_3\sin\omega t + Q_4\cos\omega t\right], \tag{3.68}$$

$$Q_3 = S_1\cos k_\beta r + S_2\sin k_\beta r + S_3\sin k_\alpha r + S_4\cos k_\alpha r, \tag{3.69}$$

$$Q_4 = -S_1 \sin k_\beta r + S_2 \cos k_\beta r + S_3 \cos k_\alpha r - S_4 \sin k_\alpha r, \tag{3.70}$$

$$Q_\phi = Q_4 + \imath Q_3 = |Q_\phi| e^{\imath \varphi_\phi}. \tag{3.71}$$

For the near-field condition ($k_\alpha r < k_\beta r \ll 1$), the amplitude of the sine over the cosine components,

$$\tan \varphi_r = \frac{Q_1}{Q_2}, \tag{3.72}$$

and

$$\tan \varphi_\phi = \frac{Q_3}{Q_4} \tag{3.73}$$

for radial and transverse components, respectively, have large values. Expanding the terms in powers of $k_\alpha r$ and $k_\beta r$, it can be shown that, assuming a Poisson medium (where $k_\beta = k_\alpha \sqrt{3}$), in the near field the terms are

$$\lim_{k_\alpha r \to 0} \left| \frac{Q_1}{Q_2} \right| = \frac{9}{k_\alpha r (1 + 6\sqrt{3})} \gg 1, \quad \to \infty \text{ with } k_\alpha r \to 0, \tag{3.74}$$

$$\lim_{k_\alpha r \to 0} \left| \frac{Q_3}{Q_4} \right| = \frac{6}{k_\alpha r (1 + 6\sqrt{3})} \gg 1, \quad \to \infty \text{ with } k_\alpha r \to 0. \tag{3.75}$$

Using equations 3.74 and 3.75, we obtain also

$$|\cos \varphi_r| \approx \frac{k_\alpha r (1 + 6\sqrt{3})}{9} \ll 1, \quad \to 0 \text{ with } k_\alpha r \to 0, \tag{3.76}$$

and

$$|\cos \varphi_\phi| \approx \frac{k_\alpha r (1 + 6\sqrt{3})}{6} \ll 1, \quad \to 0 \text{ with } k_\alpha r \to 0. \tag{3.77}$$

In equations (3.76) and (3.77), the phases φ_r and φ_ϕ of the radial and tangential components (equations (3.64) and (3.68), respectively) are calculated in the near field with the source model of a single force. Note that these angles, close to $\pi/2$, do not depend on ϕ but on kr, i.e., on frequency, formation velocity and radius. For a given frequency ω of the source, the limits are calculated for r tending towards zero. If the single-force model is assumed to represent the bit source, $r = 0$ is a singularity for displacement.

3.9.2 Near-field axial displacement

Because the real bit is not a point, we can integrate the contribution of the force distributed over its face, of radius r_b. However, the knowledge of the exact distribution of forces at the bit face is, in general, not available (Ma, Zhou and Deng, 1995). To calculate the near-field effect, we use the averaged over-bit-area force, by assuming that the source is distributed uniformly over $0 < r < r_b$. Using equations (3.74) and (3.75) to calculate the phase of the harmonic displacements, it can be shown that the ratios Q_1/Q_2 and Q_3/Q_4

may be of the order of 10^2 at $r = r_b$, for realistic bits and formation properties, and for frequencies ranging from about 15 Hz to 50 Hz. A comparable result can be obtained by calculating the axial displacement at distance h in front of the bit assumed to be flat and subject to uniform axial stress distribution, namely,

$$\sigma_0 = \frac{F_0}{A_b} \sin \omega t \tag{3.78}$$

over the bit area $A_b = \pi r_b^2$. In this approximation, we can integrate (Poletto, 2002a) the contribution of the elementary areas to obtain the integrated axial-displacement component at a point of the bit axis at distance h from the bit face (Figure 3.8) as

$$\hat{u} = \int u_z \mathrm{d}A = \int_0^{2\pi} \mathrm{d}\theta \int_0^{r_b} (u_r \cos\phi - u_\phi \sin\phi)\xi \mathrm{d}\xi = \Gamma_1 \sin\omega t + \Gamma_2 \cos\omega t, \tag{3.79}$$

where ξ and ϕ are the radial and angular variables of Figure 3.8 and the coefficients Γ_i are calculated separately for sine- and cosine-wave components as

$$\Gamma_1 = \frac{2\pi}{A_b} \int_0^{r_b} (R_0 Q_1 \cos\phi - S_0 Q_3 \sin\phi)\, \xi \mathrm{d}\xi, \tag{3.80}$$

$$\Gamma_2 = \frac{2\pi}{A_b} \int_0^{r_b} (R_0 Q_2 \cos\phi - S_0 Q_4 \sin\phi)\, \xi \mathrm{d}\xi. \tag{3.81}$$

It can be shown that

$$\lim_{h \to 0} \Gamma_1 = C_0 r_b, \tag{3.82}$$

$$\lim_{h \to 0} \Gamma_2 = -C_0 r_b \frac{k_\alpha r_b (1 + 6\sqrt{3})}{12}, \tag{3.83}$$

where $C_0 = F_0/(A_b \rho \alpha^2)$ and we have used the near-field approximation for the coefficients Q_i. Therefore, we have

$$\lim_{h \to 0} \frac{\Gamma_1}{\Gamma_2} = -\frac{12}{k_\alpha r_b (1 + 6\sqrt{3})}, \tag{3.84}$$

and

$$\lim_{h \to 0} \left| \frac{\Gamma_1}{\Gamma_2} \right| \gg 1. \tag{3.85}$$

The integrated result in equation (3.84) is obtained at the bit face ($h \to 0$), and is of the same order as those we obtain if we use $r = r_b$ in equations (3.74) and (3.75) for a point force. In other words, the axial displacement at a position very close to the bit source has a phase close to that of the force, as it is composed of a sine wave of magnitude $|\Gamma_1|$ ($\gg |\Gamma_2|$) and a cosine wave of magnitude $|\Gamma_2|$ ($\ll |\Gamma_1|$). The phase of the near-field axial displacement is

$$\varphi(\omega, \alpha, r_b) = \arctan \left[-\frac{12}{k_\alpha r_b (1 + 6\sqrt{3})} \right] \approx -\frac{\pi}{2}, \tag{3.86}$$

for seismic frequencies, i.e., $k_\alpha r_b \sim 100^{-1}$. This gives

$$\cot \varphi = -\frac{k_\alpha r_b (1 + 6\sqrt{3})}{12} \approx -0.95 k_\alpha r_b, \tag{3.87}$$

with $\cot\varphi \approx \cos\varphi$ for $k_\alpha r_b \ll 1$. The phase φ is used in the next paragraph to calculate in- and out-of phase stress-velocity relations.

3.9.3 Energy flux and near-field effects

The phase φ determines the amount of vibrations in phase and out of phase with force. This analysis is used to determine not only the radiated energy (far-field waves), but also the energy interchanged between the formation and the bit, as part of the drill string, for near-field effects. The particle velocity in terms of sine and cosine waves is given by

$$v_r = \frac{\partial u_r}{\partial t} = \omega R_0 \left[Q_1 \cos \omega t - Q_2 \sin \omega t \right], \tag{3.88}$$

$$v_\phi = \frac{\partial u_\phi}{\partial t} = \omega S_0 \left[Q_3 \cos \omega t - Q_4 \sin \omega t \right]. \tag{3.89}$$

In the near field we have

$$v_r \approx \omega R_0 Q_1 \cos \omega t, \tag{3.90}$$

$$v_\phi \approx \omega S_0 Q_3 \cos \omega t, \tag{3.91}$$

while in the far field ($k_\alpha r \gg 1$) we obtain again the propagating waves of equations (3.29) and (3.30). The integrated axial velocity is calculated from equation (3.79) as

$$\hat{v} = \frac{\partial \hat{u}}{\partial t} = \omega \Gamma_1 \cos \omega t - \omega \Gamma_2 \sin \omega t. \tag{3.92}$$

From equations (3.82), (3.83) and (3.84) we obtain the *real* axial velocity at the bit as (see Figure 3.8)

$$v_0 = \lim_{h \to 0} \hat{v} = \frac{F_0 k_\alpha r_\mathrm{b}}{A_\mathrm{b} \rho \alpha} \left[\cos \omega t + \frac{k_\alpha r_\mathrm{b} (1 + 6\sqrt{3})}{12} \sin \omega t \right]. \tag{3.93}$$

Let us calculate the instantaneous energy flux at the bit axis by

$$\sigma_0 v_0 = \frac{F_0^2 k_\alpha r_\mathrm{b}}{A_\mathrm{b}^2 \rho \alpha} \left[\cos \omega t \sin \omega t - \cot \varphi \sin^2 \omega t \right], \tag{3.94}$$

where we have used equations (3.78) and (3.87). To account for work done (expended) by the source, we multiply by the bit area, and average in time over a period $T = 2\pi/\omega$ to obtain the average power

$$A_\mathrm{b} \overline{\sigma_0 v_0} = -\frac{F_0^2 k_\alpha r_\mathrm{b}}{2 A_\mathrm{b} \rho \alpha} \cot \varphi = W_\mathrm{R}^{(\mathrm{P})}, \tag{3.95}$$

which equals the total radiated power in a Poisson medium of equation (3.53). Note that only the second term – with very small amplitude – of equation (3.94) contributes in average to energy expenditure: this is the energy of far-field wave radiation. On the other hand, the $\cos \omega t$ component of particle velocity is orthogonal to the force which is a $\sin \omega t$ time function. The contribution of this component to average power over a period T is zero. This near-field term has higher amplitude and represents the ("*blind*") power continuously and locally interchanged between source and formation. For a similar analysis with spherical pressure waves, see Kaufman and Levshin (2000, p. 179). Hence, for a given harmonic amplitude the expended power is determined by the $\cos \varphi$ factor of phase between force and particle velocity, dependent on frequency.

3.9.4 Complex impedance

We have to reformulate the definition of the integrated impedance to include the near-field effects. Because the stress in the proximity of the bit has the phase of a sine wave (see also Somerton, Timur and Gray, 1961), in order to obtain particle velocity by superposition of sine and cosine waves, a complex impedance has to be used (Lutz, Raynaud, Gastalder, Quichaud, Raynald and Muckerloy, 1972; Clayer, Vandiver and Lee, 1990).

It is convenient to rewrite particle velocity and stress in a complex form. We can use the near-field approximation of equation (3.87) to write the *complex* axial particle velocity at the bit (equation (3.93)) as

$$\tilde{v}_0 = \imath V_0 e^{\imath(\omega t - \varphi)}, \tag{3.96}$$

where

$$V_0 = \frac{F_0 k_\alpha r_\mathrm{b}}{A_\mathrm{b} \rho \alpha \sin \varphi}. \tag{3.97}$$

Moreover, the axial *complex* stress becomes

$$\tilde{\sigma}_0 = -\imath \frac{F_0}{A_\mathrm{b}} e^{\imath \omega t}. \tag{3.98}$$

The radiation impedance is therefore calculated as

$$\frac{\tilde{\sigma}_0}{\tilde{v}_0} = -\frac{F_0}{A_\mathrm{b} V_0} e^{+\imath\varphi} = -\frac{\rho\alpha}{k_\alpha r_\mathrm{b}} \left(\cos\varphi \sin\varphi + \imath \sin^2\varphi\right). \tag{3.99}$$

It has to be remembered that the impedance of equation (3.99) is calculated in the near-field approximation. The near-field integrated impedance over the bit area is

$$\mathcal{Z}_\mathrm{b} = \frac{A_\mathrm{b}\tilde{\sigma}_0}{\tilde{v}_0} = -\frac{A_\mathrm{b}\rho\alpha}{k_\alpha r_\mathrm{b}} \left(\cos\varphi \sin\varphi + \imath \sin^2\varphi\right). \tag{3.100}$$

Now we relate the near- and far-field integrated impedances. We observe that the total integrated impedance of equation (3.55) can be expressed as

$$\mathcal{Z}_\mathrm{R}^{(\mathrm{P})} = -\frac{A_\mathrm{b}\rho\alpha}{k_\alpha r_\mathrm{b}} \tan\varphi. \tag{3.101}$$

We obtain

$$\mathcal{Z}_\mathrm{b} = \mathcal{Z}_\mathrm{R}^{(\mathrm{P})} \left(\cos^2\varphi + \imath \cos\varphi \sin\varphi\right). \tag{3.102}$$

We will see in Section 3.9.5 that the result of equation 3.102 is similar to that obtained for pressure waves from a pressure source. Note also that the real part of the integrated impedance is close to the impedance at the bit $\left(\Re(\mathcal{Z}_\mathrm{b}) \sim A_\mathrm{b}\rho\alpha\right)$ as calculated by Poletto, Malusa and Miranda (2000) in the plane-wave approximation, without accounting for the complex near-field effects (Section 4.11.1).

3.9.5 Waves from a pressure source at the origin

We discuss the previous results with the following analogy. The complex impedance to radiation of spherically propagating acoustic waves is a good example for the analysis of near-field and far-field effects (see, for instance, Seto, 1994). Assume spherical harmonic pressure waves

$$p = \frac{P_0}{r} e^{\imath(\omega t - kr)}, \tag{3.103}$$

where $k = \omega/\alpha$. We obtain particle velocity from $\partial_r p = -\rho \partial_t v$ (with the notation $\partial_\psi = \partial/\partial\psi$) and write it as

$$v = \left(\frac{1}{r} + \imath k\right) \frac{p}{\imath \rho \omega}, \tag{3.104}$$

which gives the complex impedance

$$z_p = \rho\alpha \left(\frac{k^2 r^2}{1 + k^2 r^2} + \imath \frac{kr}{1 + k^2 r^2}\right). \tag{3.105}$$

Using

$$|z_p| = \rho\alpha \frac{kr}{\sqrt{1 + k^2 r^2}}, \tag{3.106}$$

and

$$\cos\varphi = \frac{kr}{\sqrt{1 + k^2 r^2}}, \tag{3.107}$$

we can also write

$$|z_p| = \rho\alpha \cos\varphi, \tag{3.108}$$

and finally

$$z_p = \rho\alpha \left(\cos^2\varphi + \imath \cos\varphi \sin\varphi\right), \tag{3.109}$$

which is similar to equation (3.102). We observe that $|z_p|$ in the far field (where $\varphi \to 0$) is $\rho\alpha$, while it tends to zero in the near field where the amplitude of particle velocity

$$v = \frac{p}{\rho\alpha}(1 - \imath \tan\varphi) \tag{3.110}$$

increases. However, the energy flux (intensity) $pv\cos\varphi$ has the same form $P_0^2/(\rho\alpha r^2)$ in the near and far field, so that the total power is the same.

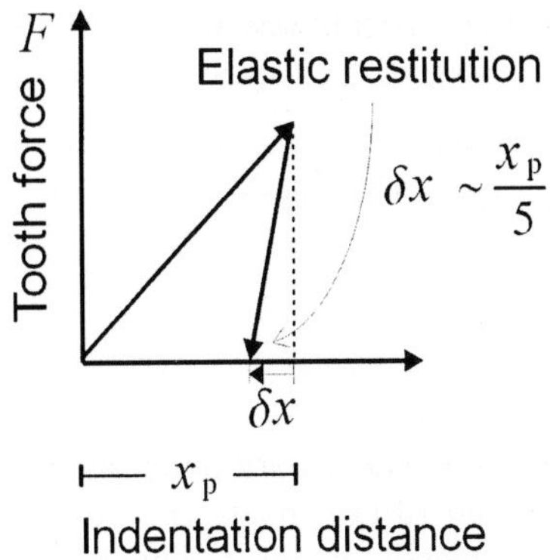

Figure 3.9: Restitution of vertical displacement after tooth penetration process (modified after Sheppard and Lesage, 1988).

3.9.6 Relation between rotary-drilling power and radiated power

Sheppard and Lesage (1988) analyzed the forces at the teeth of a drilling roller-cone bit (see Section 3.12.1). They observed that higher torque values, caused by vertical indention and ROP, correspond to weaker rocks. They observed also that a part of the work expended in indenting the vertical distance x_{p} (Section 3.6.4) is restituted in the form of elastic energy and results in a partial retraction of the rock once the force of the retracting indentor has dropped to zero. They observed a rock retraction δ_x, due to elastic restitution corresponding typically to $\approx 1/5$ of x_{p}, i.e., (Figure 3.9) that can be written as

$$\delta_x\left(x_{\mathrm{p}}\right) \approx \frac{x_{\mathrm{p}}}{5}. \tag{3.111}$$

Neglecting local permanent deformation effects, we have assumed (Poletto, 2000a) that the indention from 0 to x_{p} corresponds to the accumulation and restitution of elastic energy during the indention process. We have also assumed that the force and elastic deformation are approximately in phase. The vertical force exerted by the bit tooth is proportional to x_{p}. According to the results of Section 3.9.3, the near-field phase of particle velocity can be expressed by a $\cos\tilde{\varphi}$ factor, where the symbol $\tilde{\varphi}$ means that the phase is averaged over the range of frequencies of the signal. Hence, we can assume that the amplitude of the vibration in phase with force is a fraction $\cos\tilde{\varphi}$ of the elastic restitution δ_x at the completion of the deformation and relaxation process, which is of magnitude $\delta_x = x_{\mathrm{p}}/5$. We determine that the elastic-radiated power produced by the vertical action of the teeth can be estimated by substituting $(x_{\mathrm{p}}^2/5)\cos\tilde{\varphi}$ in place of x_{p}^2 in the rotary drilling equation (3.13). With these assumptions, we determine that the elastic power produced by the vertical force approximately equals $0.2\cos\tilde{\varphi}$ times the vertical indention power, in which a part of the rotary drilling energy is transformed. This part is calculated from rotary-drilling equation (3.14) expressed in terms of drilling parameters, in which the ratio of the indention power to the total power is $\mathrm{b}_8 R_D^{1/n}/T_D$. Finally, the power of elastic vibration produced by the vertical force in the formation can be expressed in terms of the rotary equation as a fraction of the rotary power by

$$W_{\mathrm{R}} \approx 0.2\ \frac{\mathrm{b}_8 R_D^{1/n}}{T_D}\ \frac{2\pi \cdot \mathrm{TOB} \cdot \mathrm{RPM}}{60}\ |\cos\tilde{\varphi}|\,, \tag{3.112}$$

or

$$\frac{W_{\mathrm{R}}}{W_{\mathrm{drill}}} \approx 0.2\ \frac{\mathrm{b}_8 {\mathrm{R}_{\mathrm{D}}}^{1/n}}{T_D}\ |\cos\tilde{\varphi}|\,, \tag{3.113}$$

where we have approximated the torque-rotary power as the total drilling power of equation (3.3). Equations (3.112) and (3.113) relate the average drilling conditions to local effects and the expenditure of elastic energy radiated from the bit assumed as a vertical source. The radiation model does not include transverse components, which are accounted for in the drilling equations. Moreover, the approach has the limitation that the experimental restitution value of $x_{\mathrm{p}}/5$ is obtained while drilling rock specimens in the laboratory. The restitution occuring during full-scale drilling may be very different and difficult to measure. Finally, the approach has the disadvantage that the evaluation of the drilling bit constant (b_8) may involve laborious analysis of drilling data sets and of cross plots of drilling parameters with different bit types.

Nevertheless, we assume a milled roller-cone bit with $\mathrm{b}_8 = 0.08$, $n = 2$ (as recommended by Sheppard and Lesage, 1988), ROP $= 10$ m/h, TOB $= 6$ kN m, WOB $= 98$ kN, RPM $= 120$ rev/min, and $D = 12\ 1/4$ in. We obtain from equations (3.113) and (3.84), for a 30 Hz source in a Poisson formation of compressional velocity $\alpha = 4200$ m/s ($|\tan\tilde{\varphi}| \approx 150$),

$$\frac{W_{\mathrm{R}}}{W_{\mathrm{drill}}} \approx\ 0.2 \cdot 0.0272 \cdot |\cos\tilde{\varphi}|\ \approx\ 0.36 \cdot 10^{-4}. \tag{3.114}$$

This value is in good agreement – in order of magnitude – with the result (E_{R}/E_0) calculated by Simon (1963). This comparison is done to verify the compatibility of different approaches rather than to determine the radiated energy using the results of the different energy-balance methods reported in the drilling literature, which would require much more detailed information and better definitions of boundary conditions.

3.10 Balance of the borehole and radiated power

To complete the balance of the vibration power for SWD purposes, we have to consider also the power of the waves transmitted in the drill string and in the borehole. We have

$$W_{\mathrm{vibr}} = W_{\mathrm{R}} + W_{\mathrm{l}} + W_{\mathrm{N}} + W_{\mathrm{G}}, \tag{3.115}$$

where W_{R} is the power radiated in the formation in the far field, W_{l} is the power of the waves generated by the reaction force in the drill string, W_{N} is the "blind" power temporarily stored in the formation in the near field and continuously exchanged with the drill string and W_{G} is the power radiated in the form of borehole and mud guided waves (Lea and Kyllingstad, 1996; Poletto, Malusa and Miranda, 2001).

Special considerations

In the ensuing discussion of energy balance, we emphasize the following aspects.

The near-field power term W_{N} has zero-mean value in time and does not contribute – on average – to the energy flux of the waves radiated in the formation, and transmitted in the drill string. However, the near-field effects are important in the formation close to the bit and for pilot-vibration analysis because they determine the partition of the propagating and standing drill-string waves (see Section 4.11.2 on bit/rock reflection coefficients).

The borehole interface and mud-guided waves, W_{G}, are important in borehole seismics, in particular with recording tools in the borehole (Lee and Balch, 1982; Toksöz and Stewart, 1984; Tang and Cheng, 2004). Meredith, Cheng, Toksöz and Bouchon (1990) have shown that the near-field effects of tube waves are detectable at a distance of about twenty-five times the borehole radius (about five meters). However, drill-bit wavelengths are typically higher and we assume that the presence of the mud-filled borehole does not substantially change the radiation pattern of the bit vertical force (Rector and Hardage, 1992) (see also Section 3.8.2). The tube waves are not included in the source-radiation model, and conical wave radiated energy (Rector and Hardage, 1992; Heelan, 1953) is rather considered as a product obtained at the expenses of the drill-string propagating waves (Section 4.6.1). As discussed in Section 4.3, this effect involves attenuation analysis. Therefore, for our purpose, to determine the partition of radiation and drill-string pilot waves, herein we refer to a simple model of an empty borehole in which the drill string is not in contact with the wall of the hole.

The bit vibrations induced by mud-pressure modulation are estimated in Section 3.18. In addition, we use the mud-wave pilot signals in Chapter 7. However, here we do not calculate tube-wave effects W_{G} in the energy balance of the bit source and assume an empty-hole model. This is because we are interested in the partition of the waves radiated in the formation and of the pilot waves in the drill string to determine the correlated energy. In addition, SWD measurements are performed during mechanical drilling which, however, necessarily involves mud circulation and hydraulic drilling. Drilling hydraulics involves static and dynamic aspects (see, for instance, Maurer and Heilhecker, 1969; Maurer, 1980), but is not included in the mechanical rotary-energy model we have used in equation (3.3). For this reason, a study of the interactions between mud-pressure waves, bit forces and drilling performances requires a separate analysis. This analysis should be focused on several important technological aspects discussed in the drilling literature, such as: properties of the mud circulation system, mud-flow regimes, Doppler effects, drill-string and bit geometry (in the sections exposed to mud pressure), bit-jet design as well as downhole motor performances.

Partition of radiated and drill-string powers

For simplicity in the formulation of the energy balance of the waves produced by a vertical force exerted on the drilled formation and on the drill string, we assume

$$W_{\rm vibr} \approx W_{\rm R} + W_1. \tag{3.116}$$

Herein, we assume that W_1 is the power of the axial drill-string waves and neglect the flexural waves (Carcione and Poletto, 2000). The energy balance relates the impedances of the drilled formation and of the drill string. A similar analysis is done by Lutz, Raynaud, Gastalder, Quichaud, Raynald and Muckerloy (1972).

Assume, for simplicity, that the drill string is a tube with constant properties: cross-section A_1, density ρ_1, and Young modulus Y_1. The vertical reaction force at the bit, say $F = -F_0 \sin \omega t$, produces axial-extensional vibrations propagating in the drill string with the velocity $c_1 = (Y_1/\rho_1)^{1/2}$. Therein, c_1 is the rod axial velocity in the low-frequency approximation, which holds when the wavelength is much larger than the pipe diameter (Kolsky, 1953; Carcione and Poletto, 2000). Let z be the axial string coordinate and $\sigma_0 = -F_0/A_1$ be the axial-stress magnitude in the drill string close to the bit ($z \cong 0$). In the absence of attenuation, the axial-stress wave induced in the drill string is

$$\sigma_1 = \sigma_0 \sin(\omega t + k_1 z), \tag{3.117}$$

where $k_1 = \omega/c_1$. The average energy stored for elastic deformation in a unit volume of the medium is given by (Jaeger, 1969)

$$E_1 = \frac{\sigma_0^2}{2Y_1}. \tag{3.118}$$

To obtain the average power of the stress wave in the drill string, we multiply E_1 by the volume, $c_1 A_1$, of the pipe interval traveled by the wave in unit time, which gives

$$W_1 = c_1 A_1 \frac{\sigma_0^2}{2Y_1}, \tag{3.119}$$

and, finally

$$W_1 = \frac{1}{2}\frac{F_0^2}{\mathcal{Z}_1}, \tag{3.120}$$

where we have used $Y_1 = \rho c_1^2$ and where $\mathcal{Z}_1 = A_1 \rho_1 c_1$ is the characteristic (axial) impedance of the drill string assumed as a rod (Lutz, Raynaud, Gastalder, Quichaud, Raynald and Muckerloy, 1972; Aarrestad, Tønnesen and Kyllingstad, 1986; see also Section 4.2.1). Note that, unlike for the body waves radiated in the formation by the single-force source, the power transmitted by one-dimensional stress waves in homogeneous drill pipes does not depend on frequency (in the low-frequency approximation). However, due to the fact that the composition of the actual drill string is made up of sections of different acoustic impedances, the frequency response, i.e., the transfer function, of actual drill strings is not flat. Variation of impedances, signal reflectivity, energy transmission and attenuation in actual drill strings are analyzed for instance by Paslay and Bogy (1963), Barnes and Kirkwood (1972), Drumheller (1989), Poletto, Malusa and Miranda (2001).

Moreover, drill-string stationary-vibration modes are coupled to the near-field vibration modes and can be included in the model using a complex impedance (Section 3.9.4) and reflection coefficient at the bit (Section 4.11.3).

Using equations (3.51) and (3.120), the ratio of power radiated in the formation to drill-string power can be expressed as

$$\frac{W_{\rm R}}{W_1} = \frac{\mathcal{Z}_1}{\mathcal{Z}_{\rm R}}. \tag{3.121}$$

For example, assume a 30 Hz wave radiated in a Poisson medium of P velocity $\alpha =$ 3160 m/s and density $\rho = 2400$ kg/m^3, when drilling with a drill collar of outer and inner diameters of 8.5 in. and 2.75 in., a bar velocity of 5130 m/s and a density of 7840 kg/m^3. The ratio of at-the-bit and radiated impedances is

$$\frac{\mathcal{Z}_1}{\mathcal{Z}_{\rm R}^{(\rm P)}} = \frac{0.033 \cdot 5130 \cdot 7840 \cdot 0.302\,(2\pi \cdot 30)^2}{2400 \cdot 3160^3} = 1.88 \cdot 10^{-4}. \tag{3.122}$$

In other words, we obtain that $W_{\rm vibr} \sim W_1$. This means that the expenditure of the energy in the drill string can be more relevant for drilling-energy balancing than the energy radiated in the formation. The energy in the drill string was not considered by Simon (1963) in his initial analysis of the drilling-energy balance, which focuses only on the bit action. Hence, if a no-rate-of-penetration model is assumed (i.e., $W_{\rm break} = 0$, $W_{\rm heat} = 0$) (see, also, Chin, 1994) a large part of the rotary energy is transformed into drill-string vibrations, and a very small part into radiated waves.

3.10.1 Measuring the power of axial drill-string waves

Assuming a monochromatic drill-string wave, we can calculate the impedance and use the force to measure the power of axial waves produced in the drill string and radiated in the formation. However, acceleration measures are often available and can be used in place of force. Peak-to-peak values of axial acceleration may range between a few g's (acceleration of gravity) under common drilling conditions (Cunningham, 1968; Deily, Dareing, Paff, Ortloff and Lynn, 1968; Eronini, Somerton and Auslander, 1982) and may rise to tens of g's for anomalous drill-string pendulum vibrations of high amplitude (Dykstra, Chen, Warren and Azar, 1996). Let the axial acceleration be expressed as

$$a_1 = a_0 e^{\imath(\omega t + k_1 z)}, \tag{3.123}$$

$$a_1 = \frac{\partial^2 u_1}{\partial t^2} = -\omega^2 u_1, \tag{3.124}$$

where u_1 is the axial displacement in the drill string. Values of peak-to-peak axial displacement ranging from about 0.5 to 2 in. were observed by Cunningham (1968). The axial displacement is related to the axial stress σ_1 by

$$\sigma_1 = Y_1 \frac{\partial u_1}{\partial z} = \imath Y_1 k_1 u_1. \tag{3.125}$$

Using $k_1 = \omega/c_1$, $\sigma_1 = F/A_1$ and the characteristic impedance of drill-string axial waves $\mathcal{Z}_1 = A_1\rho_1 c_1$, we obtain

$$F = -\imath \frac{\mathcal{Z}_1 a_1}{\omega}, \tag{3.126}$$

(Table 3.4). Taking the time-average value of $\Re[F]^2$, hence of $\Re[a_1]^2$, over T from equation (3.120), we have

$$W_1 = \frac{\mathcal{Z}_1 a_0^2}{2\omega^2}. \tag{3.127}$$

Note that from equations (3.121) and (3.127) we have

$$W_{\mathrm{R}} = \frac{\mathcal{Z}_1^2 a_0^2}{2\mathcal{Z}_{\mathrm{R}}\omega^2} = 0.151 \frac{\mathcal{Z}_1^2 a_0^2}{\rho\alpha^3}. \tag{3.128}$$

In other words, we obtain that the radiated power expressed by equation (3.128) in function of the drill-string axial acceleration is independent on frequency. Using equations (3.127) or (3.128), the drill-string and the radiate powers can be calculated directly from field MWD acceleration measurements in the drill pipes (see Table 3.4). These measurements can be related to the acceleration at the bit using the drill-string transfer function (Section 4.10). The drill-collar sections are typically used for estimating the drill-string impedance $\mathcal{Z}_1$ at the bit (Poletto, Malusa, and Miranda, 2000). Note that from equation (3.121) it follows that by using thinner pipe sections, the amount of energy W_1 generated in the drill string increases because the impedance $\mathcal{Z}_1 \propto A_1$ decreases.

Table 3.4 – Axial harmonic force and acceleration magnitude in drill-strings

String component	Section A_1 [cm^2]	Freq. $\omega/2\pi$ [Hz]	Force F [kN]	Acc. a [g]
8.0 in. drill collar	286.0	6	100	0.33
8.0 in. drill collar	286.0	30	50	0.84
8.0 in. drill collar	286.0	80	10	0.44
5 in. heavy-weight dp	81.1	6	100	1.18
5 in. heavy-weight dp	81.1	30	50	2.95
5 in. heavy-weight dp	81.1	80	10	1.57
5 in. drill pipe	34.1	6	100	2.80
5 in. drill pipe	34.1	30	50	7.01
5 in. drill pipe	34.1	80	10	3.74

3.11 Drill bit versus conventional seismic sources

The drilling bit, under the weight loaded by the drill string, transforms the rotary energy into the action of breaking the rock and a small percentage into radiated vibrational energy (see Section 3.7.3). Different types of drill bits – of different shapes and mechanical action – can be used to drill a well (Chapter 2). The most common are the roller-cone bits (also denoted as tricone or rock bits) and the bits with fixed cutter inserts, namely the diamond and the polycrystalline diamond compact (PDC) bits. Special types of bits are the core bits, designed to pick up plug-shaped sample cores of rock. The criteria of bit-type selection from hundreds of different drill bits for a given drilling plan is related to factors such as the expected drilling-rate performance, the cost per unit depth and the bit life (Section 2.6.7). The action of the different bit types characterizes the vibrations produced while drilling and the signal used for SWD purposes. Their working action, including conditions of bouncing, whirl and stick-slip, are herein described.

The radiation properties of the downhole drill-bit seismic source are related to the amplitude and frequency of the forces exerted by the working bit. The main vibration modes of roller-cone and polycrystalline diamond compact bits are investigated under different drilling conditions for this purpose. The analysis includes vibrations with periodic and random effects produced by teeth indention, multi-lobe patterns, bouncing, single-cutter forces, stick-slip and whirling, mud-pressure modulation forces and bit wear. Then, the drill-bit radiation properties are calculated by using the results shown in Section 3.8.2, and compared to the radiation of conventional VSP sources.

In the previous sections, we analyzed the energy requirements of the drilling process (Poletto, 2002a); the radiation properties of the drill-bit source; near-field effects, which are used to scale the energy expended by the bit in lateral gouging and vertical rock penetration, calculated by using the rotary drilling equations.

In particular, we analyzed the radiation of waves in the far field, the near-field effects and the downhole vibrations caused by a single vertical (or axial) force. In real drilling conditions, the drill-string downhole vibrations at the bit are caused by many factors, including transverse forces, beats from irregular rotation at the bit, normal and anomalous vibration modes of the drill string, and modulation caused by the fluctuation of the mud pressure with the frequency of the pump strokes.

In Section 3.12, we will analyze the drilling conditions and vibration modes that affect drilling and contribute to increasing the level of downhole string vibrations (Poletto, 2002b). In particular, we analyzed the main periodic and non-periodic random forces of the rotary bits acting as a source of vibrations radiated in the formation and induced in the drill string.

3.12 Roller-cone bit as a periodic vibration source

We use the analysis developed in the previous sections to calculate the radiation of the drill-bit source. In this calculation, we assumed that the radiation impedance to the bit-axial single force is not influenced by the lateral and torsional vibrations. We show below that the roller-bit source produces sufficient vibration energy for seismic-while-drilling purposes, which is comparable to that of conventional seismic sources (Poletto, 2002b). This is in agreement with application results (Rector and Marion, 1991; Miranda, Aleotti, Abramo, Poletto, Craglietto, Persoglia and Rocca, 1996), which show that, in general, roller-cone bits produce compressional seismic components of sufficient energy for checkshot analysis, prediction of the seismic interfaces ahead of the bit, and while-drilling imaging by migrating the multioffset SWD data in depth. When drilling hard formations with a tricone bit, two kinds of drill-string axial vibrations appear (Yang, 1990): one of them is a high-frequency vibration caused by the teeth of the cones; the other is a low-frequency vibration, caused by the lobed-pattern shape of the bottom of the hole. A third kind of axial vibration is the pulsation due to the mud pump pressure fluctuations (Cunningham, 1968).

3.12.1 Vibrations induced by teeth indention

Using laboratory tests, geometrical analysis and statistics, Ma and Azar (1985) studied the dynamics of roller-cone bits in order to determine the number of teeth in contact with the bottom hole and their instantaneous position and velocity. The tests they used indicated that the penetration depth versus the revolution of the bit is not a smooth straight line but a discontinuous line with transition steps, shaped like the profile of a staircase (Figure 3.10). However, it has to be observed that, in this analysis, the role of the boundary conditions in the drilling system, which may influence to some extent the bit vibrations and performances, is unclear. The tests showed that, after revolving a certain number of turns, when all the craters over the bottom are formed, the roller bit suddenly drops down. The drop distance equals the average-crater depth, and can be easily calculated as the product of the average-ROP-per-turn times the number of turns necessary to drop down a step. After dropping, the bit continues revolving at the new depth level until the next drop. In this process, the top of a tooth will not leave the bottom of the excavated crater until the next tooth (not necessarily of the same row) is the new instantaneous rotation center (Figure 3.11). These authors conclude that instantaneous positions and velocities at each moment of the teeth being in contact with the bottom are, in practical applications, unknown.

Sheppard and Lesage (1988) calculated the (non-linear) equations governing the force-penetration relations for a single cone, by assuming zero-torque moment around the journal-bearing axis and by balancing the total WOB loaded on all the indentors in contact with the formation. For a full tri-cone bit, two additional equations are required to balance the total force on the three cones and the torque couple moment of each axis. In real drilling conditions, it is very difficult to calculate the complete set of equations to determine the time evolution of the forces at the teeth of the bit. The bit itself could be considered as a multi-pole seismic source because one tooth is disengaged from rock contact only when another tooth is engaged. However, due to the fact that the multi-pole random effects are distributed around the bit rotation angle, they can be neglected, and a vertical source model can be used for our purposes – as pointed out above. In fact, since the magnitude F_0 of the axial force – and of the reaction force in the drill string – is different from zero, the drill-bit source cannot be represented by a simple dipole model.

In a simplified interpretation model of the tricone bit, Sheppard and Lesage (1988) assumed that only one rolling indentor per row contacts the rock at a given instant and that the total load is equally distributed on the three rows of a cone. These researchers studied in detail the periodic force at the teeth of a drilling roller-cone bit. They used a fully instrumented bit cone to measure, in the laboratory, the forces on the inner and outer rows independently, as well as to determine the torque and the single indentor position

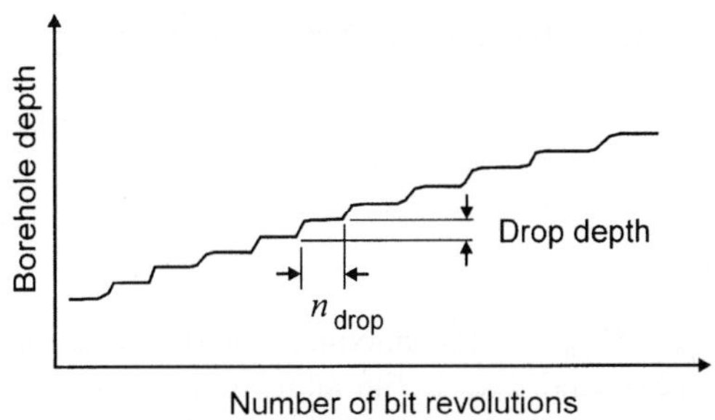

Figure 3.10: Increases in drilling depth by steps versus number of revolutions. The drop distance is drilled in $n_{\rm drop}$ revolutions (modified after Ma and Azar, 1985).

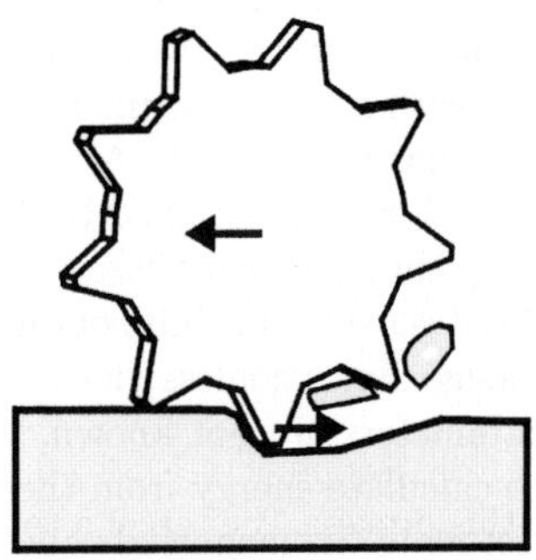

Figure 3.11: Tooth indention and gouging cutters. The tooth indenting the rock becomes the instantaneous center of rotation of the cone.

during cone rotation. They modeled and measured the vertical and horizontal forces versus vertical penetration and horizontal displacements. The vertical penetration of a single indentor tooth can be related to the vertical force by equation (3.12) (a quadratic relation can be calculated for the transverse horizontal action). Sheppard and Lesage (1988) analyzed the rolling and scraping action of the single cone and of the single indentors when drilling rock specimens. They plotted the results versus the rotation angle of the cone about the journal bearing axis (giving cycles per cone revolution) (see Chapter 2). They showed that the mean rotation rate of the cone is from 1.25 to 1.31 times the rotation rate of the bit around its vertical axis. In addition, they observed that, during its extraction phase, the indentor experiences an elastic "restitution" with slope $\arctan(\alpha_x x_{\rm p})$ where $\alpha_x \sim 5$, as determined in single indentor experiments (Figure 3.9).

The results of Sheppard and Lesage (1988) were used by Hardage (1992) to describe the periodic action of the roller-cone source. Figure 3.12a shows the vertical force of the 19-teeth outer row in a complete turn (360 degrees) versus the cone rotation angle. Figure 3.12b shows the periodic force of the 9-teeth inner row. For our purposes, these axial forces can be approximated by a periodic force of the type

$$F(t) \approx F_{\rm m} + F_0 \sin \omega_{\rm row} t, \qquad (F_{\rm m} \geq F_0), \tag{3.129}$$

with the "mean" F_m and the "dynamic" F_0 forces given by

$$F_m = \frac{F_{max} + F_{min}}{2}, \tag{3.130}$$

$$F_0 = \frac{F_{max} - F_{min}}{2}, \tag{3.131}$$

respectively, where F_{max}, F_{min} are the maximum and minimum load forces on a row, respectively (Figure 3.12a), and ω_{row} is the angular frequency of the force of the given row of teeth. Figures 3.12c and 3.12d show the forces during individual indention events. A faster variation in the force during unloading (Biggs, Cheatham and Rice, 1969) can be observed after the maximum peak of the signal of the inner row, with a slope about five times that observed during loading. As discussed before, the faster instantaneous variation of force dF/dt causes an increase in the average radiated energy. The synthetic example of Figure 3.13b is obtained with a modified sine function, in which the unloading period is five times shorter than the loading period. As can be calculated by equations (3.58) and (3.50), the average radiated energy obtained by using this force is three times the average energy of the mono-frequency harmonic force.

These researchers showed also full-scale field results obtained by using 8 1/2 in. milled-tooth bits, which are plotted versus dimensionless downhole parameters (see Section 3.6.3). As the original drilling parameters are not known, the rotary drilling equations (3.3) and (3.13) cannot be used to calculate energy from these literature data. Nevertheless, these results can be used to estimate the magnitude and the frequency of the vertical (axial) force, and to calculate the energy radiated in the far field by the source (equation (3.50)), assumed as a vertical force, with the hypothesis that the peak-to-peak dynamic force component $2F_0$ remains comparable to the maximum force magnitude F_{max}.

3.12.2 Vibrations induced by lobed patterns

Oil field measurements have shown that the cone bits work frequently in the condition of bouncing off the bottom, no matter whether rotary drilling or turbo drilling (Yang, 1990; Cunningham, 1968; Wolf, Zacksenhouse and Arian, 1985). Drill-string fluctuations have been observed at three times the rotary speed, i.e., at frequency

$$f_{LP} = \frac{\text{RPM}}{20} \quad [\text{Hz}] \tag{3.132}$$

while drilling with tricone bits. These vibrations are attributed to a three-lobed bit pattern – or, more generally, in a n-lobed pattern, where n is an integer multiple of the number of cones (Skaugen, 1987) – at the face of the rock being drilled (Figure 3.14).

The bouncing excitation condition was analyzed by Yang (1990). A lobed pattern is a particular shape of the bottom (Figure 3.14a) with excavated spots ("*cams*" on bottom) corresponding to the shape of the cone bit. Skaugen (1987) showed an example in which a six-lobed pattern produced by a tricone bit is interpreted. If a lobed pattern is formed, the bit bounces off the bottom when the axial velocity of the cone rotating in the lobed bottom pattern is higher than the maximum falling velocity of the bit according to Hooke's law, which governs the vibrations of the drill string.

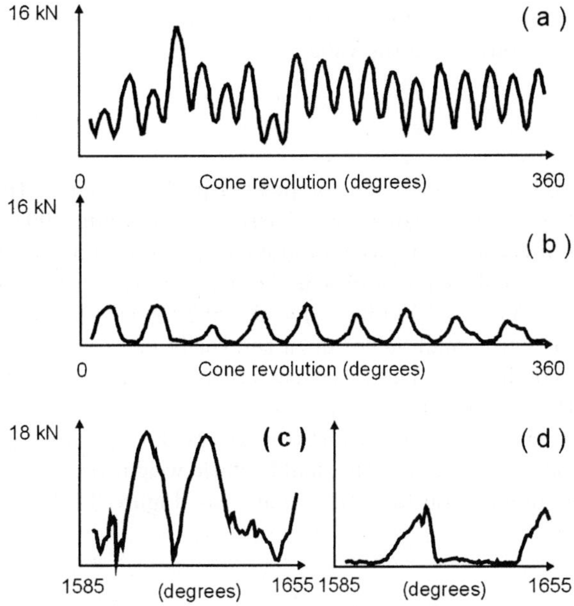

Figure 3.12: a) Outer 19-teeth-row force versus cone revolution angle. b) Inner 9-teeth-row force versus cone revolution angle. c) Particular of outer-row event. d) Particular of inner-row event. Modified after Sheppard and Lesage (1988).

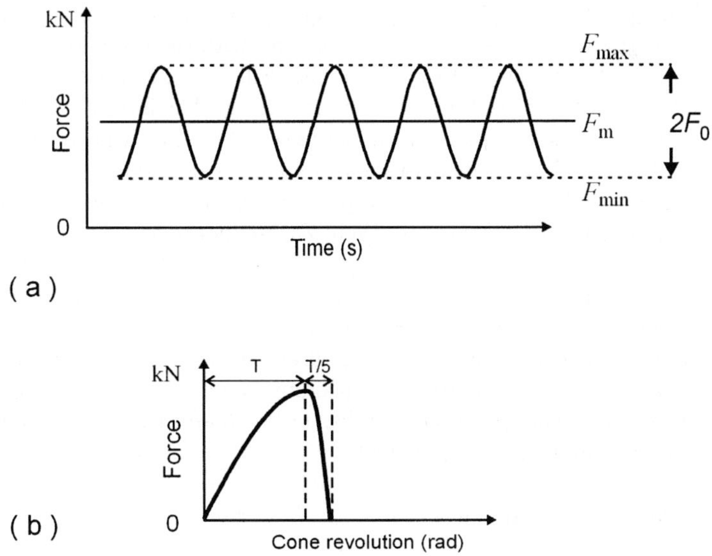

Figure 3.13: a) Maximum F_{max} and minimum F_{min}, average F_m and dynamic $2F_0$ forces. b) Model of a single event of the indention force, with the unloading period five times shorter than the loading period.

Yang (1990) showed that the bit bounces off the bottom when the first critical frequency of the forced vibration has the value

$$f_{c1} = \frac{l_c g}{\pi c_1 s_0}, \tag{3.133}$$

where l_c is the length of the drill-string compressed portion, that is, that portion below the drill-string neutral point (Adams and Charrier, 1985; Chin, 1994; Poletto, Malusa and Miranda, 2001), g is the acceleration of gravity, c_1 is the propagation velocity of the axial elastic wave in the drill string, and s_0 is the peak-to-peak displacement value of the regular lobed bottom. As $c_1 = 5130$ m/s, $g = 9.8$ m/s^2 and $l_c \sim 100$ m, from equation (3.133) it follows that even small values of s_0 can easily cause bouncing off the bottom in practical drilling conditions. The lobed pattern is a frequent condition and induces large variations in the weight on bit. Yang pointed out that as long as the tricone bit keeps steady contact with the lobed bottom, the $2F_0$ peak-to-peak value of the downhole dynamic weight on bit cannot exceed the double static weight on bit (2 WOB). Hence, at the critical frequency of equation (3.133), we have (see Figure 3.14b)

$$F_0 = \text{WOB}. \tag{3.134}$$

Larger relative amplitude variations of weight and torque have been measured in real experiments with the bit bouncing off the bottom. Deily, Dareing, Paff, Ortloff and Lynn (1968) showed examples in which the bit weight varied from 0 to 625 kN, when the nominal weight applied at the surface was of about 180 kN, and peak loads more than three and a half times higher than the load average values were observed (Figure 3.14c).

Wolf, Zacksenhouse and Arian (1985) gave a qualitative explanation of the positive feedback process linking the drill-string dynamics with the three-lobed pattern excitation. In this process, the dynamic response of the drill string forms the three-lobed pattern, which, in turn, provides excitation to the drill string. The force applied by the drill string to the formation, when in phase with the initial three-lobed pattern, deepens the three low spots. Thus, excitation increases and resonances develop.

The vibrations due to a three-lobed pattern depend on the hardness of the formation, and are larger for harder formations. Wolf, Zacksenhouse and Arian (1985) observed that this effect is excited, in general, at frequencies different than the natural resonant drill-string frequency (Dareing, 1982). After the maximum depth of the lobes that correspond to the geometry of the bit is reached, additional energy destroys the lobed pattern and the process decays abruptly. Skaugen (1987) observed that the minimum modulation period $T_{\rm L}$, in which a multi-lobed pattern of maximum vertical amplitude $a_{\rm L}$ can be generated and broken down again (Figure 3.14d), has to be greater than the time necessary to drill the peak-to-peak amplitude $2a_{\rm L}$, namely,

$$T_{\rm L} = 3600 \, \frac{2a_{\rm L}}{\text{ROP}} \quad [\text{s}]. \tag{3.135}$$

For instance, if ROP $= 8$ m/h, and $a_{\rm L} = 2$ mm, we have $T_{\rm L} = 1.8$ s. We can consider negligible any additional long-period precession effect due to the fact that more work may be done on one side, front or back, of the pattern.

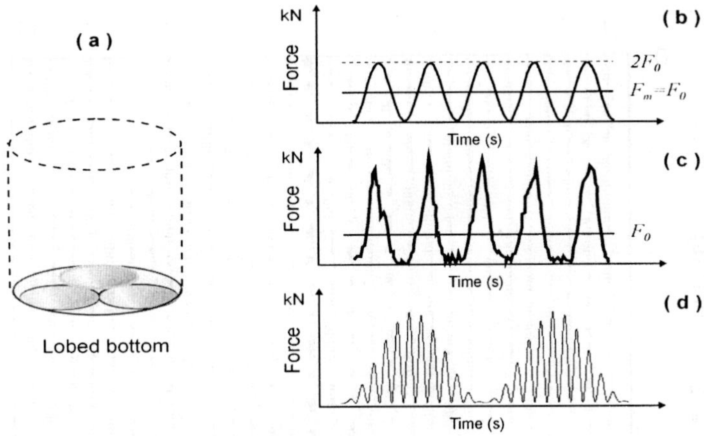

Figure 3.14: a) Lobed bottom layout, with three spot cams. b) Vibration model at the critical frequency (limit of bouncing off bottom), where the average vertical force F_m equals F_0 = WOB (where WOB is the "static"-without-rotation weight on bit) and the maximum vertical force F_{max} equals $2F_0$ = 2WOB. c) Example of bouncing off the bottom, downhole real data. Modified after Deily, Dareing, Paff, Ortloff and Lynn (1968). d) Model of lobed pattern modulation effect for periodic creation and destruction of the lobed bottom.

Analysis of SWD data have demonstrated that the three-lobed excitation makes an important contribution to SWD investigations. Figure 3.15 shows an example of good-quality axial-pilot bit signals measured at the top of the drill string while drilling a carbonate formation using a tricone bit. Figure 3.16 shows the corresponding seismic vertical-component data measured at ground level for reverse VSP purposes. The low-frequency peak at about 5 Hz in Figures 3.17 and 3.18 is interpreted as being due to the three-lobed pattern excitation (with RPM $\approx$ 100 rev/min), according to equation (3.132). Figure 3.19 shows the correlated SWD results versus offset. In general, drillers can consider large downhole vibrations either useful, for increasing the rate of penetration, or undesirable for their effects on wellbore stability (Santos, Plácido and Wolter, 1999) and drill-string fatigue problems. Recently, Hutchinson, Dubinsky and Henneuse (1995) introduced a MWD downhole-assistant-driller prototype system based on downhole measurements of forces and motion to detect in real time the undesired downhole vibrations, such as the large amplitude three-lobed patterns. This type of system can assist the driller to make adjustments to the drilling control in order to remove the undesired downhole effects and improve drilling performances. In this case, a partial reduction of the SWD drill-bit signal can be expected.

3.12.3 Pore pressure and roller-cone bit forces

An important factor that relates the roller-cone bit forces to the single-tooth displacement in the rock fracturing process is the downhole pressure. Under different pressure condi-

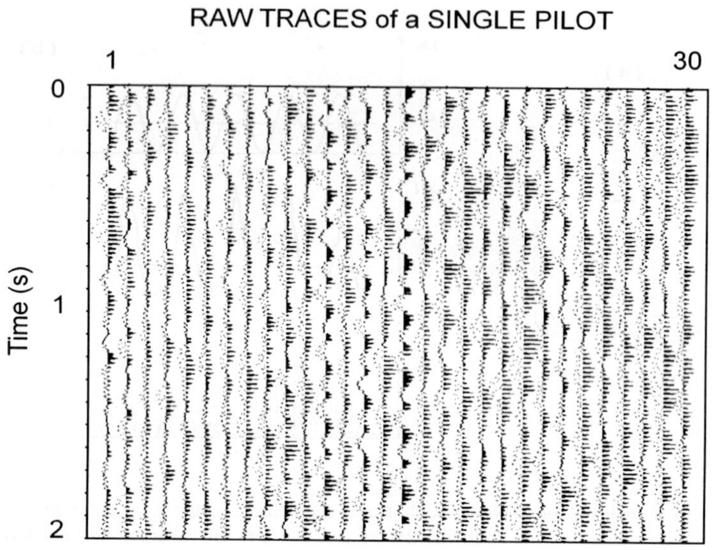

Figure 3.15: Drill-bit data measured while drilling with a roller cone at 2140 m depth. Traces of the axial pilot signal measured at the top of the drill string. The low-frequency component is interpreted as due to the trilobed pattern.

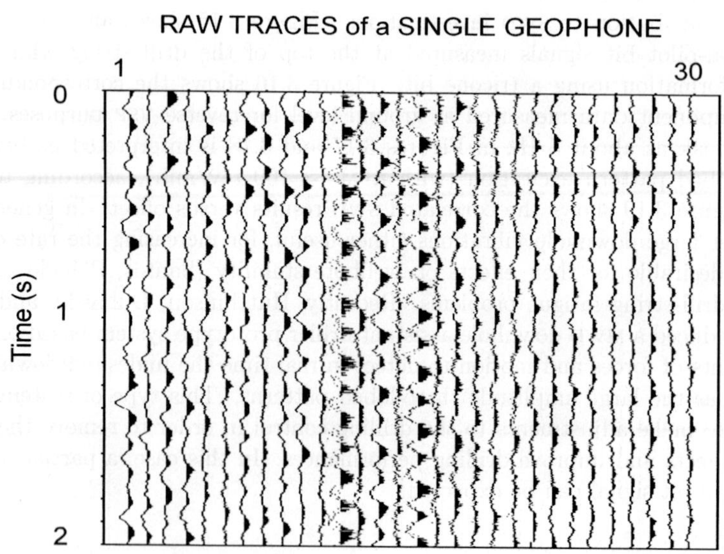

Figure 3.16: Drill-bit data measured while drilling with a roller cone at 2140 m depth. Geophone signals measured at offset of 450 m from the wellhead. The low-frequency lobe-pattern effect is clearly detectable also in the geophone data.

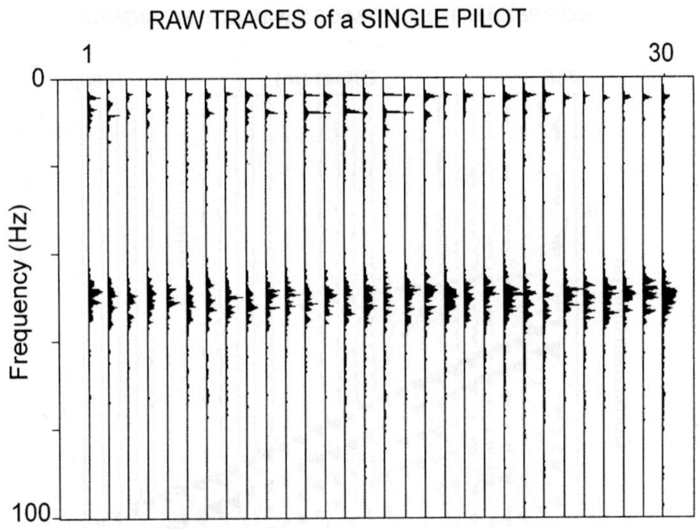

Figure 3.17: a) Amplitude spectrum of the pilot traces of Figure 3.15.

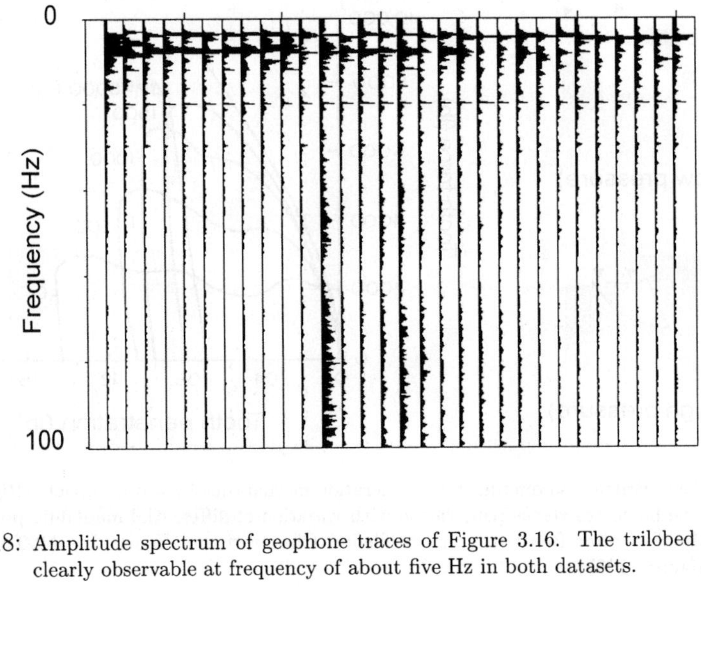

Figure 3.18: Amplitude spectrum of geophone traces of Figure 3.16. The trilobed pattern peak is clearly observable at frequency of about five Hz in both datasets.

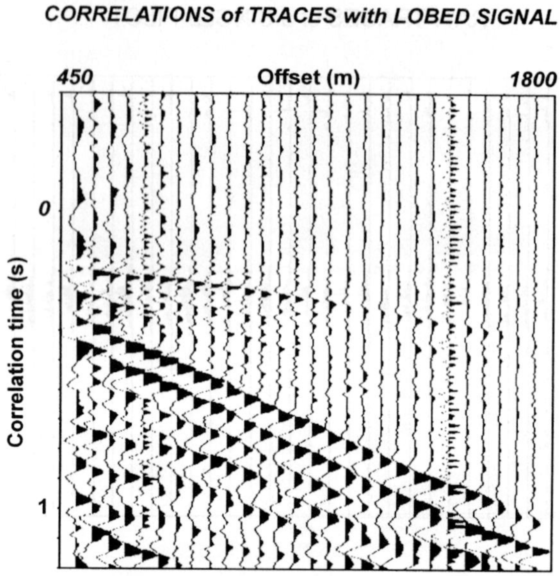

Figure 3.19: Correlations of the SWD data versus offset. The pilot signal is deconvolved. Data are not bandpass filtered.

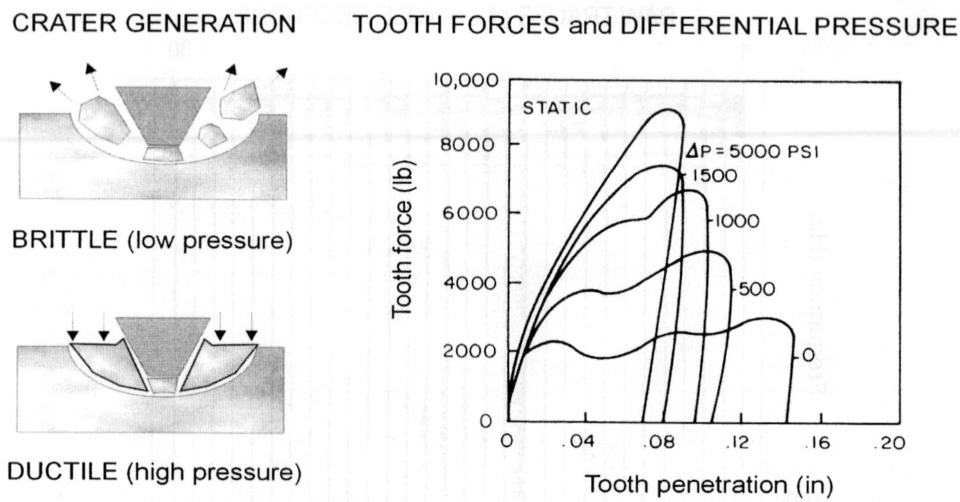

Figure 3.20: (Left) Brittle and ductile crater-generation mechanisms by tooth impact. (Right) Roller-cone bit forces versus penetration with variation of differential mud/fluid-pore pressure. Modified after Bourgoyne Jr., Millheim, Chenevert and Young Jr. (1991), and after Maurer (1965).

tions, the slope of force-displacement curve, schematically shown in Figure 3.9, changes (Maurer, 1965). This effect is due to the fact that the pressure influences the way in which the rock after failure is ejected from the crater created by a tooth. Maurer (1964) studied the conditions of *brittle* and *ductile* fracturing, i.e., deformation without change in applied stress; see, also, p. 104. He observed that the crater mechanism depends on the differential pressure between the borehole, mud, pressure p_{h} and rock-pore pressure p_{pore} (Maurer, 1965), written as

$$\Delta p = p_{\mathrm{h}} - p_{\mathrm{pore}}. \tag{3.136}$$

The mechanism of crater creation is shown in Figure 3.20a. At low differential pressure, the rock fragments beneath the bit are ejected from the crater ("brittle" fracture). At higher differential pressure, the friction between fractured surfaces increases, the rock beneath the bit is not completely ejected and deformation occurs, with formation of "pseudoplastic" craters. Figure 3.20b shows the variation in the tooth-force versus tooth-penetration curves with the differential pressure (after Maurer, 1965). As a consequence, the differential borehole– and formation–pressure may influence the amount of bit vibrations in drilling.

3.12.4 Effects of teeth wear on roller-cone vibrations

Wear of a roller-cone tooth, either milled steel or insert, is classified by levels of dull teeth wear from T0 to T8, which correspond to the status of a new tooth with no loss of the tooth height and to the total loss of the tooth height (Figure 3.21a), respectively (McGehee, Dahlem, Gieck, Kost, Lafuze, Reinsvold, and Steinke, 1992b; Gabolde and Nguyen, 1999). Teeth wear has two effects on SWD. The first is a variation in the indention energy. The second is a variation in the row radius, cone revolution speed and vibration frequency.

When a dull flat develops on the tooth wedge, a greater penetration force is required. In the presence of wear, equation (3.12) relating force per unit chisel length and tooth penetration depth becomes

$$F = \sigma_{\mathrm{p}} \left(d_{\mathrm{flat}} + x_{\mathrm{p}} \tan \Theta \right), \tag{3.137}$$

where d_{flat} is the flat distance of the dull-bit wedge (Burgess and Lesso, 1985). As a consequence, the indention coefficient b_8 increases a little in the rotary energy equation (3.14), while the gouging coefficient b_6 decreases significantly with increasing tooth flatness due to wear. Falconer, Burgess and Sheppard (1988) calculated that the indention coefficient of the worn-bit $\mathrm{b}_8^{\mathrm{worn}}$ is always less than two times the coefficient $\mathrm{b}_8^{\mathrm{new}}$ of the new tooth, i.e., $\mathrm{b}_8^{\mathrm{worn}} < 2\mathrm{b}_8^{\mathrm{new}}$. This means that, if we assume unit WOB, RPM and bit diameter, the indention energy (dimensionless torque of equation (3.10)) expended to drill at the same ROP (dimensionless rate of penetration of equation (3.11)) by a worn bit can be, at most, twice the indention energy expended by the new bit. Including also the gouging term, wear typically results in a decrease in torque and an increase in specific energy $\mathrm{E_S}$ of equation (3.7) required to drill a unit volume of rock, rather than in an increase in the average flux of energy radiated from the downhole force. Sheppard and Lesage (1988) measured the load distribution at the inner and outer rows of a new (T0 wear state) and worn-bit cone (T3, i.e., 3/8 of the T-taper wear state) in laboratory experiments.

They observed that load distribution is insensitive to lithology type, but sensitive to wear geometry. With the new bit, a total load of 8 kN was distributed as 6 kN and 2 kN on the outer and inner rows, respectively. With the bit in the T3-wear state, the same total load was distributed as 5 kN and 3 kN on the outer and inner rows. So, in the wear state, the ratio of outer over inner loads changed from 3 to 1.66, thus modifying the frequency content of the total loading force and radiation properties.

The geometry of a worn bit also changes the frequency of the row vibrations. In fact, if we assume to cover the same hole bottom at a fixed RPM around the bit axis, the cone revolution speed around the journal bearing axis increases as the wear process progressively reduces the cone radius. Jardine, Lesage and McCann (1990) studied the power spectra of the near-bit force and acceleration to determine the row-wear state from the bit signature. This method is based on the log-spectrum (cepstral) analysis of frequency peaks in data sampled as a function of the angle (cycles per cone revolution). These researchers observed that in the T3-wear state of a 29-teeth cone, the reduction in tooth-row radii increased the cone-speed factor from 1.25 to 1.30. The cone-speed factor (CSF) is defined as the ratio of the cone and bit-revolution speed. It assumes values of 1.6–1.7 for bits with high wear (T8), with an approximately linear increment of the cone-speed factor versus wear. The cone-revolution frequency can be approximated as

$$\omega_{\rm cone} = \frac{2\pi{\rm RPM}}{60}\left[{\rm CSF}_0 + 0.125({\rm CSF}_8 - {\rm CSF}_0)\jmath_{\rm wear}\right], \tag{3.138}$$

where we can have, for instance, $\mathrm{CSF}_0 \sim 1.2$ and $\mathrm{CSF}_8 \sim 1.65$, and $\jmath_{\rm wear}$ is the wear integer index (from 0 to 8). However, the calculation of the spectrum of a bit row may be more complex for a worn bit than for a new bit, because the wear state may introduce teeth irregularities. In this case, the harmonics of the cone speed, arising from the general periodic nature of the signal, become dominant peaks (Jardine, Lesage and McCann, 1990).

3.13 Roller-cone bit as a wideband seismic source

As a basic requirement, a seismic source needs to be wideband – generally, two or three octaves are required (Hardage, 1987). This is a necessary condition for obtaining short wavelets in time, either by using impulsive sources or non-impulsive sources and coherency processing, i.e., as with Vibroseis data. Thus, a basic condition for SWD is that the rotating drill-bit produces wideband signals. Two causes that contribute to making the roller-cone bit a wideband source are discussed below.

3.13.1 Unevenness of the formation, random breakage process

The drill bits, either roller cone or PDC, act by imparting a series of pulses to the rock. The laboratory tests performed by Sheppard and Lesage (1988) showed that the frequency of the pulses produced by the periodic loading and unloading of the indentor teeth can be related to the roller-cone revolution speed and geometry. Hardage (1992) used the results of these researchers to analyze the vibrations produced by the roller-cone bit as a percussive source and proposed that each tooth pulse can be considered – at least ideally – as a wideband signal, which, ultimately, contributes to making the spectrum of the

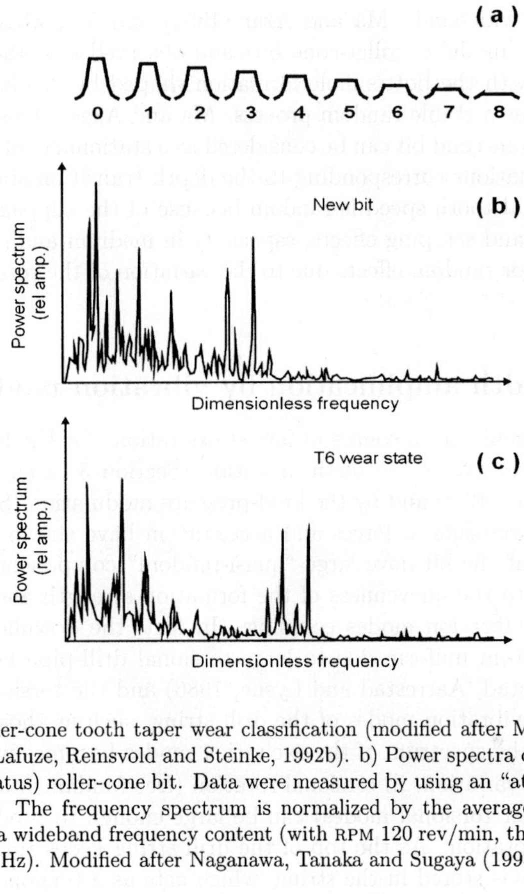

Figure 3.21: a) Roller-cone tooth taper wear classification (modified after McGehee, Dahlem, Gieck, Kost, Lafuze, Reinsvold and Steinke, 1992b). b) Power spectra of a new (T0) and a worn (T6 status) roller-cone bit. Data were measured by using an "atmospheric test machine" in lab. The frequency spectrum is normalized by the average bit rotation speed and shows a wideband frequency content (with RPM 120 rev/min, this band could correspond to 100 Hz). Modified after Naganawa, Tanaka and Sugaya (1995).

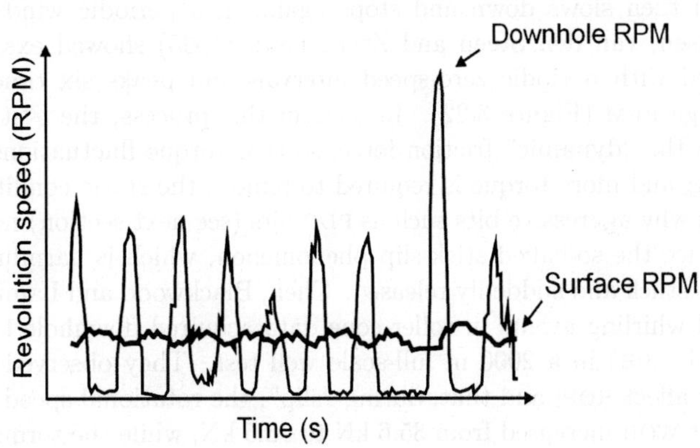

Figure 3.22: The downhole rotation speed has large fluctuations and stops (stick slip), while surface RPM remains nearly constant. Modified after Jansen, van den Steen and Zachariasen (1995).

resulting SWD signal wideband. Ma and Azar (1985) and Ma, Zhou and Deng (1995) analyzed the dynamic model of roller-cone bits and observed that the interaction of the bit teeth in contact with the bottomhole formation shaped by drilling varies with time, and can be regarded as a stable random process. Ma and Azar (1985) observed that the angular speed of the cones and bit can be considered as a stationary process – with positive or negative speed variations corresponding to the depth transition steps of Figure 3.10 – and that the tangential tooth speed is random because of the slippage of the contacting teeth due to gouging and scraping effects, especially in medium and soft formations. We also have to account for random effects due to the variation of the rotational speed of the bit (Section 3.13.2).

3.13.2 Bandwidth amplification by vibration-mode coupling

Periodic force components are a source of forced excitations for the bit axial vibrations. Axial vibrations are caused by the teeth indention (Section 3.12.1), by the bottomhole lobed pattern (Section 3.12.2) and by the mud-pressure modulation (Section 3.18). However, downhole measurements of forces and acceleration have shown that the axial and rotational vibrations at the bit have large "quasi-random" components (Skaugen, 1987). These are attributed to the unevenness of the formation strength and random breakage process, and also to vibration modes coupling. In fact, the downhole rate of drill-bit rotation is often far from uniform due to large torsional drill-pipe flexibility (Skaugen, 1987; Hasley, Kyllingstad, Aarrestad and Lysne, 1986) and the torsional modes interact with the forced axial vibration modes of the drill string, such as those due to the lobed pattern. As a result, the spectrum of the excitation modes becomes wideband.

In some cases, the variations in torsional rotation (free fundamental mode, also called "*pendulum*", and higher torsional modes) can be large enough to produce periodic stops in the downhole bit rotation. As the top of the drill string keeps rotating, the string is wound up, and energy is stored in the string, which acts as a torsional spring. Then, at increased torque, this elastic energy is released because the bit starts spinning, its speed increases and then slows down and stops again, in a periodic winding and unwinding process. Jansen, van den Steen and Zachariasen (1995) showed examples of downhole rotation speed with periodic zero-speed intervals and peaks six times higher than the surface average RPM (Figure 3.22). In fact, in this process, the "static" friction force is larger than the "dynamic" friction force, so that torque fluctuations do not damp out during sliding, and more torque is required to remove the static condition of no-rotation. This explains why aggressive bits such as PDC bits (see next section) more than roller bits tend to produce the so-called stick-slip phenomenon, which is harmful for bit failure, in that the bit is stuck and suddenly released. Chen, Blackwood and Lamine (1999) observed stick-slip and whirling events in roller-cone data acquired downhole by an instrumented recording tool (SUB) in a 2000 m full-scale well test. They observed that stick-slip did not adversely affect ROP, and that, during "slip", the rotational speed changed from 0 to 146 RPM, that WOB increased from 35.6 kN to 93.5 kN, while the torque on the bit (TOB) changed from 0.27 kN m to 1.1 kN m.

Moreover, in real drilling situations, the frictional interaction between drill string and wellbore also gives rise to torque transfer to the drill string of a more stochastic nature. Hence, in general, we can assume that unpredictable behavior of the bit action and drill-string interaction is an intrinsic part of the drilling process that produces impulsive random time-signal components as well as wideband signals.

Torsional modulation of lobe-axial vibrations

Let us consider only the free fundamental drill-string torsional mode (also called pendulum mode) of frequency ω_0 and a three-lobed pattern excitation of frequency $\omega_{\rm ax}$. The angle of the axial-displacement vibration coupled to the torsional mode at the drill bit can be expressed as (see equations (23) and (24) in Aarrestad and Kyllingstad, 1988)

$$\varphi_{\rm b} = \omega_{\rm ax} t \left(1 + A_0 \frac{\sin \omega_0 t}{\omega_0 t}\right), \tag{3.139}$$

and the axial displacement as

$$u = u_0 \sin \left[\omega_{\rm ax} t \left(1 + A_0 \frac{\sin \omega_0 t}{\omega_0 t}\right)\right], \tag{3.140}$$

where the lobe-pattern frequency is approximated by

$$\omega_{\rm ax} = 3\omega_{\rm rev} = \frac{\pi {\rm RPM}_{\rm av}}{10}, \tag{3.141}$$

being $\rm RPM_{av}$, the average rotary speed. Note that bit-rotation stopping and reverse-spinning effects will appear when $A_0 \geq 1$. Aarrestad and Kyllingstad (1988) measured $A_0 \sim 0.3$ in full-scale experiments. The pendulum frequency can be calculated as

$$\omega_0 = 2\pi f_0 = \sqrt{\frac{S}{J}}, \tag{3.142}$$

where S is the static torsional stiffness and J is the effective moment of inertia of the drill string. For a drill pipe of length $L_{\rm pipe}$ and a drill collar of length $L_{\rm DC}$, the inertia J can be approximated as (Kyllingstad and Hasley, 1987)

$$J = \rho_1 \left(\frac{I_{\rm pipe} L_{\rm pipe}}{3} + I_{\rm DC} L_{\rm DC}\right), \tag{3.143}$$

where ρ_1 is the steel density, $I_{\rm pipe}$, $I_{\rm DC}$ are the moment of inertia per unit length and mass of drill pipe and collars, respectively. The torsional stiffness S is approximated by

$$S = \frac{\mu I_{\rm pipe}}{L_{\rm pipe}}, \tag{3.144}$$

where μ is the steel shear modulus. As the fundamental frequency f_0 is of the order of magnitude of 0.1 Hz, the frequency modulation index, $A_0 \omega_{\rm ax}/\omega_0$, is typically larger than 1. The frequency-modulation effect introduces harmonic components at the frequencies $\omega_{\rm ax} \pm n\omega_0$ (n integer), as well as large changes in the downhole RPM and torque spectrum – torque is assumed to be proportional to the twist angle between the surface and bit. It was observed that during drilling the string can be twisted several turns with torque on bit between 0.5 and 10 kN m (Jansen, van den Steen and Zachariasen, 1995).

3.13.3 Roller-cone bit as a high-frequency source

High-frequency acoustic waves are generated by the rock cracking produced by the drilling process. Simon (1963) calculated that this energy can be a large part of the drilling energy, ultimately transformed in local vibrations and heat. These high-frequency events could be successfully used for short-range acoustic investigations with downhole tools for geosteering purposes (Section 8.12.5).

3.14 PDC bit as a vibration source

Polycrystalline diamond compact (PDC) bits act prevalently as shearing tools and, in general, produce poorer SWD axial pilot signals and compressional vibrations than those produced by tricone bits, which work both by gouging and percussive action. In other words, while generating vibrations is an intrinsic process of the mechanical percussive action of the roller-cone bits, the axial forces involved in PDC and diamond-bit drilling are not only lower, but are also quite constant, i.e., they drill typically with a flatter dynamic-load function, thus producing poorer radiation for SWD purposes. For instance, Falconer, Burgess and Sheppard (1988) assumed that for PDC bits, no energy term is associated with the penetrating action of the teeth, so that

$$b_{8(\mathrm{PDC})} \cong 0, \tag{3.145}$$

in the rotary-drilling equation (3.14), while, by assuming that the cutting face generates many overlapping circles which in total make about a circle with the bit radius, these authors used

$$b_{6(\mathrm{PDC})} \sim 0.25. \tag{3.146}$$

Vertical and lateral vibrations and forces are produced by the impacts of single PDC cutters and vibrations of the PDC-bit/drill-string system. Since there is no "mechanical-bit action" associated with vertical penetration of teeth in the PDC rotary-drilling process, the contribution of axial vibrations is reduced and more difficult to use for SWD purposes than that of roller-cone bits. As a particular case, bicenter-PDC bits produce more vibrations for SWD purposes, giving results more similar to those obtained with the roller-cone bits (oral communication at the Schlumberger Cambridge Conference on SWD, 1996).

In addition to the disadvantage of generating poor signal excitation, PDC bits are commonly used with downhole motors or turbines, which are other less-favorable conditions for SWD investigations (Chapter 7). As a large and increasing number of wells are drilled with PDC and turbodrilling, effective use of SWD beyond roller-bit drilling is an important aspect, which can be solved by multichannel processing and downhole near-bit measurements of PDC axial vibrations (Poletto, Miranda, Corubolo and Abramo, 1997) (Section 7.8.1).

3.15 Analysis of PDC single-cutter forces

The stress concentration induced in the elastic rocks under a diamond-bit tooth was studied by Guyen Minh Duc, Cholet and Tricot (1972). They analyzed the plane stress

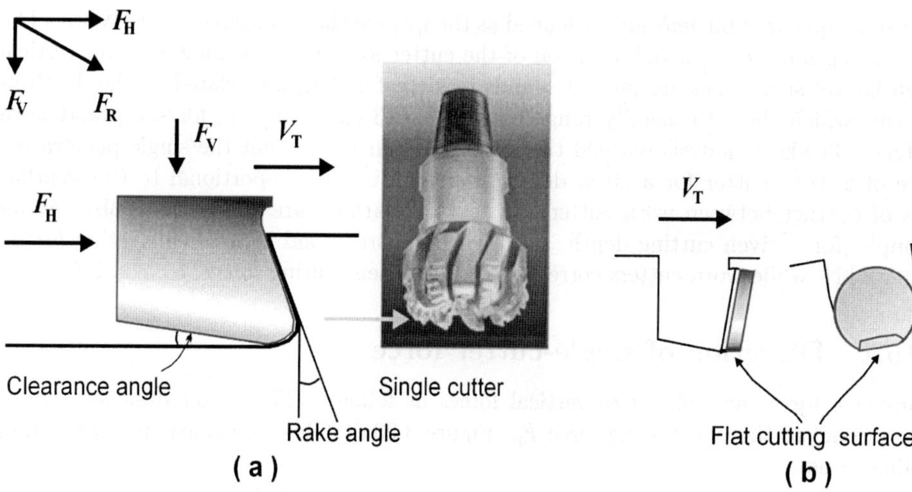

Figure 3.23: a) Vertical F_V and horizontal F_H forces of a single cutter. F_R is the resultant force from composing F_V and F_H (when there are side lateral forces, the F_p and F_d penetration and drag forces are used instead of F_V and F_H). b) Wearflat area. Modified after Guyen Minh Duc, Cholet and Tricot (1972).

conditions produced in bench tests in rocks by a tooth subjected to the vertical force, F_V, produced by the load on bit, and to the horizontal force, F_H, which is the effect of the applied torque (Figure 3.23). Their analysis was extended to fracture initiation in the vicinity of the contact area as a function of loading, by assuming the fracture model on the basis of the Griffith-Walsh criterion (see also, Section 3.7.1). Guyen Minh Duc, Cholet and Tricot (1972) analyzed the effects of the two fundamental mechanisms, cutting and grinding, with the following conclusions.

Cutting action – In the cutting mechanism, high-compression stress concentrations are induced under the tooth for clearance angles (Figure 3.23a) other than zero (Figure 3.23b). These researchers observed that the stress transmitted by the diamond bit can be assimilated with an inclined force in the model, in that, at "a certain distance away", a simple compression zone spreads in approximately the same direction as the resultant force of the single tooth, F_R, where F_R is the vectorial sum of F_V and F_H. They considered examples with applied forces $|F_H| = 2\,|F_V|$ and $|F_H| = 0.5\,|F_V|$. At the same time, high tensile stresses appear around the cutting edge.

Grinding action – In the grinding mechanism, tensile stress appears behind the tooth and high compressive stresses are propagated in front of it. In the grinding mode, for a given vertical force, friction increases tooth efficiency.

Glowka (1986) performed a fairly exhaustive study of laboratory experimental data and developed a method based on a single-cutter analysis for predicting cutter forces, temperatures, wear on PDC, and integrated bit-performance parameters such as weight on bit

(WOB), torque, and bit *imbalance* (defined as the ratio of the total side force to WOB). The penetrating force F_p – in the direction of the cutter load which is different from vertical when lateral side forces are present – and the drag force F_d are related to the depth of the cut, which should generally range between 0.013 cm to 0.2 cm for sharp and worn cutters. Glowka's analysis reached the important conclusion that the single penetrating force of a worn cutter for a given depth of cut is directly proportional to the wearflat area of contact between worn cutter and rock ("wearflat" area) (Figure 3.23b). As an example, for a given cutting depth, sharp cutters correspond to low penetrating forces, e.g., 2.2 kN, while worn cutters correspond to high penetrating forces, e.g., 11 kN.

3.15.1 Direction of single-cutter force

Transverse forces are related to vertical forces as follows. The cutter drag force F_d is proportional to the penetrating force F_p (Figure 3.23a). Glowka defined the cutter drag coefficient as

$$\nu_d = \frac{F_d}{F_p} = \tan \vartheta. \tag{3.147}$$

The drag coefficient ν_d (hence, the direction ϑ of the resulting force) is a function of the rock type, and is relatively independent of the depth of cut and wearflat area (Figure 3.23b). Values of $\nu_d \sim 0.65$ and $\nu_d \sim 0.95$ were estimated in dry tests with granite and sandstone samples, respectively. In water (mud) drilling at elevated nozzle pressures, the penetrating stresses are significantly reduced (Glowka, 1986) due to the effects of hydraulic horsepower with jet-drilling on PDC rate of penetration (ROP) (Holster and Kipp, 1984). With jet pressures of 13.8 MPa, the penetrating stresses required to cut a given depth of granite are reduced by 10% to 15% with respect to drilling without jet pressure and with the simple presence of water at the cutter/rock interface. With assistance from 31 MPa waterjets, cutter penetrating stresses are reduced by 50% to 65% (31 MPa jet causes considerable damage to rock surface). Drag forces are reduced by the presence of water but not greatly affected by jet pressure (Glowka, 1986).

PDC bits with different profiles and cutters, loaded vertically by WOB and laterally by side forces, have different distributions of forces. The total WOB is the sum of the longitudinal (axial) components. The drilling-bit torque is the sum of the moments caused by the circumferential drag forces. Glowka analyzed the WOB and rotary-speed conditions necessary to prevent thermally accelerated wear (which is of one or two orders of magnitude greater than the ordinary abrasive wear) because wearflat friction temperatures must be kept below 350 °C for standard PDC bits.

3.15.2 Influence of wear on PDC performance parameters

Wear greatly affects bit performance. Higher WOB is required for drilling with a worn bit at the same ROP of a sharp bit. For instance, to drill at ROP 9 m/h with a sharp bit, a 16 kN WOB may be required, whereas about ten times this WOB may be required for drilling at the same ROP with a worn bit. Furthermore, excessive, i.e., greater than usual for a given ROP, drilling torque is an indicator of bit-wear conditions. Wear conditions may affect balanced drilling and cause large lateral vibrations (stick slip) (see Table 3.5 and, also, Section 4.9). To avoid high non-uniform wear conditions, which may produce lateral vibrations, the density of the radial distribution of cutters and the effects of the bit profile are very important parameters. Glowka suggested two criteria for determining bit life in relation to wear. The first is the drilled footage, i.e., the drilled distance, at which thermally-accelerated wear effects begin. The second is the drilled footage at which performance parameters, such as WOB or torque or imbalance reach some selected levels. For example, a bit-life criterion for WOB is of 147 kN at ROP 9 m/h. In this analysis, wear in hard- and soft-rock drilling produces the following effects. In hard rock, wear tends to occur at an accelerating rate with respect to the footage drilled, even in the absence of thermal wear. If the formation is fractured or has a high fraction of quartz, the probability of wear in the "hard mode" increases regardless of the apparent strength of the rock. With respect to wear, expected PDC-bit life in soft rock is much greater than in hard rock.

All these conditions are possibly involved in the distribution of the PDC-bit forces and vibrations – in particular the lateral vibrations – used for SWD purposes.

3.15.3 Influence of downhole pressure on cutter forces

Another downhole condition influencing the PDC cutting action is mud pressure. The normal (penetration) and drag forces of PDC single cutters were studied in the laboratory by simulating downhole pore and total bottomhole pressures by Zijsling (1987). He used a single-cutter testing system on shale samples to work under the different pressure conditions that govern the cutting process at different drilling depths. He reported the effects of pressures on dynamic PDC forces varying in the range from about 3 to 30 kN on individual cutters at several simulated drilling depths, between 800 and 3500 m. He showed that the dilatancy of the rocks during shearing is responsible for cutter-force variation with different bottomhole pressures.

3.16 Dynamic variation of PDC-cutter forces

Behr, Warren, Sinor and Brett (1993) developed a 3D-PDC model to evaluate cutter loading for drilling in inhomogeneous rock and, in cases where the bit moves laterally, while whirling. In this model the WOB, torque and bit-imbalance forces are updated each time an individual cutter force is modified. The following cases of high drilling loads were studied using model and laboratory tests.

Tilted bed boundaries

Drilling tests were conducted to observe the performance of different PDC bits drilling through a tilted bed boundary between shale and limestone. Behr, Warren, Sinor and Brett (1993) reported that at a constant 22 kN WOB, the load on the cutters remains nearly constant (the range of variation was not specified) when drilling 100% pure shale or limestone. But when drilling through the bed boundary, the load on individual cutters varies greatly – from about 1.2 kN to about 8.5 kN – within a revolution as a result of alternatively engaged shale and limestone. In other examples, both vibration and downdip displacement of PDC bits having long-tapered profiles (Chapter 2) were observed.

Drilling nodules

When the formation is inhomogeneous, larger vibrations of PDC bits may occur. An example is a soft-medium shale containing much stronger nodules (concretions, which are formed by the deposition of calcite in the pores). The dimension of these hard nodules may be about a foot (0.3 m). Simulated drilling tests were run under constant WOB of 44.5 kN using a four-blade PDC bit. The normal load on a single cutter was about 2.7 kN, while the peak force on the cutter during engagement with the concretion was about 26 kN. In this case, the vertical force on the cutter was about 18 kN, i.e., 40% of the total load. An approximately tenfold increase in the imbalance lateral force (passing from about 3.5 kN to 38 kN) was also measured. This force may produce whirl (Table 3.5).

Whirling bits

As may happen with roller-cone bits, PDC bits may produce lobed-bottom patterns which may be the cause of severe vibrations. In this case, the lobed patterns are related to the lateral motion of the center of rotation of the bit, which precesses out of the bit-axis and whirls laterally (Brett, Warren and Behr, 1990). Behr, Warren, Sinor and Brett, (1993) observed in the laboratory the lobed-bottom precession motion of a four-blade PDC bit. In this motion, the bit formed a nine-lobed-star shaped borehole wall instead of a cylindrical one. High imbalance, i.e., the ratio of the total side force over WOB, and normal single-cutter forces with large variations and a periodicity of about eight cycles per revolution of the drill string were reported. For a bit with equally spaced blades, assuming whirl frequency per revolution f_w, the number of lobes should be

$$N_{\rm lobe} = (f_{\rm w} + 1)\, m, \qquad m = 1, 2, 3, \cdots . \tag{3.148}$$

Bit precession causes intermittent contact between the cutters and the lobed-shaped formation and, as a consequence, the normal, i.e., perpendicular to the cutter motion, tooth force varies.

Examples of normal forces on a single tooth are the following: during steady-state drilling without whirling the normal force is about 0.5 kN; during whirling the normal force varies from zero, when the tooth is off the bottom, to about 5.8 kN when the tooth is fully engaged with a lobe of the hole formation. The WOB resulting from the sum of all the single-teeth vertical forces oscillated around a mean value of 22 kN. Periodic imbalance forces (with 8 cycles per drill-string revolution) ranged from 3.5 kN to 11.1 kN, and the centrifugal force was about 4.7 kN.

Anti-whirl PDC bits – Anti-whirl bits were introduced in the early 1990s. Their design is based on the concept that the cutter is placed on the bit so that there always is a non-zero side force (Hanson and Hansen, 1995). This side force is aimed at stabilizing the bit against the wall of the well, and the result is that the instantaneous center of rotation of the PDC remains at a fixed position. It was shown by laboratory experiments

(Cooley, Pastusek and Sinor, 1992) that anti-whirl PDC bits may whirl at small cut-depths (cutter penetration) and low WOB. Larger cut-depths and higher WOB produce higher anti-whirl side-forces and stabilization in a smooth drilling condition. Hanson and Hansen (1995) defined the critical rate of penetration at which the PDC bit stabilizes as "transition ROP".

3.16.1 Dynamic models of the PDC axial vibrations

The lateral, torsional, and axial vibration modes of a PDC bit were calculated by Langeveld (1992a; 1992b), who developed a three-dimensional, fully dynamic, PDC-bit and drill-string numerical model, including formation removal. The axial mode is associated with the bit motion in the z direction, assumed to be vertical. To calculate the vertical component of the time-varying force resulting from the cutting action of the bit, the drill string is modeled as a damped mass-spring system. The equivalent string mass was calculated as the mass of the BHA plus one third of the drill-pipe mass. The axial equation of motion includes the friction force and the mud-damping force for the axial motion of the drill string.

The following dynamic results were reported by Langeveld (1992a; 1992b) using a 2350 m, 5 in. drill pipe, 150 m of 6 1/2 in. drill collars and three different 8 1/2 in. anti-whirl PDC bits. Axial vibrations occurred in the frequency range from 4 to 7 Hz (in relation to rapid fluctuations of the ROP). This axial response proved to be governed by both the drill-string properties and by the properties of the formation. Langeveld (1992) concluded that PDC bits may induce larger axial drill-string vibrations during transients in the applied load (WOB). However, the effect of the axial vibration on the bit/drill-string system is small, and the sensitivity analysis indicated that larger amplitude vibrations in the reactive force of the drill string occurred at relatively low WOB and high rotary speeds.

Langeveld (1992) reported also that standard PDC bits showed a tendency to produce lateral vibrations consisting of backward whirl, which results in reduced penetration rates due to sidewise and backward motion of the cutters. These motions may accelerate cut wear and may produce bit failure.

Based on Langeveld's model and results, Hanson and Hansen (1995) developed a numerical bit-dynamics method to predict PDC-bit performance under the conditions of a full-scale drilling simulator test. They tested 20 bit designs of different diameters of both conventional and anti-whirl type. Normal (perpendicular to the direction of the cutter motion) and tangential (in the direction of the cutter motion) single-cutter forces were calculated. This dynamic analysis showed the presence of large vertical vibrations during the whirling state, with load variations of the order of 45 kN. As the vertical load increased and the rate of penetration reached the transition-ROP value of 14.6 m/h, the axial vibrations diminished and drilling stabilized. After stabilization, some low-level axial vibrations still persisted. The spectra of the vertical vibrations (WOB) were computed at ROP rates of 7.3 m/h and 14.6 m/h and 120 RPM. They showed a large bandwidth feature, with energy distributed from 0 to 80 Hz. In particular, during the whirling of an eight-blade bit, a peak at 16 Hz clearly occurred and a consistent nine-lobed pattern was observed. On the other hand, the same analysis performed on conventional bits gave some examples for which stabilization was not achieved.

3.17 Summary of large bit-vibration modes

Table 3.5 summarizes some types of bit-excitation modes, of both roller-cone and PDC bits, in which large downhole vibrations can be produced (Zannoni, Cheatham, Chen and Golla, 1993). These large bit vibrations, i.e., bouncing, whirling and stick slip, may be harmful for bit life, induce drill-string fatigue, and cause borehole instability.

These vibrations can be controlled by feedback systems that change RPM and WOB on the basis of a while-drilling analysis of rigsite drilling parameters (Fear and Abbassian, 1994) or by using downhole measurements and mud-pulse telemetry (Hutchinson, Dubinsky and Henneuse, 1995). Seismic while drilling is also sensitive to the large bit vibrations, and may be used as an auxiliary monitoring tool to help drilling diagnostics.

3.18 Bit vibrations induced by mud pressure modulation

Another axial force exciting axial vibrations is the pulsation of the mud-pump pressure fluctuation. For double-acting pump cycles (double-acting reciprocating pumps are also denoted as duplex), the fluctuation frequency equals four times the pump frequency. Typical values of cycles (or strokes) per minute (SPM) are from 60 to 70. Since a number $N_{\rm P}$ of two to four pumps can be used simultaneously, the total frequency is

$$f_{\rm pump} = \frac{4\,{\rm SPM} \cdot N_{\rm P}}{60}, \qquad [{\rm Hz}], \tag{3.149}$$

and is typically of the order of 8–16 Hz, where we assume that the pumps are synchronized with alternating phase. Cunningham (1968) discussed examples of moderate weight fluctuation of about 67 kN, corresponding to a pump-pressure fluctuation of about 4150 kPa.

The importance of the pump pressure fluctuation on the bit weight can be appreciated by considering a static drill pipe, held rigid at both ends, with a pulsating pressure inside. With no axial motion of the drill pipe at any point, the weight fluctuation of the bit due to 6900 kPa pump-pressure fluctuation could be 29 kN for a 5 1/2 in. drill pipe (Cunningham, 1968).

Deily, Dareing, Paff, Ortloff and Lynn (1968) analyzed the forces that are produced by the fluid pressure acting at the bit and directed along the axis of the tool. Assuming the jet-reaction force due to the mud injection through the jet nozzles to be small, they gave the following equation

$$f_{\rm T} = F_{\rm P} + p_{\rm O} \Delta A_{\rm O} + p_{\rm I} \Delta A_{\rm I}, \tag{3.150}$$

where $f_{\rm T}$ is the reaction force in the tool, $p_{\rm I} \Delta A_{\rm I}$ is the force produced by the inside pressure, $p_{\rm O} \Delta A_{\rm O}$ is the force produced by the outside pressure, and $F_{\rm P}$ is the axial force in the pipe directly below the tool (Figure 3.24).

To calculate the magnitude of the pressure effects, Deily, Dareing, Paff, Ortloff and Lynn (1968) assumed $p_{\rm I} = 24\,000$ kPa and $p_{\rm O} = 14\,150$ kPa. As the difference between the

Table 3.5 – Bit-vibration modes

Vibration type	Definition	Bit type	Pattern
Bouncing	When the bit bounces completely off the bottom	Mainly roller cone	Cam lobed bottom
Whirling	Out-of-center lateral vibrations. The instantaneous center of rotation moves around the face of the bit	Mainly PDC but also roller cone	Star-shaped PDC lobed bottom
Stick slip	Torsional oscillation, in which the bit stops rotating (stick) and starts rotating (slip) at speeds much higher than surface RPM	Mainly PDC but also roller cone	

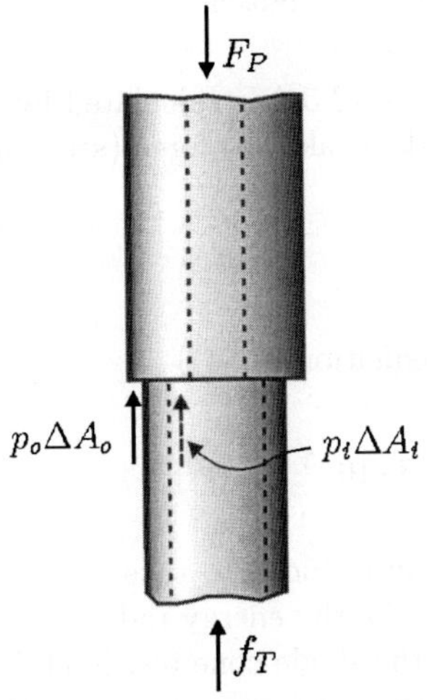

Figure 3.24: Model used to calculate the modulation of the axial force by mud pressure. The mud forces depend on downhole tool and drill-string geometry.

inner and outer sections of the tool was of the order of 35 cm^2 (Figure 3.24), these authors obtained a force induced by the differential pressure of 270 kN that was comparable to load-on-bit values and had to be subtracted from the bit weight recordings. This effect is strictly related to the geometry of the downhole tools, drill-string and BHA design.

3.19 Numerical examples of drill-bit vibrations

The following calculations of the radiated energy are performed by considering periodic bit vibrations. As discussed above, random effects in real drilling conditions may introduce sudden variations in signal amplitude and phase, which could increase the amount of the radiated power. For this reason, the values calculated in the following numerical examples have to be considered as lower limits for the energy radiated by the working bit at the given drilling and geological conditions.

Radiation caused by teeth vibration – A peak-to-peak periodic vertical force of about 16 kN was measured by Sheppard and Lesage (1988) for the 19-teeth outer row of a single cone. Rotation speed was not specified by the authors. If we assume 60 RPM as a reference value and consider only this row, the pulsation is $\omega_{\text{row}} = 2\pi \cdot 19 \cdot 1.25$, where the last number represents the average cone-to-bit revolution ratio. The rock is assumed to be a Poisson medium with Young's modulus $Y = 40$ GPa and density $\rho = 2400$ kg/m^3. The integrated-radiation impedance of equation (3.55) is

$$\mathcal{Z}_R^{(P)} \approx \frac{1}{0.302} \frac{\rho \alpha^3}{\omega_{\text{row}}^2} \approx 3.15 \cdot 10^{10}, \qquad [\text{kg/s}]. \tag{3.151}$$

The radiated power of equation (3.54) is calculated by assuming that the dynamic force F_0 is one half the peak-to-peak load force (see equation (3.131))

$$F_0 = \frac{16}{2} \qquad [\text{kN}], \tag{3.152}$$

which gives the power flux (equation (3.54))

$$W_{\text{R}}^{(1)} \approx \frac{(8\,000)^2}{2} \frac{3}{3.15 \cdot 10^{10}} \approx 3 \cdot 10^{-3} \qquad [\text{W}], \tag{3.153}$$

where the factor 3 has been introduced to take into account the average effect in loading and faster unloading. So, the energy radiated by the outer row in one hour of continuous drilling during the single-cone test is 11 J. If we assume that the load is equally distributed on all the cones, we can transfer the total downhole load force to the single-row load. The analysis of downhole data shows that the magnitude of the total loading during full-scale drilling can be of the order of the surface WOB. This force can be assumed to be much higher than the laboratory value of 16 kN (e.g., 240 kN) and the ROP frequency doubled (ROP = 120). Assuming that total and dynamic loads are comparable, we obtain values of the energy radiated for teeth indention of about nine hundred times higher than that of the experiment discussed above, and obtain radiated energy of the order of 10 kJ per drilling hour. This value increases to about 28 kJ per hour if the rock Young modulus is $Y = 20$ GPa.

Most of the energy is radiated in the form of shear waves. By using equations (3.50) and (3.52) it can be shown that, for a given P-wave velocity, that the radiated energy is very sensitive to the α/β, i.e., to the V_P/V_S, as to Poisson's ratio. Figure 3.26a shows the energy variation for V_P/V_S ranging between 1.2 and 2.4. This ratio also changes the integrated impedance at the bit which, for a Poisson medium, is given by equation (3.100).

Radiation caused by tri-lobed vibration – Let us consider a bottomhole lobe pattern in its critical condition given by equation (3.133), after which the bit bounces off the bottom. In this case, we have a peak-to-peak force equal to 2WOB and

$$F_0 = \text{WOB} \qquad [kN], \tag{3.154}$$

$$\omega_L = 3 \cdot 2\pi \frac{\text{RPM}}{60} \qquad [s^{-1}]. \tag{3.155}$$

Using WOB = 240 kN, RPM = 120, $Y = 20$ GPa, $\rho = 2400$ kg/m^3, we obtain

$$W_R \sim 0.17 \qquad [\text{W}], \tag{3.156}$$

and the low-frequency energy due to the lobed pattern which is radiated in one hour of drilling is about 600 J. Note that in this case, we have not used any factor of energy amplification to account for the "sharpness" of the loading curve as we did for teeth loading and unloading (factor equal 3), even though lobed vibrations appear to be very sharp in many plots of downhole measurements reported in the drilling literature. When drill-string torsional fundamental vibration modes induce frequency modulation, this energy value can be about 1.5 times higher, i.e., about 1 kJ. Furthermore, this value also increases if we consider the harmful condition of the bit bouncing off the bottom – in which case it may be $2F_0 = 3.5$ WOB – with the impact on the cutters when the bit returns in contact with the formation.

Composite vibrations – Assume that the cone-row rotation speeds are not equal and the rock contacts of the indentor teeth are not in phase in the three cones. Assuming $F_{\text{min}} = 0$ in equations from (3.129) to (3.131), we can write the force as

$$F(t) = \sum_i F_i \left[1 + \sin(\omega_i t + \phi_i)\right], \tag{3.157}$$

where F_i and ϕ_i are the loads distributed on the cone rows and the random rotation phases, respectively. We can use equation (3.58) to calculate the radiated energy. If we assume that the different components are uncorrelated, we obtain

$$I_f = \frac{1}{2} \sum_i (F_i \omega_i)^2 . \tag{3.158}$$

In the particular case where the load is equally distributed on the three cones, we have

$$W_{\mathrm{R}}^{(3)} = \frac{1}{2}\, 0.302 \frac{F_0^2}{3} \frac{\hat{\omega}^2}{\rho\alpha^3}, \tag{3.159}$$

where

$$\hat{\omega}^2 = \frac{1}{3} \sum_{i=1}^{3} \omega_i^2. \tag{3.160}$$

As a rule of thumb, we may assume that the factor 1/3 in equation (3.159) is compensated by the factor 3 of loading–unloading effects. In this case, equation (3.159) can be rewritten as

$$W_{\mathrm{R}}^{(3)} \approx 0.302 \frac{F_0^2}{2} \frac{\hat{\omega}^2}{\rho\alpha^3}, \tag{3.161}$$

and $W_{\mathrm{R}}^{(3)} \sim W_{\mathrm{R}}^{(1)}/3$, where $W_{\mathrm{R}}^{(1)}$ is given by equation (3.153) of the single cone.

Vibrations modulated in amplitude – Modulation effects can be introduced either by lobed-pattern forces or mud-pressure fluctuations, as well as drill-string resonances (Aarrestad and Kyllingstad, 1988) (see Section 3.13.2). It can be shown that the energy produced by the modulated force is one and a half times that of the unmodulated row force.

Roller-cone teeth wear – Sheppard and Lesage (1988) showed that the geometry of the worn bit changes the load distribution. The effect was that the total load was more equally balanced between the outer and inner rows after wear, while in the new cone the load was greater on the outer row. Even if the total load was unchanged, these examples did not report the changes in the dynamic loads produced by the shorter dull teeth, which may be of minor magnitude with the same total load. Furthermore, any uneven wear introduces low-frequency sub-harmonics, so that the spectrum of vibrations is re-distributed to a lower frequency band. Despite a possible small increase in the cone revolution speed factor, all the previous effects contribute to lowering the radiated energy. The fact that a higher indention term is calculated in the rotary drilling equation (3.14) is only an apparent contradiction. In fact, we can expect worn bits to drill more slowly and more seismic energy to be obtainable by increasing the listening time during the much longer times used to drill the same depth interval.

Vertical vibrations calculated by rotary-drilling equation – We have used rotary drilling equation (3.14) to relate drilling and radiation energies. This approach requires the knowledge of the phase factor $\cos\varphi$ between the force and velocity at the bit and knowledge of the bit constant b_8, which is reported in literature only for milled roller bits. Assuming, a constant bit parameter and phase, this equation could be used to estimate relative variations in intervals drilled with the same bit.

Estimation of radiated power from axial drill-string vibrations – The best way to calculate the radiation performances of the drill-bit source is to measure vibrations downhole near the bit. The force data, related to acceleration by equation (3.126), are derivated with respect to time and squared to calculate I_f of equation (3.58). As an example, assume a pipe with drill collars that have outer and inner diameters of 8 1/2 in. and 2 3/4 in., respectively, and acceleration of 17 m/s^2 (1.7 g). When drilling a formation with $Y = 20$ GPa, and $\rho = 2400$ kg/m^3, using equation (3.128), we obtain a radiated energy of 3.7 kJ in one hour of drilling.

3.20 Radiation properties of conventional sources

The total recording times used to acquire SWD data of a single depth level may vary from a few minutes to hours (Miranda, Aleotti, Abramo, Poletto, Craglietto, Persoglia and Rocca, 1996). This long-time listening to the power radiated while drilling makes it possible to obtain sufficient seismic energy. From previous calculations, it follows that the energy radiated by the drill-bit source in the rock during realistic SWD acquisition intervals can be compared to the working energy of conventional seismic sources, such as vibrators (mainly onshore) and air- or water-guns (onshore and offshore).

3.20.1 Vibroseis source

The seismic emission of mechanical vibrators depends on many factors, such as the frequency of the exerted force, ground properties, soil coupling, vibration plate deformation and the use of vibrator arrays. Assuming a rigid plate – in which case the displacement of the soil particles at the contact of the plate equals the plate displacement – Cassand and Lavergne (1971, Chapter 8, p. 198) analyzed the forces exerted by a single plate, by multiple plates, the coupling effects and the displacement of ground particles at a great distance from the source. In ground with high radiation impedance (discussed above and defined as the ratio of applied force to particle velocity), i.e., in hard rock, the force F_s exerted by the single plate practically equals the vibration generator force F_e. In rocks of low radiation impedance, i.e., loose ground, these forces are equal only at low frequencies, while the deviation of F_s/F_e – which defines vibration coupling – at high frequencies depends on ground and plate properties and is a matter of proper design acquisition by vibrators.

Single vertical vibrator

Radiation analysis is based on the results of Miller and Pursey (1954), who calculated the power transmitted by a single vibrating plate to a semi-infinite, elastic, homogeneous, isotropic medium with Poisson ratio 0.25. If $2F_s$ and ω are the peak-to-peak vertical force impressed on the ground and the angular frequency of the vibration, respectively, the following power distributions are obtained

$$W_\alpha^{\mathrm{vibro}} = 0.333 \frac{\omega^2 F_s^2}{4\pi\rho\alpha^3}, \tag{3.162}$$

$$W_\beta^{\rm vibro} = 1.246 \frac{\omega^2 F_{\rm s}^2}{4\pi\rho\alpha^3}, \tag{3.163}$$

$$W_{\rm surf}^{\rm vibro} = 3.257 \frac{\omega^2 F_{\rm s}^2}{4\pi\rho\alpha^3}, \tag{3.164}$$

$$W_{\rm total}^{\rm vibro} = 4.836 \frac{\omega^2 F_{\rm s}^2}{4\pi\rho\alpha^3}, \tag{3.165}$$

for compressional, shear, surface waves and total radiated power, respectively. Hence, approximately 7 %, 26 % and 67 % of the power of a single vibrating plate are transmitted in the form of compressional, shear, and surface waves, respectively. The total energy of a linear sweep of amplitude $F_{\rm s}(t)$ and instantaneous frequency $\omega(t)$ is calculated by integrating $W_{\rm total}^{\rm vibro}(t)$ from zero to the sweep-time length. Assuming $F_{\rm s}$ =constant and a linear sweep varying from the initial to the final frequency $\omega_{\rm start}$ and $\omega_{\rm end}$, respectively, as the time t varies from 0 to τ, we have

$$\omega(t) = \omega_{\rm start} + \frac{(\omega_{\rm end} - \omega_{\rm start})\, t}{\tau}. \tag{3.166}$$

Assuming a slow frequency variation during interval τ (i.e., $\tau \gg 2\pi/\omega_{\rm min}$, where $\omega_{\rm min}$ is the minimum of $\omega_{\rm start}$ and $\omega_{\rm end}$), the total compressional energy can be calculated as

$$E_\alpha^{\rm vibro} = 0.333 \frac{F_{\rm s}^2}{4\pi\rho\alpha^3} \int_0^\tau \omega^2(t) dt, \tag{3.167}$$

which gives

$$E_\alpha^{\rm vibro} = 0.333 \frac{F_{\rm s}^2 \hat{\omega}^2}{4\pi\rho\alpha^3}\, \tau, \tag{3.168}$$

where

$$\hat{\omega}^2 \;=\; \frac{(\omega_{\rm end} - \omega_{\rm start})^2}{3} + \omega_{\rm end}\omega_{\rm start}. \tag{3.169}$$

Furthermore, it can be easily shown that the radiated energy in a "vertical stack" of correlations obtained with multiple sweeps under the same conditions is simply the sum of the energies of the individual sweeps.

The compressional radiated power $W_\alpha^{\rm vibro}$ of equation (3.162) is twice the compressional power W_r of equation (3.48) radiated by an equivalent downhole bit force, in agreement with the heuristic consideration that double the energy is required to energize the entire space instead of a semi-space. This relation is not true for the shear component, which is more affected by free-surface effects when produced by a vertical vibrator. Finally, we can observe that the total power $W_{\rm R}^{\rm surf}$ calculated for the same vibration conditions using Simon's equations for surface-impact loading processes (equation (3.24)) equals the total power $W_{\rm total}^{\rm vibro}$ radiated from a single vertical vibrator (equation (3.165)).

Array of vertical vibrators

In field applications, due to the weak amplitude of the emitted signal and the presence of environmental noise, arrangements of several vibrating plates (vibrator arrays) can be used simultaneously to reinforce the compressional signal at the expenses of surface waves. In this case, the mutual admittances (impedances) of the plates vibrating in phase have to be properly accounted for in the calculation of the radiated power (Cassand and Lavergne, 1971, pp. 206–208). In fact, mutual effects are generated because the displacement under each individual vibrating plate is modified by the presence of the other vibrators. The effect is that, at least ideally, the power radiated by n very closely spaced vibrators is approximately n^2 times the power radiated by a single vibrator. This fact could be approximately and simply accounted for by substituting a force of magnitude n-times larger in equations (3.162) to (3.165). In practical applications, this condition is not true. The phase difference in function of the vibrator's distance and baseplate properties is analyzed by Baeten and Ziolkowski (1990). In their early work, Cassand and Lavergne (1971) analyzed the effects of different arrangements of vibrator arrays. For example, they showed that 3 vibrators arranged in a triangle and spaced at an optimum distance may produce compressional-wave power that equals 30 % of the total radiated power, which proves to be one and a half times the total power of a single vibrator.

Table 3.6 shows an example of acquisition values used in a conventional VSP Vibroseis application. Parameters are those of the numerical example in which the radiation from a 19-teeth single row was calculated. The total energy radiated by four vibrators energizing in five repeated sweeps is 6.7 kJ. The compressional wave energy is 0.4 kJ. It can be shown that the calculated radiation is in agreement with the theoretical radiation calculated by Baeten and Ziolkowski (1990, results scaled to models I and II).

3.20.2 Radiation from marine sources

Air-guns, based on pressured bubble expansion, and water-guns, producing waves by cavity collapse, are used for offshore conventional and also onshore VSP surveying (Hardage, 2000). Offshore, the guns are either lowered into the water from the rig to acquire zero-offset VSPs, or towed by the vessel to acquire multioffset VSPs. Onshore, the guns are used in tanks dug near the well and filled with water. The gun generates a bubble and a pressure wavefront, which radiates in the water and then in the formation. Consider that in an ideal unbounded, homogeneous isotropic medium a pressure-dilatation wave radiates with spherical geometry, with its power equally distributed and equal to its average value on the surface wavefront. On the other hand, the waves emitted by vertical forces produced by bits and vibrators have radiation lobes, and in the direction of the acting force have a maximum compressional power density, which is three times their average compressional power density.

The amount of energy effectively radiated depends on factors such as the energy of the source, bubble effects, water pressure conditions, and loss of energy due to the tank/formation reflection.

Air and water guns

The nominal energy of a gun E_{gun} is given by the product of its volume V_{gun} and working pressure p_{gun},

$$E_{\text{gun}} = V_{\text{gun}} p_{\text{gun}}. \tag{3.170}$$

For instance, a gun of volume $V_{\text{gun}} = 400$ cu.in. (0.00655 m^3) and of working pressure $p_{\text{gun}} = 2000$ psi (13 800 kPa) has a working energy $E_{\text{gun}} = 800\,000$ pound-force in. (90.4 kJ).

A simple method for determining the relative potential energy of marine sources is to observe the bubble oscillation interval and measure the potential energy from a chart, the *Rayleigh–Willis* chart, in which the energy of different sources is compared (Figure 3.25; see, also, Dobrin, 1981, pp. 129–131).

3.21 Radiation from drill-bit and conventional sources

We compare the calculated performance of the drill-bit source to those of Vibroseis and air gun. Baeten and Ziolkowski (1990, p. 3) observed that that Vibroseis has some disadvantages compared with dynamite. They reported literature examples indicating that a 2.5 kg dynamite charge can deliver about 10^7 J whereas a single vibrator delivers 1.2×10^6 J. To obtain sufficient energy, Vibroseis are typically used in arrays and with several sweeps in the same position. The radiation increases with soft formations, i.e., with low Young's modulus Y. Table 3.6 calculates the radiation of a conventional Vibroseis-array source used for VSP purposes. In this example the total radiated energy is 6.7 kJ. The numerical examples calculated in Section 3.19 for roller bits show that a comparable radiated energy, of about 12 kJ per hour, is obtained with 30 minutes of SWD listening by using a typical average row-cone frequency in equation (3.161).

Furthermore, the radiation of the roller bit at low frequencies is much higher in the presence of a lobed-bottom-pattern vibration. The numerical calculations give – with no-bouncing and including the pendulum effect – a value of 1.5 kJ in one hour, which is comparable to the total energy radiated by the group of vibrators in five repeated sweeps of 12 s. To appreciate the comparable effects at the different frequencies, Figures 3.26 and 3.27 show the compressional-power radiated by roller-cone bit and Vibroseis sources in bi-logarithmic scale. In this representation, we can appreciate the linear trends of

$$\log W_\alpha^{\text{vibro}} = 2 \log \omega + \log \gamma_{\text{vibro}}, \tag{3.171}$$

$$\log W_r = 2 \log \omega + \log \gamma_{\text{bit}}, \tag{3.172}$$

where $\gamma_{\text{vibro}} = F_s^2/(12\pi\rho\alpha^3)$ and $\gamma_{\text{bit}} = F_0^2/(24\pi\rho\alpha^3)$ for the vibrator and the roller-bit sources, respectively. This simply corresponds to comparing the constant value F_s^2 to $F_0^2/2$, which varies as a function of the frequency. These calculations and the analysis of the wideband frequency content show that seismic while drilling using roller-cone bits is, generally, a reliable, effective tool to support drilling and exploration activities.

The analysis of PDC bit dynamics has shown that vertical/axial and lateral vibrations are produced by the drilling action. In particular, large vibrations are produced during

Rayleigh-Willis Chart

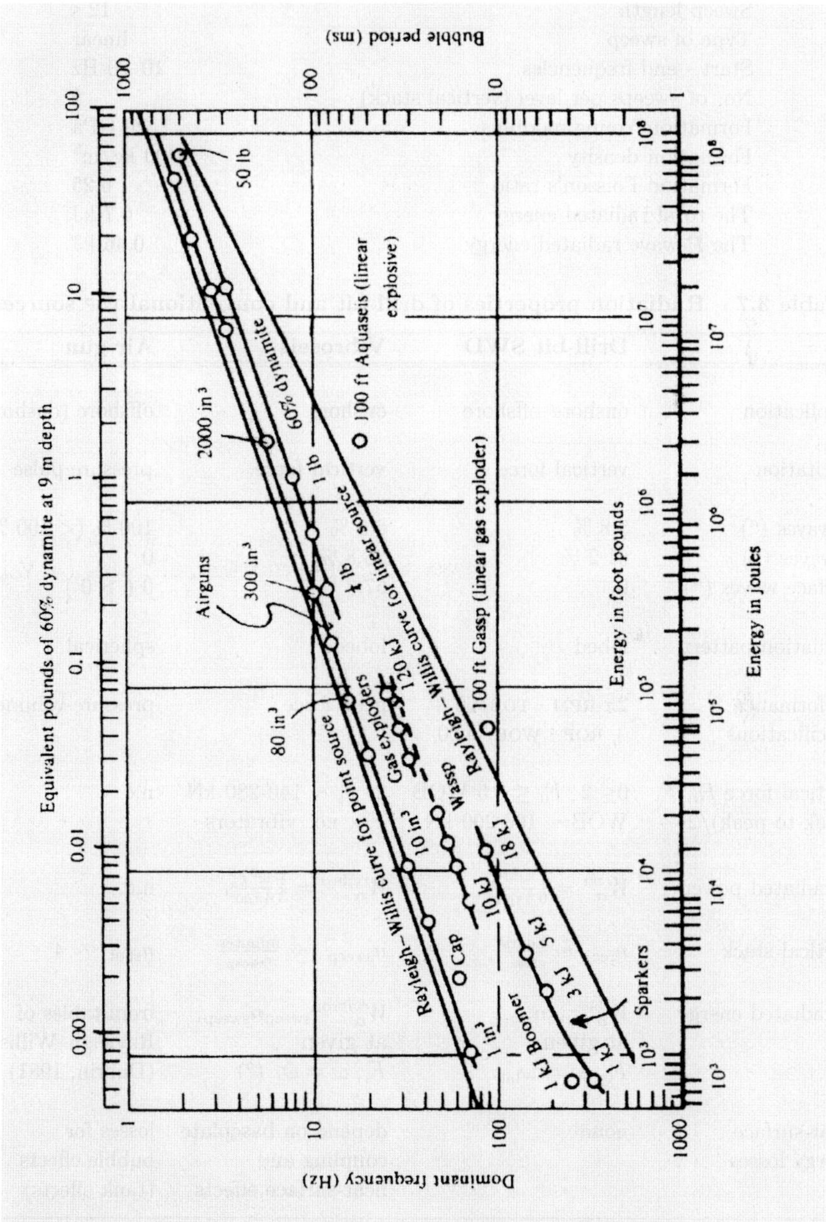

Figure 3.25: Rayleigh–Willis curve relating bubble-pulse period to potential energy for 30 ft source depth.

Table 3.6 – Example of conventional-VSP Vibroseis source

Number of vibrators	4
Peak force of single vibrator	$2F_s = 61\,000$ lbf (271 kN)
Sweep length	12 s
Type of sweep	linear
Start - end frequencies	10–70 Hz
No. of sweeps per level (vertical stack)	5
Formation Young modulus	20 GPa
Formation density	2400 kg/m^3
Formation Poisson's ratio	0.25
The total radiated energy	6.7 kJ
The P-wave radiated energy	0.46 kJ

Table 3.7 – Radiation properties of drill-bit and conventional VSP sources

	Drill-bit SWD	Vibroseis	Air gun
Application	onshore–offshore	onshore	offshore (onshore)
Excitation	vertical force	vertical force	pressure pulse
P-waves ([a])	8.8 %	6.9 %	100 % (< 100 %)
S-waves ([a])	91.2 %	25.8 %	0
Surface waves ([a])	0	67.3 %	0 (> 0)
Radiation pattern	lobed	lobed	spherical
Performance specifications	2π RPM $\cdot$ TOB/60 + + ROP $\cdot$ WOB/3600	peak force	pressure·volume
Vertical force F_0, F_s (peak-to-peak)/2	$0 \leq 2 \cdot F_0 \leq 3.5$ WOB WOB $\sim$ 100-200 kN	$2 \cdot F_s \sim$ 140-280 kN $F_s\times$ no. vibrators	n.c.
P-radiated power	$W_\alpha^{\rm bit} = \frac{1}{6}\frac{\omega^2 F_0^2}{4\pi\rho\alpha^3}$	$W_\alpha^{\rm vibro} = \frac{1}{3}\frac{\omega^2 F_s^2}{4\pi\rho\alpha^3}$	n.c.
Vertical stack	$n_{\rm rec} \sim \frac{\rm hours}{T_{\rm rec}}$	$n_{\rm sweep} \sim \frac{\rm minutes}{T_{\rm sweep}}$	$n_{\rm shot} \sim 4$
P-radiated energy	$W_\alpha^{\rm bit} T_{\rm rec} n_{\rm rec}$, at given F_0, $\omega = \omega_{\rm bit}$.	$W_\alpha^{\rm vibro} T_{\rm sweep} n_{\rm sweep}$, at given F_s, $\omega = \hat{\omega}$. ([b])	from tables of Rayleigh–Willis (Dobrin, 1981)
Near-surface energy losses	none	depend on baseplate coupling and near-surface effects	losses for bubble effects (tank effects)

[a]Values represent the mean flux of radiated energy and are calculated as the percentage of the total mean flux in the formation assumed to be a Poisson's medium.

[b]For a linear sweep, $\hat{\omega}^2 = \omega_{\rm end}\omega_{\rm start} + (1/3)(\omega_{\rm end} - \omega_{\rm start})^2$

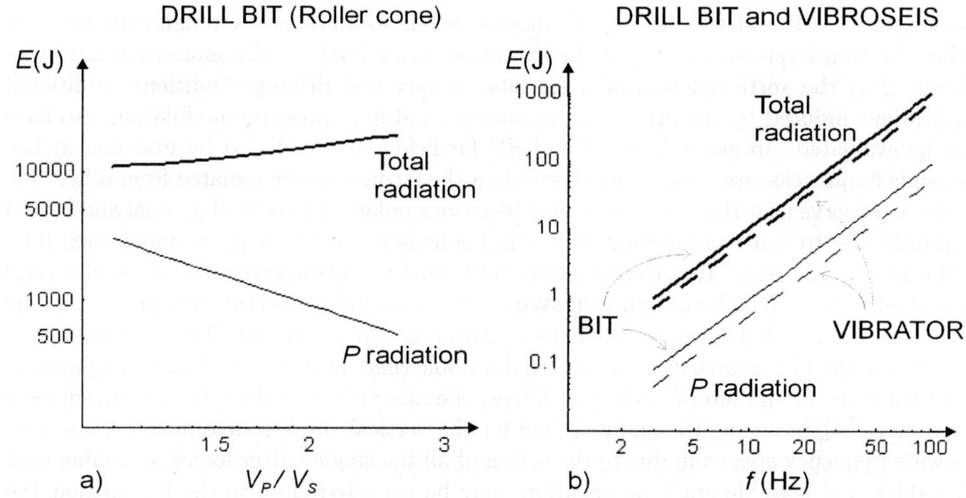

Figure 3.26: a) Amount of energy radiated in one hour by an ideal 19-teeth-cone row, represented versus V_P/V_S ratio. b) Comparison between powers of a conventional surface vibrator (energy in 1 minute) (dashed lines) and roller-cone drill-bit source (energy in 30 minutes, continuous line) at different frequencies, from 3 to 100 Hz (the calculation includes the lobed pattern, 9-teeth-inner and 19-teeth-outer-row vibrations). A total bit vertical load of 240 kN (160 kN on outer and 80 kN on inner rows), RPM 120 rev/min, cone rotation factor 1.25, Poisson formation of Young modulus 20 GPa and density 2400 kg/m^3, two vibrators of 61 klb and are assumed.

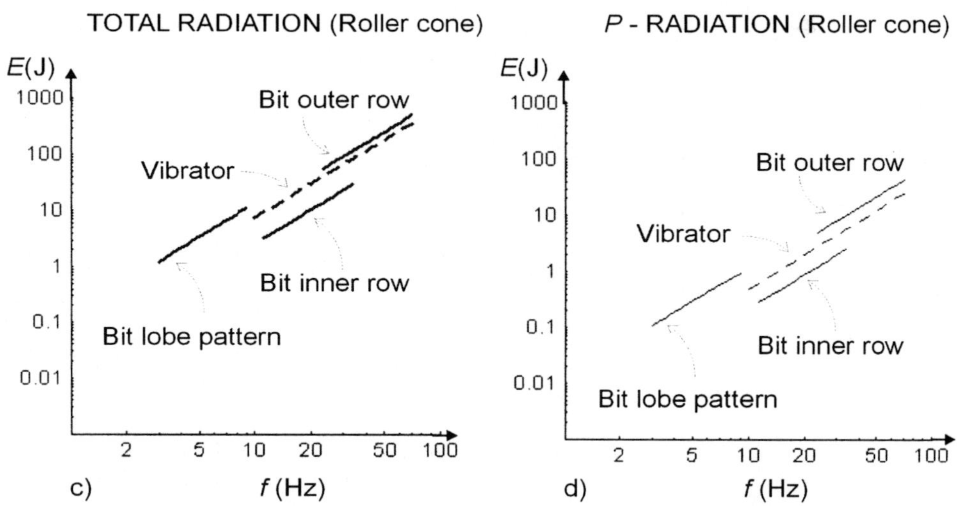

Figure 3.27: a) Total power (energy in 30 min) radiated from 19-teeth-outer, 9-teeth-inner rows and lobed-pattern forces (continuous line) are compared separately (for $60 \leq$ RPM≤ 120 rev/min) to the radiation from two vibrators (energy in 1 min, frequency from 16 to 70 Hz). b) Compressional power calculated as in the Figure (a).

transition states and the drilling of inhomogeneous formations. A rough estimation of the vibration levels indicates that the axial PDC forces are typically more than ten times lower than the vertical forces of roller bits. Under real drilling conditions, additional vibrations induced by the drill-string resonances and mud-pressure modulation also have to be evaluated. In general, the P and SV far-field waves radiated by PDC bits at low seismic frequencies are significantly lower than the far-field waves radiated from roller bits. We can observe that the PDC source may be more similar – for its level in axial and lateral vibrations – to some mini-vibrators, using loadings a few kN (e.g., nominal 2500 lbf). This is a source "size" that can be successfully used for investigating distances of several hundred meters. Results obtained in SWD applications indicates that PDC signals may be detected by downhole acquisition (Poletto, Miranda, Corubolo and Abramo, 1997).

When the PDC source is assumed as a downhole (instantaneous) polarized single force, the resultant of the lateral and axial forces, the energy radiated by the polarized force can be calculated using the analysis used for the vertical single-source model. Moreover, a wide frequency spectrum due to the action of all the single-cutter forces, including rock breaking and possible crack propagation, may be expected close to the bit, so that the PDC-bit signal could prove a good seismic source for downhole short-range high-resolution, acoustic investigations.

Table 3.7 summarizes the radiation properties of the drill-bit and conventional VSP sources.

Chapter 4

General theory: drill-string waves and noise fields

4.1 Introduction: drill-string vibration analysis

The drill-string vibrations are well known to drillers, who avoid the particular drilling parameters that cause resonance and affect the drilling efficiency. However, a large amount of these vibrations is generated during normal drilling conditions. We will see in the next chapter that – among other possible approaches – the vibrations of a working drill-bit are measured in two different ways for seismic-while-drilling surveying. The reference pilot signal is obtained by measuring the energy transmitted through the drill string. It is usually recorded by sensors placed on the drilling rig (as shown, for instance, by Rector and Marion, 1991). In particular applications, the pilot signal may also be measured near the bit. A second type of signal is recorded by seismic lines placed at the surface, laid on the sea bottom or in neighboring wells. This is the seismic signal transmitted through the formations. These two signals are crosscorrelated to get the while-drilling reverse VSP. The correlation attenuates the random noise due to the drilling-yard activity and occasional drill-string vibrations and, at the same time, reinforces the in-phase signal. However, correlation reinforces also the coherent components of undesired environmental noise (Section 4.9), drill-string vibrations and reverberations. The autocorrelation of the rig-pilot signal, and the crosscorrelations of the seismic field data have different kinds of drill-string periodicities (Section 6.13). These correspond to various propagating modes.

The complex dynamic behavior of the drill string has been investigated since the early 60's. To understand the behavior of the drill-string and rig-structure system, studies were made to obtain a theoretical model to help the drillers and supply information about the bit and drill-string wear state. Lutz, Raynaud, Gastalder, Quichaud, Raynald and Muckerloy (1972) developed a theoretical interpretation of the axial vibrations, which are transmitted to the rock and the drill system by a rotating tricone bit, to obtain information on the rock properties.

More recently, the vibrations of the drill string were investigated to evaluate the potential of elastic waves for the transmission of information between the drill bit and the drill-rig operator during drilling operations (Drumheller, 1989; 1992; 1993). Drumheller's analysis considered frequencies up to 2000 Hz. He found that the attenuation of the propagating energy depends on the frequency and the coupling between the drill string and mud, and he showed that the periodically spaced discontinuities of tool-joints cause stopbands and passbands for acoustic transmission through the drill pipes (after Barnes and Kirkwood, 1972) and change the disposition of the group-velocity (Drumheller and Knudsen, 1995). When SWD acquisition is performed by using only surface measurements, the use of the technique is restricted to low-frequency bands (say, less than 120 Hz), so that tool-joint stopband effects are not observed. But group velocity, which depends on tool-joint properties, is measured and used for data processing. A detailed discussion of the drill-string acoustic properties is given by Carcione and Poletto (2000). They solved the differential equations by a pseudo-spectral method modeling the geometrical features of pipes and coupling joints.

In its typical range of frequencies below 120 Hz, surface SWD mainly observes long-period reverberations and bottom-hole-assembly (BHA) multiples. Rector introduced a reference deconvolution method (Rector, 1989) in which the signal detected at the top of the drill string was used to develop an operator to reduce the amplitude of the drill-string multiples in the field seismic traces. A critical point in this approach is that it depends on the correct detection of the drilling dynamic conditions, i.e., the surface and downhole boundary conditions, influencing the bit/rock coupling and the downhole and surface signals. Using measurements made at the rig, Booer and Meehan (1993) developed a model-based approach to analyze the drill-string vibrations; the resulting image allowed identification of the reflection coefficient at the bit and at the intermediate points in the drill string. They used an inverse-model approach to get reflection coefficients from measurements of hookload and torque. Poletto, Malusa and Miranda (2001) interpreted the SWD wavefields by direct modeling of the drill-string waves propagating in the drill string and in the formations. They analyzed the periodic coherent events generated by the downhole signal and surface rig noise. The proposed modeling characterizes the waves propagating through the drill pipes, and the dynamic behavior of the system composed by the derrick, drill string, and rocks. Assuming, without loss of generality, that the drill-bit signals and noise sources are white, they parameterize a one-dimensional transmission-line model, using the mechanical features of the drill string to get reflection coefficients (Section 4.10.1).

4.2 Drill-string waves

Knowledge of mechanical features of the drill string is necessary to correctly understand the periodic reverberations. The drill string can be described using the field tables of Figure 2.29. It can be subdivided into drill pipes and BHA (Section 2.3.10). The most important BHA components are the heavy weight drill pipes and the drill collars. These BHA components cause important variations of impedance and strong reverberations in the pilot and seismic data.

A working drill bit and the noise sources induce axial, torsional and flexural vibrations in the drill string, which can propagate with a wide frequency band, from 1 Hz to 2 kHz (Drumheller, 1989). The flexural waves are dispersive and, owing to the presence of the mud in the borehole, cannot be easily used as a communication tool between down hole

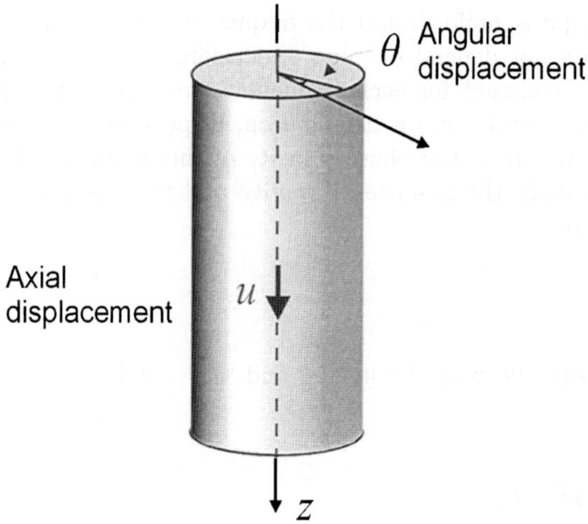

Figure 4.1: Extensional and torsional waves in a cylindrical rod.

and upper hole. On the other hand, at low frequencies, the wavelengths of the extensional and torsional components are long compared to the drill-string diameter, so they can be approximated as not dispersive.

4.2.1 Axial drill-string waves

The propagation of axial (also denoted as extensional or longitudinal) waves in a rod or pipe is governed by the following equation (Aarrestad, Tønnesen, and Kyllingstad, 1986; Drumheller, 1992; Carcione and Poletto, 2000), namely,

$$A\rho\frac{\partial^2 u}{\partial t^2} - AY\frac{\partial^2 u}{\partial z^2} + K_{\rm ax}\frac{\partial u}{\partial t} = F, \tag{4.1}$$

where z, u, A, Y, ρ, $K_{\rm ax}$ and F are the axial coordinate, the axial displacement, the rod cross-section (Figure 4.1), the Young modulus, mass density, a viscous axial damping coefficient and an external axial force, respectively. If we let be, for simplicity, $K_{\rm ax} = 0$ and $F = 0$, we have

$$\frac{\partial^2 u}{\partial z^2} = \frac{\rho}{Y}\frac{\partial^2 u}{\partial t^2}. \tag{4.2}$$

The axial wave propagation velocity in a rod is (Kolsky, 1953)

$$v_{\rm ax} = \sqrt{\frac{Y}{\rho}}. \tag{4.3}$$

The characteristic impedance for axial drill-string waves is defined as the product of mass density per unit length and propagation velocity, namely,

$$\mathcal{Z}_{\rm ax} = A\rho v_{\rm ax} = A\sqrt{Y\rho}. \tag{4.4}$$

When the drill pipe is uniform and the frequency is low, i.e., $\lambda \gg$ diameter, the axial wave velocity does not depend on pipe dimensions. In this case, the phase velocity of equations (4.3) is constant for each frequency component (no dispersion). If the pipe is not homogeneous or if there is attenuation, dispersion effects are introduced and the group velocity differs from the phase velocity of the uniform rod. Consider the damped equation ($K_{\rm ax} \neq 0$) in the absence of gravity and load forces ($F = 0$). Substituting a plane-wave solution

$$u = e^{\imath(\omega t - kz)}, \tag{4.5}$$

where ω and k are the wave frequency and wavenumber, respectively, we obtain the dispersion equation

$$\omega^2 - \imath \frac{K_{\rm ax}}{A\rho}\omega - \frac{Y}{\rho}k^2 = 0, \tag{4.6}$$

and the dispersion relation can be expressed as

$$k(\omega) = \pm\omega\sqrt{\frac{\rho}{Y}\left(1 - \imath\frac{K_{\rm ax}}{A\rho\omega}\right)} = \pm\frac{\omega}{v_{\rm ax}}\sqrt{1 - \imath\frac{K_{\rm ax}}{A\rho\omega}}. \tag{4.7}$$

Introducing the complex slowness

$$s = \frac{k(\omega)}{\omega}, \tag{4.8}$$

the phase velocity and attenuation factor are given by (Carcione and Poletto, 2000) as

$$v_{\rm p,ax} = \Re(s)^{-1}, \tag{4.9}$$

and

$$\alpha_{\rm ax} = -\omega\Im(s), \tag{4.10}$$

respectively, where attenuation is

$$e^{-\alpha_{\rm ax} z}. \tag{4.11}$$

The axial group velocity in pipes with viscous friction is

$$v_{\rm g,ax} = \left[\Re\left(\frac{\partial k}{\partial \omega}\right)\right]^{-1}. \tag{4.12}$$

4.2.2 Torsional drill-string waves

The propagation of torsional (also denoted as torque or transverse) waves in a rod or pipe is governed by the following equation

$$I\rho\frac{\partial^2\theta}{\partial t^2} - I\mu\frac{\partial^2\theta}{\partial z^2} + M_f = M, \tag{4.13}$$

where θ, I, μ, M_f and M are the angular displacement (twist angle of Figure 4.1), the moment of inertia, the shear modulus, a friction torsional damping for unit length and an external moment, respectively. M_f can be expressed as (Hasley, Kyllingstad, Aarrestad and Lysne, 1986)

$$M_f = M_C \operatorname{sign}\left(\frac{\partial\theta}{\partial t}\right) + K_{\rm tor}\frac{\partial\theta}{\partial t}, \tag{4.14}$$

where M_C is the Coulomb torque-friction coefficient and $K_{\rm tor}$ is the viscous-friction coefficient. Assuming $M_f = 0$ and $M = 0$, we have

$$\frac{\partial^2\theta}{\partial z^2} = \frac{\rho}{\mu}\frac{\partial^2\theta}{\partial t^2}. \tag{4.15}$$

The torsional wave propagation velocity is (Kolsky, 1953)

$$v_{\rm tor} = \sqrt{\frac{\mu}{\rho}}. \tag{4.16}$$

The characteristic impedance for torsional drill-string waves is defined as

$$\mathcal{Z}_{\rm tor} = I\rho v_{\rm tor} = I\sqrt{\mu\rho}. \tag{4.17}$$

For torsional waves, the moment of inertia is calculated with respect to the pipe axis. For a hollow cylindrical tube of inner and outer radius $r_{\rm I}$ and $r_{\rm O}$, respectively,

$$I = I_{zz} = \int_A r^2 dA = \int_0^{2\pi}\int_{r_{\rm I}}^{r_{\rm O}} r^3 dr d\vartheta = \frac{\pi}{2}\left(r_{\rm O}^4 - r_{\rm I}^4\right). \tag{4.18}$$

When the drill pipe is uniform and there is no friction, the torsional velocities do not depend on pipe dimensions. The phase velocities of equation (4.16) are constant for each frequency component (no dispersion in the low-frequency approximation). If the pipe is not homogeneous or in the presence of friction, dispersion effects are introduced, and the group velocity differs from the phase velocity of the uniform rod. Viscous effects can be separated from the Coulomb friction, which is assumed equal to zero in the absence of lateral forces ($M_C = 0$). Also let $M = 0$. We substitute the plane-wave solution

$$\theta = e^{\imath(\omega t - kz)}, \tag{4.19}$$

into equation (4.13) and using equation (4.14) obtain the dispersion equation and relation

$$\omega^2 - \imath\frac{K_{\rm tor}}{I\rho}\omega - \frac{\mu}{\rho}k^2 = 0, \tag{4.20}$$

$$k(\omega) = \pm\omega\sqrt{\frac{\rho}{\mu}\left(1 - \imath\frac{K_{\rm tor}}{I\rho\omega}\right)} = \pm\frac{\omega}{v_{\rm tor}}\sqrt{1 - \imath\frac{K_{\rm tor}}{I\rho\omega}}. \tag{4.21}$$

Using reasoning similar to that used above for the extensional waves (equations (4.9) to (4.12)) we can calculate the phase velocity $v_{\rm p,tor}$, the attenuation factor $\alpha_{\rm tor}$, and the group velocity $v_{\rm g,tor}$ in pipes with viscous friction.

4.2.3 Transversal and flexural drill-string waves

Let the pipe bend in the (x, z) plane and y be the axis perpendicular to the bending plane; let w and u be the x- and z-displacement components. According to the geometry of Figure 2.30a, w is the displacement in the plane of bending that corresponds to a deflection in the x direction. For a large curvature radius R, we can assume with sufficient accuracy that equation (2.27) can be rewritten as

$$R = \left(\frac{\partial^2 w}{\partial z^2}\right)^{-1}. \tag{4.22}$$

The moment equation (2.26) can be expressed as

$$M = -YI\frac{\partial^2 w}{\partial z^2}. \tag{4.23}$$

We have seen in Section 2.6.5 that, in this case, I is the transversal moment of inertia with respect to the axis parallel to y and passing through the beam centroid. Hence, for flexural bending of a hollow cylinder

$$I = I_{yy} = \int_A r^2 \sin^2 \vartheta \, dA = \int_0^{2\pi} \sin^2 \vartheta \, d\vartheta \int_{r_{\rm I}}^{r_{\rm O}} r^3 dr = \frac{\pi}{4}\left(r_{\rm O}^4 - r_{\rm I}^4\right). \tag{4.24}$$

The radius of gyration is defined as

$$R_F = \sqrt{\frac{I}{A}}. \tag{4.25}$$

The equations of motion for flexural vibrations of bars is (Kolsky, 1953; Carcione and Poletto, 2000)

$$\rho A\frac{\partial^2 w}{\partial t^2} = -\frac{\partial S}{\partial z} + f_{\rm F}, \tag{4.26}$$

where $f_{\rm F}$ is the external force and S is the shear force oriented in the x direction. It is given by

$$S = -\frac{\partial M}{\partial z}, \tag{4.27}$$

and equation (4.26) becomes

$$\rho A\frac{\partial^2 w}{\partial t^2} = \frac{\partial^2 M}{\partial z^2} + f_{\rm F}, \tag{4.28}$$

and, using equation (4.23) we obtain

$$\rho A\frac{\partial^2 w}{\partial t^2} + YI\frac{\partial^4 w}{\partial z^4} - f_{\rm F} = 0. \tag{4.29}$$

Vibrating-rope approximation

Assume that the resistance to bending and shear are negligible; that the applied axial tension force $f_F = AT$, where T is the tension, is constant for small displacements. In this approximation, the pipe is equivalent to a vibrating rope. For small vibration amplitudes, we approximate sine by tangent, and the resultant of the transversal component (in the x direction) of the tension-force over a small rope element is proportional to the variation of the slope $\partial w/\partial z$. We obtain the well-known one-dimensional wave equation (Seto, 1994)

$$\rho \frac{\partial^2 w}{\partial t^2} = T \frac{\partial^2 w}{\partial z^2}. \tag{4.30}$$

The rope-transversal wave is not dispersive and its speed is

$$v_{\text{rope}} = \sqrt{\frac{T}{\rho}}. \tag{4.31}$$

The mechanical impedance is defined as

$$\mathcal{Z}_{\text{rope}} = A\rho v_{\text{rope}} = A\sqrt{\rho T}. \tag{4.32}$$

Bending vibrations

Assume that there are no external forces, and use the plane-wave solution

$$w = e^{\imath(\omega t - kz)}. \tag{4.33}$$

We obtain the dispersion equation

$$k^4 = \frac{\rho A}{YI}\omega^2. \tag{4.34}$$

Equation (4.34) has two real roots

$$k_{1,2} = \pm \left(\frac{\rho A}{YI}\right)^{\frac{1}{4}} \omega^{\frac{1}{2}}, \tag{4.35}$$

corresponding to progressive and regressive propagation modes, and two imaginary roots

$$k_{3,4} = \pm \imath \left(\frac{\rho A}{YI}\right)^{\frac{1}{4}} \omega^{\frac{1}{2}}, \tag{4.36}$$

corresponding to static evanescent modes. These correspond to non-propagating motions with dissipation of energy (Carcione and Poletto, 2000). Hence, flexural waves are dispersive. Using the previous results, the phase velocity is

$$v_{\text{p,flex}} = \frac{\omega}{k}. \tag{4.37}$$

Substituting the expressions for the radius of gyration (equation (4.25)) and the extensional velocity (equation (4.3)), the flexural-wave phase velocity can be expressed as (Drumheller, 1992)

$$v_{\text{p,flex}} = \sqrt{R_F v_{\text{ax}} \omega}. \tag{4.38}$$

The group velocity of the flexural wave is twice the phase velocity, i.e.,

$$v_{\text{g,flex}} = \frac{\partial \omega}{\partial k} = 2\sqrt{R_F v_{\text{ax}} \omega}. \tag{4.39}$$

4.2.4 Coupled extensional and flexural drill-string waves

Drumheller (1992) showed that coupled extensional and flexural waves are governed by the following two equations. If R_F and R are the gyration and (large) curvature radii, respectively, the equations are

$$\frac{\partial^2 u}{\partial z^2} + \frac{\partial}{\partial z}\left(\frac{w}{R}\right) - \frac{R_F^2}{R}\frac{\partial^3 w}{\partial z^3} = \frac{1}{v_{ax}^2}\frac{\partial^2 u}{\partial t^2}, \tag{4.40}$$

and

$$-R_F^2\frac{\partial^4 w}{\partial z^4} - \frac{1}{R}\left(\frac{\partial u}{\partial z} + \frac{w}{R}\right) = \frac{1}{v_{ax}^2}\frac{\partial^2 w}{\partial t^2}. \tag{4.41}$$

Similar equations are discussed by Carcione and Poletto (2000).

4.3 Attenuation of extensional waves

Damping consists of the loss of vibrational energy. Aarrestad, Tønnesen and Kyllingstad (1986) treated the following causes of axial damping:

- Radiation of acoustic waves into the surrounding medium (mud and formation) due to the interaction with the drill-string radial motion (Lee, 1991). Conical head waves are an example of this loss;
- Viscous losses due to coupling and interaction with drilling mud;
- Frictional forces produced by wall contacts;
- As discussed above, the bit/rock interaction;
- Losses in the rig suspension at the top of the drill string;
- Dissipation effects in shock absorbers (Section 2.3.10).

The first three items concern damping effects distributed along the drill string. Aarrestad, Tønnesen and Kyllingstad (1986) found that the contribution of the radiation damping is much more important than viscous damping. For instance, they showed examples of radiation damping with coefficient $K_{ax} = 120$ [Pa·s] and viscous damping with coefficient $K_{ax} = 2.3$ [Pa·s] for 3 Hz harmonic extensional waves (equation (4.1)). To calculate the radiation damping assuming the drill string in fluid, they used the expression (after Squire and Whitenhouse, 1979)

$$K_{ax} = 2\pi^2\nu^2 r_o^2 \rho_f \omega, \tag{4.42}$$

where ν, r_o, and ρ_f are the steel's Poisson ratio, outer pipe radius, and fluid (water) density, respectively. The wave decay length, Λ, is defined as the distance the wave has to travel to be attenuated by a factor e. From equation (4.7) these researchers approximated

$$\Lambda = \frac{2\rho v_{ax} A}{K_{ax}}. \tag{4.43}$$

Substituting equation (4.42) into this equation gives

$$\Lambda = \frac{1}{\pi\nu^2}\left(1 - \frac{r_{\mathrm{I}}^2}{r_{\mathrm{O}}^2}\right)\frac{\rho v_{\mathrm{ax}}}{\rho_{\mathrm{f}}\omega}, \tag{4.44}$$

where r_{I} is the pipe inner radius. From equation (4.44) and drill-string tables it can be calculated that the damping in the drill collars, e.g., $\Lambda = 7000$ m at 3 Hz frequency, is much less than in 5 in. drill pipes, e.g., $\Lambda = 2300$ m at 3 Hz frequency. The attenuation is often expressed in decibel per kilometer, which gives

$$-\frac{20\log e}{\Lambda\cdot 10^{-3}} \cong -\frac{8686}{\Lambda} \qquad [\mathrm{dB/km}]. \tag{4.45}$$

This corresponds to attenuation of the wave amplitude of about 4 dB/km at 3 Hz frequency in 5 in. drill pipes. Because the attenuation in equation (4.45) increases linearly with frequency, an attenuation of about 37 dB/km is expected at 30 Hz. However, this value seems to be too large.

Attenuation of extensional waves at higher frequencies was studied by Drumheller (1993). He analyzed the signal dispersion and empirically estimated the attenuation of the extensional waves, produced by explosive sources in full-scale drill strings, for passband frequencies above 250 Hz. These passbands are the passbands for extensional waves in periodic drill strings, which are analyzed in Section 4.4.1. The attenuation values measured by Drumheller (1993) are of the order of 11–12 dB/Km in the second band centered at about 400 Hz, of 19–21 dB/km in the third band centered at about 670 Hz, and of 28–31 dB/km in the fourth band centered at about 930 Hz. Extrapolation of Drumheller's results to the first band, from 0 to about 250 Hz and centered at about 120 Hz, is difficult. However, this extrapolation should lead to an attenuation of the order of about 6 dB/km, which is significantly lower than the radiation attenuation calculated above by equation (4.43) of Aarrestad, Tønnesen and Kyllingstad (1986). A possible explanation is that, in real drilling conditions with pipes, fluid and formation coupled at the long wavelengths of the SWD seismic signals, the fluid model is not valid for the surrounding medium which should include the formation. The attenuation for radiation in fluid, expressed by equation (4.42), is a possible model for SWD drill-string signals propagating through marine-rig risers where attenuation effects could be more important.

Malusa (2001) made direct measurements of the attenuation of axial-pilot waves propagating in 5 in. drill pipe. He analyzed onshore SWD pilot traces in the 20–40 Hz seismic frequency band. His results are more in agreement with those of Drumheller (1993) and show low attenuation for propagating axial waves. Malusa (1991) used the amplitudes of the long-period drill-string multiples in drill-pipe lengths of 1000 to 2500 m (Figure 4.2). He found an average attenuation of 4 dB/km in this frequency band (Figure 4.3).

4.3.1 Attenuation of vibrations by shock absorbers

Shock absorbers are used to reduce drill-string vibrations (Section 2.3.10). Large drill-string vibrations may cause drill-string fatigue and wear and may reduce the rate of penetration. Shock absorbers have two effects (Skaugen and Kyllingstad, 1986). The first is to attenuate the vibrations generated at the drill bit. Parfitt and Abbassian

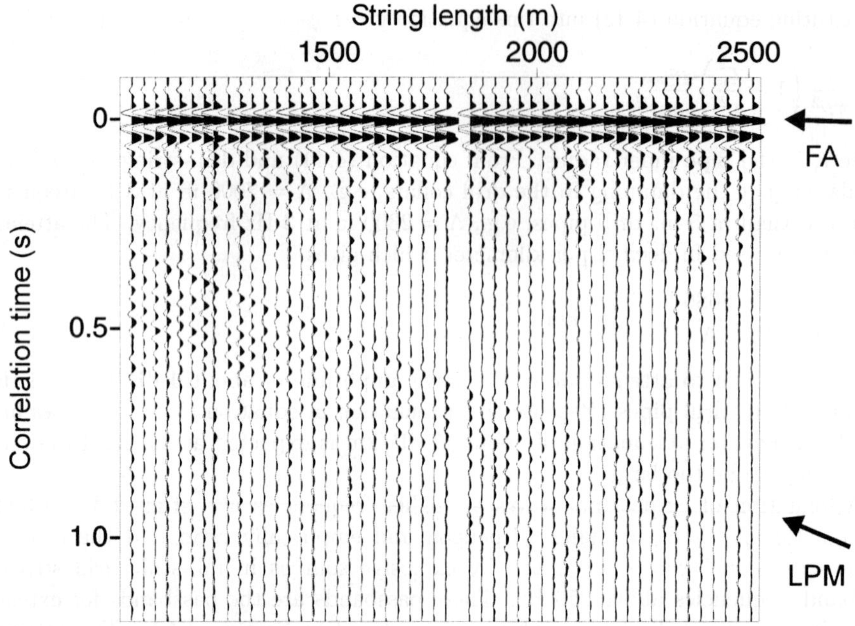

Figure 4.2: Axial drill-string pilot signal after autocorrelation (with the first arrival FA at $t = 0$ and normalized) and long-period multiple signal (dipping LPM event). Data are bandpass filtered (15/20/40/45 Hz).

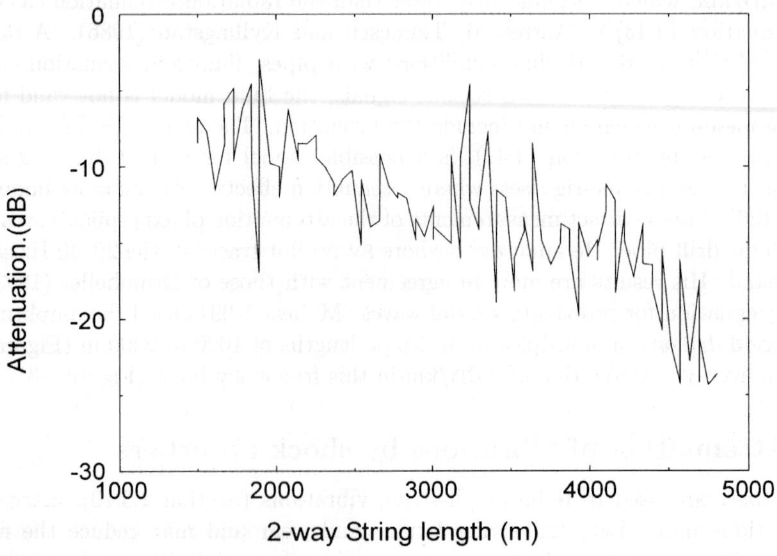

Figure 4.3: Attenuation of long-period multiples versus drill-string length.

(1995) modeled the interaction between a tricone bit and formation with axial lobed-pattern vibrations to determine the shock-absorber selection criteria. The second is to attenuate the longitudinal drill-string vibrations produced by drilling (Kreisle and Vance, 1969). Skaugen and Kyllingstad (1986) measured the performances of shock absorbers in a 1000 m deep vertical well drilled by a full-scale drilling rig. They used a downhole (lobed-pattern) exciter of known response mounted at the lower end of the drill string. The drill-string response was measured by an instrumented tool (sensor sub) located above the shock absorber mounted just above the drill bit. They measured WOB, torque, inner- and outer-mud pressure and four acceleration channels at sampling rates ranging between 25 and 100 Hz. Each absorber is modeled as a spring of constant κ and a damper of damping coefficient b in parallel, by using the equation

$$m\frac{\partial^2 u}{\partial t^2} + b\frac{\partial u}{\partial t} + \kappa u = 0, \tag{4.46}$$

where m is the suspended-string mass and u is the axial displacement. This equation corresponds to the equation of viscous damping in which b is constant. These researchers determined that because the viscous damping is small, the more important effects are produced by structural damping – related to elastic material hysteresis – and non-linear Coulomb friction. Structural damping is related to frequency by

$$b = \frac{\kappa}{\omega}\gamma, \tag{4.47}$$

where γ is a constant. The damped solution of equation (4.46) is of the type

$$u = e^{-\frac{b}{2m}t \pm \iota\omega_n\sqrt{1-[b/(2m\omega_n)]^2}\,t}, \tag{4.48}$$

where $\omega_n = \sqrt{\kappa/m}$ is the natural resonance frequency of the system.

Internal Coulomb's friction force F for the metal sliding against metal is independent of velocity magnitude but depends on its sign. The motion equation becomes

$$m\frac{\partial^2 u}{\partial t^2} + \text{sign}\left(\frac{\partial u}{\partial t}\right) F + \kappa u = 0. \tag{4.49}$$

Measurements performed by these researchers showed that shock absorbers have hysteresis loops in the force/displacement domain, with variation of the spring stiffness κ at different load amplitudes. In absorber tools with rubber components, κ depends also on temperatures. In general, large non linearity of the shock absorber performances in damping and elastic mechanisms was reported. Under constant WOB and small oscillation, κ constant (linear spring behavior) was observed. For some tools γ depended on WOB. For all the considered shock absorbers, γ was nearly independent of vibration frequency.

Reported results of measurements performed with and without shock absorbers indicate that bouncing effects disappeared. The analysis of vibrations show attenuation – to about half – of axial loads and acceleration in the frequency range from 0 to 6 Hz. However, assuming only the structural damping described by equation (4.47), the attenuation term is $e^{-\kappa\gamma/(2m\omega)\,t}$ and damping effects should be less important at higher frequencies. Measured values of γ ranged between 0.5 and 1.2. Spring-stiffness ranges were from 5 to 100 kN/mm for different shock absorbers.

The use of shock absorbers in SWD was investigated by Naville, Layotte, Pignard and Guesnon (1994). They suggested the shock-absorber tool be used while measuring the drill-string vibrations using downhole pilot sensors in order to improve the noisy conditions at the recording point close to the bit.

4.4 Waves in *periodic* and *non-periodic* drill strings

Real drill strings are not homogeneous nor exactly periodic. The bottom-hole assembly contains elements of different dimensions and materials (for instance, some parts can be made of rubber or aluminum). Drill pipes can be formed of sections with different type of tool joints and may have different lengths. However, the pipes, in many cases, can be approximated as a periodic system of individual elements each formed of an elongated body and the pipe joints (tool joints) at its ends. Pipe periodicity introduces important impacts for wave propagation. One such impact is the passbands and stopbands in the frequency spectrum. Another is the wave dispersion, and the changes in the group velocity, also at low frequencies. Drill-string signals are used to measure the pilot delay and to correct the arrival time of the data acquired at the surface by a deployed seismic line. Because the correction is based on the group velocity of the guided waves, it is important to determine with accuracy this velocity, which depends on the acoustic properties and the geometry of the drill string.

4.4.1 Wave propagation in *periodic* strings

Several authors have investigated the behavior of guided waves in drill strings. Barnes and Kirkwood (1972) analyzed the passband and stopband effects caused by the presence of tool joints in the drill pipes of an idealized (periodic) drill string. In this analysis, the drill string is assumed to be an infinitely long cylindrical steel pipe with identical coupling joints at equal intervals, as in the example of Figure 4.4.

They calculated the transmission conditions, which can be expressed as

$$\left|\cos\frac{\omega}{v}l_1\cos\frac{\omega}{v}l_2 - \mathcal{M}\sin\frac{\omega}{v}l_1\sin\frac{\omega}{v}l_2\right| \leq 1, \tag{4.50}$$

where v is the propagation velocity given by equations (4.3) and (4.16) for extensional and torsional waves, respectively; l_1 and l_2 are the lengths of the pipe body upset and tool joints, respectively, and where the coefficient $\mathcal{M}$ – representing the wave reflection (and transmission) effects – is

$$\mathcal{M} = \frac{1}{2}\left(\frac{A_2}{A_1} + \frac{A_1}{A_2}\right), \tag{4.51}$$

$$\mathcal{M} = \frac{1}{2}\left(\frac{I_2}{I_1} + \frac{I_1}{I_2}\right), \tag{4.52}$$

for axial and torsional waves, respectively, where $A = \pi(r_{\rm O}^2 - r_{\rm I}^2)$ and $I = \pi(r_{\rm O}^4 - r_{\rm I}^4)/2$ are the tube cross-section and moment of inertia, respectively. Reject or stopbands correspond to values of frequency for which the term in equation (4.50) exceeds 1. The explanation

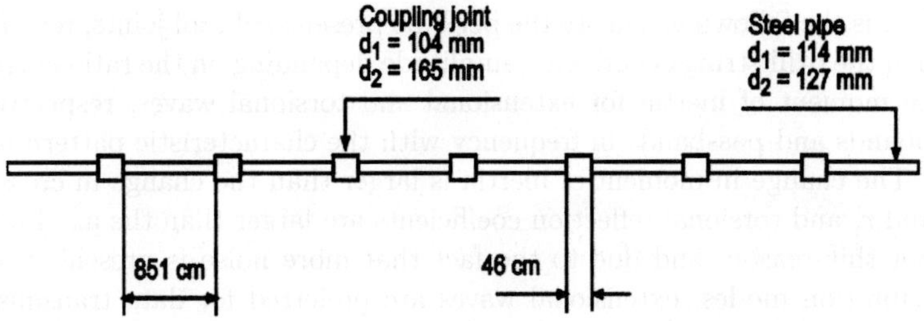

Figure 4.4: Periodic drill-string pipe.

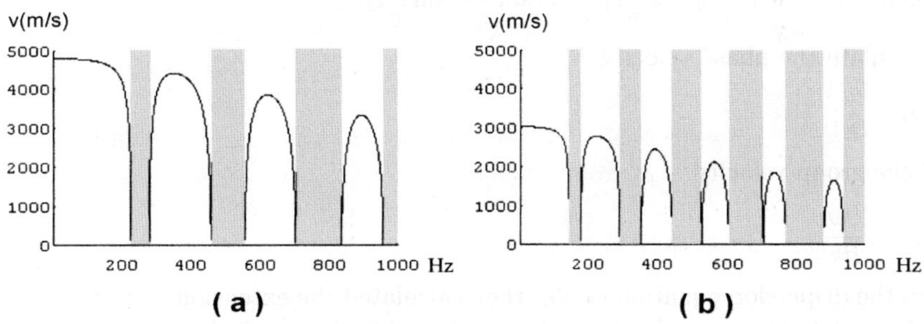

Figure 4.5: Passbands and stopbands in frequency for a) extensional group velocity b) torsional group velocity (modified after Drumheller and Knudsen, 1995).

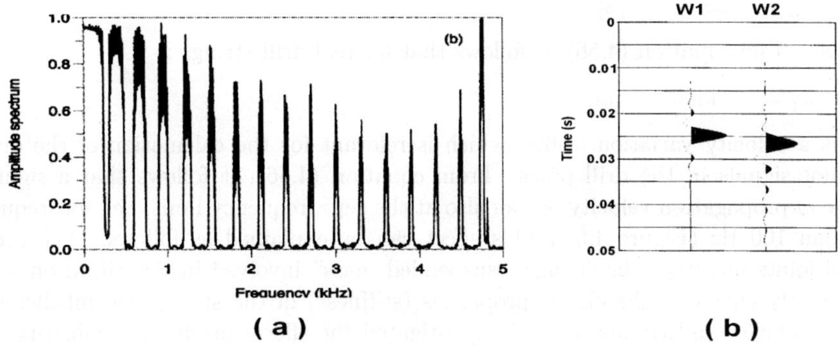

Figure 4.6: a) Passband and stopband frequencies for extensional waves propagated by full-wave modeling (after Carcione and Poletto, 2000). b) Waves calculated by numerical simulation, and filtered in the low seismic frequencies. Wavelets in rods without (W1) and with (W2) tool joints have different propagation delays, which, if not accounted for, produce important errors for SWD signal correction.

for this effect is the following. Due to the periodic presence of tool joints, repeated internal reflections in the drill string occur, with amplitude depending on the ratio of cross-sections and on the moment of inertia for extensional and torsional waves, respectively. These cause stopbands and passbands in frequency with the characteristic pattern of a periodic structure. The change in moment of inertia is larger than the change in cross-section for given r_{O} and r_{I} and torsional reflection coefficients are larger than the axial ones (Section 4.10.1). For this reason, and due to the fact that more noise is present in the form of torsional vibration modes, extensional waves are preferred for data transmission in the drill string (Drumheller, 1993). It was also pointed out that periodic distribution of tool joints is a favorable condition for data transmission, at least in the passbands calculated by equation (4.50), which are different for axial and torsional waves.

Drumheller and Knudsen (1995) used the same dispersion relation, namely,

$$\cos k(l_1 + l_2) = \cos \frac{\omega}{v} l_1 \cos \frac{\omega}{v} l_2 - \mathcal{M} \sin \frac{\omega}{v} l_1 \sin \frac{\omega}{v} l_2, \tag{4.53}$$

to calculate the phase velocity

$$v_{\mathrm{p}}(\omega) = \frac{\omega}{k}, \tag{4.54}$$

and the group velocity

$$v_{\mathrm{g}}(\omega) = \frac{\partial \omega}{\partial k}. \tag{4.55}$$

From the dispersion equation (4.53), they calculated the extensional group velocity in the low-frequency approximation, i.e., for ω in the zero frequency limit, as

$$v_{\mathrm{g}}(0) = v_{\mathrm{p}}(0) = v \frac{l_1 + l_2}{\sqrt{l_1^2 + (A_1/A_2 + A_2/A_1) l_1 l_2 + l_2^2}}. \tag{4.56}$$

For steel, $Y = 206$ GPa, $\rho = 7840$ kg/m^3 can be assumed; for extensional waves in a steel rod, we have

$$v = \sqrt{Y/\rho} = 5126 \quad \text{[m/s]}. \tag{4.57}$$

However, from equation (4.56), it follows that for real drill strings it is

$$v_{\mathrm{g}}(0) = v_{\mathrm{p}}(0) = 4800 \qquad \text{[m/s]}. \tag{4.58}$$

This is a velocity variation of 6%, which is relevant for the calculation of the delay of the pilot signals in the drill pipes. From equation (4.56), it follows that a significant change of propagation velocity occurs also at the zero frequency limit, i.e., for frequencies less than 100 Hz (Figure 4.5). This effect can be explained as follows. The presence of tool joints increases the string's "suspended mass" involved in the vibration without significantly changing the elastic properties (stiffness) of the string. Drumheller (1992) and Drumheller and Knudsen (1995) investigated the effects produced by fluctuations of drill-string length and calculated the group velocity of the extensional waves for different lengths and cross-sections of real drill pipes and tool joints in periodic systems. In Section 4.4.3, we will see that the average low-frequency effect can be generalized and calculated for arbitrary periodic and non-periodic strings with stationary properties (Poletto and Carcione, 2001).

4.4.2 Wave propagation in *non-periodic* strings

Wave motion in arbitrary periodic drill strings was calculated by Lous, Rienstra and Adan (1998) using frequency-domain algorithms. In order to model wave-propagation in real drill strings of arbitrary composition, either periodic or non-periodic, using a full-wave modeling algorithm, Carcione and Poletto (2000) solved the differential equations describing wave propagation through the drill string. They computed, with a fourth-order Runge-Kutta algorithm, the waveforms of the extensional, torsional and flexural waves by modeling the geometrical features of the coupling joints, including piezoelectric sources and sensors. Their equations include a relaxation mechanism simulating the viscoelastic behavior of the steel, dielectric losses and any other losses, such as those produced by the presence of the drilling mud, the casing and the formation. This algorithm simulates the passbands and stopbands due to the presence of the coupling joints, and pulse distortion and delay due to non-uniform cross-section areas (Figure 4.6). These codes are used to calculate wavelets, signal content and time delay of group-velocity propagation with sources and receivers in arbitrary positions along a drill string including BHA, pipe and rig suspension (traveling block).

4.4.3 Group velocity in *non-periodic* string

A method to compute the group velocity of extensional and torsional waves in a drill string composed of several elements of arbitrary acoustic properties and geometrical characteristics was proposed by Poletto and Carcione (2001). The method is based on averaging the forces, the mass and the moment of inertia of the different components of the drill string. The condition of stationarity is required; that is, in a given length of drill string the proportion of each element (e.g., drill pipes, tool joints, etc.) remains constant. The resulting equations for the group velocity are not restricted to periodic systems.

A drill string is composed of several sections and elements of different weights, diameters and mechanical properties. The upper part is composed of several sections of drill pipes and coupling joints, and the lower part is the BHA, which includes the heavy-weight drill pipes and the drill collars. The BHA is heavier and may be of complex composition. The drill string is normally made of steel with uniform elastic properties, but parts of the BHA can be made of aluminum and hard rubber.

Each element of drill string is described by a density ρ, a length d, an internal radius $r_{\rm I}$, an external radius $r_{\rm O}$, a Young modulus Y, and a shear modulus μ. At seismic frequencies, the wavelengths of the guided waves are long compared to the drill-string lateral dimensions, and the drill string can be approximated as a rod. As shown in Figure 4.1, z is the axial coordinate of the drill string, u the axial displacement and θ the angular displacement. The axial and angular strains are

$$\epsilon = \frac{\partial u}{\partial z} \quad \text{and} \quad s = \frac{\partial \theta}{\partial z}, \tag{4.59}$$

respectively. The strains (4.59) vary in the drill string according to the properties of the different elements. In the low-frequency approximation, which holds for wavelengths that are much longer than the length of the drill-string sections, these sections can be viewed as homogeneous rods, with constant diameter and elastic properties and where constant axial and torsional stresses generate constant axial and angular strains, respectively.

The phase and group velocities are given by equations (4.54) and (4.55) where $\omega = \omega(k)$ (or $k = k(\omega)$) is the dispersion relation. Instead of using the exact dispersion relation, which would require a lengthy and complicated numerical calculation, Poletto and Carcione (2001) invoke the long-wavelength approximation and replace the non-uniform drill string by an effective rod of similar average properties. Then, the group velocity at seismic frequencies can be easily calculated from these average properties.

Let $m_{\rm ext}$ [kg/m] and $M_{\rm tor}$ [kg·m] be the mass per unit length and the moment of inertia per unit length, respectively. They are given by

$$m_{\rm ext} = \rho A \quad \text{and} \quad M_{\rm tor} = \rho I, \tag{4.60}$$

where A and I are the cross section and the moment of inertia per unit mass, given by $I = \pi(r_{\rm O}^4 - r_{\rm I}^4)/2$ of the "effective average" rod. In a uniform rod, the group velocities of extensional and torsional waves equal the respective phase velocities; that is, $v_{\rm g(ext)} = (Y/\rho)^{1/2}$, and $v_{\rm g(tor)} = (\mu/\rho)^{1/2}$ (Kolsky, 1953), and the group velocities of the effective rod can be calculated as

$$V_{\rm g(ext)} = \sqrt{\frac{YA}{m_{\rm ext}}}, \quad \text{and} \quad V_{\rm g(tor)} = \sqrt{\frac{\mu I}{M_{\rm tor}}}, \tag{4.61}$$

where YA and μI are the average axial force per unit strain and the average torque per unit strain, respectively.

4.4.4 Average drill-string properties

Let the drill string be composed of N elements, of axial lengths d_i and cross sections A_i, $i = 1, \ldots, N$. We calculate the average of a given property or field variable a by

$$\langle a \rangle = \left[\sum_{i=1}^{N} d_i a_i\right] \Big/ \sum_{i=1}^{N} d_i. \tag{4.62}$$

Average dynamic properties

The axial force (Newton's law) and the torque moment per unit length are (Kolsky, 1953)

$$F_i = A_i \rho_i \frac{\partial^2 u_i}{\partial t^2} \quad \text{and} \quad M_i = I_i \rho_i \frac{\partial^2 \theta_i}{\partial t^2}, \quad i = 1, \cdots, N. \tag{4.63}$$

The average axial force and torque moment per unit length are then given by

$$F = \langle F_i \rangle, \quad \text{and} \quad M = \langle M_i \rangle, \tag{4.64}$$

respectively. Assuming statistical independence (Ochi, 1990) between u_i and $A_i\rho_i$ and between θ_i and $I_i\rho_i$, we have

$$F = \left\langle A_i \rho_i \frac{\partial^2 u_i}{\partial t^2} \right\rangle = \langle A_i \rho_i \rangle \left\langle \frac{\partial^2 u_i}{\partial t^2} \right\rangle \quad \text{and} \quad M = \left\langle I_i \rho_i \frac{\partial^2 \theta_i}{\partial t^2} \right\rangle = \langle I_i \rho_i \rangle \left\langle \frac{\partial^2 \theta_i}{\partial t^2} \right\rangle. \tag{4.65}$$

(for instance, since u_i, in the long-wavelength limit, varies linearly and very slowly from element to element, we may assume, omitting the time derivative, $\langle A_i\rho_i u_i\rangle = \langle A_i\rho_i\rangle\langle u_i\rangle$, even if $A_i\rho_i$ varies by large fractions from element to element).

Then, setting

$$m_{\rm ext} = A\rho = \langle A_i\rho_i\rangle \quad \text{and} \quad M_{\rm tor} = I\rho = \langle I_i\rho_i\rangle. \tag{4.66}$$

we have

$$F = A\rho\frac{\partial^2\langle u_i\rangle}{\partial t^2}, \quad \text{and} \quad M = I\rho\frac{\partial^2\langle \theta_i\rangle}{\partial t^2}. \tag{4.67}$$

Average static properties

Let F be a constant axial force and M a constant torque moment acting on the drill string and let ϵ and s be the average axial and torsional strains, respectively. The strains are related to the axial force and torque by

$$\epsilon = \frac{F}{YA} \quad \text{and} \quad s = \frac{M}{\mu I}, \tag{4.68}$$

where YA and μI are the averaged properties.

The average value of the axial and torsional strains are

$$\epsilon = \langle \epsilon_i\rangle \quad \text{and} \quad s = \langle s_i\rangle, \tag{4.69}$$

respectively. Because

$$\epsilon_i = \frac{F}{Y_iA_i} \quad \text{and} \quad s_i = \frac{M}{\mu_iI_i}, \tag{4.70}$$

we obtain,

$$\langle \epsilon_i\rangle = F\left\langle\frac{1}{Y_iA_i}\right\rangle \quad \text{and} \quad \langle s_i\rangle = M\left\langle\frac{1}{\mu_iI_i}\right\rangle. \tag{4.71}$$

Comparison of equations (4.68) and (4.71) and the use of (4.69) and (4.70) yields

$$YA = \left\langle\frac{1}{Y_iA_i}\right\rangle^{-1} \quad \text{and} \quad \mu I = \left\langle\frac{1}{\mu_iI_i}\right\rangle^{-1}, \tag{4.72}$$

respectively. Equations (4.72) are the reciprocal of the averages of the reciprocal acoustic properties. They are similar in form to the effective elasticity constant c_{33} of a thinly layered medium in the long-wavelength limit (Bruggeman, 1937; Postma, 1955; Helbig, 1958; Backus, 1962; Carcione, Kosloff and Behle, 1991).

4.4.5 Group velocity at low frequency

Substituting equations (4.66) and (4.72) into (4.61) yields the following group velocities for extensional and torsional waves in the long-wavelength (low-frequency) approximation

$$v_{g(\mathrm{ext})} = \left[\left\langle \frac{1}{Y_i A_i} \right\rangle \langle A_i \rho_i \rangle \right]^{-1/2} = \left(\sum_{i=1}^{N} d_i \right) \left[\left(\sum_{i=1}^{N} d_i \rho_i A_i \right) \left(\sum_{i=1}^{N} \frac{d_i}{Y_i A_i} \right) \right]^{-1/2} \tag{4.73}$$

and

$$v_{g(\mathrm{tor})} = \left[\left\langle \frac{1}{\mu_i I_i} \right\rangle \langle I_i \rho_i \rangle \right]^{-1/2} = \left(\sum_{i=1}^{N} d_i \right) \left[\left(\sum_{i=1}^{N} d_i \rho_i I_i \right) \left(\sum_{i=1}^{N} \frac{d_i}{\mu_i I_i} \right) \right]^{-1/2} . \tag{4.74}$$

These equations are not restricted to periodic systems and provide a good approximation in the frequency range used for while-drilling investigations.

If the density and the Young modulus are constant, and the drill string is composed of a periodic system of pipes and coupling joints, equation (4.73) becomes the group velocity obtained by Drumheller and Knudsen (1995). Moreover, if the shear modulus is constant, equation (4.74) simplifies to

$$v_{g(\mathrm{tor})} = \sqrt{\frac{\mu}{\rho}} (d_1 + d_2) \left[d_1^2 + \left(\frac{I_1}{I_2} + \frac{I_2}{I_1} \right) d_1 d_2 + d_2^2 \right]^{-1/2} , \tag{4.75}$$

where d_1 and d_2 and I_1 and I_2 are the lengths and moment of inertia of the pipes and coupling joints, respectively.

Example of group velocity

We consider wave propagation through the drill collars (part of the BHA) for two different cases. They are referred to as systems 1 and 2 in Table 4.1, where d indicates the length of each element of the drill collar. In system 1, the radii are quite uniform, and in system 2 the radii have a larger variation. Each element has the same mechanical properties: ρ = 7840 kg/m^3, Y = 206 GPa and μ = 78.5 GPa. These values give rod velocities of 5126 m/s and 3164 m/s for the extensional and torsional waves, respectively. The group velocities obtained from equations (4.73) and (4.74) are 5123 m/s and 3157 m/s and 5022 m/s and 2902 m/s for the extensional and torsional waves corresponding to systems 1 and 2, respectively. As expected, the values for system 1 are close to those of the uniform rod because its mass distribution is more uniform.

Let us consider now the drill pipe/coupling joint system above the BHA, with dimensions r_{O} = 5 in., r_{I} = 4.275 in., d = 9.2 m (pipes), and r_{O} = 6.625 in., r_{I} = 2.75 in., d = 0.5 m (tool joints). Assuming a periodic system, the extensional and torsional group velocities are 4727 m/s and 2860 m/s, respectively. Note that the wave velocities in the drill collars for system 2 – with an uneven distribution of masses – are between the velocities of a uniform rod and the velocities of the drill pipe/coupling joint system.

These results can be used to estimate the delays of low-frequency acoustic signals in the different drill-string sections.

Table 4.1 – Dimensions of the drill-collar elements

System 1			System 2		
r_O [in.]	r_I [in.]	d [m]	r_O [in.]	r_I [in.]	d [m]
6.500	2.8125	0.47	6.500	2.8125	0.97
6.500	2.8750	27.58	6.500	2.8125	28.11
6.250	2.0500	10.33	6.500	2.8125	1.10
6.500	2.8750	75.40	8.000	2.8125	18.43
6.250	2.8125	1.89	8.000	3.0000	5.96
6.500	2.8750	18.65	8.000	3.0000	5.50
6.750	2.8125	1.48	8.000	2.8125	83.73
6.750	2.8125	1.48	8.000	2.7500	9.43
6.500	2.8750	9.46	12.125	2.7500	0.90
6.250	2.8125	0.66	8.000	2.8125	9.04
8.500	0.1000	0.25	12.125	2.7500	0.95
			8.000	2.7500	6.14
			12.125	2.7500	1.0000
			12.250	0.1	0.32

$v_{1,(ext)}$	$v_{1,(tor)}$	$v_{1,g(ext)}$	$v_{1,g(tor)}$	$v_{2,(ext)}$	$v_{2,(tor)}$	$v_{2,g(ext)}$	$v_{2,g(tor)}$
5126	3164	5123	3157	5126	3164	5022	2902

4.5 Drill-bit mud waves

Mud waves are coupled to the pipes and the formation, and they play an important role in borehole signal propagation. The acoustic properties of drilling mud affect the propagation velocity of these guided waves. Poletto, Lovo and Carcione (2000) modeled the compressional-wave velocity of drilling mud with different compositions of low- and high-density (also called low- and high-gravity) solids and a given drilling plan (Section 2.6.3). They took into account water- and oil-based drilling muds, and the presence of formation cuttings. This theory was used to determine the velocity of the borehole guided waves, which – in addition to the extensional waves traveling through the drill string – can be used as pilot signals to obtain SWD seismograms, and information about the drilling conditions. These waves are used when there is no rotation of the drill string while drilling (sliding mode), and the static friction between the drill pipes and the borehole wall attenuate the extensional waves in the pipes. Another reason to analyze the mud-wave velocity is that the low-velocity events due to guided waves can be often confused with extensional multiples. A linear-moveout analysis of delay versus string length may be required to correctly interpret the mud waves or axial multiples in the SWD correlations. These researchers computed the velocities of the tube wave – with and without casing – and of the guided wave traveling in the mud inside the drill string (pipe wave). The results indicated that the pipe wave constitutes a reliable pilot signal in the absence of drill-pipe rotation.

Guided waves in SWD data

Elastic wave propagation in a fluid-filled borehole is characterized by dispersive modes which can be used for SWD measurements. Two main guided waves are observed in SWD experiments. These are: the low-frequency Stoneley wave traveling between the mud and the formation (tube wave) and the wave traveling inside the drill pipes, filled with drilling mud (mud-pipe wave). The velocity of these guided waves depends on the elastic properties of the surrounding formation and on the borehole radial dimensions, as well as on the velocity of the compressional waves in the drilling mud.

Coupled mud waves in SWD experiments were discussed by Poletto and Miranda (1998). The SWD correlation results with these slow pilot mud waves are discussed in Section 7.8.3.

The basic equations for synthetic acoustic logging in a fluid-filled borehole are given, for instance, in Cheng and Toksöz (1981), who investigated the dispersion curves of the Stoneley mode (tube waves). In their examples – hard formations – these waves show very little dispersion in the seismic frequency range, and both the phase and group velocities increase from about 0.9 $c_{\rm m}$ at low frequencies (tube-wave limit) to about 0.96 $c_{\rm m}$ at high frequencies, where $c_{\rm m}$ is the body-wave velocity in the drilling mud. The other factors affecting the tube-wave velocity are the compressional and shear velocities of the formation and the borehole lateral dimensions. In their analysis, Cheng and Toksöz (1981) used a mud velocity $c_{\rm m}$ = 1830 m/s, which according to the calculations of Poletto, Lovo and Carcione (2000) is too high for actual drilling muds.

More recently, Rama Rao and Vandiver (1999) analyzed the acoustic properties of a water-filled borehole with pipes, calculating the axisymmetric propagation of the different modes for frequencies less than 1000 Hz, with soft and hard formations. In their analysis, the authors use a velocity α_f = 1558 m/s, which is closer to estimations of Poletto, Lovo and Carcione (2000) than that obtained by Cheng and Toksöz's.

The sound velocity of drilling mud saturated with reservoir gas is modeled by Carcione and Poletto (2000). They give the basic equations to calculate α_f for water- and oil-based drilling mud versus solid content and gas saturation, and different temperature and pressure conditions. On the basis of these results, we analyze the variation of the compressional velocity of the drilling mud, for different solid proportions (bentonite and barite), water and oil mixtures and formation cuttings. The calculations are based on a realistic drilling plan, which takes into account the variation of the pressure-temperature conditions with depth.

4.5.1 Acoustic properties of drilling mud

The composite density of the drilling mud is simply the weighted average of the densities of the constituents. Assuming that no component is in solution, the mud density is given by

$$\rho_{\rm m} = \sum \rho_i \phi_i = \phi_q \rho_q + \phi_b \rho_b + \phi_w \rho_w + \phi_o \rho_o + \phi_c \rho_c, \tag{4.76}$$

where ϕ_q, ρ_q and ϕ_b, ρ_b are the volume fractions and densities of the clay particles (bentonite, modeled with the quartz properties) and barite (high-gravity solids), respectively;

ϕ_w, ρ_w and ϕ_o, ρ_o are the volume fractions and densities of water and oil; and ϕ_c, ρ_c are the volume fractions and densities of the cuttings. We have assumed that the drilling fluid is a mixture of water- and oil-based drilling mud (oil emulsion mud). By definition,

$$\sum \phi_i = \phi_q + \phi_b + \phi_w + \phi_o + \phi_c = 1. \tag{4.77}$$

The volume fractions of bentonite are fixed, while those of water and barite are calculated by compensating the mud-weight versus depth predicted by the drilling plan (Carcione and Poletto, 2000). Wood's model is used to obtain the bulk modulus of the composed mud. This model gives the properties of fluid suspensions and mixtures (Mavko, Mukerji and Dvorkin, 1998). The model gives the average bulk modulus as the reciprocal of the average of the reciprocals of the individual bulk moduli (isostress assumption), namely,

$$\frac{1}{K_{\rm m}} = \frac{\phi_q}{K_q} + \frac{\phi_b}{K_b} + \frac{\phi_w}{K_w} + \frac{\phi_o}{K_o} + \frac{\phi_c}{K_c}, \tag{4.78}$$

where K_q, K_b, K_w, K_o, and K_c are the bulk moduli of bentonite, barite, water, oil, and cuttings, respectively (see Table 4.2). The oil properties are given, as an example, at atmospheric conditions, and we consider an API gravity of 40 degrees. The volume fraction of the cuttings in the outer mud is given by

$$\phi_c = 16.67\,\pi r_{\rm b}^2 \frac{\rm ROP}{{\rm FLOW}_{\rm m}}, \tag{4.79}$$

where $r_{\rm b}$ [m] is the radius of the bit, ROP is the rate of penetration [m/h], and $\text{FLOW}_{\rm m}$ is the flow of the mud [liter/min] (Gabolde and Nguyen, 1999).

4.5.2 Velocities of the acoustic mud waves

The mud wave velocity is then given by

$$c_{\rm m} = \sqrt{\frac{K_{\rm m}}{\rho_{\rm m}}}. \tag{4.80}$$

This velocity is generally lower than the wave velocity of the pure fluid – water in this case – because the increase in density due to the presence of solids is not compensated for by the increase in bulk modulus. Differentiating equation (4.80), we obtain

$$\frac{\delta c_{\rm m}}{c_{\rm m}} = \frac{1}{2}\left(\frac{\delta K_{\rm m}}{K_{\rm m}} - \frac{\delta \rho_{\rm m}}{\rho_{\rm m}}\right), \tag{4.81}$$

where δ denotes the increment (positive or negative) in the respective quantity.

4.5.3 Sensitivity analysis for acoustic mud velocity

A velocity sensitivity analysis was performed by Poletto, Carcione, Lovo and Miranda (2002). Increasing the density by adding solids decreases the sound velocity, but an increase in bulk modulus increases the velocity. Hence, it is necessary to study how the sound velocity varies with the addition of solids to the drilling mud. For simplicity, we

Table 4.2 – Material properties

Material	Density [kg/m³]	Elastic properties [GPa]
Bentonite	2650	36.00+
Barite	4200	55.00+
Water	1000	2.25+
Oil‡	826	1.50+
Cuttings	2000	23.00+
Pipe	7840	(0.29†) 206.00*

+ Bulk modulus
∗ Young modulus
† Poisson's ratio

‡ at atmospheric conditions and API gravity of 40 degrees

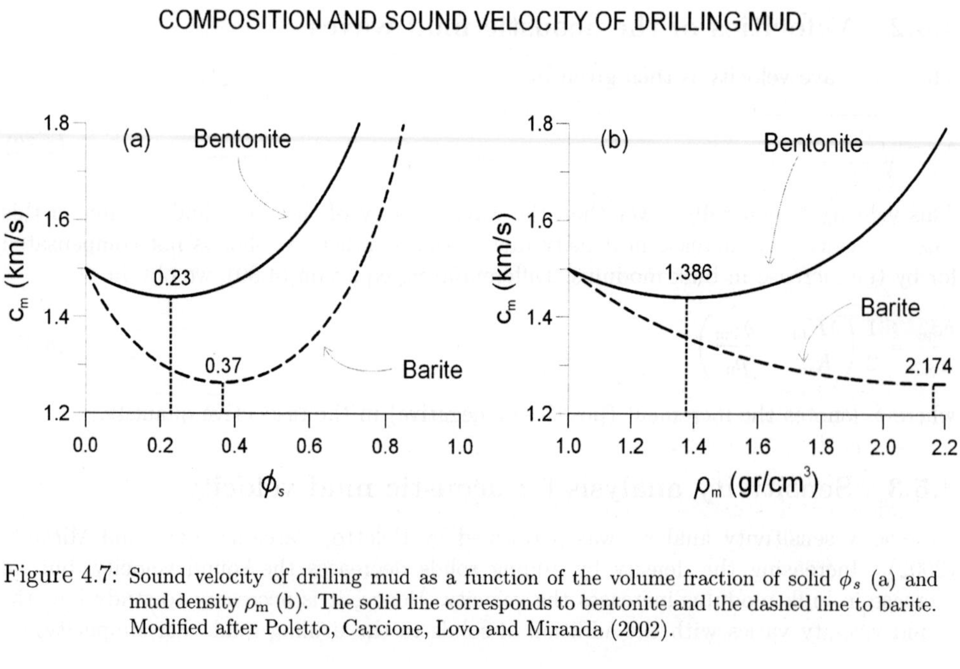

Figure 4.7: Sound velocity of drilling mud as a function of the volume fraction of solid ϕ_s (a) and mud density ρ_m (b). The solid line corresponds to bentonite and the dashed line to barite. Modified after Poletto, Carcione, Lovo and Miranda (2002).

assume a single additive solid, with volume fraction $\phi_s = \phi_1$ or $\phi_s = \phi_2$. Differentiating the bulk modulus and the density with respect to the volume fraction ϕ_s, we obtain

$$\frac{\delta c_{\rm m}}{c_{\rm m}} = \frac{1}{2}\left[\frac{K_s - K_f}{K_s - \phi_s(K_s - K_f)} - \frac{\rho_s - \rho_f}{\rho_f + \phi_s(\rho_s - \rho_f)}\right]\delta\phi_s, \tag{4.82}$$

where the subscripts s and f denote the solid and the fluid, respectively. The value $\phi_s = \phi^*$ at which the right side of equation (4.82) equals zero – and $c_{\rm m}$ has its minimum value – is

$$\phi^* = \frac{1}{2}\left(\frac{K_s}{K_s - K_f} - \frac{\rho_f}{\rho_s - \rho_f}\right). \tag{4.83}$$

Using the material properties given in Table 4.2, we obtain $\phi^* = 0.23$ for bentonite and $\phi^* = 0.37$ for barite, corresponding to $\rho_{\rm m} = 1386$ kg/m^3 and $\rho_{\rm m} = 2174$ kg/m^3, respectively. Figure 4.7 shows the sound velocity of drilling mud as a function of the volume fraction of solids ϕ_s (a) and mud density $\rho_{\rm m}$ (b).

The solid line corresponds to bentonite and the dashed line to barite. The sound velocity of drilling mud is lower than the sound velocity of water for realistic values of the volume fraction. In real applications, barite is generally used in the drilling plans to obtain high values of mud weight. As can be seen, barite-saturated mud has a lower velocity than bentonite-saturated mud. For this reason we expect sound velocities lower than 1500 m/s.

The volume fraction of the cuttings in the outer mud (equation (4.79)) becomes significant for high rates of penetration (ROP $>$ 30 m/h) and large bit diameters, since the presence of cuttings increases the density of the mud and lowers its sound velocity.

4.5.4 Velocities of the guided waves

The velocity of the pipe waves is given by

$$c_P = \left[\rho_{\rm m}\left(\frac{1}{K_{\rm m}} + \frac{1}{\mathcal{K}}\right)\right]^{-1/2}, \tag{4.84}$$

where

$$\mathcal{K} = \frac{Y(r_{\rm O}^2 - r_{\rm I}^2)}{2[(1+\nu)r_{\rm O}^2 + (1-\nu)r_{\rm I}^2]}, \tag{4.85}$$

$r_{\rm O}$ and $r_{\rm I}$ are the outer and inner radii of the pipe, and Y and ν are the Young modulus and Poisson ratio of the pipe (White, 1965, p. 150). This approximation holds at low frequencies, and assumes that the outer medium is a vacuum.

On the other hand, the velocity of the Stoneley mode in the absence of a drill string is obtained from equation (4.84) by taking $\mathcal{K} = \mu$, where μ is the shear modulus of the formation; that is

$$c_S = \left[\rho_{\rm m}\left(\frac{1}{K_{\rm m}} + \frac{1}{\mu}\right)\right]^{-1/2}. \tag{4.86}$$

The presence of casing in the upper sections of the borehole affects the tube wave velocity. If we include the steel casing, the tube-wave velocity is

$$c_T = \left[\rho_\mathrm{m} \left(\frac{1}{K_\mathrm{m}} + \frac{1}{\mu + Y \frac{r'_\mathrm{O} - r'_\mathrm{I}}{2r'_\mathrm{I}}} \right) \right]^{-1/2}, \tag{4.87}$$

where r'_O and r'_I are the outer and inner radii of the casing (White, 1965, p. 155; a slightly different equation is given by Marzetta and Schoenberg, 1985). Equation (4.87) assumes that there is no mud between the casing and the formation, but that the corresponding interface is lubricated. Note that in equation (4.87), we do not consider the presence of a drill string in the borehole.

The accuracy of low-frequency approximations (4.84), (4.86) and (4.87) have been verified by comparison to the exact expressions provided by Rama Rao and Vandiver (1999) and Lea and Kyllingstad (1996). Rama Rao and Vandiver (1999) show that, at low frequency, the difference between the approximated and the exact pipe-wave velocities is 0.07 % and 0.35 % for hard and soft formations, respectively. The differences for the Stoneley mode are 3.5 % and 7 % for the hard and soft formations, respectively. On the other hand, the differences for the pipe and tube waves compared to the values obtained by Lea and Kyllingstad (1996) for a soft formation, are 0.04 % and 6 %, respectively. Unlike the tube waves, the pipe waves are not very sensitive to the properties of the formation.

4.5.5 Sensitivity analysis for velocity of mud guided waves

Let us analyze the sensitivity of the pipe-wave velocity in terms of the bulk modulus and density of the drilling mud, and in terms of the pipe modulus $\mathcal{K}$. We consider the variation of $\mathcal{K}$ for new- and worn-pipe conditions. Differentiating equation (4.84), we obtain

$$\frac{\delta c_p}{c_p} = \frac{1}{2} \left[\left(\frac{c_p}{c_\mathrm{m}} \right)^2 \frac{\delta K_\mathrm{m}}{K_\mathrm{m}} - \frac{\delta \rho_\mathrm{m}}{\rho_\mathrm{m}} + \left(\frac{\rho_\mathrm{m} c_p^2}{\mathcal{K}} \right) \frac{\delta \mathcal{K}}{\mathcal{K}} \right]. \tag{4.88}$$

Let us define the relative-wear coefficient

$$\delta_w = \frac{\delta r_\mathrm{O} - \delta r_\mathrm{I}}{r_\mathrm{O} - r_\mathrm{I}}. \tag{4.89}$$

If $\delta r_\mathrm{O} = -\delta r_\mathrm{I}$, $r_\mathrm{O} + r_\mathrm{I}$ is constant. If $0 \leq \delta_w \leq 0.2$, the denominator in equation (4.85) is nearly constant. Hence, we have

$$\frac{\delta \mathcal{K}}{\mathcal{K}} \cong \delta_w. \tag{4.90}$$

Assuming one saturating solid, we can recast equation (4.88) in terms of the volume fraction of the solid. Using equation (4.90), we obtain

$$\frac{\delta c_p}{c_p} = \frac{1}{2} \left\{ \left[\left(\frac{c_p}{c_\mathrm{m}} \right)^2 \frac{K_s - K_f}{K_s - \phi_s (K_s - K_f)} - \frac{\rho_s - \rho_f}{\rho_f + \phi_s (\rho_s - \rho_f)} \right] \delta \phi_s + \left(\frac{\rho_\mathrm{m} c_p^2}{\mathcal{K}} \right) \delta_w \right\}. \tag{4.91}$$

4.6 Coupled pipe-mud-formation guided waves

Lee (1991) analyzed the axially propagating waves in an infinitely-long, uniform and cylindrically multi-layered waveguide. His results reveal that there are three principal modes of propagation: the pipe mode dominated by stress waves in the drill string and the both inner- and outer-mud waves dominated by acoustic waves (Lee, 1991). After Lee, Lea and Kyllingstad (1996) investigated the coupled modes in the drill-string/borehole system, including the formation and the inner and outer drilling mud. They modeled the drill string and borehole as a cylindrical multi-layered wave guide with three distinct wave modes (Lee, 1991). Each mode has a characteristic velocity and involves physical variations in all the layers. The radial flexibility of the drill string causes communication between the pressure of inner and annulus mud (Figure 4.8). These coupled modes are particularly important in the drill pipes and less important in the drill collars, which are much thicker. The basic coupled-wave equations, in the low-frequency approximation of wavelengths much larger than the radial borehole dimensions, relate the pipe axial displacement u to the inner- and outer-mud (annulus) pressure p, and are

$$A_P \rho_P \frac{\partial^2 u}{\partial t^2} - A_P Y \frac{\partial^2 u}{\partial z^2} = 2\nu A_I \frac{\partial p_I}{\partial z} - 2\nu (A_I + A_P) \frac{\partial p_O}{\partial z}, \tag{4.92}$$

and

$$A_0 \rho \frac{\partial^2 u}{\partial t^2} - A_0 B \frac{\partial^2 u}{\partial z^2} = B \frac{\partial A}{\partial z}, \tag{4.93}$$

for the drill string, and fluid regions (inner and outer mud), respectively. In equation (4.92), A_P, ρ_P are pipe cross-section and density, p_I, p_O, and A_I are inner and outer mud pressure and inner cross-section, respectively. In equation (4.93), ρ is mud density, A_0 and A are the original cross-section of the fluid region and the section including dynamic variation, respectively, and B is a coupling parameter. When mud pressure vanishes, equation (4.92) is the classical wave propagation for extensional waves in a rod. The borehole boundary condition is modeled by relating the radial displacement and dynamic area variation to the formation shear modulus. Lea and Kyllingstad (1996) solved the system of three coupled equations and computed the eigenvectors and velocities of the different wave modes of the inner mud – a little slower than the mud acoustic velocity – the pipe steel – a little slower than extensional velocity – and the annulus mud – significantly slower than mud velocity – for different strings and formations of different shear moduli. They assumed a sound velocity of 1304 m/s for the inner and outer mud. They concluded that wave coupling is important, mainly between the fluid modes (inner and annular pressure waves).

More recently, in order to account for the coupling effects for SWD borehole waves, and assess the impact of formation properties on drill-bit pilot signals, Tinivella, Poletto and Carcione (2002), starting from the results of Lea and Kyllingstad (1996), solved the propagation matrix equations numerically (using a Runge-Kutta algorithm of the fourth order) and calculated the one-dimensional propagation of coupled waves along a wellbore with a typical full-scale drill string and borehole. The properties of the inner mud, the drill string, the annular mud, and the surrounding formation, are described by variables along the wellbore axis. They used a mesh with variable spacing, so that the one-dimensional

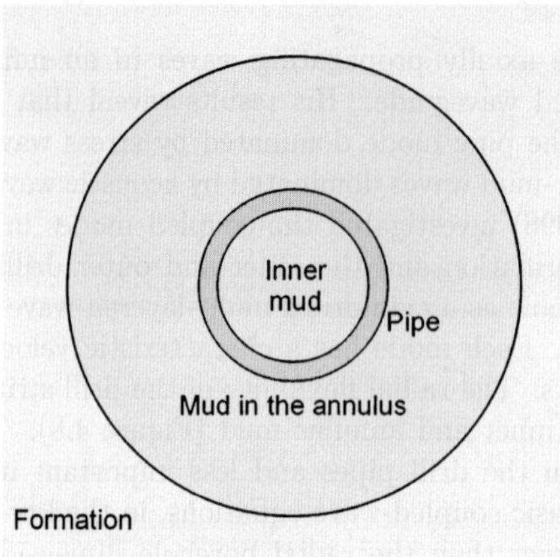

Figure 4.8: Borehole cross-section, with external formation, mud annulus, steel pipe, and inner pipe mud (modified after Lea and Kyllingstad, 1996).

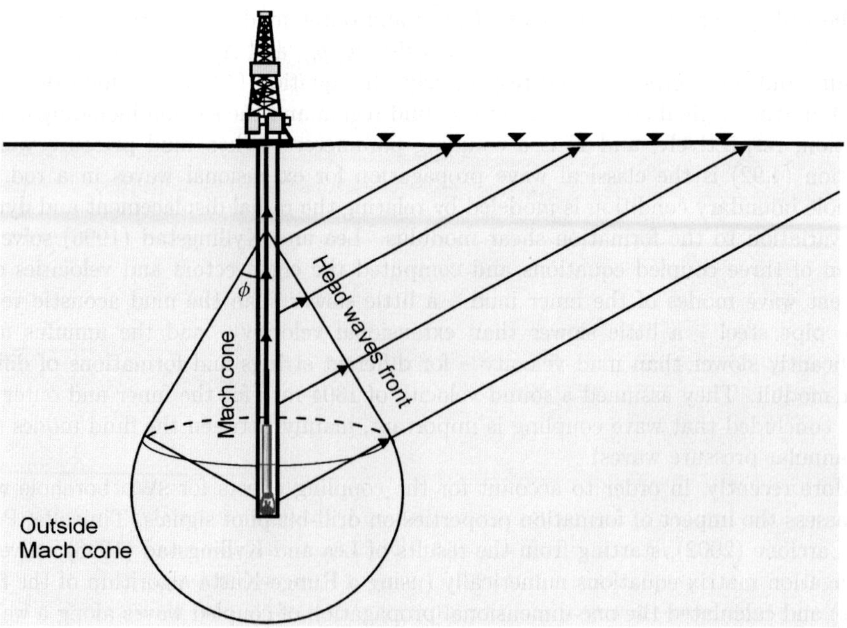

Figure 4.9: Conical drill-bit head waves (modified after Meredith, 1990; after Rector and Hardage, 1992).

coupled system of inner mud, drill string – including tool joints and BHA – annulus mud and boundary formation can be described in detail along the axial coordinate of the well. Reflections of the annulus-mud waves due to changes of the formation shear velocity were observed.

4.6.1 Conical head waves in the formation (borehole radiation)

Because tube waves and borehole waves are radiated in the formation with loss of axial-pilot energy (Sections 4.3, 4.5.4 and 4.6), noise is produced for SWD. In other words, in this section we consider noise produced from "pilot" axial and mud waves. In Sections 4.8 and 4.9, we discuss noise produced by other types of vibrations that can generate in the borehole and at the surface.

Assume an axially-symmetric geometry. Two types of similar conical waves propagating from the borehole can be important for SWD recordings. The first are the shear head waves, also called Mach waves, calculated by Meredith (1990). The second are the compressional head waves produced in the presence of the drill string (Rector, 1992; Rector and Hardage, 1992).

Mach shear waves

When the tube-wave velocity, $c_{\rm T}$, is higher than the shear velocity of the formation, β, conical shear waves (head-shear waves) are produced. This condition is common because the effect of the casing is to increase the tube-wave velocity (Meredith, 1990). The Mach number is defined by Meredith (1990) – making use of the aerodynamics analogy used for an airplane traveling at supersonic velocity – as the ratio

$$\mathrm{M} = \frac{c_{\rm T}}{\beta}. \tag{4.94}$$

The cone angle is calculated as (Figure 4.9)

$$\sin\phi = \frac{1}{\mathrm{M}}. \tag{4.95}$$

If the borehole is filled only by fluid, it is very rare that supersonic conditions ($\mathrm{M} > 1$) will be satisfied for compressional waves, which are faster.

Drill-string head waves

In the presence of the drill-string in the borehole, the velocity of the axially propagating pipe waves $v_{\rm ext}$ can be higher than the compressional velocity of the surrounding formation, α. Hence, because the pipe-mud-formation coupling, conical head waves are generated while the pilot signal propagates through the drill string. These are the conical head waves propagating with cone angle given by the same Mach equation (4.95) calculated for compressional and drill-string waves (Rector and Hardage, 1992)

$$\phi = \arcsin\frac{\alpha}{v_{\rm ext}}, \tag{4.96}$$

if the "supersonic" condition $M_\alpha = v_{ext}/\alpha > 1$ is satisfied. Drill-string head waves result as an additional noise measured in SWD correlated wavefields. These waves have a moveout and frequency content similar to that of the drill-bit direct arrivals and their separation is sometimes difficult for shallow data (Section 5.14.2).

4.7 Summary of drill string waves

Tables 4.3 to 4.6 summarize the main drill-string and borehole waves we have analyzed. Table 4.3 shows the phase velocity of the main rod-wave modes in the low-frequency approximation ($\lambda \gg$ radius). Tables 4.4 and 4.5 summarize the group velocities and the mud-wave velocities. Table 4.6 shows the characteristic impedance (also integrated impedance or mechanical impedance) which is given by the ratio of force over cross-section A to particle velocity.

Table 4.3 – Drill-string waves (homogeneous pipe)

Wave type	Phase velocity
Extensional	$v_{ext} = \sqrt{\frac{Y}{\rho}}$
Torsional	$v_{tor} = \sqrt{\frac{\mu}{\rho}}$
Transverse (rope)	$v_{rope} = \sqrt{\frac{f_T}{A\rho}}$
Flexural (bending)	$v_{p,flex} = \sqrt{R_F v_{ext} \omega}$

Table 4.4 – Drill-string waves

Wave type	Group velocity (low-frequency approximation)
Extensional (non-homogen. pipe)	$v_{g(ext)}(0) = \left(\sum_{i=1}^N d_i\right) \left[\left(\sum_{i=1}^N d_i \rho_i A_i\right) \left(\sum_{i=1}^N \frac{d_i}{Y_i A_i}\right)\right]^{-1/2}$
Torsional (non-homogen. pipe)	$v_{g(tor)}(0) = \left(\sum_{i=1}^N d_i\right) \left[\left(\sum_{i=1}^N d_i \rho_i I_i\right) \left(\sum_{i=1}^N \frac{d_i}{\mu_i I_i}\right)\right]^{-1/2}$
Flexural (homogeneous pipe)	$v_{g,flex} = 2\sqrt{R_F v_{ext} \omega}$

Table 4.5 – Mud wave and coupled modes

Wave type	Phase velocity
Mud acoustic	$\alpha_{\mathrm{m}} = \sqrt{\frac{K_{\mathrm{m}}}{\rho_{\mathrm{m}}}}$
Inner-pipe mud	$\sim c_p = \left[\rho_{\mathrm{m}}\left(\frac{1}{K_{\mathrm{m}}} + \frac{1}{\mathcal{K}}\right)\right]^{-1/2}$
Outer mud (annulus)	$\sim c_T = \left[\rho_{\mathrm{m}}\left(\frac{1}{K_{\mathrm{m}}} + \frac{1}{\mu + Y\frac{r'_{\mathrm{O}} - r'_{\mathrm{I}}}{2r'_{\mathrm{I}}}}\right)\right]^{-1/2}$

Table 4.6 – Characteristic (integrated) impedance (homogeneous pipe/tube)

Wave type	Characteristic impedance
Drill-string extensional	$A\rho v_{\mathrm{ext}} = A\sqrt{Y\rho}$
Drill-string torsional	$I\rho v_{\mathrm{tor}} = I\sqrt{\mu\rho}$
Transversal (rope)	$A\rho v_{\mathrm{rope}} = \sqrt{A\rho f_{\mathrm{T}}}$
Mud acoustic	$A\rho_{\mathrm{m}}\alpha_{\mathrm{m}} = A\sqrt{K_{\mathrm{m}}\rho_{\mathrm{m}}}$

4.8 Surface/rigsite noise wavefields

Many kinds of coherent and random noise components are produced at the rig and yard (Figure 4.10). This noise is due to human and drilling activity. The operation of supply trucks (or vessels) and cranes, and the movement of equipment at the rigsite are common sources of surface noise. They are rather variable in location and timing, and ultimately may introduce prevalent random noise components into the SWD data (Figure 4.10a).

On the other hand, diesel rig-power engines, mud shakers, and pumps perform continuously in fixed positions during drilling, and generate stationary noise wavefields (Figure 4.10b). In particular, the diesel engines are used to generate the three-phase 60 Hz electric power, and run with RPM frequency close to 1200 rev/min (20 Hz) (see Section 2.3.3). Even when the engines are protected by acoustic isolators for human safety, this noise is strongly present in the raw seismic traces measured in the proximity of the well. In some cases, the activation of an additional motor, e.g., as when passing from 3 to 4 active en-

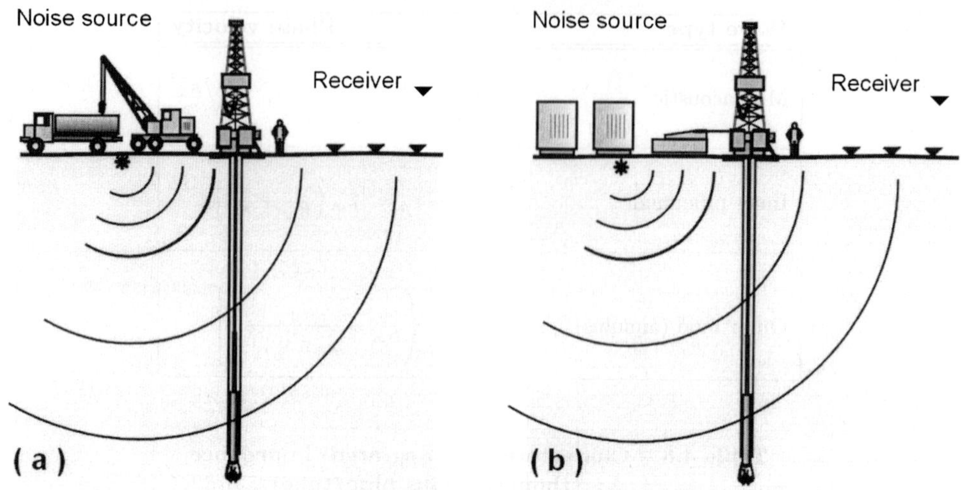

Figure 4.10: Rigsite random and coherent noise sources.

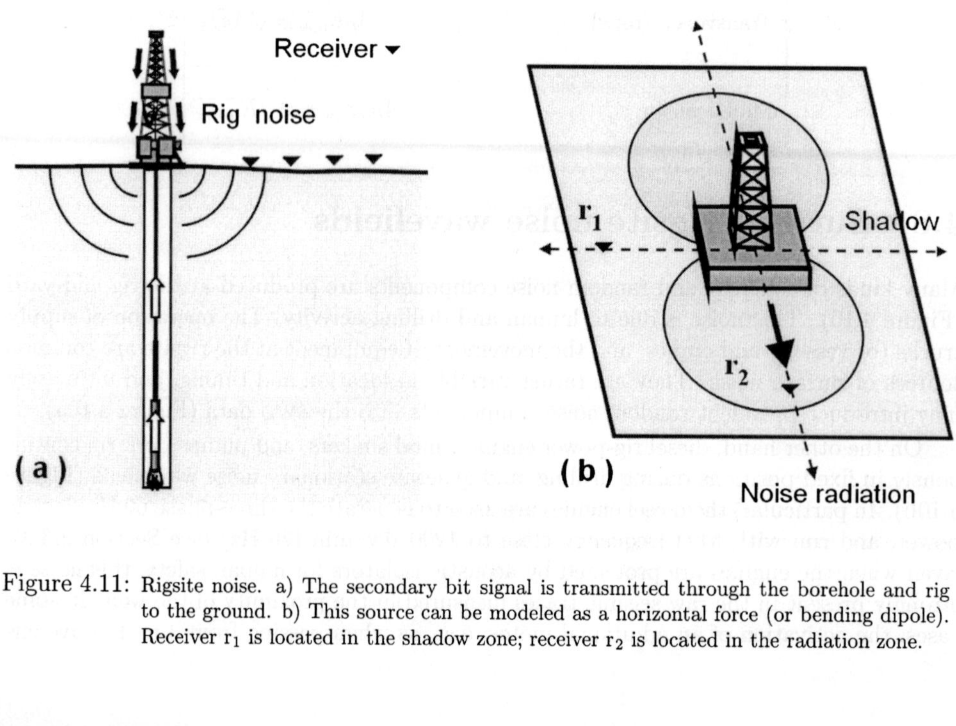

Figure 4.11: Rigsite noise. a) The secondary bit signal is transmitted through the borehole and rig to the ground. b) This source can be modeled as a horizontal force (or bending dipole). Receiver r_1 is located in the shadow zone; receiver r_2 is located in the radiation zone.

gines, changes the signal-to-noise ratio in SWD data. The design of the optimum patterns of the receiver groups of the seismic line is one of the important tasks in the preparation of a SWD survey (Section 5.14).

Measurements of the main rigsite environmental noise sources have been introduced (Angeleri, Persoglia, Poletto and Rocca, 1996) with the purpose of obtaining reference noise traces and separate signal and noise wavefields. Separation based either on orthogonalization or statistical independence is discussed in Chapter 7, dedicated to SWD signal processing.

In general, several rigsite noise vibration sources are more strongly present in seismic geophone data than in pilot data measured on the rig – above the drilling pad – where these noise waves are transmitted only in small amounts (Rector, 1992). Other parts of the surface rigsite noise, like the low-frequency "colored" noise produced by pump strokes, is transmitted to the drill string and modulates the bit vibrations (see Section 3.18). At the same time, the drill-bit signal itself is transmitted to the drill-string structure and to the ground (Rector, 1992). In this way, the rig derrick itself acts as a secondary noise source (Figure 4.11a).

In more recent applications of 3D-SWD (Petronio, Poletto, Carcione, Seriani, Luca and Miranda, 2001), it has been shown that the model of the drilling-noise source of rig vibrations could be better simulated by using a horizontal force model (or bending dipole), rather than a vertical force. This effect is probably related to the structural asymmetry of the vibrating derrick. In this analysis, lobes of maximum amplitude and zones of no radiation of the rig noise (Figure 4.11b) are clearly observed on the seismic lines spread out circularly around the well (Section 8.10.3).

4.9 Drill-string noise and borehole interactions

Noise is also produced by drill-string vibrations and contacts with the borehole. Drill-string and borehole interactions have the double effect of producing secondary-radiation points and coherent noise in SWD data and, at the same time, of creating conditions that adversely affect drill-bit data transmission. The following main effects are important for seismic while drilling, and may adversely affect the rig pilot signals and/or the seismic line data.

Surface rig suspension vibrations

A source of surface noise affecting the drill-string waves is the "jumping" of the traveling block and surface-rig equipment. The shaking and jerks of the hoisting wireline introduce non-linear effects, and broadband ("white") components of surface noise in the measured data. Measurements of vibrations in the rig traveling-block suspension and their effect on drill-string vibrations have been analyzed by Aarrestad and Kyllingstad (1993). These researchers studied the coupled axial-drill-string and transverse-cable vibrations, including the rig low-frequency resonances. They observed that coupling is characterized by non-linear behavior that significantly influences the surface drill-string vibrations (Figure 4.12a). These surface vibrations are introduced in the rig pilot signal, and may be transmitted to the formation as surface noise traveling through the drill string.

Lateral swivel vibrations

Lateral vibrations polarized in the direction perpendicular to the plane containing the bail – the heavy-metal arm that connects the swivel to the hook of the traveling block, which is free to rotate about its pins (Figure 4.12b) – affect the reference pilot signal measured at the top of drill string when a rotary table with the swivel system is used (Rector, 1992). The noise in the signal measured at the top of the drill string is practically not correlated with the geophone signals. However, it produces distortions in the pilot-deconvolution operators (Section 6.5) and it can be subtracted from the disturbed drill-bit signal using either polarization or independence methods (Sections 7.5 and 7.6).

Lateral bending vibrations

Large drill-string lateral bending vibrations are a major risk for drill-string failure, even if only axial and torsional oscillations can be detected at the surface (Chin, 1988). This loss-of-transmission effect is due to the fact that the variation of axial tension along the drill string creates "traps" for the bending vibrations, which typically remain confined below the drill-string neutral point (drill-string loads are described in Section 2.3.10). We can explain this effect using the example of a rope vibrating under tension. The lateral disturbances propagate at wave speed, namely,

$$c_b = \sqrt{\frac{T}{\rho}}, \tag{4.97}$$

where $T = T(z)$ and ρ represent *tension-stress* and *mass density*, respectively (see also equation (2.3), where the *tension force* and the *mass per unit length* of the rope line are used). Hence, bending-wave impedance is

$$z_b = \rho c_b = \sqrt{T\rho}. \tag{4.98}$$

Therefore, independent of the frequency $\omega = \omega(k)$, the impedance vanishes at the load neutral point where $T = 0$. Each harmonic component generated downhole is, therefore, reflected back at the neutral point where the propagation vanishes too. As a consequence, energy is trapped downhole and may grow up with destructive effects. Hence, surface measurements can not easily detect downhole lateral vibrations. However, the large bending motions generate coupled axial disturbances in the drill string that are detectable at the surface (Chin, 1988). Furthermore, large noise waves may be also expected in the formation, as when lateral contacts with the formation occur at intermediate string positions.

Close, Owens and MacPherson (1988) measured downhole BHA vibrations using a MWD recording system in the drill collars. They observed that most of the severe vibrations occurred while reaming. When a reamer was used (Chapter 2), rapid changes in rotational speed, and large lateral accelerations were observed (130 m/s^2 or 13 g), while axial acceleration was of 2.5 g. During rotation at constant velocity, lateral acceleration was of 0.5 g and axial acceleration was of 0.1 g.

These researchers also observed vibrations due to probable occasional contacts of the collars with the casing. In this case, a peak of lateral acceleration of 25 g (245 m/s^2) and

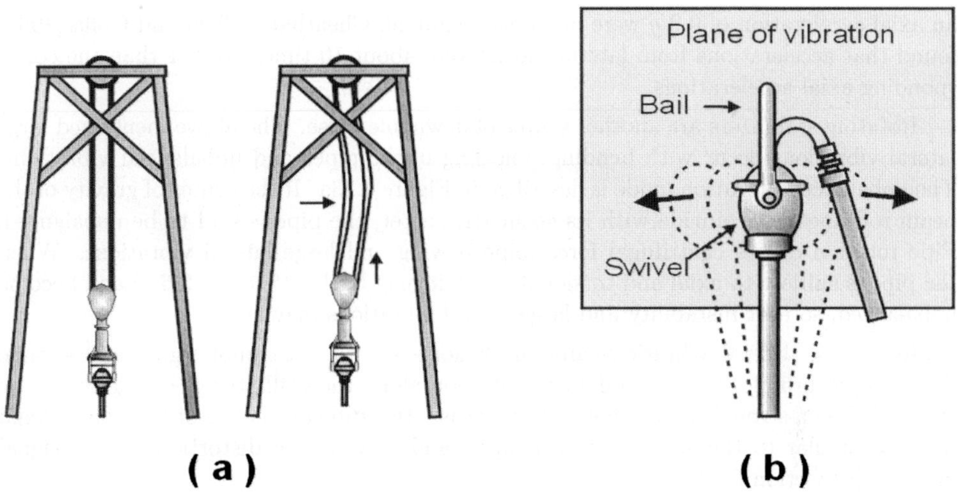

Figure 4.12: Sources of rigsite surface noise. a) Rig suspension shaking. b) Vibrations of the swivel polarized in the direction perpendicular to the plane containing the bail (modified after Rector, 1992).

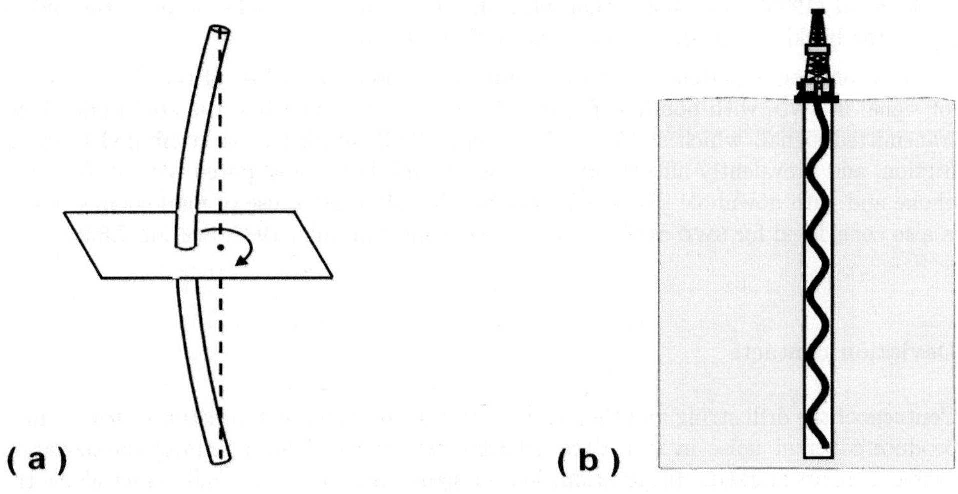

Figure 4.13: a) Unbalanced beam. b) Pipe buckling.

an axial acceleration of 0.2 g were observed. Zannoni, Cheatham, Chen and Golla (1993) found that accelerations from lateral shocks were about 10 times greater than the corresponding axial accelerations.

Rotating stabilizers are another source of downhole noise. The above mentioned large lateral vibrations occur with bending, buckling of the pipes and unbalanced vibrations. The unbalanced-vibration mode is described in Figure 4.13a. If the center of gravity of the beam rod does not coincide with its geometric center, the pipe is said to be unbalanced. Pipe rotation causes centrifugal force, pipe bowing and large lateral vibrations. When the pipe is subject to axial and torsional loads, it may buckle (Section 2.6.6) and become unbalanced, so that instability and large lateral vibrations may start.

In general, lateral vibrations and shock noise affects both pilot signal and seismic data. When produced with random distribution along the drill string, the noise can be considered as a strong random noise. In any case, the autocorrelation of the noise's axial waves is similar to the signal autocorrelation and may induce distortions in the signal processing (Section 6.13).

Pipe buckling and static friction effects

Buckling of the pipes may be produced for axial loading and torque stresses (Section 2.6.6). Buckling (Figure 4.13b) is influenced by the friction forces that are different in the *static* and *dynamic* friction conditions for pipes. Static and dynamic friction forces are produced when drilling in the sliding and kinetic rotary mode, respectively (Mitchell, 1995). Buckling itself changes friction forces, with non-linear effects. He, Hasley and Kyllingstad (1995) have shown that when drilling with a downhole motor in the sliding mode, the buckling threshold lowers and buckling is easier.

It is common experience that the sliding mode produces a loss of the drill-string pilot signal in SWD, with non-linear threshold effects with respect to the amplitude of the transmitted signal, which suddenly disappears. This signal loss is attributed to static friction, and prevalently affects surface pilots. Good data are acquired with surface geophone and with downhole pilots (Chapter 5). The alternative use of mud-coupled waves is also considered for SWD applications (Poletto and Miranda, 1988; Section 7.8.2).

Deviation contacts

Contacts of the drill string and the borehole during deviation and directional drilling may produce coherent noise in SWD data, as noise emission in fixed points of the deviation profile is more marked. In addition, loss of signal energy in the drill string along the tangential contact, where the pipe contact is due to the lateral weight force, is significant (Figure 4.14b). Directional drilling is noisy for SWD. This behavior is similar to that described for buckling and rotating stabilizers. This noise affects both pilot signals measured at the surface and the seismic traces.

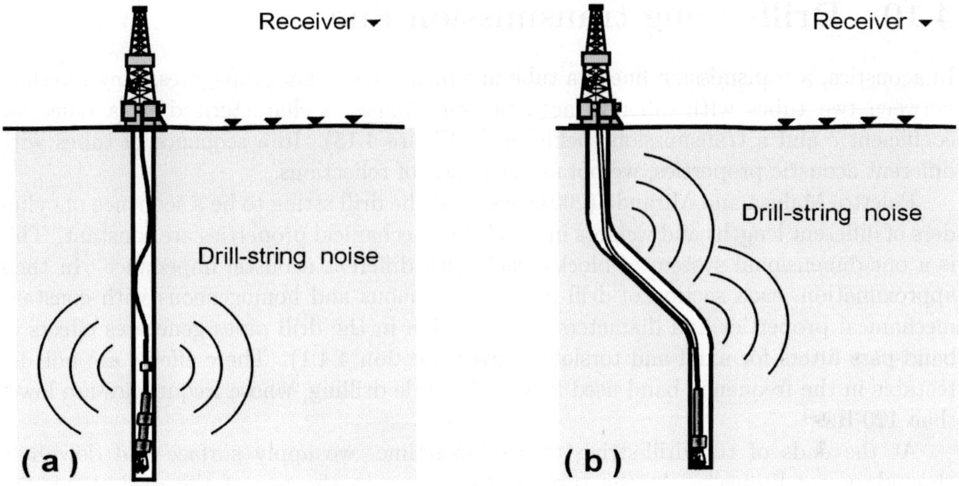

Figure 4.14: Downhole drill-string noise due to a) unbalanced vibrations, reaming, stabilizers and b) directional drilling contacts.

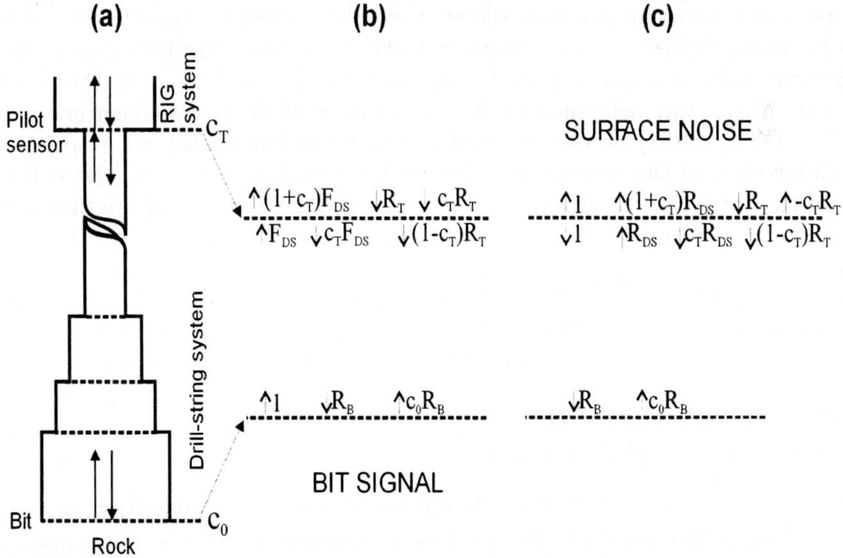

Figure 4.15: Boundary conditions for the drill-string and rig transmission lines (a). The waves in the drill string are coupled to waves in the rock by means of the reflection coefficient c_0 and to the waves in the rig system by means of the reflection coefficient c_T. The upgoing and downgoing waves produced by a unit signal at depth (b) and the upgoing and downgoing waves generated by a unit noise pulse at the surface (pilot location) (c) (after Poletto, Malusa and Miranda, 2001).

4.10 Drill-string transmission line

In acoustics, a transmission-line is a tube in which a vibration propagates. Any interface between two tubes with different acoustic impedances is characterized by a reflection coefficient c and a transmission coefficient t (Figure 4.15). In a sequence of tubes with different acoustic properties, we obtain a cascade of reflections.

Poletto, Malusa and Miranda (2001) assumed the drill string to be a sequence of cylinders of different lengths and weights in which the mechanical properties are constant. This is a one-dimensional system of blocks, each with different acoustic impedance. In their approximation, each section of drill pipe is continuous and homogeneous with constant mechanical properties and diameters. Propagation in the drill pipes generates effects of band-pass filters for axial and torsional waves (Section 4.4.1). These effects are not detectable in the frequency band used in seismic while drilling, whose frequencies are lower than 120 Hz.

At the ends of the drill-string transmission line, we apply surface and downhole "boundary conditions". At the surface, the waves in the top of the drill string are coupled to the waves in rig structure by means of a reflection coefficient c_T. The rig line is described by its reflection coefficients, and assumes the simplified condition that a wave E_W is transmitted from the rig to the ground without reflections ("escaping" wave). Downhole, the drill-string waves at the bit are coupled to the waves in the formation by a reflection coefficient c_0. This coefficient depends on the variation of the acoustic impedance between bit and rock and on the drilling conditions.

A recursive algorithm describes the propagation of acoustic waves in a transmission line (Claerbout, 1976). Let the propagation in the transmission line be represented by the Z-transform of the time series, defined as $a_0 + a_1 Z^{-1} + \ldots + a_j Z^{-j} + \ldots$ for the time-series $\{a_j\}$. Let the one-way delay in each discrete element of the transmission line be equal to $Z^{-1/2}$. Let c_j and t_j be the reflection and transmission coefficients, respectively, of the j-th interface of the transmission line. It holds that $t_j = 1 + c_j$. The well-known propagation matrix, which allows us to compute the downgoing and upgoing waves at each interface, is given by

$$M_K = \frac{Z^{K/2}}{\prod_{i=1}^{K} t_i} \begin{bmatrix} F_K(Z) & Z^{-K} G_K(Z^{-1}) \\ G_K(Z) & Z^{-K} F_K(Z^{-1}) \end{bmatrix}, \tag{4.99}$$

where

$$\begin{cases} F_K(Z) = F_{K-1}(Z) + c_K Z^{-1} G_{K-1}(Z) \\ G_K(Z) = c_K F_{K-1}(Z) + Z^{-1} G_{K-1}(Z) \end{cases}. \tag{4.100}$$

Recursions (4.100), together with the initial conditions $F_1 = 1$ and $G_1 = c_1$, give the functions $F_K(Z)$ and $G_K(Z)$ from reflection coefficients c_k, with $k = 1, \ldots, K$. This algorithm allows us to calculate the propagating signal in any position of the transmission line after the introduction of a codified signal, e.g., an impulse or a narrow-band or "colored" signature. In the general approach, we use two propagation matrices, M_H^{String} and M_L^{Rig} to describe the propagation in the drill-string and rig systems, respectively. The matching conditions between upgoing and downgoing waves at the pilot sensor location and at the bit are shown in Figure 4.15.

Absorption effects are included by using $Z = e^{\imath\omega'}$ with a complex frequency $\omega' = \omega - \imath\pi f\Delta/Q$, where f is the frequency in Hz and Δt is the sampling interval [s]. Attenuation of extensional waves is discussed in Section 4.3.

4.10.1 Reflection coefficients in the drill string

The reflection coefficients for axial and torsional waves in strings formed with tubes of different materials and properties are computed, for instance, by Poletto, Malusa and Miranda (2001). Let z be the axial coordinate. Assume two string blocks with density, cross-section, Young and shear moduli $(\rho_1, A_1, Y_1, \mu_1)$ and $(\rho_2, A_2, Y_2, \mu_2)$, respectively. The matching conditions at the interface are

$$\begin{cases} u_1 = u_2 \\ Y_1 A_1 \frac{\partial u_1}{\partial z} = Y_2 A_2 \frac{\partial u_2}{\partial z} \end{cases} \tag{4.101}$$

for axial waves and

$$\begin{cases} \theta_1 = \theta_2 \\ \mu_1 I_1 \frac{\partial \theta_1}{\partial z} = \mu_2 I_2 \frac{\partial \theta_2}{\partial z} \end{cases} \tag{4.102}$$

for torsional waves. The symbols u_1 and u_2, θ_1 and θ_2 denote the relative axial and angular displacements while I_1 and I_2 denote the moments of inertia – with respect to the tube axis. For a tube with inner and outer radius $r_{\rm I}$ and $r_{\rm O}$, respectively, $I = (\pi/2)(r_{\rm O}^4 - r_{\rm I}^4)$.

For our purposes, the string can be considered as a rod. When the wavelength of the propagating wave is large with respect to the lateral dimensions of the rod, the theoretical velocities of axial (or extensional) and torsional waves are given by equations (4.3) and (4.16), respectively. Using equation (4.60), where $m_{\rm ext}$ [kg/m] is the linear density of the mass in the string, and $M_{\rm tor}$ [kg m] represents the moment of inertia of the string mass for unit length, we obtain

$$v_{\rm ext} = \sqrt{\frac{Y}{\rho}} = \sqrt{\frac{YA}{m_{\rm ext}}}, \tag{4.103}$$

for axial waves and

$$v_{\rm tor} = \sqrt{\frac{\mu}{\rho}} = \sqrt{\frac{\mu I}{M_{\rm tor}}}, \tag{4.104}$$

for torsional waves. Assuming the values ρ = 7840 kg/m^3, Y = 206 GPa and μ = 78.5 GPa, the theoretical velocity in a steel rod is $v_{\rm ext}$ = 5126 m/s for axial and $v_{\rm tor}$ = 3164 m/s for torsional waves.

We insert a plane wave $e^{\imath(\omega t - kz)}$ with phase velocity $v = \omega/k$. Because the frequency ω is the same at the interface, from equations (4.103) and (4.104), we obtain discontinuity of the wavenumber k with the relationship

$$\left(\frac{k_1}{k_2}\right)_{\rm ext} = \sqrt{\frac{Y_2\rho_1}{Y_1\rho_2}} \tag{4.105}$$

for extensional waves and

$$\left(\frac{k_1}{k_2}\right)_{\text{tor}} = \sqrt{\frac{\mu_2\rho_1}{\mu_1\rho_2}} \tag{4.106}$$

for torsional waves. From equations (4.105) and (4.106) it follows that the discontinuity of wavenumbers depends only on the elastic and not on the geometrical properties of the rod. For example, if the two blocks are made of steel of constant elastic properties, the wavenumber ratio is 1 at the interface. However, the reflection coefficients differ from zero because of the difference in the radial rod dimensions. Using these relations and substituting the plane-wave solution in equations (4.101) and (4.102), it can be shown that the reflection coefficient at the interface is

$$c = \frac{a-1}{a+1}, \tag{4.107}$$

where a is given by

$$a_{\text{ext}} = \frac{A_1}{A_2}\sqrt{\frac{Y_1\rho_1}{Y_2\rho_2}}, \qquad a_{\text{tor}} = \frac{I_1}{I_2}\sqrt{\frac{\mu_1\rho_1}{\mu_2\rho_2}} \tag{4.108}$$

for axial and torsional waves, respectively. In practical applications, we use a simplified approach and assume that the Young and shear moduli of the steel and the density are constant in the string. In this case, the cross-section is proportional to the linear-mass density m_{ext}, and in equation (4.107) we use

$$a_{\text{ext}} = \frac{A_1}{A_2} = \frac{m_{\text{ext},1}}{m_{\text{ext},2}}, \tag{4.109}$$

and

$$a_{\text{tor}} = \frac{I_1}{I_2} = \frac{M_{\text{tor},1}}{M_{\text{tor},2}}. \tag{4.110}$$

More directly, the reflection coefficients for displacement can be calculated as

$$c = \frac{\mathcal{Z}_1 - \mathcal{Z}_2}{\mathcal{Z}_1 + \mathcal{Z}_2}, \tag{4.111}$$

where $\mathcal{Z}$ is the string's (rod's) characteristic impedance, given by equations (4.4) and (4.17) for axial and torsional waves, respectively.

In a similar way, we obtain the reflection coefficients for the waves in the rig by evaluating the mass distribution of its main blocks with reflection coefficients close to 1 at the top of the rig and at the kelly-bush interface. A discussion about rig vibration modes is provided by Aarrestad and Kyllingstad (1993).

Table 4.7 – Rigidity and maximum HWDP/DP reflection coefficients

Formation type	α_{cross}	Radial displacement (maximum) [in.]	$\lvert c_{\max}\rvert$ (extensional)	$\lvert c_{\max}\rvert$ (torsional)
Soft	5.5	3.43	0.59	0.80
Hard	3.5	2.25	0.46	0.66

Drill-string rigidity and reflection coefficients

In Chapter 2 we discussed limitations for changes of bending rigidity of drill strings. The recommended constraints for bending properties also limit the internal drill-string reflectivity. In fact, the recommended values of α for equation (2.31) correspond to limits for the maximum reflection coefficients ($c_{\max}$) for extensional and torsional waves. In Table 4.7 the coefficients $c_{\max}$ are calculated for cross-sections between 5 in. heavy weight drill pipes (HWDP) and drill collars (DC), assuming that the internal diameter is constant, $D_{\text{I}} = 3$ in., by assuming constant steel properties. It can be also shown that the cross-section between 5 in. drill pipes and 5 in. heavy-weight drill pipes satisfy the condition given by equation (2.31) because the variation is limited to only the internal diameter.

4.11 Bit/rock reflection coefficient

The drill-string axial vibrations, produced when drilling a well, are transmitted to the surrounding rock at the drill-bit interface where a reflection coefficient is calculated. The reflection coefficient at the bit/rock interface can be used while drilling to determine the elastic properties of the drilled rocks. Lutz, Raynaud, Gastalder, Quichaud, Raynald and Muckerloy (1972) first developed a theoretical interpretation of the vibrations measured at the top of the drill string to obtain information on the rock properties (SNAP log). More recently, other authors analyzed the drill-string transmission line and the drill-string vibrations in the drill-bit seismograms for SWD purposes. For instance, Booer and Meehan (1993) used a model-based approach to interpret rig measurements and determine the bit/rock reflection coefficient. This approach is by inversion of the drill-string reflectivity. Below, we consider two different approaches, both assuming an ideal flat bit without drilling effects.

- ◇ The first is to calculate a (real) bit/rock reflection coefficient in the plane-wave approximation. This method was proposed by Poletto, Malusa and Miranda (2000).
- ◇ The second is to include near-field effects and to calculate a complex bit/rock reflection coefficient (Poletto, 2002a, and see the calculations of this book).

In both these approaches, we use a simplified model to calculate and interpret the bit/rock reflection coefficient. We consider the case of an idealized rod and borehole that are coupled at the drill-bit/rock interface (Figure 4.16). In this way, we calculate the reflection-coefficient for drill-string extensional waves.

4.11.1 Bit/rock reflection coefficient (plane-wave approximation)

In the plane-wave approximation, these waves are incident at the bit and transmitted through the bit/rock interface to the rock where they are radiated in the form of plane compressional waves. Lateral bending effects and conversions to shear and flexural modes are not considered. The calculated reflection coefficient depends on the cross-sections of the rod and borehole, and on the elastic properties of the rock. A relation between the rock impedance and the reflection coefficient is found.

The constitutive equations (bit/rock contact)

Let us consider an unbounded medium with a semi-infinite borehole and a rod in the borehole. We assume the low-frequency approximation, in which the minimum investigated wavelength is much larger than the borehole radial dimensions. In this approximation, the choice of rectangular coordinates does not detract from the generality of the results. The vertical axial dimension is z, while x and y are the horizontal co-ordinates, orthogonal to the well-axis. We assume also that the well is empty of fluid, without drilling mud, and the rod is laterally-free, coupled to the formation only at the bit interface. For the calculation of the coefficient, the bit is assumed as not-working, and its ideal surface flat. Its section, A_b, equals the cross-section of the borehole. The rod cross-section is A_1 and equals the cross-section of the drill-collars in the BHA.

Let σ_x, σ_y and σ_z be the normal-stress components, and u, v and w be the displacements in the x, y, z directions, respectively, which are assumed to be the principal axes of the stress. The stress-strain relations (Jaeger, 1969) are

$$\begin{aligned} Y\varepsilon_x &= \sigma_x - \nu(\sigma_y + \sigma_z), \\ Y\varepsilon_y &= \sigma_y - \nu(\sigma_z + \sigma_x), \\ Y\varepsilon_z &= \sigma_z - \nu(\sigma_x + \sigma_y), \end{aligned} \tag{4.112}$$

where the strains ε_x, ε_y, ε_z are defined as

$$\varepsilon_x = \frac{\partial u}{\partial x}, \quad \varepsilon_y = \frac{\partial v}{\partial y}, \quad \varepsilon_z = \frac{\partial w}{\partial z}. \tag{4.113}$$

Let us consider an extensional stress wave in the rod, incident at the bit/rock interface and partially transmitted to the formation. The stress conditions are different for the rod and in the formation. Let w be the axial displacement in the formation and w_1 the vertical displacement in the rod, ρ and ρ_1 the density, and Y and Y_1 the Young modulus of the formation and of the drill collars, respectively, and ν the Poisson ratio of the formation. For the rod, laterally-free in the borehole, $\sigma_{x1} = \sigma_{y1} = 0$ and

$$Y_1\varepsilon_{z1} = \sigma_z. \tag{4.114}$$

In the formation the condition for the horizontal stress is different, since the rock coupled to the bit is surrounded by other rock. Using the hypothesis of isotropy and homogeneity, in general, stress is $\sigma_x = \sigma_y \neq 0$ in the formation. We consider a compressional plane wave propagating in the vertical direction, and assume that in the formation close to the bit-interface the presence of the surrounding rock produces a reaction stress which compensates for the horizontal relative displacements, such that $\varepsilon_x = \varepsilon_y = 0$ at the bit depth. Hence, in the formation we have

$$\sigma_x = \sigma_y = \frac{\nu}{1-\nu}\sigma_z, \tag{4.115}$$

where σ_z is the vertical stress produced in the formation at the bit interface, and we obtain

$$Y\varepsilon_z = \frac{(1+\nu)(1-2\nu)}{1-\nu}\sigma_z. \tag{4.116}$$

Real bit/rock reflection coefficient (plane waves)

We assume that the bit is coupled to the formation at the bit interface ($z = 0$). The continuity condition for the displacement at the drill-string and formation contact gives $w_{z1} = w_z$. The continuity condition for the vertical force at the interface gives

$$A_1 Y_1 \varepsilon_{z1} = A_b Y \varepsilon_z \frac{1-\nu}{(1+\nu)(1-2\nu)}. \tag{4.117}$$

It can be shown that, using the compressional plane-wave solutions,

$$w_1 = e^{\imath(\omega t - k_1 z)} + c_0 e^{\imath(\omega t + k_1 z)}, \tag{4.118}$$

$$w = t_0 e^{\imath(\omega t - k_\alpha z)}, \tag{4.119}$$

in the rod and formation, respectively. Thus, we obtain the (real) bit/rock reflection coefficient

$$c_0 = \frac{a-1}{a+1}, \tag{4.120}$$

where a is given by

$$a = \frac{A_1 Y_1 k_1}{A_b Y k_\alpha} \frac{(1+\nu)(1-2\nu)}{1-\nu}. \tag{4.121}$$

From the continuity of the phase (and angular frequency ω) at $z = 0$, the ratio of the wavenumbers in the rod and formation is

$$\frac{k_1}{k_\alpha} = \frac{\alpha}{c_1}, \tag{4.122}$$

where c_1 and α are the velocity of propagation of the extensional wave in the rod, and of the compressional wave in the formation, respectively. The velocity of extensional waves in the rod in the low-frequency approximation (Kolsky, 1953) is $c_1 = (Y_1/\rho_1)^{1/2}$. The compressional velocity in the formation can be expressed in terms of its elastic properties as

$$\rho\alpha = \sqrt{\frac{1-\nu}{(1+\nu)(1-2\nu)}}\sqrt{Y\rho}. \tag{4.123}$$

We obtain

$$\frac{k_1}{k_\alpha} = \sqrt{\frac{(1-\nu)}{(1+\nu)(1-2\nu)} \frac{Y}{Y_1} \frac{\rho_1}{\rho}}, \tag{4.124}$$

and

$$a = \frac{A_1}{A_b} \frac{\sqrt{Y_1 \rho_1}}{\alpha \rho}, \tag{4.125}$$

which, using equation (4.120) and $Y = \rho c_1^2$, gives the reflection coefficient

$$c_0 = \frac{A_1 \rho_1 c_1 - A_b \rho \alpha}{A_1 \rho_1 c_1 + A_b \rho \alpha} = \frac{\mathcal{Z}_1 - \mathcal{Z}_{Ab}}{\mathcal{Z}_1 - \mathcal{Z}_{Ab}}. \tag{4.126}$$

Bit/rock coefficient and formation impedance

The result of equation (4.126) also estimates the impedance of the drilled formation as

$$\rho \alpha = \frac{A_1}{A_b} \rho_1 c_1 \frac{1 - c_0}{1 + c_0}, \tag{4.127}$$

(where we typically use $Y_1 = 206$ GPa and $\rho_1 = 7840$ kg/m^3 for the drill-string Young modulus and density, respectively). Previous results gives only a plane-wave approximation of the bit/rock reflection coefficient, which, as discussed before is a complex coefficient depending on frequency. Examples of formation evaluation using equation (4.127) are given in Section 8.5.2.

4.11.2 Complex bit/rock reflection coefficient (near-field approximation)

To account for the near-field effects discussed in Section 3.9.4, we have to substitute the complex impedance of equation (3.102) to the real impedance $A_b \rho \alpha$ calculated above. This is an important point to understand. The drill string in the low-frequency approximation (say with wavelengths greater than five times the pipe diameter) is a one-dimensional system (rod) and, when isolated, does not have any "near-field" effect on the wave propagation. However, we have to calculate the response of the drill-string/formation system to the near-field effects occuring in the formation. We start from the impedance results of Section 3.9.4.

4.11.3 Drill-string waves and near-field effects

In Section 3.9.1 we have calculated the near-field phase of sine and cosine waves in the formation in contact to the bit – assumed flat and in perfect contact to the rock – with a harmonic force at the bit. If the drill string – uniform and of infinite height above the bit – were isolated, the harmonic force would generate simple harmonic waves in the drill-string itself. However, the coupling to the formation subject to the near-field vibrations introduces both in-phase and out-of-phase components in the string.

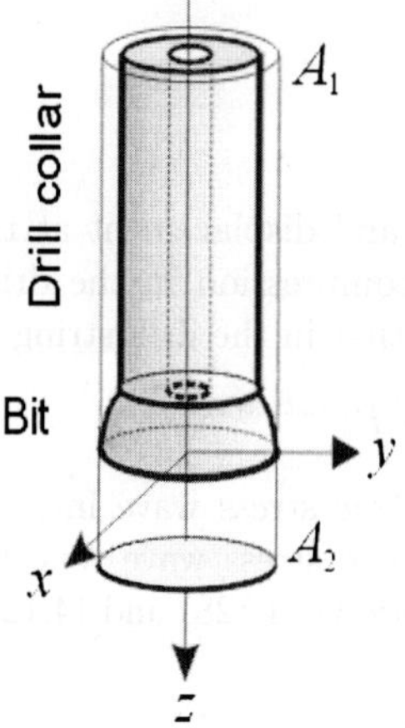

Figure 4.16: Simplified bit model.

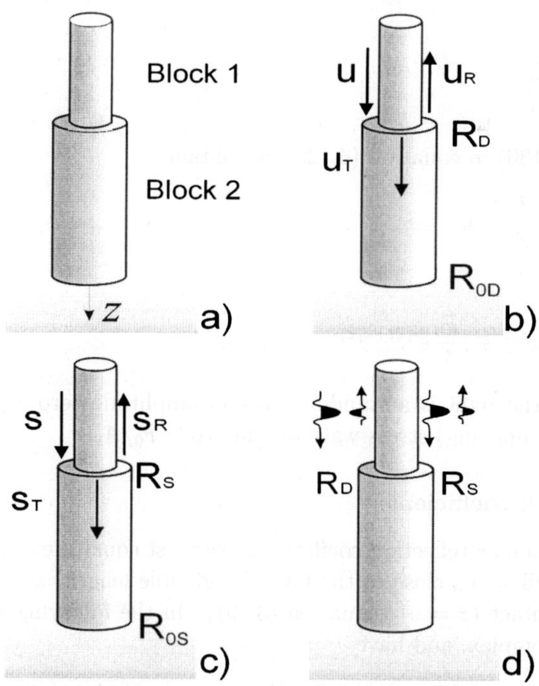

Figure 4.17: We assume two string blocks of different geometrical and mechanical characteristics (a). A displacement wave u (b) and a strain wave (c) are incident on the interface in block 1 where are a reflected back in block 1 and transmitted in block 2. From the continuity conditions at the interface, we obtain opposite reflection coefficients for strain and displacement waves (d).

Let us assume that the harmonic axial stress is uniform over the pipe cross-section A_1 and that the stress is given at the near-bit/rock interface depth ($z = 0$) as the real part of

$$\sigma_1 = \imath \frac{F_0}{A_1} e^{\imath\omega t}. \tag{4.128}$$

We assume continuity of stress and displacement at the bit/rock contact $z = 0$. From equation (4.128), we obtain a compression in the drill string if a positive bit force is assumed. Moreover, we assume that in the drill string it is

$$\sigma_1 = A\sigma_D + R\sigma_U = Ae^{\imath(\omega t - k_1 z)} + Re^{\imath(\omega t + k_1 z)}, \tag{4.129}$$

where σ_D is a downgoing harmonic stress wave incident at the bit/rock interface from above, σ_U is the upgoing harmonic stress wave reflected at the interface, A and R are complex coefficients. From equations (4.128) and (4.129) simplified for $e^{\imath\omega t}$, we have

$$A + R = \imath \frac{F_0}{A_1}, \tag{4.130}$$

and the relations for the real and imaginary parts

$$\Re[R] = -\Re[A], \tag{4.131}$$

$$\Im[R] = -\Im[A] + \frac{F_0}{A_1}. \tag{4.132}$$

Using equation (4.130) in equation (4.129), we obtain

$$\sigma_1 = A\sigma_D - \left(A - \imath \frac{F_0}{A_1}\right)\sigma_U, \tag{4.133}$$

from which we have

$$\sigma_1 = -2\imath A \sin(k_1 z) e^{\imath\omega t} + \imath \frac{F_0}{A_1} e^{\imath(\omega t + k_1 z)}. \tag{4.134}$$

The first term on the right is a standing wave of amplitude zero at the bit; the second one is a regressive (upgoing) stress wave of amplitude F_0/A_1.

Complex bit/rock coefficient

To calculate the complex reflection coefficients, we must equal the complex axial displacements u_1 in the drill-string close to the bit (of negligible length) and $\tilde{u}_0$ in the formation at the bit/rock contact ($z = 0$) (equation (3.79)). In the following we omit the symbol over $\tilde{u}$ to denote complex, and have

$$u_1 = u_0 = \lim_{h \to 0} \hat{u}. \tag{4.135}$$

From

$$\sigma_1 = Y_1 \frac{\partial u_1}{\partial z} \tag{4.136}$$

and from $Y_1 k_1 = \omega \rho_1 c_1$, we have

$$u_1 = A \frac{\sigma_D}{-\imath\omega\rho_1 c_1} + R \frac{\sigma_U}{\imath\omega\rho_1 c_1} \tag{4.137}$$

and from equation (4.130)

$$u_1 = \imath \frac{A}{\omega\rho_1 c_1} (\sigma_D + \sigma_U) + \frac{F_0 \sigma_U}{\omega A_1 \rho_1 c_1}. \tag{4.138}$$

From equation (3.96) and from $v_0 = \imath\omega u_0$, we have in the formation

$$u_0 = -\frac{F_0 r_b}{A_b \rho \alpha^2} (\imath - \cot\varphi)\, e^{\imath\omega t}, \tag{4.139}$$

with

$$\cot\varphi = -\frac{k_\alpha r_b (1 + 6\sqrt{3})}{12}. \tag{4.140}$$

From previous equations and equation (4.130), we obtain

$$A = \frac{F_0}{2\imath A_1} \left[-\frac{k_\alpha r_b}{A_b \rho\alpha} (\imath - \cot\varphi)\, A_1 \rho_1 c_1 - 1 \right], \tag{4.141}$$

and

$$R = \frac{F_0}{2\imath A_1} \left[\frac{k_\alpha r_b}{A_b \rho\alpha} (\imath - \cot\varphi)\, A_1 \rho_1 c_1 - 1 \right]. \tag{4.142}$$

The complex bit/rock reflection coefficient (for stress) is obtained as

$$c_0 = \frac{R}{A}, \tag{4.143}$$

which, after some calculations, becomes

$$c_0 = -\frac{A_1\rho_1 c_1 + A_\alpha \rho\alpha (k_\alpha r_b)^{-1} \left(\cos\varphi \sin\varphi + \imath \sin^2\varphi\right)}{A_1\rho_1 c_1 - A_\alpha \rho\alpha (k_\alpha r_b)^{-1} \left(\cos\varphi \sin\varphi + \imath \sin^2\varphi\right)}. \tag{4.144}$$

Note that we obtain the same result if we use the drill-string impedance $\mathcal{Z}_1 = A_1 \rho_1 c_1$ and the integrated complex impedance $\mathcal{Z}_b$ of equation (3.100) to calculate the coefficient for stress

$$c_0 = -\frac{\mathcal{Z}_1 - \mathcal{Z}_b}{\mathcal{Z}_1 + \mathcal{Z}_b}. \tag{4.145}$$

The reflection coefficient for displacement is opposite to that of stress because stress and displacement are dual fields (see Section 4.12). The coefficient c_0 can be used to calculate the partition (frequency dependent) of stationary and regressive waves in the drill string. Moreover, it can be related to the drilling conditions by using $Z_1 = A_1 \rho_1 c_1$ and equation (4.140). We can interpret the results in the following way. In the drill string, there is a traveling wave, which is usually assumed in SWD. However, there is also a standing wave sensitive to near-field vibrations of the rock. The partition is related to the phase angle φ. In general, $\varphi = \varphi(\alpha, \beta, r_b, \omega)$. This explains why, for instance, the detection of frequency resonances in relation to drilled rock can be used for while-drilling acoustic logging purposes (Lutz, Raynaud, Gastalder, Quichaud, Raynald and Muckerloy, 1972).

4.12 Dual fields in the drill string

A wavefield can be separated into different partial wavefields consisting, e.g, of up- or downgoing waves. A technique for separation is to use dual sensors. The basic concept of dual-sensor technology (Section 5.6) is to observe the waves produced from a given source by measuring for the same physical process, i.e., at the same time and in the same location, at least two physical quantities having opposite reflection coefficients. These are the dual fields. This concept is explained, for instance, by Loewenthal and Robinson (2000). In reflection seismic, dual fields are obtained by measuring particle velocity (using geophones) and pressure (using hydrophones). The fact that these quantities have opposite reflection coefficients is a concept well known in acoustics: it is easily understood by considering that particle velocity diminishes while pressure increases when passing from a soft to a hard medium. These seismic dual fields are combined and subtracted to remove undesired multiple reflections, for instance the ghost reflections in marine OBC data, from primary events. Loewenthal and Robinson (2000) developed an analysis of dual-seismic signals and proposed a method that involves the separation of upgoing and downgoing waves, followed by application deconvolution of the upgoing field by using the downgoing operator, and dynamic deconvolution of the deconvolved upgoing data.

The basic concept can be explained as follows. Consider a one-dimensional system (like a drill string), where we have only downgoing, i.e., propagating in the $+z$ axial direction, and upgoing waves, i.e., propagating in the $-z$ axial direction. Let the displacement be represented as

$$u(z,t) = U_{\rm dw} f\left(t - \frac{z}{V}\right) + U_{\rm up} g\left(t + \frac{z}{V}\right), \tag{4.146}$$

where $U_{\rm dw}$ and $U_{\rm up}$ are constants and V is the propagation speed. Using particle velocity $v(z,t) = \partial u/\partial t$ and strain $s(z,t) = \partial u/\partial z$, we can separate the downgoing and upgoing waves as

$$f'\left(t - \frac{z}{V}\right) = \frac{v(z,t) - Vs(z,t)}{2U_{\rm dw}} \tag{4.147}$$

and

$$g'\left(t + \frac{z}{V}\right) = \frac{v(z,t) + Vs(z,t)}{2U_{\rm up}}. \tag{4.148}$$

The dual-field method is proposed by Poletto (2002) to separate the drill-string reflections in the pilot signals used for seismic-while-drilling (SWD) correlation. We can use, for instance, acceleration – but also velocity or displacement – and strain – but also stress or force – drill-string waves. These quantities have different waveforms and have to be modified (waveshaping) in order to be successfully combined. Assume a progressive (downgoing) displacement plane wave propagating in a uniform drill string as

$$u = e^{\imath(\omega t - kz)}, \tag{4.149}$$

where ω is the angular frequency, and $k = \omega/V$ is the wavenumber. Acceleration and strain are related to displacement by

$$a = \frac{\partial^2 u}{\partial t^2} = -\omega^2 u, \tag{4.150}$$

$$s = \frac{\partial u}{\partial z} = -\imath k u, \tag{4.151}$$

respectively. For a regressive (upgoing) wave

$$u = e^{\imath(\omega t + kz)}, \tag{4.152}$$

we have

$$s = \frac{\partial u}{\partial z} = \imath k u. \tag{4.153}$$

The operator $F_{\mathrm{s,a}}(\omega)$ converting strain into acceleration is

$$\mathcal{F}_{s,a}(\omega) = \mp \imath \frac{\omega^2}{k(\omega)} = \frac{a}{s}, \tag{4.154}$$

and

$$a = \mathcal{F}_{s,a}(\omega) s. \tag{4.155}$$

In the absence of dispersion, i.e., when $\omega/k = V$ = constant, we have

$$\mathcal{F}_{s,a}(\omega) \propto \mp \imath \omega, \tag{4.156}$$

so that the filter converting the strain waveform into acceleration waveform becomes a simple time derivative, of negative or positive sign. If we consider force instead of strain, according to equation (3.126), we see that the filter relating force to acceleration is a time derivative scaled by the drill-string characteristic impedance

$$\mathcal{F}_{f,a} = \mp \frac{\partial_t}{\mathcal{Z}_1}, \tag{4.157}$$

where the negative and positive signs hold for progressive and regressive waves, respectively.

4.12.1 Dual (displacement and strain) reflection coefficients

In Section 4.10.1 we compute the reflection coefficients for axial waves in strings formed with tubes of different materials and properties. Assume two string blocks, denoted as 1 and 2 (Figure 4.17a), with density, Young and shear moduli, cross-section, and moment of inertia per unit length $(\rho_1, Y_1, \mu_1, A_1, I_1)$ and $(\rho_2, Y_2, \mu_2, A_2, I_2)$, respectively. The continuity conditions at the interface, assumed at $z = 0$, are discussed in Section 4.10.1. Similar conditions and results can be obtained for torsional waves by substituting the relative angular displacement θ, the shear strain $\partial\theta/\partial z$, the shear modulus μ, and the moment of inertia I in place of the axial displacement u, the axial strain $\partial u/\partial z$, the Young modulus Y and the cross-section A, respectively.

Reflection coefficients of displacement waves

Let us consider a unit plane wave of axial displacement incident on the interface in block 1 (Figure 4.17b). It is

$$u = e^{\imath(\omega t - k_1 z)}. \tag{4.158}$$

The displacement wave reflected back in block 1 is

$$u_{\mathrm{R}} = R_{\mathrm{d}} e^{\imath(\omega t + k_1 z)}. \tag{4.159}$$

The displacement wave transmitted forward in block 2 is

$$u_{\mathrm{T}} = T_{\mathrm{d}} e^{\imath(\omega t - k_2 z)}. \tag{4.160}$$

Here, R_{d} and T_{d} are the reflection and transmission coefficients, respectively, of the displacement at the interface between block 1 and block 2. Using continuity of displacement, we obtain immediately the well-known relation between the transmission and reflection coefficients of displacement, namely,

$$T_{\mathrm{d}} = 1 + R_{\mathrm{d}}. \tag{4.161}$$

Using the definition of strain given by equation (4.151) and equations from (4.158) to (4.160), the equation of continuity of the force can be expressed as

$$-\imath Y_1 A_1 k_1 (1 - R_{\mathrm{d}}) = -\imath Y_2 A_2 k_2 T_{\mathrm{d}}. \tag{4.162}$$

Observe that the coefficients of the products $Y_i A_i$ in equation (4.162) are the amplitudes of the strains corresponding to the propagation of a unit displacement wave, incident and reflected in block 1 and transmitted in block 2. To solve for R_{d} in equations (4.161) and (4.162), we can use the discontinuity of wavenumber given by equation (4.105), and have immediately

$$R_{\mathrm{d}} = \frac{\mathcal{Z}_1 - \mathcal{Z}_2}{\mathcal{Z}_1 + \mathcal{Z}_2}, \tag{4.163}$$

and

$$T_{\mathrm{d}} = \frac{2\mathcal{Z}_1}{\mathcal{Z}_1 + \mathcal{Z}_2}, \tag{4.164}$$

where $\mathcal{Z} = A\sqrt{Y\rho}$ ($\mathcal{Z} = I\sqrt{\mu\rho}$) is the characteristic impedance for axial (torsional) waves in the drill string (Sections 4.2.1 and 4.2.2).

Reflection coefficients of strain waves

Consider now a strain plane wave incident on the interface in block 1 (Figure 4.17c). This is

$$s = S_0 e^{\imath(\omega t - k_1 z)}, \tag{4.165}$$

where S_0 is a constant scaling factor. The strain wave reflected back in block 1 is

$$s_{\mathrm{R}} = R_{\mathrm{s}} S_0 e^{\imath(\omega t + k_1 z)}. \tag{4.166}$$

The strain wave transmitted forward in block 2 is

$$s_{\mathrm{T}} = T_{\mathrm{s}} S_0 e^{\imath(\omega t - k_2 z)}. \tag{4.167}$$

R_{s} and T_{s} are the reflection and transmission coefficients, respectively, of the strain at the interface between block 1 and block 2. Assuming continuity of the force at the interface, we have

$$Y_1 A_1 (s + s_{\mathrm{R}}) = Y_2 A_2 s_{\mathrm{T}}. \tag{4.168}$$

We can choose the scaling factor $S_0 = -\imath k_1$, in order to match the amplitude of the strain corresponding to the unit displacement wave. Hence, at the interface we obtain from equations (4.165) to (4.168),

$$-\imath Y_1 A_1 k_1 (1 + R_{\mathrm{s}}) = -\imath Y_2 A_2 k_1 T_{\mathrm{s}}. \tag{4.169}$$

Equations (4.162) and (4.169) both represent the continuity of the force at the interface. Comparing the left-side terms of equations (4.169) and (4.162) gives (see Figure 4.17d)

$$R_{\mathrm{s}} = -R_{\mathrm{d}}, \tag{4.170}$$

while comparing the right-side terms gives

$$T_{\mathrm{s}} = \frac{k_2}{k_1} T_{\mathrm{d}}. \tag{4.171}$$

In the particular case of constant elastic properties, we have $k_2/k_1 = 1$ and $T_{\mathrm{s}} = T_{\mathrm{d}}$. In other words, in pipes made only of steel, the strain reflection coefficient is the negative of the displacement reflection coefficient and the strain transmission coefficient equals the displacement transmission coefficient. This is not a contradiction. In fact, we have discontinuity of strain at the interface, and the strain transmission and reflection coefficients (see equation (4.169)) are related by

$$T_{\mathrm{s}} = \frac{Y_1 A_1}{Y_2 A_2} (1 + R_{\mathrm{s}}). \tag{4.172}$$

This means that if we acquire dual waves by using acceleration and strain sensors located at different positions along a string made only of steel, we need to use only one scaling factor to combine them, assuming that the sensors at the different locations have the same instrumental response.

Transmission and reflection coefficients for displacement and strain torsional waves can be calculated in a similar manner. Acquisition and processing of dual drill-string fields are discussed in Sections 5.6 and 6.15, respectively.

4.12.2 Dual fields in the drill-string transmission line

Consider the recursive algorithms of Section 4.10. We can use the equations (4.99) and the recursions of equations (4.100) to calculate the acceleration waves in the drill-string, rig and layered formation systems. The acceleration drill-string reflection and transmission coefficients are given by equation (4.163) and (4.164), respectively. When calculating the strain waves, we use the opposite reflection coefficients of equation (4.170), and, in the product of equation (4.99), we can use the transmission coefficients

$$T_{\mathrm{s},j} = \frac{k_{j+1}}{k_j} T_{\mathrm{d},j}. \tag{4.173}$$

Equation (4.173) gives

$$\prod_{j=1}^{J} T_{\mathrm{s},j} = \frac{k_{J+1}}{k_1} \prod_{j=1}^{J} T_{\mathrm{d},j}. \tag{4.174}$$

A property of the propagation-matrix of equation (4.171) is that the polynomials $F_J(Z)$ and $G_J(Z)$ contain an even and an odd number of reflection coefficients (Claerbout, 1976), respectively. For this reason, $F_J(Z)$ is unchanged while $G_J(Z)$ changes its sign when passing from acceleration to strain waves.

Chapter 5

Acquisition of SWD data

5.1 Introduction

The acquisition of seismic-while-drilling data involves the technical aspects and operating procedures both of mudlogging and of seismic surveys.

The SWD data are collected all through the well-drilling activity, as is required when measuring the mudlogging drilling parameters. This means that the drill-bit VSP data acquisition must be planned in relation to the well-drilling conditions. It is important to avoid, or minimize, any loss of data; to achieve that, the signals are sampled at prefixed depth levels at regular intervals, during surveys performed day and night. A careful plan (feasibility study) is necessary to prepare acquisition. The information gathered in this phase is useful for while-drilling processing. The well rigsite information (drilling-yard and drilling-plant layouts), the drilling plan and the timing of the well need to be collected. Acquisition is scheduled day by day, based on the daily drilling plan and operations. However, often the drilling plan is subjected to unexpected variations because of the drilling decisions. In this case, the acquisition process has to be adapted quickly to the new conditions, which are continuously monitored for quality-control purposes. At the wellsite, the SWD system is connected to the drilling-operation control system (usually the mudlogging), from which it receives the drilling parameters necessary to drive the automated acquisition.

The technical aspects of seismic surveys are also very important during the acquisition of SWD data. The recording procedures for SWD, with the source at depth and geophones at the surface, are for some aspects more similar to those of surface seismic, with geophones at the surface, than conventional/normal VSP, with the source at the surface and geophones in the well. In fact, SWD may use many (say, several tens or hundreds) seismic channels around the well. When many seismic recording channels are used around the well, continuous maintenance and testing activity by field operators is required to prevent any possible environmental or technical problem producing failure of the acquisition/communication system (telemetric line). Similar to conventional surface-seismic surveys, it is necessary to collect all the logistic, topographic and geologic information of the well area and target – including, when onshore, a preliminary surface-refraction survey for determining static corrections.

It has to be emphasized that SWD-data acquisition is subjected to variable drilling conditions and requires reliable, well tested technology and operational consistency. Reliability in acquisition is necessary because drill-bit data recording is not repeatable for any depth level. Any loss of data is not recoverable, whereas continuous and regular sampling in depth is one of the basic requirements for effective processing of VSP reflected waves.

5.2 Signal recognition and acquisition layout

In Chapter 3, we showed that the drill-bit source can produce seismic energy comparable to that of conventional seismic sources during the drilling of "short" depth intervals, i.e., few meters. We have observed that the source signal contains energy over a wide frequency band and has a random character in time. Consequently, it can be processed, by cross- or auto-correlation or pulse-compression methods, to obtain a signal of wide bandwidth (see also Chapter 6, Sections 6.3, 6.5 and 7.12).

As mentioned in Chapter 1, both the recognition of the signal of the bit and the determination of the delay of the reference drill-bit signal may be problematic during SWD. It is also necessary to shape the signal of the bit into an impulsive and interpretable signal. Each of these challenges are discussed below.

The recognition of the signal of the bit – The time variation of the drill-bit source signal is unknown and the signal has to be identified or extracted from all the environmental noise fields: drilling-rig noise, cultural noise as well as source-generated noise, such as borehole and scattered waves. A solution to recognize the bit signal is to measure its signature in a location allowing good signal-to-noise ratio, for example at the drill string where the cultural noise produced by the rig activity and the geological response are minimized. This method provides a primary (reference or pilot) signal. Other solutions are to simulate the primary bit-source signal by means of a parametric modeling based on drilling parameters, or to synthesize a reference signal by processing gathers of recorded traces (Chapter 7). This processing can be based on wavefield analysis, propagation delays (Sections 6.8.2) or statistical assumptions (Section 6.5). Some other methods, based on correlation and migration (Section 7.12.1) or polarization analysis, do not make use of a reference signal. The use of the pilot signal only for SWD prediction is discussed in Section 7.12.2.

The determination of the delay of the reference drill-bit signal – This step is required to determine an origin time for the source relative to the time of the signals received at the surface-seismic geophones. The drill-bit signal can be recorded continuously in time intervals of several minutes or hours. An arbitrary "reference start time" can be assigned to the zero-time in order to fix the time origin in an ideal signal measured using a receiver in a position close to the bit. This near-bit receiver must have a known and accurate clock. This ideal signal has no propagation delay. However, if the drill-bit reference signal (or signature) is measured at some distance from the bit, it has a propagation delay which must be determined accurately. This delay is then subtracted from the measured reference signal in order to shift its "zero-time" to that of the bit source (shift of the reference signal to the bit position). These time shifts are necessary because the times of the seismic events –

that is, direct arrivals and reflections – obtained using reference correlation with the geophone signals must represent the seismic-propagation times from the bit source and have to be accurate to give a reliable interpretation of the SWD seismograms.

The zero time is defined as the initial time at which a given signal, assumed known, is produced at the bit. The measured scales of the seismic signals, which result after the propagation of this "known bit signal" in the formation, have to be shifted in order to be referred to the zero time. The correction of the zero time in the original raw-field traces is not necessary if the correction is done after correlation (Section 6.3). However, it is, in general, necessary to calculate the time delay from the bit of the recorded waves in order to process the drill-bit seismograms. This delay can be estimated in different ways (Chapter 7).

Shaping the bit signal – The bit signal is a random-in-time and non-impulsive signal that must be shaped into a coherent, impulsive and interpretable signal.

The solutions used to accomplish the above tasks determine the SWD-data acquisition methods. All the three challenges presented above can be accomplished using only the signals from vibration sensors (geophones, accelerometers, etc.) located at the surface. However, the SWD method can be significantly improved by using downhole reference-signal measurements. Such downhole measurements are now in an initial (non-standard) demonstration phase.

In Sections 1.4 and 1.5, we introduced different approaches that have been proposed to make the seismic drill-bit source usable. These involve using vibration sensors of different types, located at different positions at the rig and on the ground. The solutions and technologies adopted to sample the signal and noise fields may differ in part by using the different recording configurations. As a consequence, the SWD methods used by different operators may differ in effectiveness in acquiring and processing the data, minimizing the distortions and satisfying the conditions of source repeatability in the resulting seismograms. Some of these approaches are listed below.

- ◇ Most of the proposed approaches are based on correlation (say comparison) of signals acquired – generally but not always – at different locations. These correlations are between seismic signals or with reference (pilot) signals or, also, between reference signals. The more common approach is to correlate the seismic signals with the rig-pilot signals.

- ◇ Calculating adaptive filters only between seismic traces acquired by geophones or hydrophones in different positions with respect to the well. In this case, the seismic response is calculated from the ratio of the adaptive filters after convergence (Widrow, 1990).

- ◇ Using only polarization analysis of raw-uncorrelated data recorded by using sensors located in nearby boreholes (Asanuma and Niitsuma, 1992) for obtaining geophysical information in the proximity of the well. However, this method does not solve the problem of determining the zero seismic time. Therefore, it does not seem to be effective for while-drilling checkshot purposes (Section 8.2).

Reference correlation may be performed by using the traces recorded by seismic sensors and a signal of a pilot transducer located on the rig (Staron, Arens and Gros, 1988; Rector and Marion, 1991), or by using a pilot signal obtained from an optimally weighted stack of seismic traces only. The latter is sometimes called digital grouping (see Section 6.8.2). The recorded seismic geophone or hydrophone traces are cross correlated and processed to remove non-random noise. As the traces of each surface seismic sensor has been individually recorded, these can be grouped together digitally using an adaptive filter process called beam forming. This procedure allows the elimination of unwanted events from noise sources such as the rig and provides optimal deconvolution filter (Miller, Haldorsen and Kostov, 1990; Section 6.8.2).

A statistical method for extracting the drill-bit signature for correlation is based on using a plurality of reference traces measured in the rig and in the drilling yard from which the pilot signal is obtained by independent signal and noise separation (Angeleri, Persoglia, Poletto and Rocca, 1996; Section 7.6.1). Another approach – not involving the use of rig-pilot sensors – is to migrate the autocorrelograms and/or crosscorrelograms of the traces measured by the seismic receivers which contain drill-bit signal reflections (Katz, 1984; Yu and Schuster, 2001; Section 7.12.1). On the other hand, Malusa, Poletto and Miranda (2002) use only the autocorrelations of the measurements recorded by a drill-string pilot sensor to process the reflections in the formation (Section 7.12.2). We describe and compare the different processing methods to get the signal from acquired data in more detail in Chapters 6, 7 and the SWD applications in Chapter 8.

5.3 Pilot sensors and transducers

Reference-pilot and land-seismic recordings provide complementary measures of the drill-bit signal for SWD purposes. These measures are complementary because they allow us to monitor the partition of vibrational energy produced in the drill string (or borehole) and in the formation (Section 3.10). As pointed out in Section 5.2, some authors suggested a method of SWD surveying by measuring *only* the waves radiated in the formation using seismic sensors. However, the use of auxiliary pilot sensors at the rigsite or in the borehole is an important and technically recommendable aspect of SWD applications. There are more reasons to use the reference-pilot sensors. One reason is to improve the signal-to-noise ratio in SWD data. Another reason is that the delay of the pilot signals can be measured or calculated with a good approximation. Consequently, the reference correlation provides traces which can be easily processed and interpreted to obtain time-depth information (checkshots) and prediction ahead of the bit (Section 8.2). Moreover, rig measurements provide additional information that can be used to characterize the drill-bit forces and vibrations and, ultimately, the source performance. A similar approach has been taken for Vibroseis sources where acceleration and force sensors are commonly used to measure the reaction-mass acceleration and the ground force. These feedback signals are used together with the baseplate-motion signals to control and, if necessary, to synchronize the actual sweep and the phase of surface vibrator sources (Sallas, 1984; Landrum, Brook and Sallas, 1994).

The seismic measurements of SWD are typically performed to detect velocity or pressure waves using multi-component geophones or hydrophones, respectively. The properties of these types of transducers are well characterized (Pieuchot, 1984; Evans, 1997).

SWD involves reference pilot receivers, which are used to monitor the signal transmitted through the drilling plant (drill string, drilling mud, as well as casing, etc.) in the form of extensional, torsional and mud guided waves (see Chapter 3), as well as sensors used to measure both the signal and noise vibrations at the rig. Measuring shocks and vibrations on the rig, in the borehole and in the drill string involves various types of transducers like accelerometers, strain and pressure gages (which may be less familiar to geophysicists).

5.3.1 Accelerometers

An accelerometer is a transducer whose output is proportional to acceleration. For example, a mass-spring system acts as an accelerometer when it operates below its natural frequency (see, for example, Pieuchot, 1984 or Sheriff, 1999). The damped mass-spring system used as an accelerometer is schematically represented in Figure 5.1, where m is the mass suspended by the spring of stiffness k, and c is the coefficient of viscous damping. The response of the mass-spring transducer to a harmonic acceleration, $a(\omega)$ is given by the well-known equations (see, for example, Chu, Eller and Whittier, 1995, in Harris, 1995)

$$\frac{|\delta(\omega)|}{|a(\omega)|} = \frac{1}{\omega_n^2} \frac{1}{\sqrt{\left[1 - \left(\frac{\omega}{\omega_{\rm n}}\right)^2\right]^2 + \left(2\zeta \frac{\omega}{\omega_{\rm n}}\right)^2}}, \tag{5.1}$$

and

$$\theta = \arctan \frac{2\zeta \frac{\omega}{\omega_{\rm n}}}{1 - \left(\frac{\omega}{\omega_{\rm n}}\right)^2} - \pi, \tag{5.2}$$

where $a = \partial^2 u / \partial t^2$ denotes acceleration, θ is the phase angle between displacement δ and acceleration a, $\zeta = c/c_{\rm c}$ denotes the fraction of critical damping, $c_{\rm c}$, given by

$$c_{\rm c} = 2\sqrt{km}, \tag{5.3}$$

and

$$\omega_{\rm n} = \sqrt{k/m}, \tag{5.4}$$

is the undamped resonance, or natural, frequency (see also, Section 4.3.1).

When the natural frequency of the transducer is high with respect to the excitation frequency, the transducer substantially acts as an accelerometer. In fact, if the frequencies are much lower than the natural frequency, the amplitude response of the mass-spring system of equation (5.1) becomes flat and the phase response of equation (5.2) tends to $-\pi$. For undamped transducers, this condition approximately holds when (Chu, Eller and Whittier, 1995, in Harris, 1995)

$$\frac{\omega}{\omega_n} < 0.2. \tag{5.5}$$

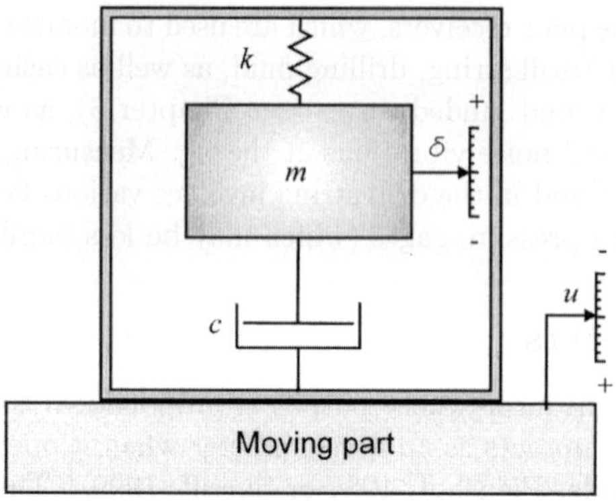

Figure 5.1: The mass m is supported by the spring of constant k, and c is the coefficient of the viscous damper (damping force = - $c\partial\delta/\partial t$). The ground (moving part or "frame") displacement is u and the relative displacement between mass and case is δ. Modified after Harris (1995).

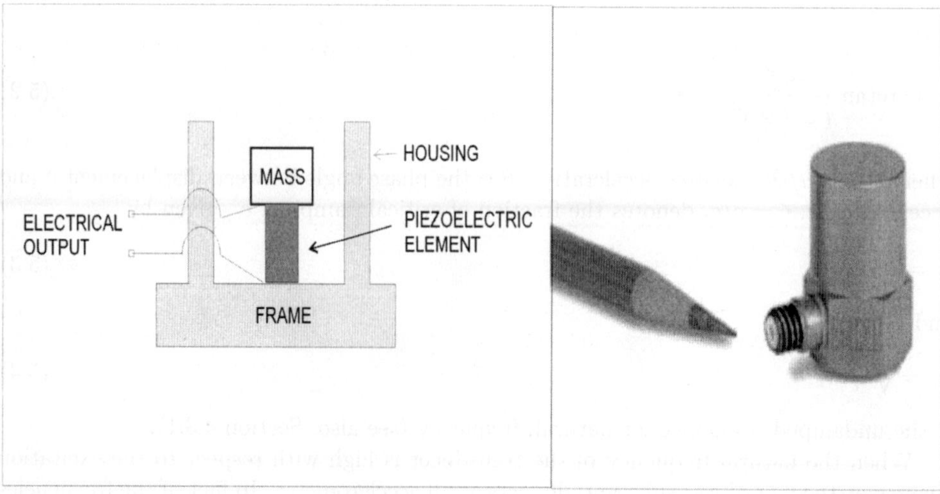

Figure 5.2: Left: Diagram of a piezoelectric accelerometer. The mass m is supported by the piezoelectric element, which acts as the spring of stiffness k in Figure 5.1, and is fixed to the housing frame of the transducer. The dashpot c of Figure 5.1 represents the viscous damping between the internal mass and the frame. Right: Picture of a compression-type piezoelectric accelerometer.

High values of ω_n are obtained by increasing the stiffness of the spring (high k) and diminishing the mass m supported by the spring. From equation (5.1) it follows also that the deflection δ for a given excitation frequency and acceleration magnitude is inversely proportional to ω_n^2. For this reason, the electric signal produced by the transducer may be very small, so that a large amplification is necessary to transform it to a recordable electric signal. Commercially available piezoelectric accelerometers are high-impedance sensors with integrated amplification circuits, which require an external power supply. When used for SWD purposes, these sensors have to be protected for explosion risk by following the safety rules of the wellsite – because these sensors are used at the rigsite risk Zone-2 (see Section 2.3.1).

We can observe that accelerometers are suitable to measure shocks of the drilling plant. In fact, it is required that the sensor and recording system have a good low-frequency response, as the shock and related motions characteristically contain low-frequency components. "*Shock pulses can be measured accurately only using instrumentation whose response is flat down to the lowest frequencies*" (Chu, Eller and Whittier, 1995, in Harris, 1995).

Main characteristics of accelerometers

The frequency response of the sensor is the ratio between output voltage and input acceleration at a given frequency. The following important properties are defined to describe the characteristics of an accelerometer (see, for example, Chu, Eller and Whittier, 1995, in Harris, 1995).

Sensitivity – The ratio of electrical output to mechanical input. It is expressed in terms of voltage per unit acceleration. When the instrument requires an external voltage, and the output is proportional to the voltage, the sensitivity is expressed in output voltage per unit of supplied voltage, per unit of acceleration, for example, millivolt per volt per g of acceleration ($1\ g = 9.8\ \text{m/s}^2$).

Resolution – The smallest change in input acceleration for which a change in electrical output can be discerned. Resolution depends on the noise level in the recording system.

Transverse sensitivity – The maximum sensitivity of a transducer is obtained for input excitation acting along the axis of its maximum sensitivity. If, ideally, the X axis of the transducer is the maximum sensitivity axis, then the cross-axis sensitivity along the Y direction normal to X should be 0. In practice, the ratio of these sensitivities ranges between 0.01 and 0.05 (from 1 to 5 %). The measurement errors due to cross-axis sensitivity have to be taken into account when analyzing orthogonal three-component drill-string vibrations.

Amplitude linearity and limits – Amplitude linearity exists when the ratio of voltage output to acceleration input, i.e., the sensitivity, is constant within specified limits. Amplitude limits bracket the acceleration range for which the linearity holds.

Frequency range – The range of frequencies over which the sensitivity of the transducers vary less than a stated percentage from a reference sensitivity. Low-frequency limits depend on the amplification system. High-frequency limits depend on the damping characteristics and resonance frequency of the transducer (Figure 5.3).

Other factors – Other factors, such as temperature, humidity, acoustic noise, and strain sensitivity may affect the measurements and generate spurious outputs. Temperature may change the damping and sensitivity values. However, generally, temperature operating ranges are quite large for accelerometers. Physical dimensions of the accelerometers are important only when the mass of the instrument becomes comparable to those of the vibrating structure. In that case, the vibration properties of the structure are different from that of the system composed of the structure and sensor fixed to it. This is not a practical problem in drilling applications!

5.3.2 Piezoelectric accelerometers

A material which generates a voltage across it in response to a stress (strain), and vice versa, is defined as piezoelectric (or "electrostrictive", Sheriff, 1999). This physical effect is due to the fact that when a piece of piezoelectric material is strained by a stress there is a deflection of the crystal lattice and a displacement of the charges. One of the most sensitive and stable piezoelectric materials is crystalline quartz. The piezoelectric material can be made of natural quartz – more stable with temperature – or reprocessed material – man-made polycrystalline piezoceramics.

A piezoelectric accelerometer is a linear transducer using a piezoelectric element to produce an electrical charge proportional to acceleration. The schematic diagram of a piezoelectric accelerometer is shown in Figure 5.2. The relative displacement of the mass compresses the piezoelectric element, acting as the spring, by the amount δ, in agreement with equations (5.1) and (5.2). The charge and electric voltage are proportional to the displacement δ and, ultimately, to acceleration at low frequencies with respect to ω_n. In the ideal system represented in Figure 5.2, we assume that the spring has zero mass, that the stiffness of the mass and of the frame are infinite and that damping exists only between the mass and the frame (Chu, Eller and Whittier, 1995, in Harris, 1995).

A typical frequency response curve of a piezoelectric accelerometer is shown in Figure 5.3. The output is given in millivolts per g of acceleration. The low-frequency limit and the attenuation curve ("roll off") is determined by the characteristics of the preamplifier that follows the accelerometer. The high-frequency limit is defined relative to the acceptable deviation with respect to the mean value of the response. For a deviation that varies not more than 0.5 dB (6 %), this limit is assumed as $f_n/5$ in terms of the natural resonance frequency f_n. However, the values of the usable frequency range may vary and depend also on the type of mounting of the sensor.

Piezoelectric transducers made of polarized ceramic, called zirconium titanate, are the most commonly used because of their low cost and their usable temperature ranges between from -100 °C to $+288$ °C. Polarized ceramics in bismuth titanate, and quartz or ferroelectric ceramics are also used.

Many different constructions are also used. A compression-type accelerometer consists of a piezoelectric disc and a mass on a frame. The mass is preloaded on the piezoelectric disc in order to measure both compression and tension vibrations along its cylindrical axis. Shear-type accelerometers use flat piezoelectric plate elements to detect shear waves.

These sensors are preloaded too, using different methods, and have a very low cross-axis response.

The physical characteristics of accelerometers also vary. Commercial accelerometers are typically of cylindrical shape, small (few centimeters) and relatively light in weight (from a few grams to a few hundred grams) (see Figure 5.2). Usually, larger accelerometers correspond to higher sensitivity and lower resonance frequency, which can be 100 000 Hz, or more. The fractional damping ζ is negligible for most piezoelectric accelerometers, i.e., less than 0.1. Therefore, from equation (5.1), it follows that at low frequencies the acceleration response is flat and the phase distortion of the transducer is quite negligible.

A part of the signal conditioning before amplification is included in the sensor by using microelectronics to perform conversion of the high-impedance voltage signal of the piezoelectric element into a readable, low-impedance, output voltage. For this reason, these sensors require a power source.

5.3.3 Damped geophones as accelerometers

Geophones are mass-spring systems with a transducer that produces a voltage output proportional to the relative velocity of the frame/suspended mass system. Geophones with high damping – say, damping factor ζ equal three – act as accelerometers. They have a frequency response with a bandwidth of some octaves centered in their natural undamped frequency. Damping can be achieved either electronically (using a "shunt" of low resistivity), or with oil of suitable viscosity in the magnetic gap.

5.3.4 Strain gages

The strain gage is a transducer used universally to measure strain for stress (and force) analysis and is broadly applicable to detect also the shocks and vibrations that accompany stress in a body (Wilson, 1995, in Harris, 1995). This means that this type of transducer can be used not only to measure the magnitude of the strain, but also the time-history of the related vibrations.

The electrical-resistance strain gage essentially consists of a foil or fine wire that has a change in resistance proportional to the imposed mechanical strain. An elongation of the thin foil or wire in consequence of an imposed strain corresponds also to a contraction of its cross-section, and the overall effect is a change in its resistance. The equation relating strain and resistance is

$$\frac{\Delta L}{L} = \frac{1}{K}\frac{\Delta R}{R}, \tag{5.6}$$

where L and R are the initial length and the resistance of the wire or foil, ΔL and ΔR are the wire or foil length and resistance variations, respectively. The ratio $\Delta L/L$ is the unit strain. K is defined as the "gage factor" of the wire or foil, and determines the sensitivity of the transducer. Not all materials exhibit this sensitivity to strain. For a discussion about strain-gage materials see Wilson (1995) in Harris (1995). Gage factors of strain gages typically range between +2.0 and +3.5.

The foil used in a strain gage must be very fine and sufficiently long to have an adequately high initial resistance R. This resistance usually ranges between $R = 60\ \Omega$

and $R = 350\ \Omega$. Since it is difficult to handle such a thin foil, it is provided with a support or "backing material". This is usually a piece of paper, plastic or other material. The backing provides easier handling and simplifies the application of the foil. A typical construction of a foil strain-gage is shown in Figure 5.4. Notice that a sufficient amount of material is left on the edges (loop ends) of the foil gage, to reduce its sensitivity to transverse strain. Low transverse sensitivity is a desirable feature for a foil-type strain gage. Its value may vary between 0.5 to 0, depending on gage construction.

Temperature and gage selection

Temperature affects the gage factor because of thermal expansion and dependence of resistivity of the foil and gives an apparent strain indication (this is denoted as the "apparent strain" effect). Strain gages can be used for static or dynamic measurements, and are made of different materials. Typically, static gages have a minimum change of resistance with temperature. At the same time, these transducers are typically less sensitive (low gage factor K). Dynamic gages have a much greater change of resistance with temperature, and are more sensitive (high gage factor K). The first can be selected to measure absolute values, the latter to detect rapid vibrations and variations that are not influenced by the relatively slow changes in temperature. Other effects, such as humidity, hydrostatic pressure may produce strain-gage errors also.

Strain gage response

Strain gages have a small mass and a linear response over a large frequency range, from 0 to more than 50 000 Hz. When using strain gages, it has to be taken into account that their life is typically a given number of deformation cycles. For example, wire gages may have 2 million cycles of fatigue life for strains of the order of 0.001 (Warring, 1985).

Strain gage installation

Installation is another important aspect for the strain-gage applications. In fact, the correct functioning of a strain transducer completely depends on the bonding of the transducer to the structure on which the measurements have to be performed. Manufacturers recommend special cements, and the structure surface has to be carefully cleaned. For example, installation of strain gages bounded directly on the derrick-rig structure can be relatively easy. The same application on the external part of the drill string is a difficult operation that requires care and the use of protective treatments and supports. A better solution is to install the strain gages in a probe measurement tool, which is prepared in the laboratory and is then mounted as a part of the rotating drill string. This is done in order to measure strains and stresses (load, torque and bending forces) in the drill pipes, e.g., to measure the WOB and axial loads in the drill string for MWD purposes.

5.3.5 Force and pressure transducers

Given that force equals mass per acceleration and also given that stress equals force per unit area, a force transducer uses the same basic technology and properties of the

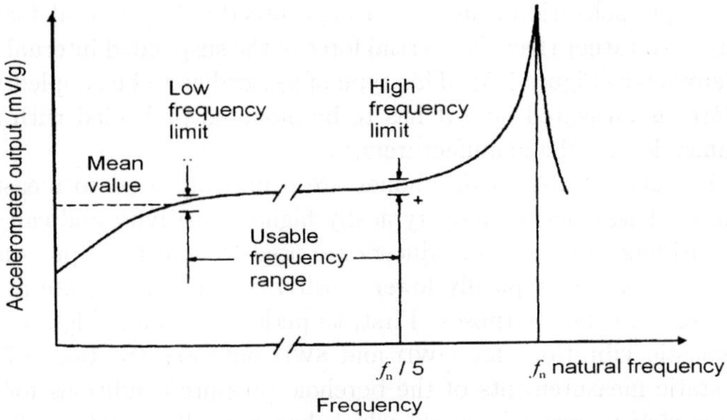

Figure 5.3: Response of piezoelectric accelerometer. The usable frequency range is determined by the low- and high-frequency limits, the later about 5 times below the natural frequency f_n (modified after Harris, 1995).

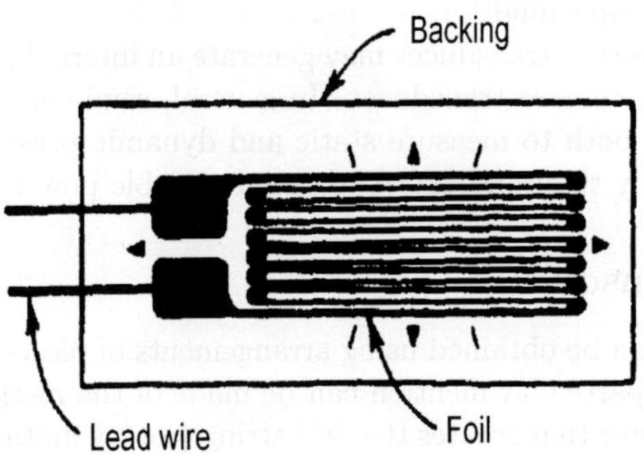

Figure 5.4: Typical strain-gage construction (modified after Harris, 1995).

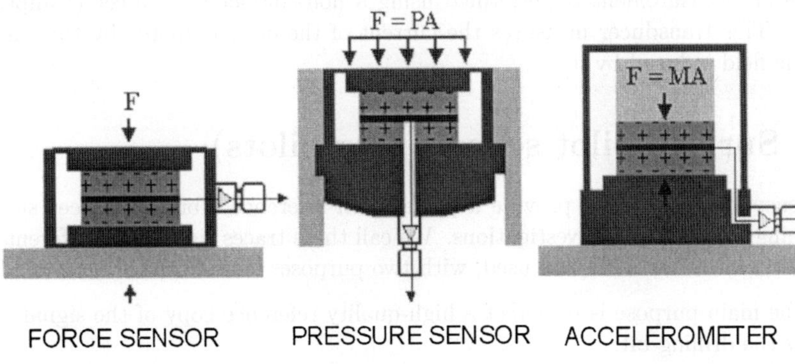

Figure 5.5: Configuration of force, pressure and acceleration piezoelectric sensors (modified after PCB-Specifications, 2002).

piezoelectric accelerometer. These sensors may differ very little in internal configuration. In this case, a piezoelectric element measures directly the external force loaded on the sensor (load cells), rather than the inertial force of the suspended internal mass as happens in the accelerometer (Figure 5.5). This type of sensor has to be coupled with the body in which the force is measured, and it has to be mounted preloaded with a reference force value recommended by the manufacturers.

In seismics, piezoelectric hydrophones are commonly used to measure the pressure waves in water. These sensors have typically higher sensitivity and can operate at lower pressure. In drilling, pressure transducers are used to detect the pressure in the drilling mud. These sensors have typically lower sensitivity and can operate at higher pressure. They can be used for two purposes. First, to perform dynamic high-frequency measurements of acoustic vibrations for MWD and SWD purposes (Section 5.7) and, secondly, to perform static measurements of the borehole pressure conditions for drilling control. These sensors may consist of a body which has a small-diameter cylindrical hole of a known cross-section. This hole ends with a diaphragm connected to a strain gage or with a piezoelectric material to detect the force due to the pressure over the exposed and known area. These tools are typically used as plugs inserted and sealed in openings in contact either with the inner or outer mud.

In some cases, the pressure transducer may generate an internal reference pressure and operate as a differential pressure transducer. In general, semi-conductive (piezoelectric) transducers can be used both to measure static and dynamic pressure variations with a natural frequency of many thousand Hz. They need a stable power supply.

5.3.6 Torque transducer

A torque measurement can be obtained using arrangements of piezoelectric or strain-gage sensors. Furthermore, a particular mention can be made of the method of measuring the current of the electric motor that rotates the drill string (rotary motor) (see Section 2.3.7). The signal of torque is proportional to the current i_a of the primary-induced circuit in the electric motor (equation (2.5)). In this way, the electric rotary motor substantially acts as a transducer sensitive to the mechanical bit-impact response (see Section 3.6). This torque measurement is performed using a non-contact transducer ("amperometric pinch"). This transducer measures the current of the electric motor by the variation of magnetic field induced by it.

5.4 Surface pilot sensors (rig pilots)

Pilot sensors at the rigsite provide a plurality of reference vibration traces suitable for performing while-drilling investigations. We call these traces rig pilots. Different types of rig-pilot measurements may be used, with two purposes.

- ◇ The main purpose is to detect a high-quality reference copy of the signal produced by the drilling bit.
- ◇ The auxiliary purpose is to measure the disturbances produced by the drilling activity (Sections 4.8 and 4.9). These measurements are used either to subtract coherent

environmental noise from the disturbed data or, in some cases, only to interpret the noise in the disturbed data. In fact, the correlated drill-bit signal contains several types of coherent-noise events, such as multiple reflections, tube and guided waves converted in the drill string. To correctly diagnose the signature and the delays of these components is important in data-quality control. The aid of the rig auxiliary measurements, e.g., such as the mud pressure, can be very important in the correct interpretation of these "noise" waves.

Due to the mechanical and hydraulic complexity of the drilling plant and borehole, various types of transducers are used to perform the pilot measurements in different locations: on the top of the drill pipe, on the rig derrick, in the drilling yard, in the mud line and at the rotary motor. Several of these measurements are already provided by the mudlogging cabin unit or by the drilling-control measurements in the form of sampled drilling parameters, but in a low-frequency band with a sampling rate of a few samples per second and with insufficient precision for SWD high-frequency analysis.

Among the wide selection of pilot sensors that can be used to perform an SWD acquisition, the sensors fitted to the top of the drill string and drilling-rig structure are primarily used to record the signal produced by the bit. Given the critical role of these elements within the SWD system, particular care must be taken when choosing and positioning them on the derrick so that a high-quality signal can be recorded.

5.4.1 Pilot sensors on the top of the drill string

Swivel system

If the well is drilled with a rotary kelly system, the top of the drill string bears the swivel. It is hung on the hook and is connected to the mud hose (see Section 2.3.7). Pilot sensors can be mounted on the swivel and connected electrically by using wires bound to the flexible mud-hose line (Figure 5.6a). As the drill string (square kelly pipe) rotates, the torsion is transmitted through the rotary table, while the swivel is a "free end" with respect to torque, i.e., torque is zero at the swivel. Hence, the swivel pilots can detect axial vibrations transmitted through the string, not torque loads or torsional waves. Multicomponent (vertical and horizontal) acceleration transducers are suitable for the purpose of measuring axial drill-bit vibrations and noise. Conversions between lateral shaking and axial modes can be a source of noise generated at the surface. The surface noise of the swivel is partially polarized in the plane of the bail vibrations (Rector, 1992; see Section 4.9). It can be reduced by processing three-axial acceleration recordings at the swivel (Chapter 7).

Analysis of the lateral swivel vibrations is related to the flexural vibration of the kelly and hoisting system at the top of the drill string. It has variable height above the drilling floor during the excursion of the mobile hoisting. The tensioning and the resonance frequency of cables are given by equations (2.2) and (2.3), respectively.

Top-drive system

A top drive is a rotary motor that rotates the drill string and is directly mounted to the top of the drill string (Section 2.3.7). When a top drive is used, the "kelly board" (rotary

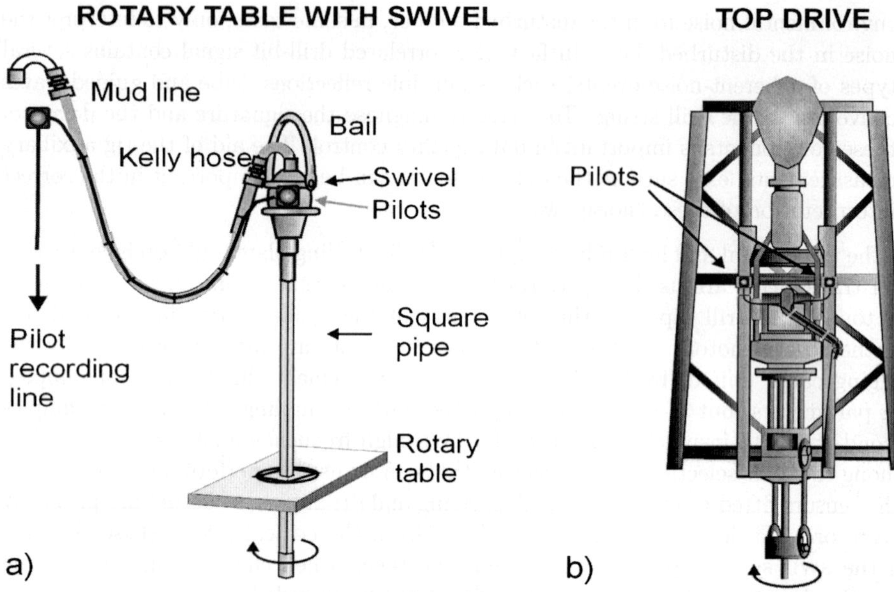

Figure 5.6: Examples of a) rotary with swivel and b) top-drive systems instrumented with SWD pilot sensors.

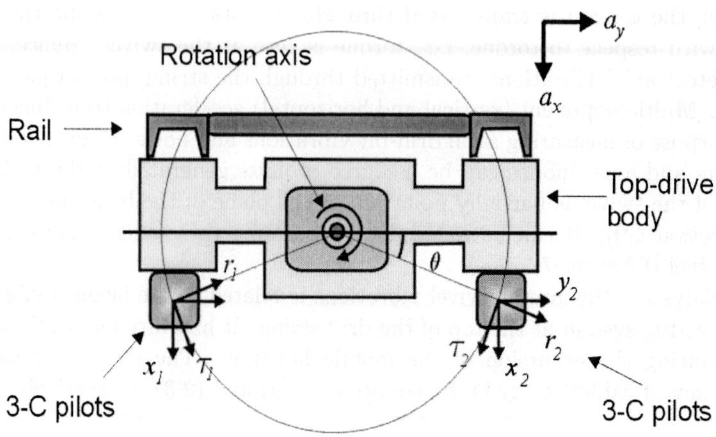

Figure 5.7: Layout in the horizontal plane of the pilot locations on the top drive (not to scale).

table) is not active. In this case, axial tension and torsional loads and vibrations are both transmitted to the top drive. Top drives are normally provided with openings to access the bearing at the top of the drill string. Figure 5.6b shows the layout of an instrumented top drive. Two sets of three-axial accelerometers are applied to the top-drive solid structure. The top drive is in close contact with the drill string. In this way, axial, torsional and transversal vibrations of the drilling-rig structure can be measured. These measurements are used to identify the vibrations produced by the impacts of the bit on the formation.

The signals of the six accelerometers can be combined and separated into the individual vibrating modes. Axial, torsional and lateral vibrations are analyzed by using the modal analysis of Sections 4.1 and 6.10 (Poletto, Malusa, and Miranda, 2001).

A cross-section of the top drive is shown in Figure 5.7, in which z is the drilling axis perpendicular to the plane. Pilots a_z and a_τ represent axial and torsional acceleration, respectively. Pilots a_x and a_y represent horizontal lateral vibrations, which can be processed by instantaneous polarization analysis to extract the instantaneous lateral flexural vibration mode, a_η. If, for simplicity, the layout is symmetric and the sensitivity of the sensors is assumed to be unity for all the axes, we calculate these pilot combinations as

$$a_z = \frac{a_{z1} + a_{z2}}{2}, \tag{5.7}$$

$$a_x = \frac{a_{r1} + a_{r2}}{2\cos\theta}, \tag{5.8}$$

$$a_\tau = \frac{a_{\tau 1} - a_{\tau 2}}{2} - \frac{a_{r1} + a_{r2}}{2}\tan\theta, \tag{5.9}$$

$$a_y = \frac{a_{\tau 1} + a_{\tau 2}}{2\cos\theta}, \tag{5.10}$$

where

$$a_{\tau j} = a_{xj}\cos\theta_j + a_{yj}\sin\theta_j, \qquad (j = 1, 2), \tag{5.11}$$

and

$$a_{rj} = -a_{xj}\sin\theta_j + a_{yj}\cos\theta_j, \qquad (j = 1, 2), \tag{5.12}$$

are calculated in positions 1 and 2 by using $\theta_1 = \theta$ and $\theta_2 = -\theta$, respectively. The individual traces a_z, a_τ and a_η (resultant of a_x and a_y) are then deconvolved to obtain wideband "white" spectrum signals and combined by independence analysis for noise separation (Poletto, Rocca and Bertelli, 2000; Section 7.6).

5.4.2 Surface pilot sensors in the rotating drill string

Pilot sensors can be installed in the rotating pipe by using a probe tool in the top of the drill string and a radio (or telephone) link for data transmission (Ng, DiSiena and Bseisu, 1990). A practical aspect is that the batteries used to supply power to the electronics of the high-impedance sensors require maintenance interventions during drilling pauses. The battery package has to be substituted after a given number of working hours. Using a different solution, the rotating systems can also communicate via electromagnetic

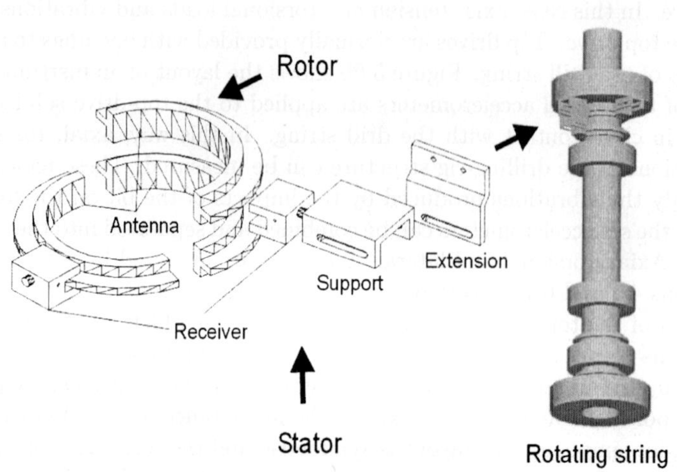

Figure 5.8: Layout of the telemetric system which allows signals to be detected from sensors installed on rotating strings (drill-string rotating pilots). It has the following components: rotating antenna, stator with the fixed-receiver system connected to the central amplification/control unit (installation of Manner technology; courtesy of A. Schleifer, 1998).

induction using AC and transformers to supply the power to the rotating sensors. The rotating sensors can be inserted into a protected probe installed in the pipe (Schleifer, 1998; Figure 5.8). Telemetric systems of this type are designed for measurements in rotating rods (Manner, 1996).

Field results show that surface measurements on the rotating drill string give, substantially, the same axial acceleration information as the signals measured at the top of the drill string (top drive or swivel). Nevertheless, they make it possible to measure the axial strain vibrations by using strain gages in the probes. They may also provide a more effective measure of high-frequency torque vibrations while drilling with a top drive or, with a swivel and kelly bush, by inserting the probe tool in the string below the rotary table level.

5.5 Downhole pilot sensors

Downhole sensors have been used extensively to monitor drilling conditions. Early applications were made about 40 years ago (Deily, Dareing, Paff, Ortloff and Lynn, 1968; Cunningham, 1968). These early applications used arrangements of sensors similar to those used in the more recent tools, however, with analog magnetic tape instead of digital recording. After retrieval, the acquired data were converted to "analog type oscillograph display, and, in some cases, were digitized for analysis purposes." Modern tools have local-processing capabilities and can use downhole transmission technology. Early systems used strain gages to measure WOB, torque and bending moments, accelerometers to measure axial, angular and radial vibrations, pressure meters to measure inner and outer mud pressure. In these early studies, a large number of wells were monitored and the main types of downhole vibrations were investigated.

Later, the measurement technologies used for drilling diagnostic technology were improved in order to perform the while-drilling control of the drilling conditions at the surface (Besaisow, Jan and Shun, 1985; MacPherson, Mason and Kingman, 1993) and downhole (Wolf, Zacksenhouse and Arian, 1985; Zannoni, Cheatham, Chen and Golla, 1993). The physical parameters measured downhole today are about the same as those of the early downhole experiments. However, the progress in digital electronic technology (integrated circuits) has allowed the collecting and storage of downhole data in local memory and allowed it to be processable in near-real time. These data can be used while drilling to control the drilling conditions and transmit information to the surface by mud-pulse telemetry (Section 2.3.18). This information is used either to drive directional drilling or to diagnose dangerous conditions and change, in consequence, the drilling parameters (Hutchinson, Dubinsky and Henneuse, 1995). Magnetometers to measure rotation angle, and "wall proximity" sensors are also used in downhole tools to detect lateral vibration modes and the position of the bit in the borehole. Temperature may be recorded for monitoring the drift of the local clock used by the recording system.

Recently, downhole measurements have been successfully extended also to drill-bit seismic while drilling. The SWD pilot data are acquired downhole and transmitted to the surface by specially wired pipes (Naville, Layotte, Pignard and Guesnon, 1994), or stored downhole and downloaded after tool retrieval after the bit trips (Miranda, Poletto, Abramo, Craglietto and Bernasconi, 1999). The use of a downhole recorded and stored reference signal and a shock absorber to limit large drill-string vibrations is discussed by Naville, Serbutoviez, Throo, Batini, Dini and Cecconi (2000).

Downhole data are shown in Section 7.8.1 and compared to surface data. In most cases, the signal-to-noise ratio and bandwidth of the data obtained with downhole sensors is significantly higher due to pilot-signal attenuation and noise. Actually, downhole drill-bit SWD tools do not seem to be mature yet, in terms of performances and reliability, for use as a standard technology for SWD purposes. However, testing is ongoing with promising results. Such tools make it possible to measure dynamic drilling conditions and the radiation performance of the drill-bit source (see Chapter 3). Figure 5.9 shows downhole-instrumented bits. Figure 5.10 shows a near-bit downhole tool while data downloading and testing between two runs on the field.

These downhole SWD tools may be available with local storage memory of several hundred Megabytes, and downhole processing capability. A technical problems of the recording with downhole tools may be related to local-clock synchronization and stability – a method based of time synchronization based on electromagnetic initial triggering was developed by Soulier (1999) – the rate of communication with the surface, and temperature limitations for electronics. If the power of this tool is not supplied by turbine power generator or wireline, limitation of the battery-package life has also to be considered in the management of the resources. A possible solution is to use a downhole power turbine to supply downhole electronics, i.e., sensors and integrated circuits for acquisition and local data processing.

a)

b)

Figure 5.9: Instrumented PDC bits used for downhole tests. a) Bit before and b) bit after insertion into the drill string. (Courtesy of ENI E&P Division)

Figure 5.10: Downhole tool in a phase of data downloading and tool re-programming (courtesy of ENI E&P Division).

Table 5.1 – Range specifications for surface and downhole accelerometers

Location	*Measurement ranges*	*Resolution*	*Linearity*	*Nat. Freq.*
Surface drill string	$\sim \pm 5\ g$ to $\pm 10\ g$	$0.0004\ g$	2 kHz (5 %)	> 15 kHz
Downhole tool	$\sim \pm 50\ g$ to $\pm 200\ g$	$0.002\ g$	3 kHz (10 %)	> 12 kHz

Table 5.1 indicates the typical specification ranges for axial accelerometers that are used at the surface and downhole. At the top of the drill string, the vibration amplitude is lower. In the drill string near the bit, the vibration amplitude is higher and the acceleration peaks may reach some tens of g's at the investigated frequencies.

5.6 Use of dual sensors in drill strings

The seismic signals obtained by correlating drill-string pilot and geophone measurements include the drill-string reverberations, because the pilot waves are reflected at each interface between string sections of different acoustic impedances (Section 4.10.1). Removing these reflections by using a reference-pilot deconvolution may cause signal distortion in the presence of additional noise due to drilling operations (Sections 6.5, 4.8 and 4.9). To overcome this problem, Poletto (2002) proposed using dual-sensor measurements in the drill string, both to remove the reflections of the drill-bit waves and to improve pilot deconvolution. The idea is the following. We can measure acceleration and strain in the drill string. These are dual fields, which have opposite reflection coefficients, and, in a drill string of constant elastic properties, the same transmission coefficients (Section 4.12). These quantities are scaled to fit the amplitude of the direct arrivals, summed to remove the events produced by an odd number of reflections in the drill string and in the rig, and may be deconvolved to characterize the reflection coefficient between drill bit and formation. Synthetic numerical examples and real measurements acquired downhole in a location close to the bit show that upgoing and downgoing drill-string pilots can be separated using dual fields and be jointly used to improve the SWD seismograms (Section 6.15).

Figure 5.11 is a layout indicating some pilot locations that can be used to measure the drill-string waves. Sensors can be placed on the rig, on the top of the drill string (top drive or swivel), by a tool inserted in the rotating drill string at the surface or downhole, close to the bit or at intermediate positions in the BHA. The surface-rig and downhole conditions and drill-string mechanical features determine the transmission modes and the quality of the pilot signal, so that it is often preferable to measure it by using the downhole instrumented tools inserted in the drill string. These tools include accelerometers, used to measure large axial and torsional vibrations, and strain gages, used to measure string loads, WOB and torque vibrations. This technology is suitable to obtain dual signals.

In fact, two classes of signals have to be considered. All the signals of each class are equivalent, in the sense that they have the same reflection coefficient. The first is composed of displacement, velocity and acceleration. The second is composed of strain, stress and force. In order to realize a dual measurement in the drill string, we simply

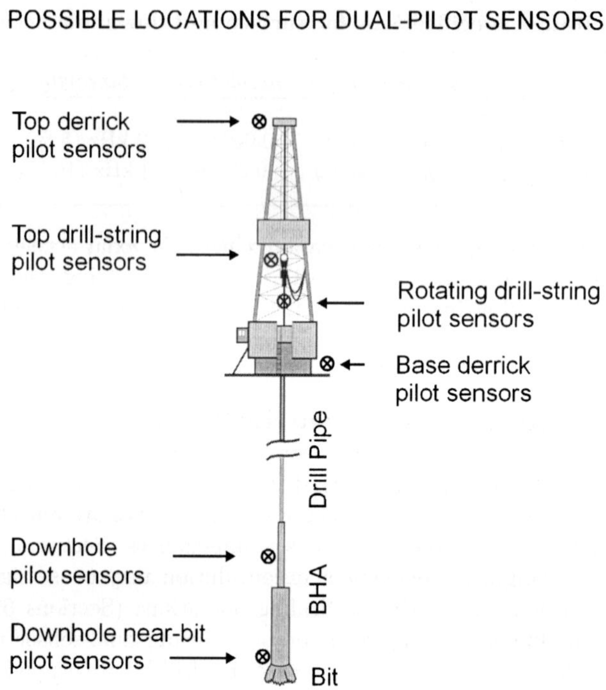

Figure 5.11: The dual sensors can be placed in different locations. In a first approximation, the drill-string transmission line can be considered as an assembly of tubes with different properties and impedances.

Table 5.2 – Examples of rig pilots

Type	Quantity	Response
Accelerometer	acceleration [m/s^2], [g]	flat in the seismic frequency
Pressurometer	pressure [psi], [bar]	flat in the seismic frequency
Strain gage	strain	flat in the seismic frequency
"Amperometric pinch"	torque [kN m]	low frequency

have to use one signal of the first and another signal of the second class. That is, one class has derivative with respect to time, the other has derivative with respect to space. For instance, we can obtain dual measurements in the drill string by using acceleration and strain. In this way, it is possible to separate the drill-string waves into up- and downgoing dual fields as discussed in Section 4.12, and to improve the pilot signal by performing downgoing deconvolution using the upgoing operator. Effective separation of signal, reflections and noise requires the analysis of the up- and downgoing wavefields in relation to both the position of the receiver and of the source of excitation in the drill string – namely, the drill bit for signal; or the upper top, for rig-vibration noise, as well as any intermediate contact points, for lateral shocks caused by the moving pipes.

5.7 Other pilot sensors at the rig

The drill-bit vibrations are transmitted through the drill string and the rope lines to the rig derrick (Rector, 1992). Rector and Marion (1990) proposed processing the rig sensor signal to obtain as a pilot signal after separating noise induced in the ground from bit vibrations. Angeleri, Persoglia, Poletto and Rocca (1992) used a plurality of yard sensors such as the rig, mud and torque sensors for drill-bit independent signal separation. Load vibrations measured at the top of the derrick, in the proximity of the crown block, were used by Booer and Meehan (1993) for drill-string imaging purposes. In general, rig-derrick sensors are used as auxiliary SWD measures of drill-bit signal and drill-string vibrations modes.

Derrick pilot sensors

Vertical (axial) component geophones and accelerometers are used at the same recording location on the derrick. Figure 5.12 shows two examples of the positioning of pilot sensors on the top of a drill string and derrick with rotary and top-drive systems respectively. These pilot sensors are used in connection with roller-cone drilling. For drilling with polycrystalline-diamond-compact (PDC) bits or for downhole-motor drilling in sliding mode (see Chapter 2), also the use of reference downhole measurements is recommended. Table 5.2 summarizes the principal types of sensors that can be used in the rigsite. A main difference to the conventional seismic sensors is that some of these tools require a conditioning and power supply unit. Table 5.3 shows an example of the location of the rig sensors in Figure 5.14.

Mud-pressure pilot

Pressure gages (Section 5.3.5) are placed in the mud delivery and return lines to measure the high-frequency pressure variations caused by acoustic waves propagating in the fluid. These instruments measure guided waves propagating through the drilling mud. Guided waves can be used to measure the delay of the pilot signal when drilling with a downhole motor in a sliding condition. This information is also used to identify the mud waves in the pilot correlations – often these events mask the drill-pipe multiples. Propagation of coupled mud and pipe waves are discussed in Section 4.6.

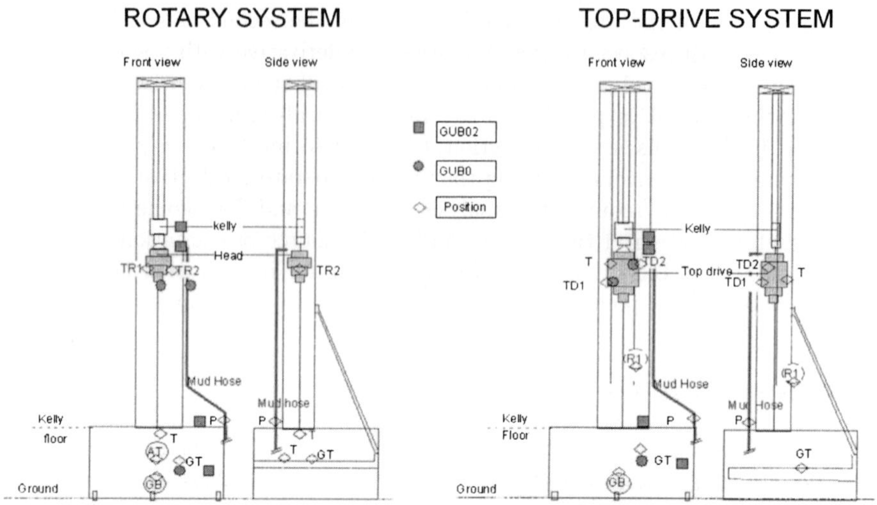

Figure 5.12: Schematic layout of instrumented rigsite (after Dordolo and Calligaris, 1993).

Figure 5.13: Example of fire-proof acquisition boxes and connections for the pilot line on the rig.

Table 5.3 – Example of location of pilots on the rig

Symbol	Location	Sensor
G_T, A_T	Top of the rig	Geophone, accelerometer (Z axis)
T_{D1} or T_{R1}	Top drive or swivel (1)	Extensimetric acc. (3 axis) + 1 piezo. acc. (Z axis)
T_{D2} or T_{R2}	Top drive or swivel (2)	Extensimetric acc. (3 axis) + 1 piezo. acc. (Z axis)
P (in)	Mud-hose in	HI-Sens. pressure transducer
P (out)	Mud-hose out	HI-Sens. pressure transducer
T	Electric-engine circuit	Torque transducer
G_B	Derrick support	3-geophone string (Z axis)
R_1	Top-drive support	Accelerometer + hor. geophone (torque arm) (Y axis)

Torque pilot

A torque pilot signal is based on the current flowing through the rotary motor. It is derived from the "amperometric pinch" (see Section 5.3.6). This signal is better with direct-current rather than alternate-current motors. However, in the current signal measured with a direct-current rotary motor, the disturbances introduced by the silicon controlled rectifiers (SCR) are observable. A wideband signal is used for SWD purposes.

Yard noise

The SWD data contain ambient noise. It is important to use sensors on the drilling yard dedicated to measuring this noise. Geophone sensors are used in the drilling yard to measure vibrations introduced by diesel engines (electric generators), mud pumps and other operating equipments, such as the mud shakers, which are also monitored by noise sensors. These signals are used to characterize the noise in the geophone recordings of the seismic line. This noise can then be removed using multi-channel filtering techniques. This method uses the sensors dominated by noise to identify the characteristics of this noise and remove it from the disturbed data (see Sections 7.5 and 7.6).

5.7.1 Connections of rig pilots to the recording system

The signals of the pilot sensors can be either transmitted via radio (or telephone) link or by wireline connection. Radio transmission uses local batteries and may require more maintenance on site. When wirelines are used, the pilot sensors and the wirelines located in the hazard wellsite zones (Section 2.3.1) have to be placed inside flameproof boxes and cables or connected by intrinsically safe circuits protected using diode barriers. In the first case, all the connection sheaths and boxes are made of reinforced flameproof material, in compliance with safety standard requirements governing the use of electrical conductors in places potentially subject to gas leaks (Figure 5.13).

5.8 SWD data-acquisition system

A general requirement for a SWD data-acquisition system is to acquire and perform preliminary processing (preprocessing) of data (Section 6.2). These steps provide a seismic signal that can be further processed at the wellsite as a reverse VSP. We take as an example the SEISBIT ® system, to give an overview of the SWD acquisition technology. Other tools may adopt different solutions.

The basic module of the system consists of a PC with a data acquisition card capable of handling a Sercel telemetric line using digital recording units SN348 and SN368. This PC-unit controls the power supplied to the telemetric line by an external appliance. A signal distribution unit (SDU) is used to supply and amplify the analog pilot sensors at the rigsite. The SDU output signals are conditioned as input to the standard Sercel recording units. In this way, both the signals from the seismic line and from the pilot sensors are recorded via the telemetric line. A workstation drives the acquisition of a variable number of powerful PCs, each with an acquisition card able to manage up to 200 seismic channels at 2 ms sampling period (1 ms or 4 ms are also available).

The acquisition card is able to initialize, i.e., to "form" the line by setting the channels "ready" to acquire and communicate, and test a telemetric line equipped with Sercel SN368 and SN348 receiver units in variable configurations (Figure 5.15). The signal acquisition, quality control and data-preprocessing stages are automated using a special software package that is connected to the drilling-operations-control system. Usually this port is provided by the mudlogging unit (Section 2.3.17). This unit continuously transmits the parameters of the drilling operation to the SWD data acquisition system. The drilling-operation-control data are processed and stored together with the drill-bit seismic data, in order to better define the data gathering conditions; they are also useful during the preprocessing stage. The SWD-data preprocessing (Section 6.2) is performed during the acquisition phase. It is essentially based on the identification and validation of one (or more) pilot signal(s) emitted by the bit, correlation of said signal(s) with those measured at the geophones, and the subsequent summing of the correlations relating to a given depth level. These and the other preprocessing steps are discussed in Chapter 6.

In practice, data recording is not performed continuously while drilling, but by acquiring a continuous sequence of seismic records of prefixed time length. The typical individual record length is some tens of seconds. These records have waiting intervals of a few seconds between consecutive records. The interval between records is determined by the time required by the system to restart a new recording phase after line test and activation ("formation"); this time may depend on the number of active recording channels. The total acquisition duration or maximum penetration for a depth level and the individual record length are decided in relation to the planned drilling speed (ROP) and acquisition conditions. Factors such as aliasing, expected formation properties, target depth, source resolution and signal-to-noise conditions are evaluated for this purpose. Performance of the system in terms of monitoring and management of the status of the seismic line are very important. In fact, failure of a single channel of the telemetric line – due to electronic problems or to external causes – may compromise the acquisition of a entire depth level. If the acquisition is activated in automated mode, this risk is higher. In part, this risk is reduced by using systems in which the failure of a part of the line does

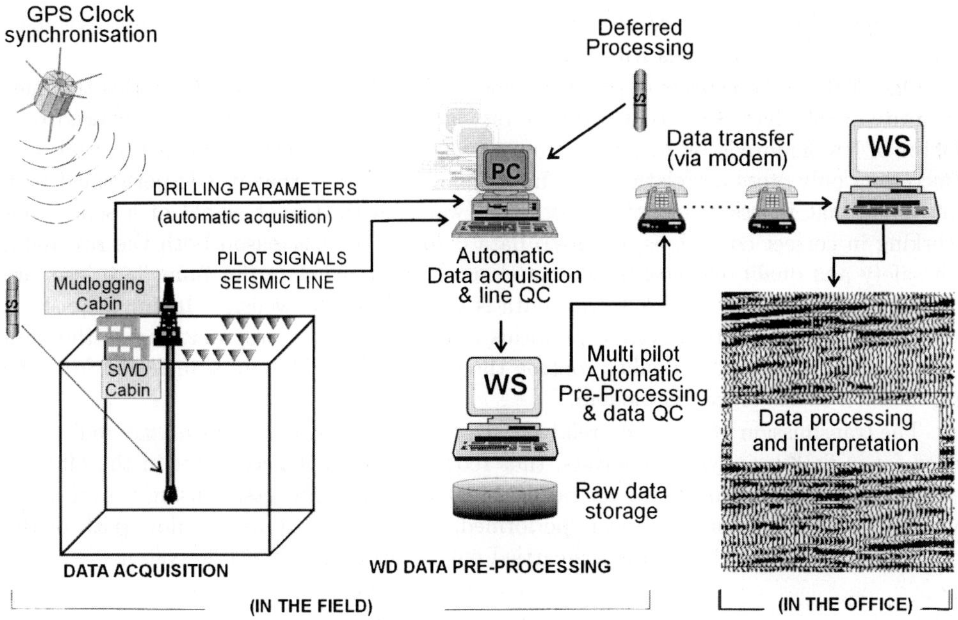

Figure 5.14: Layout of the SWD data-acquisition system. The system uses also an instrumented sub (IS) for downhole pilot data recording and storage.

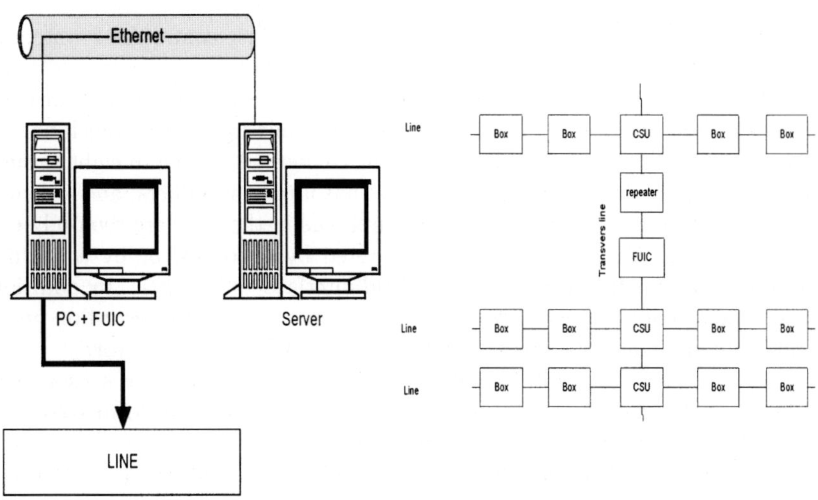

Figure 5.15: Layout of the SWD PC-acquisition system.

not compromise the entire record because the system is able to automatically reset itself after excluding the channels with problems.

One of the main aspects of preprocessing of the SWD data is that it reduces the huge quantity of raw data used to identify the bit signal. Normally, the data are reduced by a factor of few hundreds by means of cross correlation and record summing (Section 6.3). Recording only cross correlation may have the disadvantage that if it is performed with any pilot that is less than optimal or, even worse, with a pilot signal of a sensor not working in correct conditions, the SWD data is lost. For this reason both the raw data, for safety and quality-control reasons, and cross correlated data, for immediate use, are stored on disk. On-line access to raw data is available at the wellsite. In the acquisition work flow, the acquisition system performs automatically the preprocessing of the data (Section 6.2). Then, after quality control, the preprocessed data are transferred from the rig location to the office via modem.

The transmission of the precorrelated data via modem to headquarters may reduce the need for in-field geophysical analysts, thus reducing costs. Hence, either in the office or at the rigsite – according to the operational requirements – processing using conventional VSP packages and interpretation is performed. Figure 5.14 shows the flow path of the data gathered and of the operations carried out by the SWD SEISBIT ® system.

5.8.1 Other SWD commercial systems

For the sake of completeness, the following notes concern the other commercial SWD systems available worldwide from the different service companies.

TOMEX seismic-while-drilling field system (Baker Atlas)

The Baker Atlas'SWD system called TOMEX ® was the first commercial system available on the market. This system can be used onshore and offshore. The TOMEX ® field system consists of up to four field multiplex units (FMU) that convert the geophone or hydrophone signals to a digital multiplexed data stream for transmission over as much as 10 000 feet of cable to the recording unit. Each FMU can be configured for one to eight seismic inputs from geophones or hydrophones. Each channel has filters, amplifiers, and a dedicated 14-bit analog-to-digital converter. Filters and amplifiers in the FMU are remotely configured from the recording unit. The FMUs are powered by batteries, or by a thermoelectric generator for cold environments, via the field-interface unit (FIU). In addition to the electrical insulation of battery power, the FIU is connected to the recording unit by fiber optic cables to eliminate damage to the recording unit by electrical storms.

The rig multiplex unit (RMU) converts the pilot sensor and depth sensor inputs from the drilling rig into digital multiplexed data. The RMU uses transformer-coupled power and optically data lines to maintain electrical insulation.

The computer interface unit(CIU) supervises data communications and provides data and status to the computer of the recording unit. The computer supervises system operation, buffers data for recording to tape, correlates the signal of the pilot sensor with the field traces, provides quality-control displays, and performs seismic processing to produce MWD logs and data displays on the plotter. The data stream from each multiplex unit is decoded and formatted for the computer in the CIU. The waveforms from two channels can be selected for monitoring on the monitor scope. The CIU provides the signals required to operate several field systems in a master/slave configuration to expand recording capacity to over 300 channels. The system features are:

◇ 16-channel real-time correlation at 1, 2, or 4 ms.

◇ Remote-system configuration from operator's terminal.

◇ System-test function for end-to-end system verification.

◇ High-capacity tape and system-battery backup for continuous operation.

◇ Compact recording units for remote locations.

◇ Specialized software for full TOMEX® processing capability in the field.

◇ Seismic processing for MWD logs, simple images and drilling reports produced on site.

Drill-bit seismic – DBSeis system (Schlumberger)

The DBSeis[1] system utilizes 12–36 geophones and/or hydrophones in an array which is usually arranged to point from the rig with the first sensor located at 200 to 300 meters from the wellhead and connected to a conventional Sercel line. Three accelerometers are used as pilot sensors placed on the rig. The system is subdivided in a frontend PC, which performs acquisition, correlation, stacking, quality control, and data storage. Data are transferred to a secondary PC where signal processing is performed. Digital-grouping techniques can then be used to separate the bit generated signal from the ground roll. The secondary PC performs also standard VSP processing.

The signals of the pilot accelerometer and of the seismic sensors follow the standard processing discussed in Chapter 7. The pilot-accelerometer signal undergoes a drill-string signal processing in order to determine the drill-string travel time by identifying the "bit event" in the "drill-string image." This gives the one-way travel time in the drill string directly and avoids errors that can occur by using a velocity time-depth analysis method. The signals from the seismic sensors, after crosscorrelation and drill-string multiple removal, undergo a digital grouping step. The 12–36 geophone and/or hydrophone signals undergo a digital beam-forming process used to separate the bit generated signal from the noise. Two types of adaptive time domain beam formers (Section 6.8.2) are applied, depending on the noise environment (Meehan, Miller, Haldorsen, Kamata and Underhill, 1993). These are: a multiple input/single output, linearly constrained, minimum variance beam former; or a multiple input/multiple output, unconstrained beam former based on the adaptive interference canceling technique. A source sensor compensation and a shift to one-way time complete the preprocessing part before delivering processed VSP data for conventional processing and interpretation.

[1]Mark of Schlumberger

5.9 SWD-data acquisition and drilling control

The drilling-operation-control system provides the drilling parameters. This control may be performed by drillers, by MWD where it is present and, in the majority of cases, by the mudlogging unit (Section 2.3.17). Thus we call these parameters mudlogging parameters. The drilling parameters are communicated from the mudlogging unit present on the rig to the SWD unit through a serial line (Section 2.3.20). These parameters are necessary to perform automatic data acquisition and are used for quality control (QC) purposes. The most important mudlogging parameters, summarized in Table 4.4, are:

- ◇ Depth of the borehole [m]. This is the length of the borehole along its straight or curved trajectory. Depth is calculated by measuring the length of the drill string and subtracting the length of its external part emerging above the drilling floor. This part of the drill string has a variable height relative to the drilling floor (kelly position). The depth approximation is discussed in Section 5.10.
- ◇ Depth of the bit [m]. This is the depth of the bit along the borehole curve. It gives the position of the bit off bottom, during a trip or during mud circulation without drilling.
- ◇ Bit distance [m]. This is the difference between depth and bit depth. While drilling, it equals zero.
- ◇ Vertical depth [m]. This is the vertical depth (true vertical depth) of the borehole. This parameter is used if the well is deviated.
- ◇ Hourly drilling rate (ROP) [m/h].
- ◇ Position of the lower end of the mobile hoisting (kelly position) [m]. This gives the position above the kelly floor of the top of the drill string. This position changes while drilling.
- ◇ Surface rotary speed, at top drive rotor or at kelly, in number of revolutions per minute [rev/min] (RPM).
- ◇ Downhole rotary speed (rotary speed of positive–displacement–motors or turbines) [rev/min]. It is calculated from mud flow, for example, using equations (2.6) and (2.8).
- ◇ Total rotary speed [rev/min]. Sum of downhole and surface rotary speeds.
- ◇ Hookload [ton]. This is the load, measured at the hook, of the suspended drill string (F_{H} of equation (2.11)).
- ◇ Weight on bit (WOB) [ton]. This is the weight on bit obtained by using equation (2.11) and measured hookload force F_{H}, after calculating the total string weight and mud buoyancy effect by equation (2.1).
- ◇ Drill-pipe torque ([kN m] or relative scale). This is usually calculated from the torque measured by the absorbed current of the rotary motor (equation (2.5)).

◇ Number of strokes per minute of mud pumps (SPM).

◇ Mud-in pump delivery flow [l/min]. This is calculated from SPM using equations (2.16) and (2.17).

◇ Mud-out return flow [l/min].

◇ Mud injection pressure [kPa].

◇ Mud weight. This is the mud density [g/l].

Note that some parameters listed above are not strictly necessary as they are calculated in combination with other parameters. The surface mudlogging parameters are measured for the purposes of drilling and are of low frequency when compared to SWD data. A typical time sampling interval is one or two seconds. This low-frequency band makes them useful for average control of seismic-while-drilling conditions. However, a direct comparison of seismic and drilling measurements performed conventionally is, in most cases, not possible in a higher frequency band. This limitation is not only due to the fact that the sampling rate used by the drilling-operation-control recording system is lower; it could simply and easily be increased. It is mainly due to the fact that some drilling measurement systems use sensors that work with a sampling rate that is insufficient for seismic purposes.

Table 5.4 – Main surface drilling parameters

Parameter	Typical unit	Parameter	Typical unit
Date		Hour	
Depth	[m]	Vertical depth	[m]
Bit depth	[m]	Kelly position	[m]
ROP	[m/h]	RPM (surface)	[rev/min]
RPM (downhole motor)	[rev/min]	RPM (total)	[rev/min]
WOB	[ton]	Hookload	[ton]
Torque	[kN m]	Press. Pump	[kPa]
Mud weight in	[gr/l]	Mud weight out	[gr/l]
Flow in	[l/min]	Flow out	[l/min]
Pump tot	[stroke/min]	Bit distance	[m]
Tooth wear		Rig status	
String Length	[m]	Bit power	
BHA run		Bit run	

Table 5.5 – MWD downhole drilling parameters (if used)

Parameter	Typical unit	Parameter	Typical unit
WOS	[kN]	Torque	[kN m]
Vibration level	[g]	Directions	

For example, rotation speed of the drill string may be sampled by few (a pair of) magnetic markers connected to the rotating pipe. A pipe revolution is completed typically in half a second, so that, when using this measurement system, the instantaneous rotary speed variation will not be detectable in time intervals of a few milliseconds. Again, magnetic markers are used to measure the rotation of the crown block and to calculate the elongation of the rope lines. This value gives the position of the hook (kelly position) which, in turn, is necessary to calculate the borehole depth. This measurement can have a precision of 0.25 m, i.e., a sample is taken every 0.25 m. The time required to drill 0.25 m can be several minutes. Also, in this case, depth and rate of penetration – obtained from variation of depth in time – are updated at time intervals that are much longer than the times of variation of the seismic data.

Typical surface drilling parameters are shown in Table 5.4.

In some cases a MWD sub-assembly probe is used to measure downhole drilling parameters, which are sampled and transmitted via mud-pulse telemetry (see Section 2.3.18) and/or stored downhole. In this case, the low sampling rate is related to local memory resource or transmission-channel limits (few bits per second). The main downhole MWD parameters are:

- ◇ Weight on bit sub assembly (WOS) [kN].
- ◇ Torque on bit sub assembly (TOS) [kN m].
- ◇ Data processed from traces of multi-axial accelerometers used to detect vibrations, magnetometers, inclinometers and proximity sensors. These may give information about downhole rotation speed, and dangerous drilling modes.

Some downhole drilling parameters are shown in Table 5.5.

5.9.1 Automatic SWD acquisition by using drilling parameters

Automatic acquisition is programmed with the selection of the starting depths and of the maximum interval. The system starts the acquisition when this depth is reached and stops when the programmed interval is acquired (or a given number of records is acquired). With downhole pilot recording, the acquisition can be programmed in time. The recording is performed only when conditions of active perforation are matched. If, for instance, the WOB is zero because an extra drill pipe is being added, the system stops recording and starts again when the active-drilling condition is again satisfied. The DBSeis[2] system, the SWD system marketed by Schlumberger (Section 5.8.1), performs drilling recognition through estimation of spectral difference between drilling and non drilling conditions.

Selection of data by drilling parameters

A requirement for the SWD method is to reduce the noise in the seismic signal measurements of the bit. The signal-to-noise ratio can be improved using the technique of summing the contribution of a large amount of data collected by the measuring probes

[2]Mark of Schlumberger

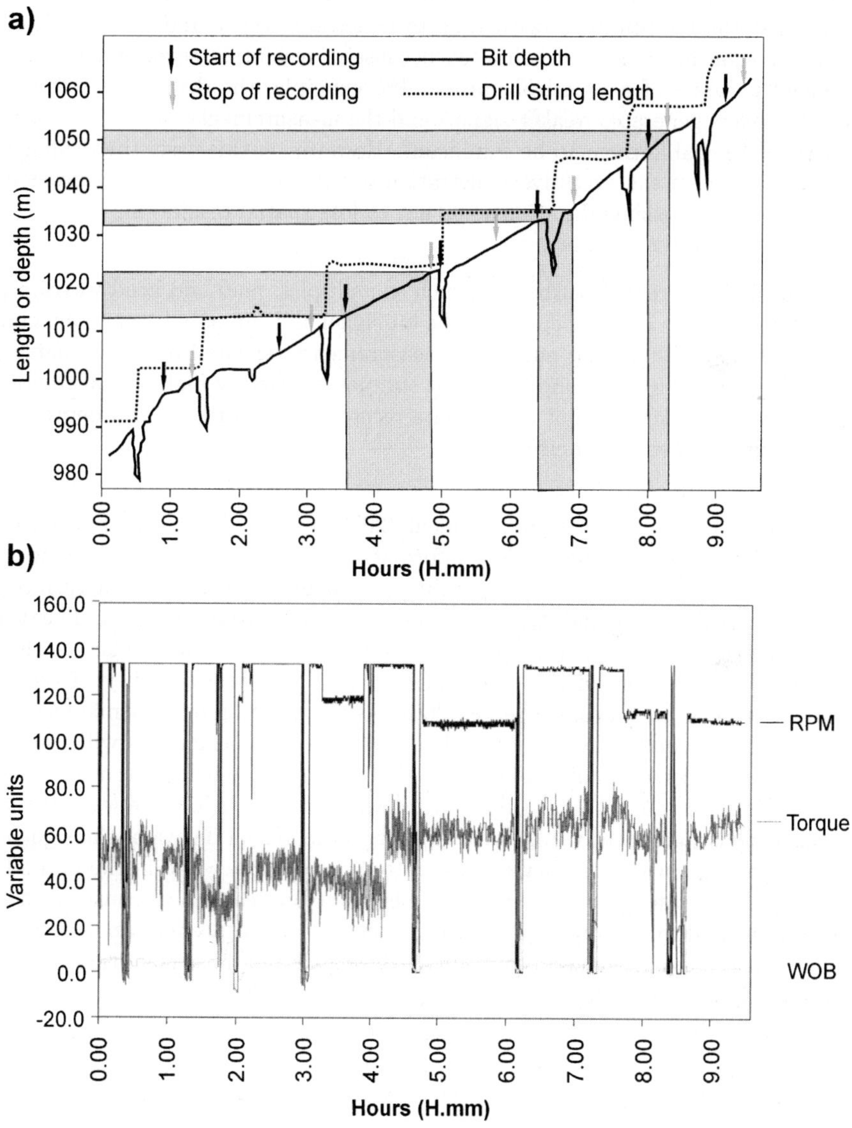

Figure 5.16: Surface drilling parameters plotted versus bit depth. a) Bit depth and drill string length. The addition of new pipes can be observed by the notches in the record. The string lengths remains constant between two consecutive additions. b) WOB, RPM and TORQUE and ROP are plotted in a relative-amplitude scale.

over a "reasonable" period of time. In this period, the drill-bit source produces sufficient energy to prevail over the random noise (see Section 7.3). However, even this method has its drawbacks. In fact, period over which the data can be summed up is limited by the fact that the bit continues with time to penetrate deeper and deeper. The allowed spatial extension of the source is limited because of resolution requirements. There are, consequently, limits of a practical nature. For example, the drilling of softer – or less compact – layers generates weaker signals and the measurements should be continued for longer times to enable them to be significant. However, in this case, the drilling speed is faster and, owing to the excessive penetration of the bit during the measurement, there are limitations in the time duration so as not to lose spatial resolution.

In addition, coherent disturbances (such as multiples, mud and head waves, rig vibrations) often prevent a clear reception of the bit signal and can only be partially eliminated with the method of summing numerous measurements repeated over reasonable periods of time; the result can be an improved, but still not sufficiently high, signal-to-noise ratio. A key point is the fact that not all the data recorded while drilling at a depth level have the same signal-to-coherent-noise ratio.

Miranda, Abramo, Poletto and Comelli (2000) proposed a method to improve the signal-to-noise ratio by using drilling information available on-site. Their approach is based on the fact that the mudlogging drilling parameters measured on the surface provide a description of the static and dynamic conditions of the drilling process. The conditions described by these parameters are related to the bit signal, which is linked to the drilling conditions at the bottom of the hole and influenced by the variability of the dynamic process and type of rock drilled (Section 3.6). As the duration of a seismic trace may, for example, be 24 seconds, the average value of the drilling parameters will, therefore, be obtained over the corresponding 24 seconds. In this way there will be a correct description of the whole signal trace of 24 seconds, during which the average value was calculated. In fact, this time is significantly long with respect to the arrival delays of the signals, which are propagated in the drilling plant from the bit to the surface and in the ground. These delays are in the order of a few seconds. It is, therefore, correct to consider the seismic measurements and the average drilling parameters as being contemporaneous. Hence, it is possible to use these parameters to examine the signal and noise conditions and improve the seismic signal of the bit.

The selective processing method is described in Section 7.4 and consists of a weighted sum of the contribution of numerous data collected by the measurement probes over a period of time. This is achieved by using weighting coefficients that only amplify records having higher signal-to-noise ratio and attenuate the records that contain prevalently noise. The weighed sum is affected by analysis of the data selected and reordered according to various drilling parameters measured during the collection of the bit signal. This leads to a separation of the bit signal and of coherent noises. Typical noises that are consistently registered in the final seismograms are the noise of the work-site activities and derrick vibrations.

5.10 Drilling depth and seismic depth

The main tasks of the SWD method are to sample the drill-bit seismic signals for well-bit-source depths, provide checkshot data while drilling and provide prediction ahead of the bit. This information is compared to the surface-seismic sections and used to update, while drilling, the casing points (Section 2.6.2) and, when necessary, correct the well trajectory. These results follow from the fact that the surface-seismic images and the well target are calibrated in depth/space, as the geological depth model is updated while drilling.

Vertical well

Let us assume, for simplicity, a vertical well. It is generally assumed that the drilling depth z_{drill} equals the true depth z_{true} (Figure 5.17a). However, this assumption is only an approximation. Let

$$\Delta L = z_{\text{true}} - z_{\text{drill}}. \tag{5.13}$$

A large part of the drill string (i.e., the 5 in. pipe) is under tension, which is variable in depth and increases as drilling depth proceeds (Section 2.3.10). Assume a given WOB and BHA composition. The increment of the hook load F_{H} of equation (2.11) is of the order of 250 kN for each kilometer of pipe (including buoyancy effects in mud). Assuming 4000 m of drill pipes in the absence of drag friction forces, an average tension force of 500 kN is exerted along the drill pipe. This force is increased by the weight of the BHA and reduced by the WOB, for which we can assume an additional tension force of 200 kN. For a pipe cross-section $A = 0.00339$ m^2 and a pipe Young's modulus $Y = 206$ GPa, we obtain the static elongation of the pipe

$$\Delta L_{\text{trac}} = \frac{700\,000 \cdot 4000}{0.00339 \cdot 206 \cdot 10^9} \approx 4 \qquad [\text{m}]. \tag{5.14}$$

Furthermore, thermal dilatation effects have also to be accounted for in the calculation of length. The linear thermal-expansion coefficient of steel is

$$\epsilon = 1.2 \cdot 10^{-5} \qquad [^{\circ}\text{C}^{-1}]. \tag{5.15}$$

With an average temperature gradient of 60 °C/km, as in high-temperature wells, the elongation of the 4000 m pipe for thermal effects is

$$\Delta L_{\text{ther}} = 1.2 \cdot 10^{-5} \cdot 60 \cdot 4000 \approx 2.9 \qquad [\text{m}]. \tag{5.16}$$

If not accounted for, these elongation effects could lead to an error of the order of 0.2% in depth evaluation, as

$$\Delta L_{\text{trac}} + \Delta L_{\text{therm}} \approx 7 \qquad [\text{m}] \tag{5.17}$$

with 4000 m pipe. In general, this error is smaller than the seismic spatial resolution. Another cause of apparent elongation effects could be buckling of the pipes. With helical

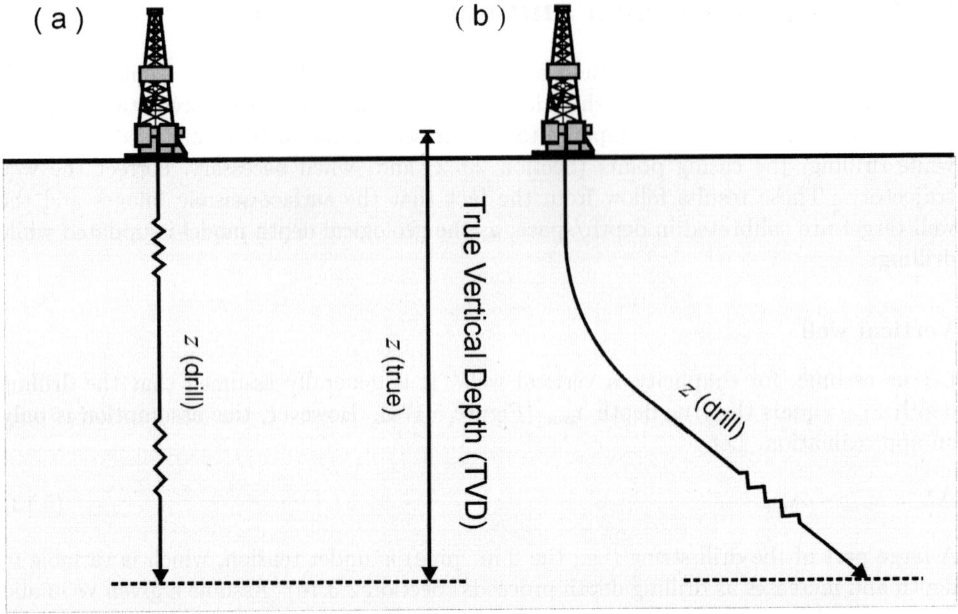

Figure 5.17: a) Depth of a vertical well. Because of the elastic properties of the drill string, the string length measured in air may differ from true depth. b) String length and depth of a directional well. Depths are often referred to the *kelly-bush* (KB) level

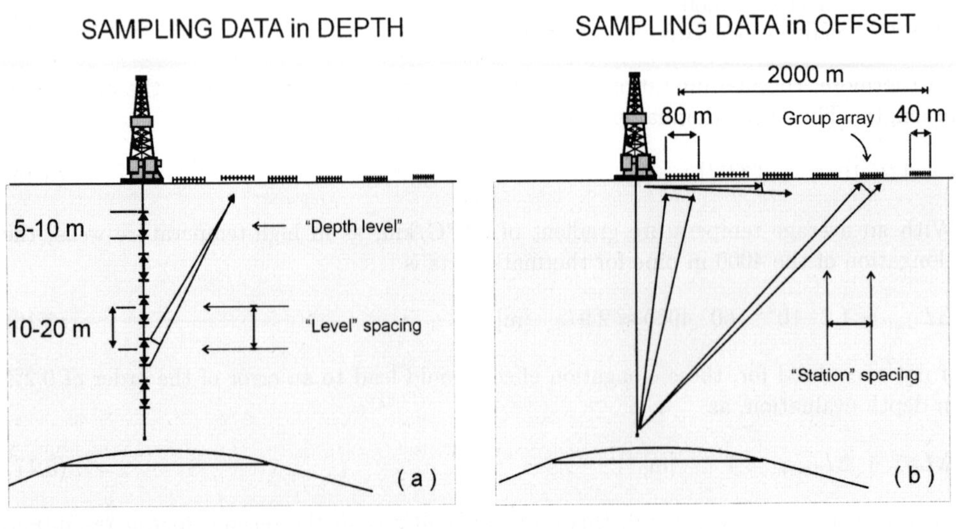

Figure 5.18: Sampling of SWD data a) in depth and b) in offset.

buckling, the elongation relative to true depth – which is, in this case, shorter – can be expressed as

$$\frac{z_{\mathrm{buck}}}{z_{\mathrm{true}}} = \sqrt{1 + \left(\frac{2\pi R_{\mathrm{c}}}{p}\right)^2}, \qquad (5.18)$$

where z_{buck} is the measured length of the bucked pipe, R_{c} and p are the clearance radius and constant pitch parameter (Figure 2.31). R_{c} is of the order of 0.1 m, and p represents the depth interval required to compete a helical turn. Assuming that p equals 30 m, the apparent contraction is of only 0.2 m per km. Therefore, it can be neglected. Possibly, pipe buckling could be of more importance for apparent elongation in deviated or horizontal drilling (as with coil tubing).

The length of the pipes forming the drill string is measured in the field. Usually, the individual lengths of the individual pipes fluctuate about an average value of 9.5 m. There are small deviations of units (or tens) of centimeters with respect to the average length of the individual pipes. The individual 9.5 m pipes (or 28.5 m joints of three pipes, see Figure 2.7c) are measured separately by drillers, and added up. The measurement errors add up as random numbers, and the total error is negligible. If the individual measurements have the RMS error of a centimeter, and 100 measures of 28.5 m individual pipe-lengths are performed, a total error of about only 10 cm can be expected (error of 0.003%).

Deviated well

In deviated wells (Figure 5.17b), the "drilling depth" (or length of the well curve) differs from the vertical depth, which is calculated using the deviation data (see Section 2.5).

5.11 Spatial sampling of SWD signals

Aliasing in multidomain SWD gathers is analyzed, for instance, by Mazzucchelli and Rocca (2001). Here we analyze the sampling of common-receiver and common-source gathers.

Common-receiver gathers

Let V be the velocity of propagation in the formation. It is well known that in a vertical well, the Nyquist sampling interval of a zero-offset VSP of maximum frequency f_{max} is

$$\Delta z_{\mathrm{max}} = \frac{V}{2 f_{\mathrm{max}}}. \qquad (5.19)$$

In offset VSP, the apparent interval velocity of the signal is higher and the condition of equation 5.19 still holds but, in this case, is too restrictive. Assuming a P wave of $f_{\mathrm{max}} = 100$ Hz and $V_{\mathrm{P}} = 3000$ m/s, we obtain $\Delta z_{\mathrm{max}} = 15$ m. Assuming an S wave of $f_{\mathrm{max}} = 60$ Hz and $V_{\mathrm{S}} = 1800$ m/s, we obtain again $\Delta z_{\mathrm{max}} = 15$ m. At parity of other acquisition factors, higher SWD frequency content is typically related to drilling harder rock, in which waves travel faster. Hence, in general SWD practice with only surface measurements, depth intervals ranging between 20 m and 10 m are commonly used (Figure 5.18a). Data of downhole tools showed higher signal-to-noise ratio and frequency content than surface data (see Section 7.8.1). In that case, shorter sampling intervals have to be used. A particular characteristic of SWD data may be that they are not sampled exactly at regular depth intervals, but only "on average", due to possible contingent operational requirements related to drilling and seismic-line maintenance.

Common-source gathers

The maximum offset of SWD traces typically increases with the target depth. Sampling in offset can be achieved with large trace spacings in order to acquire large-offset data with a number of offset VSPs sufficient to obtain the desired multiple coverage for reflections. The trace spacing can be of the order of 50 m or 100 m. The use of onshore group arrays shorter than the trace spacing may cause spatial aliasing for direct arrivals from the bit and refracted surface noise in the common-shot gathers (Figure 5.18). Provided that each individual offset SWD VSP is properly sampled in depth to avoid spatial aliasing of signal in depth, the offset-aliasing effect can be considered as an acceptable shortcoming in many field applications. The advantage is in the fact that, with a given number of traces, under-sampling along the line may make it possible to spread the available trace groups with large intervals at wide extended offset.

5.12 SWD-source pattern with bit deepening

The SWD data corresponding to a depth level are measured while drilling short depth intervals of a few meters in order to collect an amount of seismic energy comparable to that of conventional seismic sources (Section 3.21). The extension of the source in depth introduces source-pattern effects in the correlated data. If the SWD correlated data are corrected for the pilot delay as discussed in Section 5.2 or, if the pilot delay remains constant in the depth interval, the only variation of delay is due to the propagation of the signal in the formation. If we assume an ideal zero-offset reverse VSP in a constant-velocity layer, the filtering effects introduced by the spatial extension of the source are the same for the direct waves (reverse-VSP upgoing wavefield) and primary reflections (reverse-VSP downgoing wavefield). Figure 5.19a represents the descent steps of the bit source and the rise steps of the virtual bit source reflected by an interface. The depth intervals of the bit descent and of the virtual-specular-bit rise are denoted by 1 and 2, respectively. If the drill-bit emission has constant properties while drilling these steps, the source-pattern filter can be expressed as

$$\mathcal{S}(\eta) = \frac{\sin \eta}{\eta}, \tag{5.20}$$

with

$$\eta = \frac{2\pi f \, \Delta z}{V}, \tag{5.21}$$

and

$$\Delta z = 3600 \, \frac{\mathrm{ROP}}{\Delta T}, \tag{5.22}$$

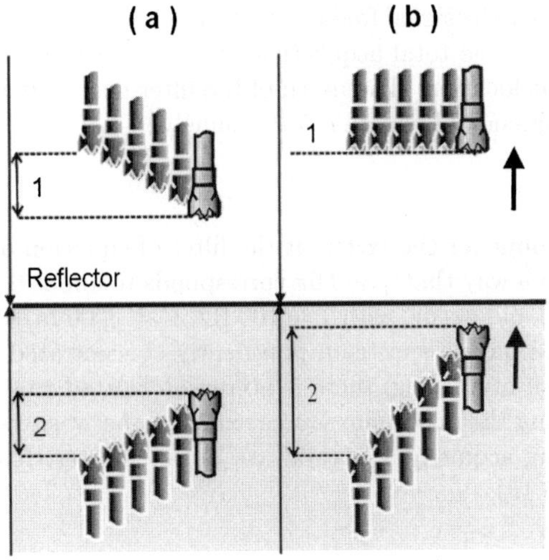

Figure 5.19: a) Effect of source pattern on upgoing and downgoing waves. b) Effect of trying to correct the source movement.

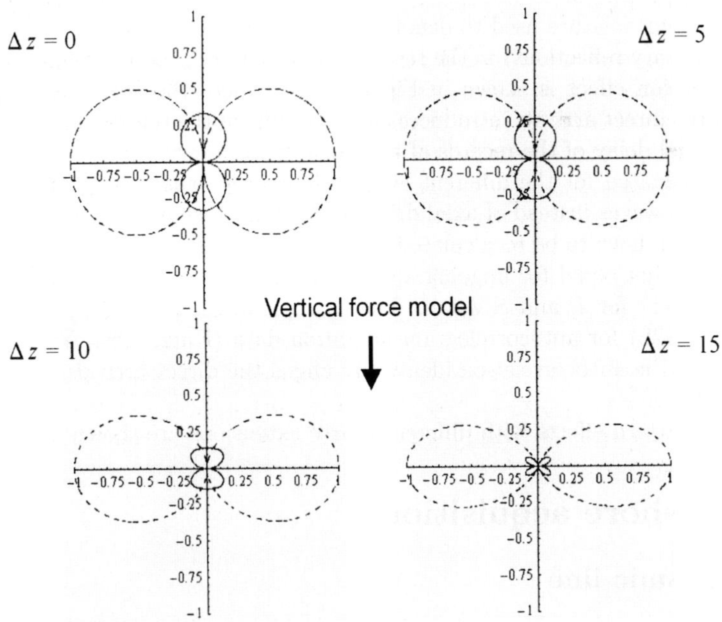

Figure 5.20: Effect of bit deepening on source pattern. Lobes of P (continuous lines) and S (dashed lines) waves are shown. The bit is assumed as a vertical force. The vertical extension of the source is Δz [m]. Velocity of P and S waves is 3000 and 1800 m/s, respectively. Frequency of P wave is 100 Hz, of S wave is 60 Hz. Δz of the order of 1/3 the Nyquist interval ($\Delta z_{\max} = 15$ m) produces a good approximation of the single-point pattern ($\Delta z = 0$).

where f, Δz, and ΔT are the signal frequency, the thickness of the interval drilled with rate of penetration ROP, and the total acquisition time, respectively, and V is the formation velocity at the drill-bit location. The notch of the filter occurs at $\eta = \pi$. Here, we obtain the same Nyquist frequency of equation 5.19, namely,

$$f = \frac{V}{2\,\Delta z}. \tag{5.23}$$

However, we may account for the fact that the filter of equation 5.20 attenuates the frequency components in a way that $\eta = 0.6\pi$ corresponds to about 6 dB of attenuation. For example, this effect should occur, with $f = 100$ Hz, $V = 3000$ m/s, for $\Delta z = 9$ m. Hence, when the original data have a spectrum prevalently concentrated at the low frequencies without significant risk of aliasing, there is no reason related to signal energy to further attenuate them by using the maximum source extension between two levels points. In this case, the active drilling acquisition interval can be set shorter than the interval between depth levels (Figure 5.18).

Data without correction to zero time

The apparently reasonable idea of correcting the effect of the source extension in the direct-arrival data solves the problem of lack of resolution for the *upgoing* VSP wavefield, but it worsens the source-extension pattern filtering effect for the *downgoing* reflections. Here we adopt the nomenclature used to denote upgoing (with direct arrivals) and downgoing fields (with primary reflections) in the reverse VSP geometry (Section 7.9.3). The concept of source-extension effect is shown in Figure 5.19b, where the corrective alignment by time shift of the direct arrivals introduces a double shift in the reflections. Any correction of the differential delay of the records of the same depth interval (say depth level) has to be carefully evaluated for the different events. Moreover, when using different types of pilots (e.g., mud waves instead of axial drill-string signals), source pattern and resolution in correlated data have to be recalculated. In general, the effects (source steps 1 and 2 of Figure 5.19) are not equal for upgoing and downgoing waves, and the radiation-pattern effects are different for P and S waves. The example of Figure 5.19b just describes the filtering effects (2b) for autocorrelogram migration data (Katz, 1984; Yu and Schuster, 2001), in which the autocorrelation identically aligns the direct arrivals at zero time (in Figure 5.19 1b).

Radiation pattern effects with different source extensions are shown in Figure 5.20.

5.13 Onshore acquisition

5.13.1 Seismic line

Onshore, the seismic line is formed by telemetric cables, analogic/digital (A/D)-converter recording-station units, repeaters, and geophone groups. The line is supplied by the central recording unit. However, when the length of a branch of the telemetric line is longer than a given length (for example, approximately three kilometers with more than 40 channels connected with 75 m cables in the SEISBIT® system), additional power supply units may be required, thus implying additional line-maintenance activity.

The laying down of the seismic line is principally guided by the geological conditions and by the drilling needs, and it follows a scouting of the area. A refraction survey provides the static corrections to apply to the data. The seismic line starts near the well and can be several kilometers long, with variable configuration and spacing between the geophones. Typically, the line is aligned with the dip direction of the geological structure. Crossing lines, i.e., in the "strike" direction with respect to structure, may be also used to study lateral velocity variation. The layout of the seismic line is designed by taking into account such aspects as the well targets, the theoretical radiation diagram of the bit (Rector and Hardage, 1992) and the noise fields. Figures 5.21 and 5.22 show a time window and the frequency spectrum, respectively, of onshore field data recorded while drilling carbonates with a roller-cone bit at about 2150 m depth. The first 19 traces are pilot signals. The remaining traces, numbered from 20 to 83, are geophone signals of the seismic line. Amplitude is normalized. The different responses of the geophone traces used at different locations – having different ground properties and ambient noise – and with different array lengths can be observed.

The following considerations and guidelines are important:

◇ A valuable characteristic of 2D SWD drill-bit surveys is the availability of both a large number of seismic channels and of source depth levels in the well. These data can be organized in common-receiver gathers (Figure 5.23a), i.e., as a single-offset VSP, or in common-source gathers, i.e., all the traces for a given bit depth (Figure 5.23b). Analysis of data in both the common-source and common-receiver domains is much more effective. It is easier to discern signal and noise arrivals both from their moveout-versus-offset and delay-versus-bit-depth properties: for example, some noise is stationary. This means that the noise has the same delay at different bit depths. Stationarity is an important criterion for noise interpretation and SWD data validation (see, also, Section 5.14.3).

◇ There are at least five reasons to use SWD offset lines with a medium-large length L (say several hundred meters or some kilometers) spreading from near the well. First, the resolution of the velocity analysis, which uses the moveout of the direct-arrival curves, generally increases with the length of the spread line (Yilmaz, 2001). This effect is important, especially for deep-drilling signals and deep reflections. Second, the maximum offset L determines the maximum distance from the well at which the reflections are investigated; that is $L/2$ if we assume a vertical well and a geological model with flat horizontal layers. Third, large spreads, extended at a distance of several hundred meters from the well, help the discrimination of the signal and of the noise, as discussed above. Fourth, large offsets also allow us to effectively record S waves. Finally, the use of long seismic lines with many offset channels increases the multiple coverage and improves the processing of the reflections ahead of the bit.

◇ The theoretical radiation diagram of a tricone bit has been discussed in Section 3.8.2 (see, also, Hardage, 1992). If the roller-cone bit source is assumed as a vertical point force in a vertical well, the radiation patterns are proportional to the $\cos\phi$ and $\sin\phi$ functions for P and SV waves, respectively (Figures 3.6 and 5.29). That is, the

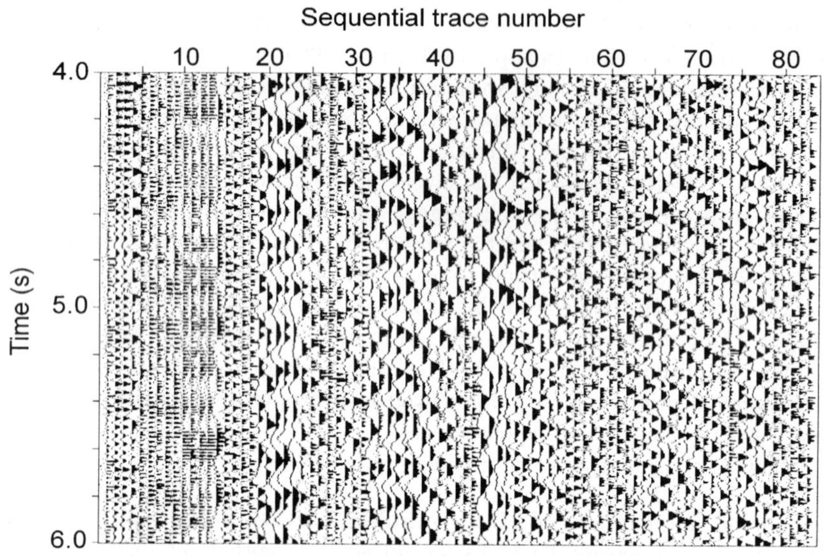

Figure 5.21: Raw field data. Traces from 1 to 19 are rig pilot signals. Traces from 19 to 83 are geophone signals sampled along the offset.

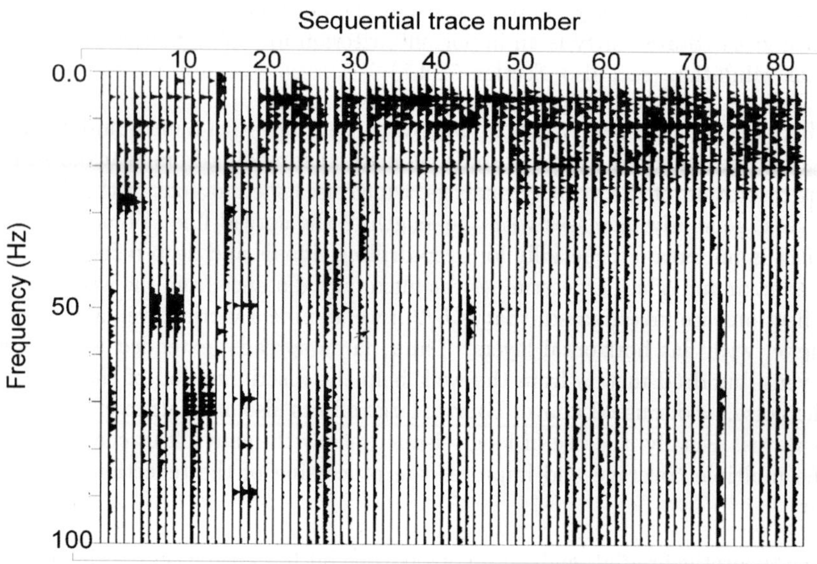

Figure 5.22: Frequency spectrum of raw field data of Figure 5.13.1. Traces from 1 to 19 are rig pilot signals. Traces from 20 to 83 are geophone signals sampled along the offset.

P-wave intensity is high at short offset and low P at long offset with respect to bit depth. The opposite holds for S waves. In a Poisson medium, the total P radiated power is about 0.1 times the total $(P + SV)$ radiated power. The P intensity in the direction of maximum radiation ($\phi = 0$) is 0.3 times the total $(P + SV)$ average intensity.

The theoretical radiation model includes horizontally-polarized, SH, shear waves (Rector and Hardage, 1992). These components are detectable in SWD experiments, and have a radiation pattern – approximately a $\sin\phi$ function – similar to that of the vertically-polarized SV shear waves. The SH pattern is due to differential transverse action between inner- and outer-row teeth for roller-cone (Section 3.12.1), and to shearing action of cutters for PDC bits (Section 3.14). If a vertical well is assumed, shear waves are more easily detectable at relatively large lateral offset with respect to source depth. Three-axial geophones oriented in the radial, tangential and vertical direction are used to record the P, SV and SH waves.

- ◇ Of particular importance is the selective use of the receiver arrays to suppress the ambient noise. The noise pattern of a working roller-cone bit is discussed by Rector (1992). A general description of the noise fields related to drilling is given in Sections 4.8 and 5.14.2. The seismic line in the proximity of the well (distance less than 300–400 m) is highly disturbed by ground roll and head-wave noise coming from the site through the uppermost layers. Long receiver arrays – oriented in the radial direction from the well – are used at short offsets; they significantly reduce the organized noise and do not significantly affect the signal. In general, the array length is reduced at increasing offset, because the array reduces the signal at oblique inclination.

- ◇ In most cases, the seismic line configuration is maintained fixed throughout a survey. This is typically done to assess the continuity of the signature in VSP common-receiver gathers of each channel. This solution is preferred to that of changing the configuration of the seismic line because of the variation of the bit depth, which may introduce more problems than benefits. When using a fixed line all through the survey, its design is, in most cases, the result of a compromise done to optimize the recording of both shallow and deep data. In some cases, there may be the need of repositioning the seismic line after interruptions due to environmental problems, such as for seasonal farming works. These activities may produce loss of data in some of the SWD-VSP channels. Therefore, some redundancy in the data of a multiple-coverage survey is beneficial.

- ◇ For all the above reasons, seismic lines of length of half or one third the maximum well depth are commonly used.

- ◇ Either twenty-four or twelve geophones per trace group has shown to be satisfactory. Receiver-trace spacing (intertrace) along the seismic line is, in general, determined by the characteristic of the instrumentation, such as the maximum length of the telemetric cables and the number of channels, and by the maximum offset. Often the survey is designed using large intertrace values, for example 75 m, which can be more than the extension of the geophone array of a single trace. This over-spacing

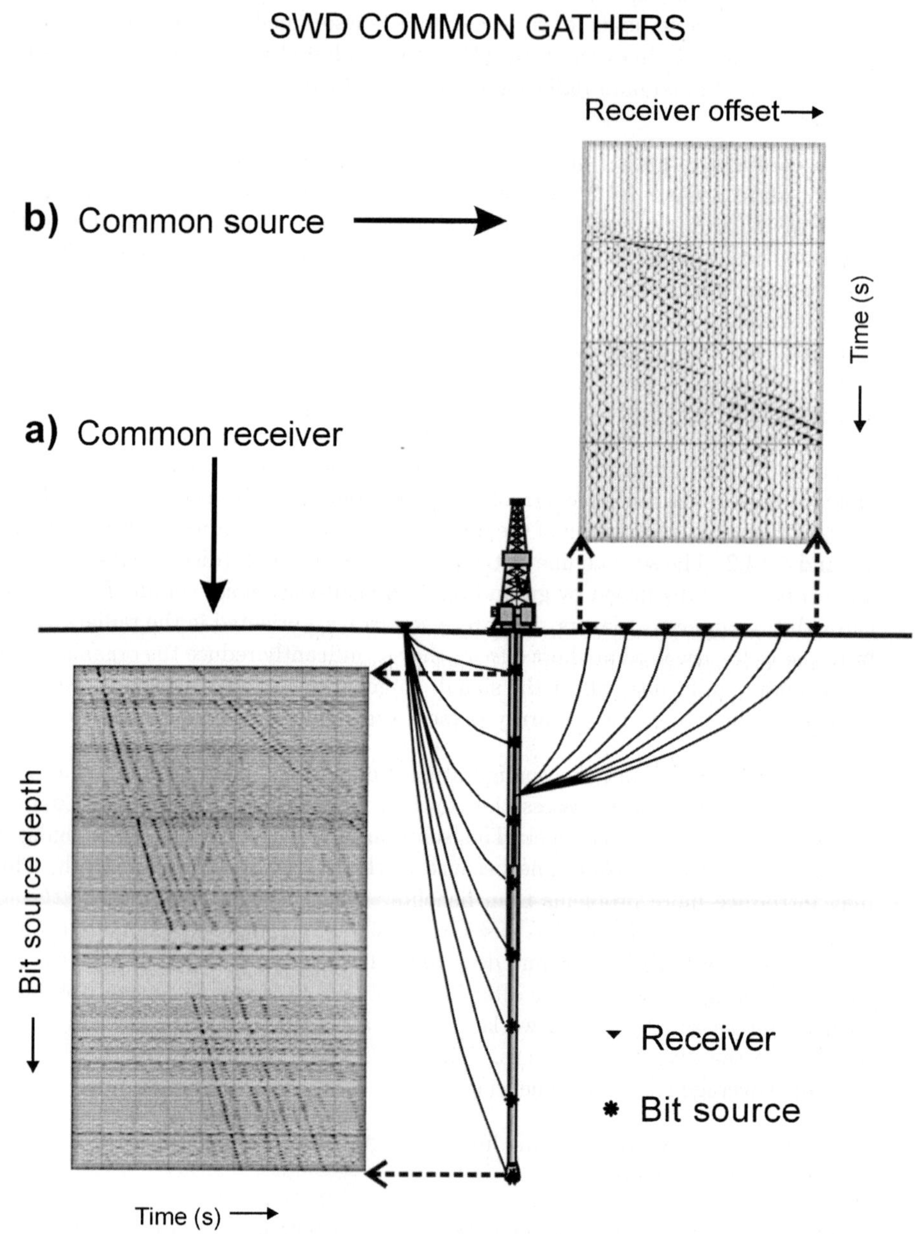

Figure 5.23: Correlated a) common-receiver (i.e., single offset VSP) and b) common-source SWD gathers.

may cause under sampling in offset and aliasing effects for the direct arrivals in the common-source gathers. However, this effect can be acceptable, provided that the aliasing in the VSP reflection data is avoided by properly sampling the acquisition in depth.

5.14 Receiver arrays in SWD

Conventional wireline-VSP typically utilizes borehole tools in which each recording channel is connected to a single geophone. SWD uses the reciprocal-VSP geometry, in which the source is in the borehole and the receivers are laid down at the surface. In this case, groups of surface receivers may be used effectively to spatially filter the waves recorded as an individual trace. This is an important advantage for SWD surveying, because the receiver groups may have array configurations that reinforce the signal and reduce the noise components.

The problem is that when we reject the noise using spatial arrays, we might also filter away a part of the signal. Indeed, what we do not want to do is to "throw away the baby with the bath water". Therefore, we have to determine the "optimum array configuration" that maximizes the signal-to-noise ratio. If we assume a group array, i.e., an array in which the traces are hard wired together and not delayed or weighted, such an optimum-array configuration does not always exist for a given pattern of signal and noise. Beam steering, that is the weighted stack of the delayed traces of a spatial array, is also used instead of group arrays to get the optimum signal-to-noise ratio (Section 6.8.2).

Arrays are used to prevent signal degradation by the presence of coherent and random noise. Because the drill-bit signal is measured over time intervals of several minutes, the drilling and environmental coherent noise adds up to a non-negligible level in the geophone (hydrophone) correlated traces. At the same time, the receiver arrays also measure random noise. This noise is produced by the scattering, at the near-surface, of the drill-bit signal and coherent-noise waves coming from drilling activity. In addition, random noise may be produced by any other random-noise source present at the surface in the proximity of the seismic line, as well as by direct and scattered borehole tube and head waves. Coherent and random noise are reinforced in the SWD correlated seismograms with different rates with respect to the total recording time. The processing, based on pilot correlation for reducing these noise components, is discussed in Section 6.3.5.

As shown below, the SWD optimum receiver arrays are calculated to spatially filter the coherent noise components and preserve the drill-bit signal in measurements with variable random-noise levels. For simplicity, we analyze the theoretical response of linear arrays, i.e., group arrays with equispaced receivers disposed along a straight line. We assume that all the receivers of the array have the same sensitivity, ground-coupling and directivity properties (Aldridge, 1989). In this case, the linear array layout is determined by two parameters. For instance, by:

- ◇ The number M_g of receivers (say geophones) in the array group.
- ◇ The total length L of the array.

The array-output trace is the sum of the individual receiver measurements. The ratios of signal to coherent noise and to random noise in the output trace both depend upon the number of receivers per group and array length. In fact, the arrays also reduce spatially random noise when the same noise is measured, locally, by more receivers of the same array.

Random noise

The effect of geophone arrays on random noise depends on the definition of that noise (Levin, 1989). Sheriff (1999) defines the random noise in receiving channels as energy that does not exhibit correlation between different receiving channels (Case 1). However, this definition of spatially random noise can be assumed as only partially true (Case 2) for the individual geophones of arrays. In fact, independence can not be assumed for noise recorded at different geophones when the individual receivers measure noise with wavelengths comparable to the receiver distance.

Case 1 – Assume that the input noise has the same intensity at each geophone. If the input noise of each geophone of the array group is uncorrelated with the noise of each other geophone of the same array ("pure random" noise), we can assume that the array-output noise N is proportional to $\sqrt{M_g}$. If the input signal of all the individual geophones of the array is the same, the array output signal S is proportional to M_g. For these reasons, geophysicists often assume that the random noise-to-signal ratio, N/S, in the array is proportional to $1/\sqrt{M_g}$. If, in particular, each input noise has unit energy, we simply obtain the output noise

$$N = \sqrt{M_g}. \tag{5.24}$$

Case 2 – Levin (1989) analyzed the array response to the partially-correlated random noise of the Aki-Denham type (Levin, 1989). This noise model consists of plane surface waves of wavelength λ – of the ground-roll type – arriving at a vertical geophone scattered from any azimuthal direction with equal probability (this model is partially true for borehole waves scattered in a limited angular aperture). In this case, the noise in the input at a receiver of the array is not independent of the noise at any other receiver. For this type of noise, the receiver responses are correlated by a function that depends of the ratio Δ/λ, where

$$\Delta = \frac{L}{M_g - 1} \tag{5.25}$$

is the geophone separation, and λ is the noise wavelength. This wavelength is the same in all the horizontal directions of propagation of the noise plane waves. In the Aki-Denham model, these directions are variable and make any azimuthal angle with the array axis, defined as the straight line passing through the receivers of the array. Assume that each geophone has a unit noise-to-signal ratio. The array noise-to-signal ratio (N/S) for random-scattered noise can be written as (Levin, 1989)

$$N/S = M_g^{-1} \left[M_g + 2 \sum_{i=1}^{M_g} \sum_{j>i}^{M_g} J_0(2\pi(j-i)\Delta/\lambda) \right]^{1/2}, \tag{5.26}$$

where $J_0(2\pi(j-i)\Delta/\lambda)$ is a Bessel function of order zero (see Figure 5.24a). The output of the array of length L to input noise waves of unit amplitude and frequency ω, i.e., the noise frequency response of the array, is

$$N(L, M_g, \omega, V_{rs}) = \left[M_g + 2 \sum_{i=1}^{M_g} \sum_{j>i}^{M_g} J_0 \left(\frac{(j-i)L\omega}{(M_g-1)V_{rs}} \right) \right]^{1/2}, \tag{5.27}$$

where $V_{rs} = \lambda\omega/2\pi$ is the horizontal propagation velocity of the random scattered waves. Equations (5.26) and (5.27) are for vertical geophones. Similar equations hold for horizontal receivers (Levin, 1989). N/S of equation (5.26) equals 1 for $\Delta/\lambda = 0$. In other words, Aki-Denham random noise with large wavelengths is insensitive to receiver spacing. Conversely, for values of Δ/λ greater than zero and approaching one, N/S approaches $1/\sqrt{M_g}$. Figure 5.24a shows N/S of equation (5.26) with $M_g = 24$. The maximum N/S attenuation is intermediate between the attenuation expected for pure random noise $(1/\sqrt{M_g})$, and the maximum attenuation of the envelope of the coherent noise $(1/M_g)$.

Coherent noise

Arrays are commonly used to attenuate organized noise, i.e., noise which bears coherence between the receivers of the array. Assume that only one time-harmonic plane wave of noise propagates with wavelength $\lambda = 2\pi V_a/\omega$ in the axial direction of the array, V_a being the apparent velocity in the array axial direction – that is, the phase velocity on the array depending on incidence angle. If the input signal is the same in all the geophones, say unit signal of apparent velocity equal ∞, the output signal $S = M_g$. We obtain the familiar equation for attenuation of monochromatic (single frequency) coherent noise by linear arrays (Figure 5.24b), namely,

$$N/S = M_g^{-1} \left[\sin^2(\pi M_g \Delta/\lambda) / \sin^2(\pi\Delta/\lambda) \right]^{1/2}. \tag{5.28}$$

The first zero ("notch") occurs at $\lambda = M_g\Delta$. The Nyquist wavelength is at $\lambda = 0.5\Delta$. The maximum attenuation of the envelope of the coherent noise-to-signal ratio is $1/M_g$ (dashed line in Figure 5.24b). If the signal has an apparent velocity different from ∞, it is filtered to some extent by the array, even if the array is designed to attenuate primarily the noise. The frequency response of a linear array of length L to an input unit plane wave of apparent velocity V_a is

$$A(L, M_g, \omega, V_a) = \left| \frac{\sin M_g\xi}{\sin \xi} \right|, \tag{5.29}$$

where

$$\xi = \frac{L\omega}{2(M_g-1)V_a} = \frac{\pi\Delta}{\lambda}. \tag{5.30}$$

In the more general case, equation (5.29) is used, as a first approximation, both for signal (with $V_a = V_s$) and coherent-noise waves (with $V_a = V_c$).

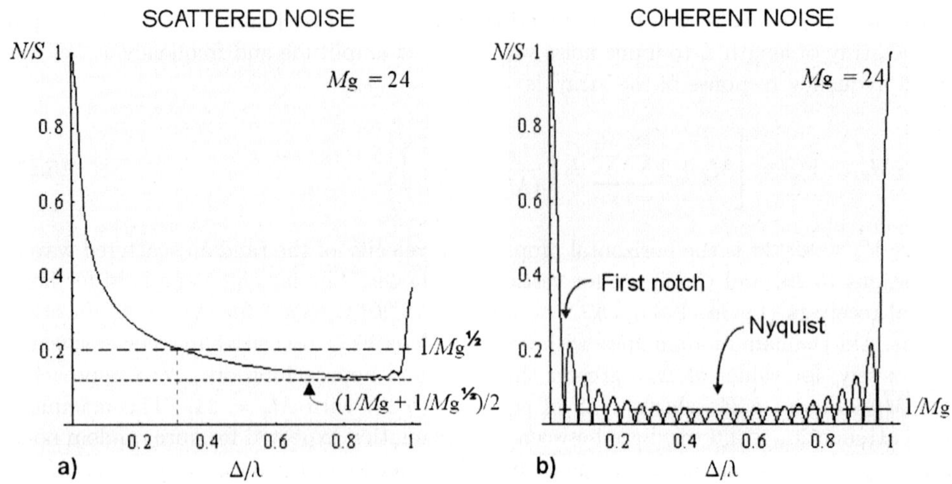

Figure 5.24: a) N/S response for scattered random noise of Aki-Denham type in a linear array of M_g vertical geophones. The "independence" distance between receivers is at about $\Delta = \lambda/3$. At this distance, the scattered-noise curve crosses the line of the pure-random-noise response (dashed line) of $N/S = 1/M_g^{1/2}$. The scattered-noise curve has no notches and periodicity is observable at $\Delta/\lambda = 1$. b) N/S for coherent noise. Maximum attenuation of the envelope is $1/M_g$. The first notch occurs at $\lambda = M_g\Delta$. The Nyquist wavelength is at $\Delta/\lambda = 0.5$ (modified after Levin, 1989).

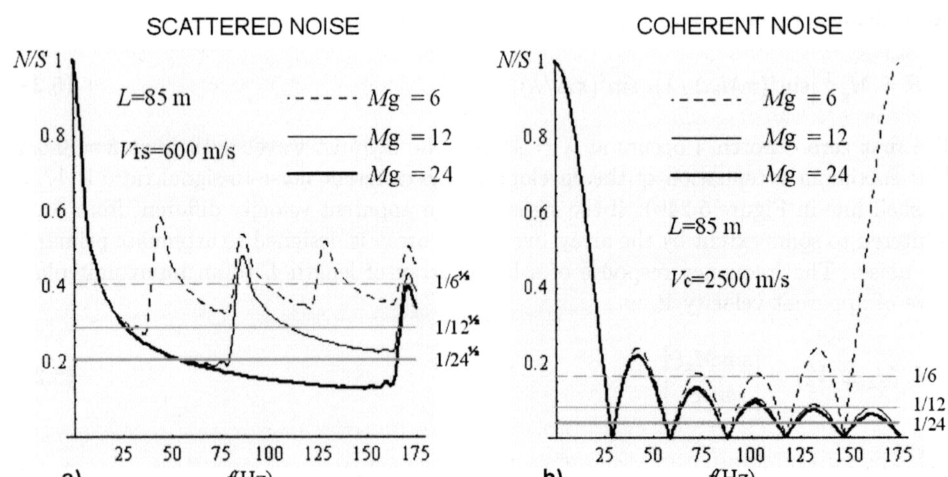

Figure 5.25: Number of receivers in the array and N/S response for a) scattered (black) and pure (gray) random noise. Periodicity in the response for the Aki-Denham scattered-noise model is observable at frequencies $f = V_{rs}/\Delta$. b) Coherent noise with the maximum envelope attenuation for 6, 12 and 24 geophones is represented by the gray lines.

Number of receivers and noise attenuation

Normalized noise-to-signal ratios obtained with a different number of receivers M_g are shown in Figure 5.25. Attenuation curves are calculated by using equations (5.26) and (5.28) for a) surface random noise of Aki-Denham type and b) time-harmonic coherent noise, respectively. Linear arrays of length $L = 85$ m, with 24, 12 and 6 receivers are used in these examples. N/S is plotted versus frequency. The frequency response is calculated by assuming horizontal propagation velocity V_{rs} of 600 m/s for random noise of scattered surface plane waves, and apparent velocity V_c of 2500 m/s for coherent noise.

We observe that at frequencies below 25 Hz, the same attenuation is obtained by using 24, 12 or 6 receivers. At frequencies between 25 Hz and 75 Hz, attenuation obtained by 24 and 12 receivers is comparable, both in the a) scattered and b) coherent-noise curves. On the other hand, 6 receivers do not attenuate random noise effectively above 25 Hz. Above 75 Hz, the attenuation obtained by using 24 geophones is much more effective than when using 6 or 12 geophones. In addition, Figure 5.25a shows the expected normalized-N/S levels (horizontal gray lines) for pure random noise with 24, 12 and 6 receivers.

This example shows that a low number of receivers ($M_g = 6$) does not provide the spatial sampling to attenuate noise. In particular, a higher number of receivers is necessary if the scattered-random-noise components have to be attenuated effectively.

5.14.1 Analysis of optimum arrays

In SWD, the receiver arrays are used to attenuate waves of the spatially organized noise recorded in the geophone traces that are time-correlated with the rig-pilot signal (Section 6.3.5). Therefore, SWD arrays are oriented usually towards the rig, in order to filter primarily those waves of spatially coherent noise that are reinforced in the correlation and result as time-coherent noise. Obviously, the filtering properties of an array do not depend on the pilot-signal properties as well as on correlation processing. However, the amplitudes of the processed signal and noise – amplitudes which ultimately determine the selection of the array configurations – do.

Use of receiver arrays in SWD is discussed by Hardage (1992). He gives the relationship between the array length L and the array moveout time. Array moveout time is defined as

$$T_{mo} = \frac{L}{V_a}, \tag{5.31}$$

where the apparent velocity can be expressed as

$$V_a \approx \frac{ZV}{X}. \tag{5.32}$$

V, Z and X are the formation velocity, bit depth and receiver offset, respectively. A wave with a period equal to the array moveout has a phase difference of one period between the ends of the array. This difference corresponds to the first notch of the array response at the frequency $1/T_{mo}$. In general, arrays filter both signal and noise waves. The maximum signal resolution is related to the signal bandwidth, from zero to the notch frequency. The array moveout time (or notch frequency) for the signal may be fixed as a rule to determine the maximum array length.

In the SWD practice using signals recorded at short/medium offset and medium/large depth, the array-notch frequency for signal waves is higher than the allowable signal bandwidth. In this case, the signal-to-noise ratio S/N is used as a criterion to design an array. The calculation of S/N in SWD arrays is performed with respect to both coherent and random noise. Let the surface coordinates – azimuth and offset from the wellhead – of the center of the receiver array be given. We assume that we know the apparent velocity V_s of the drill-bit signal, the apparent velocity V_c of the coherent noise and the horizontal-propagation velocity V_{rs} of surface-scattered random noise. Let $S(\omega)$, $N_c(\omega)$ and $N_{rs}(\omega)$ be the Fourier transforms of the signal, coherent noise and surface scattered noise, respectively. The optimum arrays are obtained by determining the array length that maximizes the signal-to-noise ratio S/N in a given frequency interval (ω_1, ω_2), $\omega_1 < \omega_2$. We have

$$S/N = \frac{\mathcal{S}(L, M_g, \omega_1, \omega_2, V_s)}{\sqrt{\mathcal{N}_c^2(L, M_g, \omega_1, \omega_2, V_c) + \mathcal{N}_{rs}^2(L, M_g, \omega_1, \omega_2, V_{rs}) + \epsilon^2 M_g}}, \tag{5.33}$$

where the root-mean-square (RMS) amplitude of the output signal

$$\mathcal{S} = \left[\int_{\omega_1}^{\omega_2} |S(\omega)|^2 |A(L, M_g, \omega, V_s)|^2 d\omega/(\omega_2 - \omega_1)\right]^{1/2}, \tag{5.34}$$

and the RMS amplitude of the output coherent noise

$$\mathcal{N}_c = \left[\int_{\omega_1}^{\omega_2} |N_c(\omega)|^2 |A(L, M_g, \omega, V_c)|^2 d\omega/(\omega_2 - \omega_1)\right]^{1/2}, \tag{5.35}$$

are calculated using the array-response of equation (5.29). Equation (5.35) can be readily extended to the case of additional coherent-noise components. The scattered-random-noise RMS amplitude is

$$\mathcal{N}_{rs} = \left[\int_{\omega_1}^{\omega_2} |N_{rs}(\omega)|^2 |N(L, M_g, \omega, V_{rs})|^2 d\omega/(\omega_2 - \omega_1)\right]^{1/2}, \tag{5.36}$$

where the array-response $N(L, M_g, \omega, V_{rs})$ to scattered noise is given by equation (5.27). We assume that coherent- and random noise types are uncorrelated in equation (5.33), in which ϵ^2 is the energy of pure random noise. Figure 5.26a shows $\mathcal{S}$, $\mathcal{N}_c$ and $\mathcal{N}_{rs}$ curves plotted versus array length L. Integration in equation (5.36) can be approximated also, for array-lengths L between 0 and 200 m, by

$$\mathcal{N}_{rs}(L, M_g, \omega_1, \omega_2, V_{rs}) \approx M_g \left[\frac{a}{\omega_2 - \omega_1} \log\left(\frac{a + \omega_2}{a + \omega_1}\right)\right]^{1/2}, \tag{5.37}$$

where

$$a = \frac{2\pi V_{rs}}{\Delta} \Bigg/ \left[\left(\frac{2M_g}{1 + \sqrt{M_g}}\right)^2 - 1\right]. \tag{5.38}$$

with $\Delta = L/(M_g - 1)$. Examples of these approximation curves are shown in Figure 5.26b.

We can observe that most of the power of the array-response curves is concentrated at Δ/λ close to zero, i.e., at low frequency. Therefore, we can expect different results if the frequencies close to zero are included in the interval of analysis. Let $V_s > V_c$. An analysis of optimum arrays calculated by using equation (5.33) leads to the following guidelines:

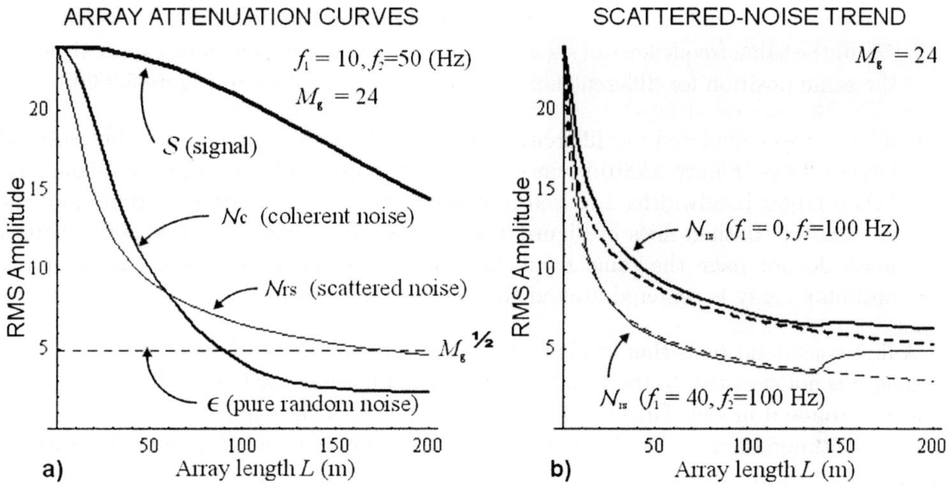

Figure 5.26: Model with $V_s = 11070$ m/s, $V_c = 1800$ m/s and $V_{rs} = 600$ m/s. a) RMS signal and noise array-attenuation trends, b) RMS scattered noise approximation (dotted line) versus array length L. Note the difference when periodic array-sampling effects occur at approximately $L = 150$ m. M_g is the number of geophones in the array.

◇ First we assume signal and coherent noise only of the same bandwidth, including the zero frequency. We assume $|S(\omega)| = |N_c(\omega)| = 1$ for $\omega < \omega_{max}$, and $|S(\omega)| = |N_c(\omega)| = 0$ for $\omega > \omega_{max}$. In this case, S/N has no maximum. It increases with L and tends, with slow rates, to a constant value for large L (thin continuous line in Figure 5.27a). That is, an optimum array does not exist. In practice, L has to be limited because of resolution requirements. In fact, a large L corresponds to a severe filtering of the high frequencies of both the signal and noise. With even a small amount of random noise, the signal may be not recoverable if the array filter is too severe. Contrarily, with only scattered noise in the same bandwidth, S/N shows a maximum (continuous line of medium thickness in Figure 5.27a).

◇ If signal and coherent noise have the same frequency bandwidth *not* including the zero frequency, the S/N curve has a maximum .

In this case, we assume $|S(\omega)| = |N_c(\omega)| = 1$ for $0 < \omega_{min} < \omega < \omega_{max}$, and $|S(\omega)| = |N_c(\omega)| = 0$ otherwise (dashed lines in Figure 5.27a). This maximum is very sensitive to the frequency content of coherent noise but not to that of scattered noise (Figure 5.27b). In fact, as can be appreciated in Figure 5.24, the coherent-noise response has a notch.

◇ When random (not scattered) noise is added to coherent noise, the S/N has a maximum. With coherent noise N_c of variable energy and constant random noise ϵ, the optimum arrays are obtained for shorter L with lower N_c. With random noise ϵ of variable energy and constant coherent noise N_c, the optimum arrays are obtained at shorter L for higher random noise ϵ (dashed lines in Figure 5.28a). Scattered noise N_{rs} in the same frequencies of signal produces a maximum, which has approximately the same position for different noise levels (continuous lines in Figure 5.28a).

◇ S/N curves calculated for different offsets show that shorter arrays are obtained at larger offsets (Figure 5.28b) because of the lower apparent velocity of the signal, V_s. When larger bandwidths are used in order to improve resolution, optimum arrays are shorter (dashed lines in Figure 5.28b). However, when the signal and coherent noise do not have the same bandwidth, it can be shown that the length of the optimum array is centered around the notch-length of the noise.

A first consideration is that random and scattered noise have different effects. The array length is not sensitive to the variation of scattered noise in the bandwidth investigated with the proposed model. On the other hand, the power of the random noise changes the length of optimum arrays. Furthermore, arrays are sensitive to coherent-noise bandwidth but not significantly to random-noise bandwidths.

As general guidelines from the analysis of these simplified models, when low-frequency coherent noise is expected – this is a typical condition for the SWD drill-bit wavefields – long arrays are preferred. Shorter arrays can be used when lower coherent-noise amplitude and larger signal bandwidth are expected in the SWD data. For instance, this condition may be achieved using downhole pilots for correlation (Section 7.8.1). Moreover, use of shorter arrays is preferred when coherent noise is weak and random noise is strong. This condition generally holds for traces acquired at large offsets, far from the wellsite source of organized noise. In general, arrays at larger offsets are shorter because of the lower apparent velocity V_s of the signal (see equation (5.32)) (Figure 5.28a).

5.14.2 Coherent and random noise in SWD geophone arrays

In the previous array analysis, comparable values of the SWD signal and of the noise have been used. It has to be emphasized that the above analysis is for geophone traces before correlation. The extension of the results to correlated data has to be evaluated carefully. Coherent waves in arrays may contribute to random noise in SWD correlations (Section 6.3.5). For this reason, the determination of the amplitude of the different types of noise is an important task in SWD. To check organized noise waves, very short arrays have to be used in order to preserve wavefields for the noise analysis tests.

Different types of organized noise are present in SWD recordings (Rector and Hardage, 1992; see, also, Sections 4.8 and 4.9). This noise can be recognized in a common-source-gather of seismic line traces obtained by crosscorrelating with a pilot sensor signal. As shown in Figures 5.30, 5.31, and 5.32 there are several noise types:

◇ Waves with low-velocity (typically from 200 to 600 m/s) are surface waves generated by rig vibrations and human activity in the well area. Like the ground roll in conventional common-shot gathers, this noise is characterized by low frequencies (typically < 15 Hz). A low-cut filter can effectively remove the ground roll in the processing phase. In view of the different spectra of signal and surface waves, this type of coherent noise does not represent a particular problem.

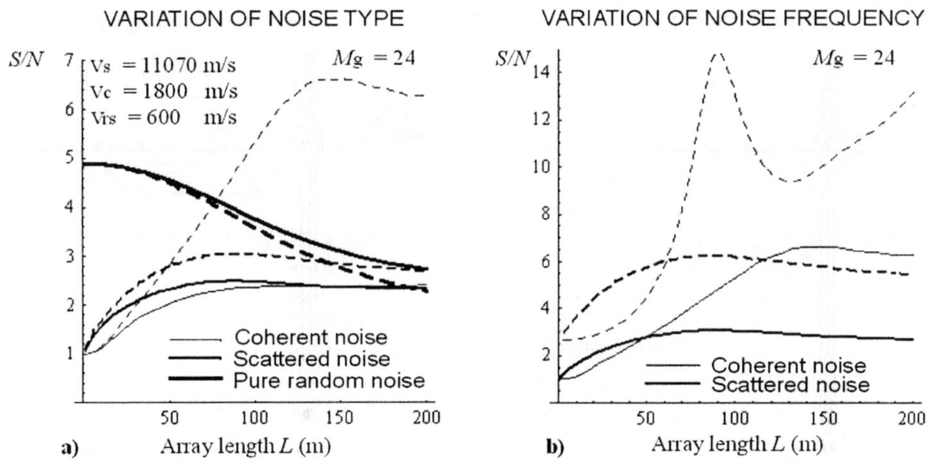

Figure 5.27: S/N curves each with only one noise type. a) Signal, coherent and scattered noise have the same bandwidth. Curves are calculated in the frequency bandwidth from 0 Hz to 80 Hz (continuous lines) and from 10 Hz to 80 Hz (dashed lines). b) S/N curves each with only one noise type. This signal has bandwidth from 10 Hz to 80 Hz. The noise has the same bandwidth (continuous lines), or a bandwidth from 15 Hz to 25 Hz (dashed lines) to simulate diesel engines. The maximum position is sensitive to noise frequency for S/N calculated with coherent noise, which has a notch at about $L = 90$ m, but not with scattered noise. The number of geophones in the array is $M_g = 24$.

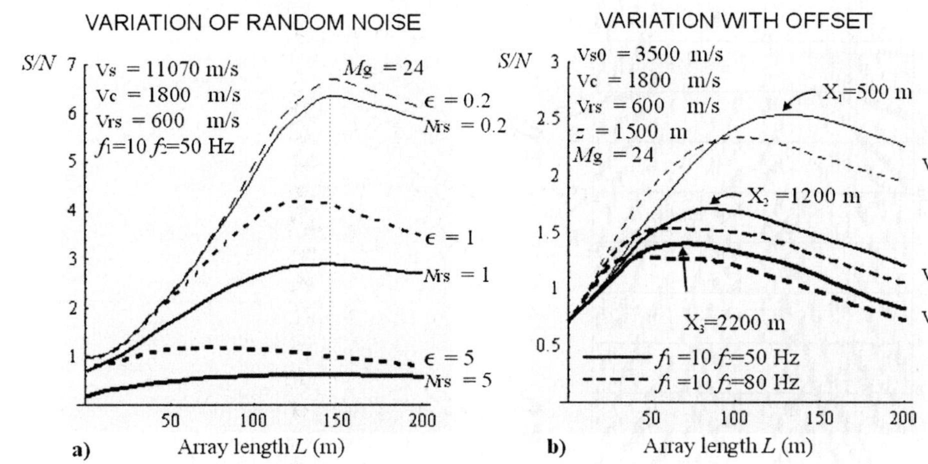

Figure 5.28: a) S/N with constant coherent noise to which is added scattered noise (continuous lines). The position of maximum for different levels of scattered noise is substantially unchanged. When pure random noise is added (dashed lines), longer optimum arrays correspond to the lower pure random noise level. b) S/N versus offset with signal and noise of equal frequency bandwidth. Bit depth is $z = 500$ m and formation velocity is $V_{s0} = 3500$ m/s. Apparent velocities are $V_{s1} = 11070$, $V_{s2} = 5600$, $V_{s3} = 4240$ m/s at offsets of 500, 1200 and 2200 m, respectively. With larger bandwidth for higher resolution, optimum arrays are shorter and S/N lower (dashed lines).

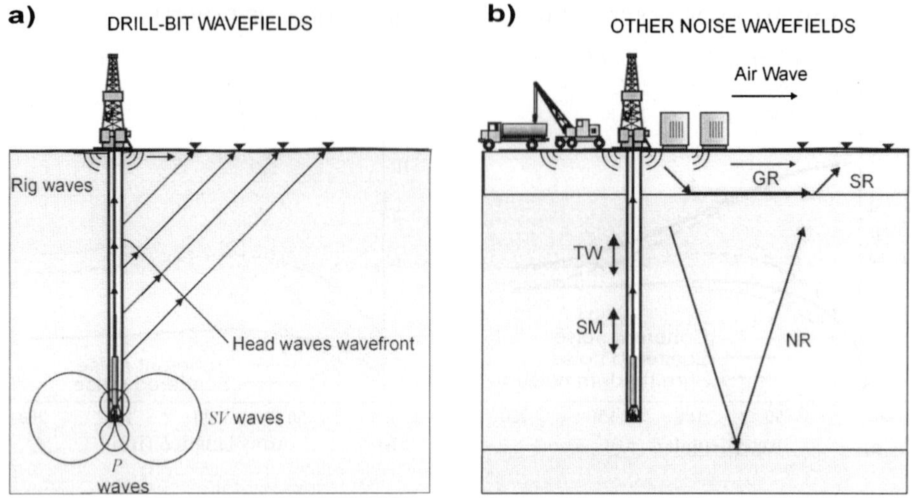

Figure 5.29: a) Main drill-bit waves (modified after Rector, 1991) and b) rig noise wavefields.

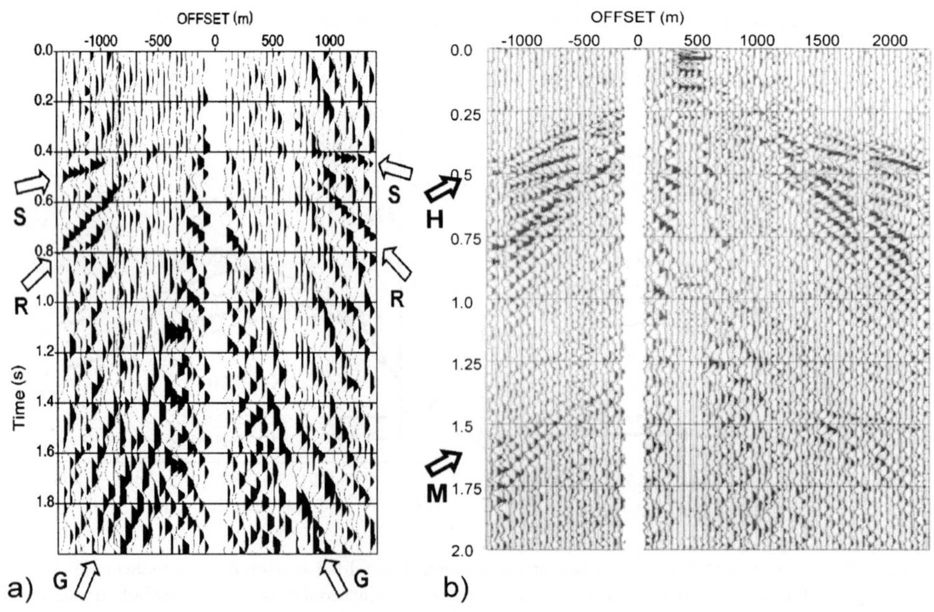

Figure 5.30: a) Correlated common-source gather acquired at 2480 m bit depth. Ground-roll (G), direct arrivals from drill bit (S) and shallow refractions (R) are indicated. b) Another example of correlated common-source depth gather on the same scale of Figure (a). These data are low-cut filtered (15 Hz). We observe head waves (H) and pilot multiples (M) of signal and noise.

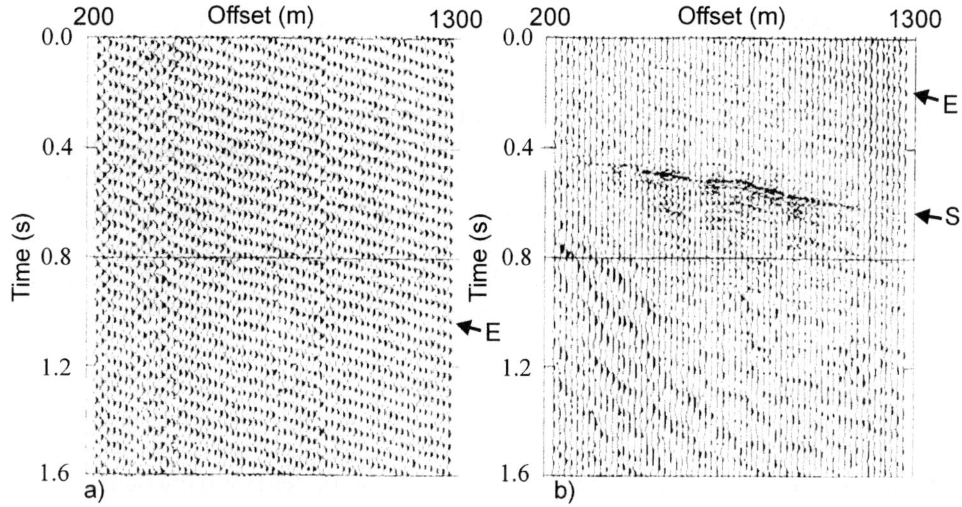

Figure 5.31: Correlated common-source gathers acquired a) without and b) with weight on bit (WOB). Pilot signal is a rig-derrick sensor. In the data without WOB (a), the 20 Hz refracted diesel-engine noise (E) is clearly observable. In the data with WOB (b), the signal (S) is clearly detectable at 0.4 s, with some persisting residual rigsite environmental noise (array length L = 25 m, M_g = 24).

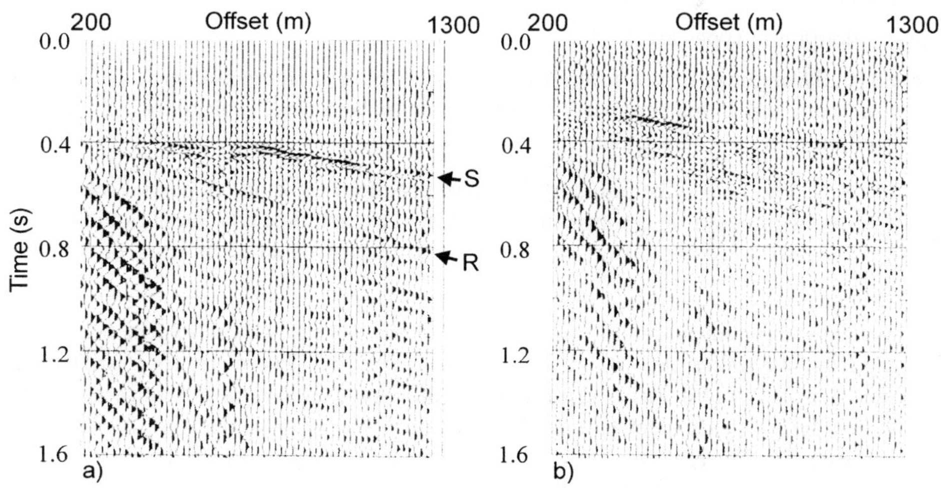

Figure 5.32: Two examples of common-source gathers acquired for different bit depths and correlated with the rig-derrick pilot signal. Refracted noise (R) propagates from the rigsite, with the same moveout of the environmental noise (E) of Figure 5.31. Note that the amplitude of the signal (S) is higher for the gather in (a) than for the gather in (b). In (a), also the amplitude of the rigsite noise (R) is higher. This is because this noise consists of secondary drill-bit waves radiated from the derrick (Rector, 1989).

- ◇ The direct (S) and refracted (R) waves, generated at the rig, travel through the uppermost layers. In common-receiver gathers, these events are independent of the drill-bit depth and, therefore, can be considered as a stationary noise. In common-source gathers, the apparent velocity of these waves is related to the velocity of the surface layers (typically from 2000 to 4500 m/s). The energy and the frequency content of this noise are comparable with those of the drill-bit signal. Therefore, in many data sets, a large decrease of the signal-to-noise ratio due to the interference of shallow refractions and signal, can be observed.

  Analysis carried out on data acquired during earlier surveys (Rocca, Persoglia, Poletto and Craglietto, 1990), showed rigsite noise with a predominant frequency of about 20 Hz.

- ◇ Conical head-waves (Section 4.6) are radiated by the drill string when the propagation velocity in the surrounding formations is lower than the extensional velocity in the drill pipes (for example 4750 m/s). This noise also interferes, with similar moveout of the signal, but is typically weaker than the surface refractions.

- ◇ Strong coherent noise can be generated during deviated drilling because of the increased contacts of the drill string with the walls of the well. This noise is not stationary during the whole drilling phase, and its removal during the processing is more difficult.

- ◇ Other coherent noise components are the bottom-hole-assembly (BHA) and drill-string multiples, affecting both signal and noise (Poletto, Malusa and Miranda, 2001).

Time-random-noise types are:

- ◇ Spatially organized noise filtered by arrays. A part of this noise, such as the noise from cars and human activity, is not stationary and results ultimately as a time-random noise in correlations.

- ◇ Pure random noise, like rain and the noise produced by wind. A partial solution to this problem could be to bury the receivers. This solution can lower random noise and, therefore, larger arrays can be used to improve spatial filtering of coherent noise.

- ◇ Scattered random noise. In the proximity of the wellsite we can expect noise coming from the well with vertical and inclined directions. Therefore, the Aki-Denham model of horizontal waves coming from different azimuths is not fully adequate. Probably, we can expect to have in SWD a scattered-noise condition intermediate between the Aki-Denham noise and coherent noise propagating with various inclinations with respect to the well.

5.14.3 Experiments with receiver arrays in onshore SWD

In seismic while drilling, most of the coherent noise comes from sources located at the well site. Therefore, the most common and effective solution is to lay out linear arrays in the radial direction, i.e., in-line arrays with equispaced receivers.

Single three-component geophones are used for polarization analysis, which can attenuate certain polarizations and improve the signal-to-coherent-noise ratio. However, single 3C geophones provide no filtering of the spatial wavelengths and lower attenuation of random noise than array groups. Effective array simulation using three-component geophones in SWD would require close spacing of many receiver traces (Hoffe, Bland, Margrave, Manning and Foltinek 1999). Use of cross-lateral (orthogonal to in-line) and linear (in-line), i.e., array groups with two orthogonal arrays, or two-dimensional arrays is recommended when there is reason to assume that high scattered or organized lateral noise is present. This is the case for instance, in the proximity of the well, i.e., at an offset distance comparable to the wellsite dimension (100–200 m). Otherwise, SWD field tests show that in-line arrays in the radial direction are better than cross-line arrays for pure rigsite noise. The main task is to decide the optimum array length for the less favorable condition at variable bit depths and in function of the expected signal-to-noise ratio.

Example 1 – A survey line of about two kilometers was used with receiver-station spacing of 50 m. Simulation with synthetic records showed an improvement in S/N ratio for SWD data, including shallower arrivals (e.g., from 1500 m bit depth), by using longer linear group arrays of geophones (say, with a length approximately equal to twice the station spacing), in the first half of the survey line and shorter (equal to the station spacing) in the second half. The actual effect of the array adopted may be assessed by comparing the data acquired with a clustered geophone group. Figure 5.33 shows the correlations of signals acquired, when the bit was between depths of 2120 and 2903 m, with a linear group array of twenty-four geophones and a length of 100 m, measured at an offset of 400 m, and those of the signals obtained using a clustered group of three geophones located at the same offset. The figure also shows a comparison between the corresponding amplitude spectra of the traces; note that the signal-to-noise ratio is higher in the data recorded with the 100 m array pattern than in the data obtained with the single geophone location, where the monochromatic noise components are stronger.

Standard onshore SWD acquisitions use seismic lines 2-3 km long. From 40 to 80 channels are spaced with a trace-interval of 50 m to 75 m, using groups of twenty-four (10 Hz) geophones. With this acquisition geometry, the length of the P-wave arrays may range between 100 m and 50 m, or shorter, changing with offset. Using a higher number of traces with shorter space intervals instead of long group arrays, makes it possible to simulate the arrays or to beam steer the signals in the processing phase. This solution may require a large number of channels and it is expensive with long-offset lines.

The use of long linear arrays allows strong attenuation of the low-velocity surface waves. The main function of the array is to reduce the amplitude of refracted waves masking the direct arrivals and reflections and to obtain a good compromise between noise amplitude reduction and signal bandwidth preservation. In fact, the length of the pattern is chosen in relation to the apparent dominant wavelength of the noise (N) and, possibly, to notch monochromatic components such as the 20 Hz power-generator noise (Section 2.3.3). Thus, in the first phase of drill-bit surveys, walkaway noise tests are acquired to determine the apparent wavelength of any stationary noise. This noise-velocity analysis is performed using signal and noise pilot sensors. Along the line, variable arrays are designed taking into account the depth of the target, the signal emergence angle for each offset and the geometry of the well path.

Example 2 – The filtering effects of three linear group arrays, respectively 100, 80 and 50 m long, are compared with that of a clustered-geophone group – for simplicity, we indicate it by 0 m – deployed with good approximation in the same location. The number of geophones per group is twenty-four, and the offset from wellhead is 300 m. Strong attenuation of the ground-roll waves generated at the wellsite was observed in the correlated traces, depending on the group array length. The signal-to-noise ratio corresponding to the 100 m pattern increased several times (6–12 dB) with respect to that of the clustered group. This result is in agreement with the calculated response of the signal-to-noise response of the array, which has high-cut filtering effect on the signal. Figure 5.34 shows two common-receiver gathers, acquired with the 100 m group array and the clustered geophone group (0 m). The data of the clustered group (top) show, below 1400 m bit depth, a horizontal event at about 0.25 s correlation time. This is rigsite noise. As can be appreciated, its delay time is constant. We define this type of noise as "stationary". This effect occurs because the source of the noise, e.g., the surface rig, and the receiver does not change the relative position for different bit depths. In fact, this wave is a rigsite wave refracted from a layer at approximately 80 m depth. It masks the direct arrival. It has also a frequency content similar to that of the signal. The spatial filtering effect of the 100 m array on this organized noise can be seen in the lower section, where the signal is recognizable in the entire section. A different level of random noise can also be seen. The array filtering effect is better appreciated if we consider that, in correlated data, the trends of signal and noise in depth are similar (Section 6.3.3), and noise separation by time-depth filters may distort signal.

In summary, noise produced by rig vibrations and by human activity connected with drilling operation, travels in the uppermost layer as surface, direct and refracted waves, as well as random noise. These noise-events interfere with the signal (direct and reflected waves) and produce a signal-to-noise reduction. The experience gained during SWD data-acquisition, performed in numerous sites, implies that the use of receiver arrays represents a powerful tool to enhance the signal-to-coherent-noise ratio. Topographic elevations and shallow geological variations along the seismic line cause different apparent velocities of the coherent noises. Random static effects have to be accounted for as well. In order to determine the characteristics of the stationary-noise signals and design the optimum receiver array, walkaway noise tests should be performed before or in the first phase of the SWD data acquisition (Figure 5.35).

A long radial (in-line) geophone array allows good attenuation of coherent noise but affects the signal bandwidth. In order to recover the original spectrum of the signal, an inverse filter is applied (Section 7.11).

5.15 Acquisition of shear and converted waves

Usually, the group arrays are designed to enhance the signal-to-coherent noise ratio of direct and reflected P waves. These configurations can seriously deteriorate the amplitude of the S-wave and converted components at large offsets if the array design is not proper.

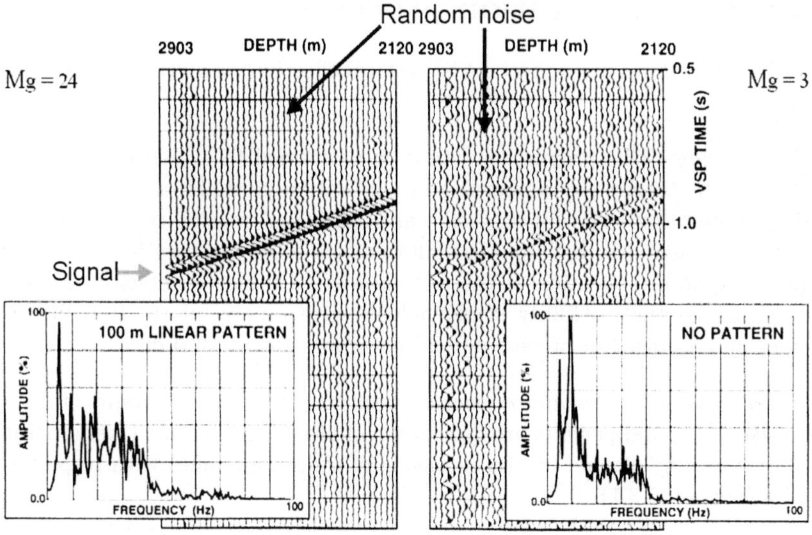

Figure 5.33: Correlated drill-bit seismograms acquired by 100 m group array and clustered group at the same 400 m offset location. Improvement of signal/random noise ratio can be appreciated. M_g is the number of geophones in the array. Modified after Miranda, Aleotti, Abramo, Poletto, Craglietto, Persoglia and Rocca (1996).

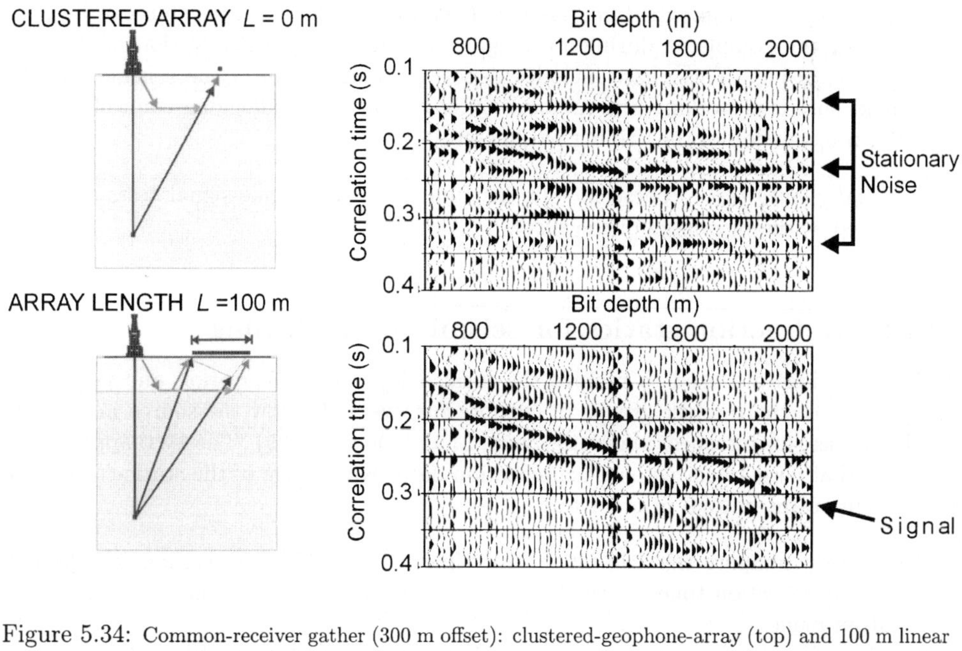

Figure 5.34: Common-receiver gather (300 m offset): clustered-geophone-array (top) and 100 m linear array (bottom). Reduction of stationary and organized noise, as well as a partial reduction of random noise, can be seen.

5.16 Survey preparation procedures

The preparation procedures are important to perform effectively a SWD survey and include the following steps:

1. *Feasibility study.* This preliminary study is performed to evaluate the survey object, decide the survey requirements and application parameters and to collect the existing information that will be used while drilling. A feasibility study collects all the seismic, logistic and drilling-plan information. This is a very important step in the planning and application of the seismic-while-drilling technology. The results collected during the feasibility study help to give timely information for drilling decisions and are used in processing and interpretation of SWD data.

2. *Permitting and environmental impact assessment* (EIA). An environmental impact assessment (EIA) and permits are typically included in the planned well operations. Specifications may depend on the extension of the seismic lines with respect to the well site (Section 2.3.1).

3. *Line positioning, topographic survey and surface refraction survey when onshore.* A refraction survey is important in onshore surveys where the SWD travel times have to be corrected for surface seismic statics to validate the checkshot-data measurements and interpretation. Refraction data are used to determine the apparent velocity of shallow layers and to design the optimum receiver arrays.

4. *Mobilization of equipment and crew and sensor installation.* This step can be performed and completed during a casing cementing phase, typically that of the first surface casing. This phase includes deposition and testing of the seismic lines to make sure it is working and preparation of the on-site mudlogging connections for the transmission of the drilling parameters.

5. *Initial quality tests.* These include instrumental and seismic-signal tests are conducted at the beginning of the survey.

5.16.1 Refraction statics for seismic while drilling

Static corrections are used to remove the local delays due to elevation and geological structure below the receiver groups in onshore surveys and below the source in conventional VSP (see, for instance, MacBeth, Zeng, Li and Queen, 1995). The SWD approach is similar to that used to correct surface reflection data. Knowledge of the static-correction times is important in SWD in order to:

- ◇ Correctly compare the drill-bit seismic times, that is, the bit-position (checkshot) and the reflection times, with the time sections of the surface seismic referred to a datum plane.
- ◇ Make the processing of multioffset data acquired in complex areas more effective.

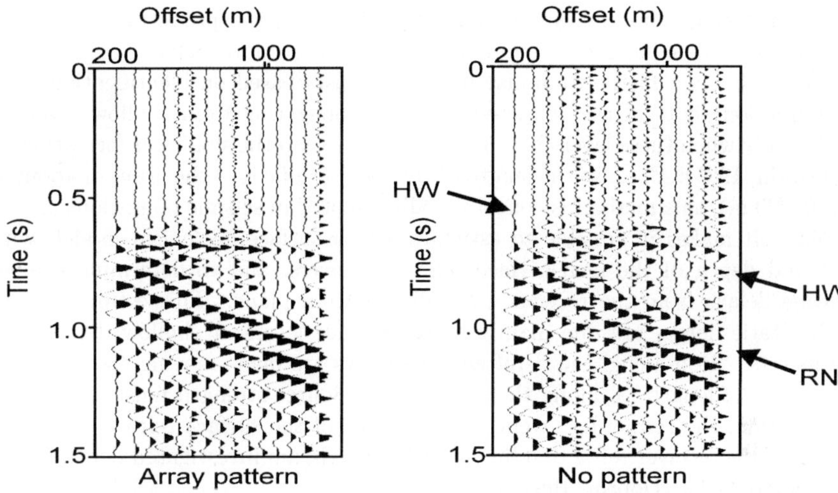

Figure 5.35: Common-bit-depth (2653 m) crosscorrelated gathers (see Chapter 6) acquired using traces with twenty-four geophones per group at different offsets from wellhead. Linear group arrays (left panel) and clustered geophone groups (right panel) were used at the same positions. Longer arrays (80 m) and clustered groups have different S/N in the near-offsets traces of the two gathers. Shorter arrays (about 35 m) at larger offset show more comparable S/N. RN denotes refracted rig noise. In the clustered line (right), head waves (HW) can be better detected. Traces are not filtered.

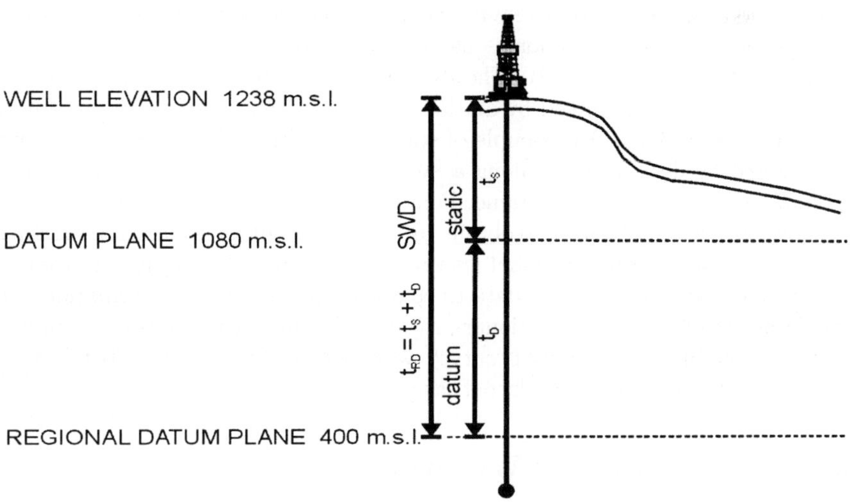

Figure 5.36: Correction of SWD data to datum plane (modified after Petronio, 1999).

For this purpose, a refraction survey aimed at detecting the properties of the shallow layers is performed along the SWD line as a typical procedure before any onshore SWD acquisition. This is a method familiar to geophysicists based on identification (picking) in the refraction shots of the refracted arrivals propagating from shallow layers. It is applicable only when these layers have higher velocity than the shallower formations above them (Dobrin, 1981). Because the length of the receiver line is of the order of about 100 m (e.g., 150–200 m), this method is conventionally denoted in seismic operations as a "*short refraction*". It makes it possible to estimate a near-surface geological model, with the velocity and depth of the near-surface refracting layers, and to determine the seismic travel time delays from the receiver to a given datum plane. This estimate is correct when the static delays are due mainly to the surface layers at shallow depths of some tens of meters. "Roll-along" short-refraction surveys are made moving the receivers of a refraction line of length L_R along the seismic line. Source strength must be adequate to get signals from the refracting layers. Performance of the surface refraction source and the length of the receiver refraction line limit the maximum investigated depth z_{ref}, which can be related to L_R (Dobrin, 1981) by

$$z_{ref} \sim \frac{L_R}{2}, \tag{5.39}$$

which is related to the velocity contrast. The dip of the refracting interface is determined by locating shots at both the ends of the receivers line ("reversed shooting").

When SWD acquisition is performed in complex areas where deeper layers cause significant static delays and correction to a deeper regional datum is used for surface-seismic sections, the length of the surface refraction line has to be increased and the source strength must be increased to get adequate signals from deeper layers (Petronio, 1999). An alternative approach is to use SWD rigsite coherent noise – including environmental noise and rig-radiated bit vibrations – as auxiliary refraction information (this idea was firstly suggested by Sinceri, 1992). The rig noise propagates as refracted wave from the drilling rig and can be reinforced by stationary – i.e., without correction of pilot delay – stack of SWD correlations. An example of stationary stacked (focused) rig noise progressively refracted by deeper layers along a SWD seismic line of about 2.4 km is shown in Figure 5.37 (Miranda, Aleotti, Abramo, Poletto, Craglietto, Persoglia and Rocca, 1996). Rigsite noise arrivals can be used to obtain deeper seismic information, as shown by Figure 5.38 in which a surface-refraction shot survey is continued by the rigsite surface refracted noise. This method requires other data to complete the "reversed" information. Integration of surface refraction data, stationary rigsite noise and surface seismic data available before the beginning of a new SWD survey was used by Petronio (1999) to correct the static delays of both shallow and deeper layers.

5.16.2 Field statics in SWD arrays

A detailed micro-spread investigation, providing local static delays an apparent velocity of refracted noise, may help array design. This is even more important for SWD than for surface seismic surveys. In surface reflection seismic, the receiver arrays are commonly used to attenuate the waves propagating in the horizontal direction, like ground-roll and

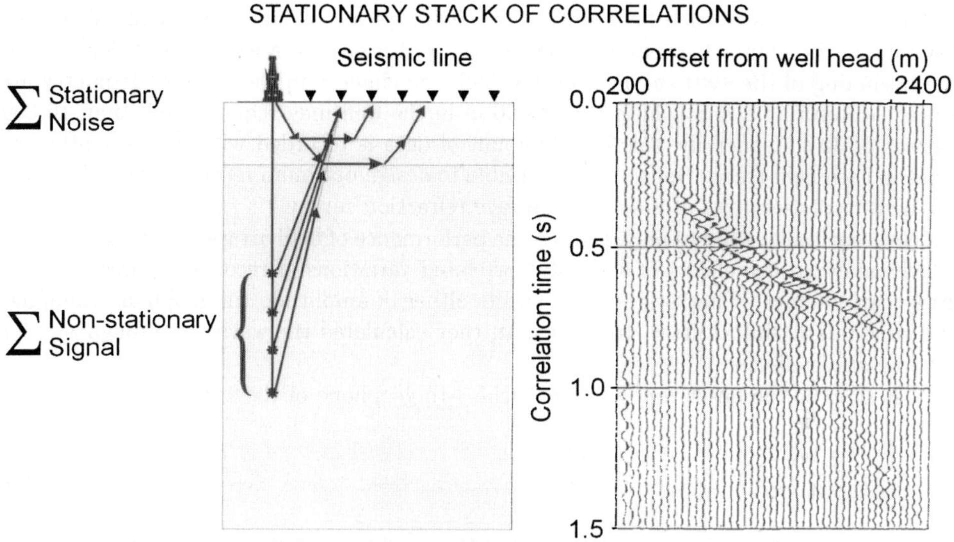

Figure 5.37: Stack of organized stationary noise (after Miranda, Aleotti, Abramo, Poletto, Craglietto, Persoglia and Rocca, 1996). This noise is mainly due to refractions of rigsite waves. It is reinforced by stacking the correlations obtained at different bit depths without pilot-delay correction. On the other hand, the non-stationary drill-bit signal is attenuated by the stack of the correlations with different source points (source array) in the well.

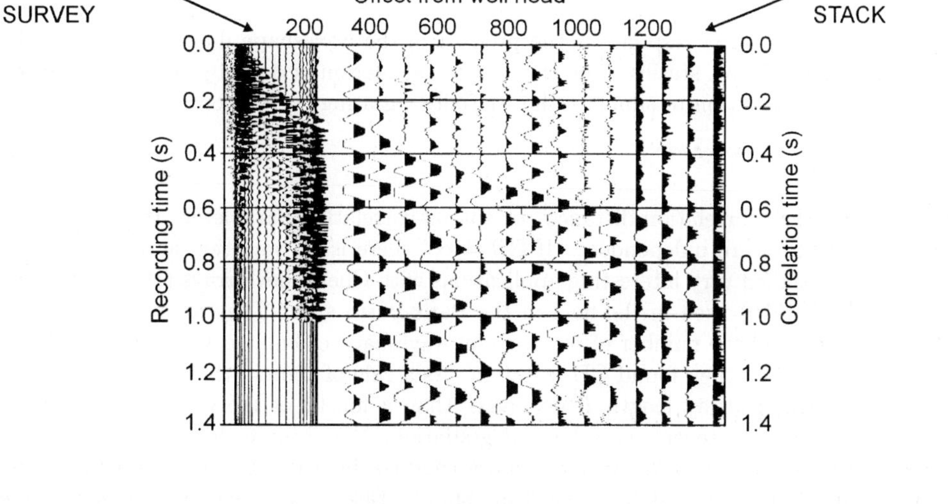

Figure 5.38: The coherent rigsite noise is used to increase the investigation depth in seismic surveys. On the left, the traces of a surface refraction profile; on the right, the correlated rigsite surface noise, are shown. Modified after Petronio (1999).

airwave. In this case, each field station acquires typically a hundred seismic shots, depending on the planned coverage. In the SWD acquisition, the seismic line is deployed at the beginning of the SWD survey and, normally, no changes in the receiver geometry are made. Monitoring a depth interval of 3500 m in the well may require more than 10 000 records of 30 s. Generally, this large amount of data is recorded with strong stationary noise in each field trace. Hence, it is advisable to design optimum receiver arrays by using the high-resolution data of a detailed surface refraction survey.

Near-surface anomalies may degrade the performance of field arrays. Berni and Roever (1989) analyzed the effects of spatially correlated variations of recording parameters on array performance. They studied the effects either of amplitude (in geophone coupling) and time (static shifts) jitter. In particular, they calculated the waveshape diagnostic for random static effects.

Let $t_e(j)$ be the onset-time variate at the j-th geophone of the array, and

$$R_s(i-j) = \frac{1}{N}\sum_{e=1}^{N}\left[t_e(i)t_e(j)\right], = \overline{t_e(i)t_e(j)}, \tag{5.40}$$

where the bar denotes expected value and N the number of observations used to calculate the statistic average. $R_s(0)$ is the mean-square time-shift fluctuation. Amplitude variation is not important for waveshape (Berni and Roever, 1989). To study the effects of statics shifts, we assume geophones of the same sensitivity and constant amplitude weights. If only time jitter is assumed (without variation of coupling amplitude), the waveshape diagnostic function calculated by Berni and Roever (1989) to obtain the group-array response for static effects is

$$\mathrm{WS}(\omega) = M_g^{-2}\sum_{i=0}^{M_g-1}\sum_{j=0}^{M_g-1} e^{-\omega^2[R_s(0)-R_s(i-j)]}. \tag{5.41}$$

That is, the total array response is the product of the waveshape diagnostic and of the array response without statics. The theoretical calculation is done by the authors using the exponential function to simulate the spatially variating correlation as

$$R_s(i-j) = R_s(0)e^{-|i-j|\Delta/\mathrm{CD}}, \tag{5.42}$$

where CD is the "correlation distance" or spatial-correlation parameter. If CD $= 0$, the time-shift variates are independent. If CD $= \infty$, there is no variation across the array. The authors studied very interesting applications with Chebyshev arrays (Holzman, 1963; Schoenberger, 1970) used in the field in the Paris-Basin area. They analyzed the effect of array length and the number of receivers on waveshape calculated with various CD. In field measurements, they found CD $\approx$ 180 m. As a rule of thumb, these authors suggest the use of a group-array length L no longer than a fourth of the correlation distance (CD/4), in order to prevent waveshape degradation. We observe that, in any case, the overall effect for S/N in SWD arrays is more complex because the input coherent noise is also statistically variated by static time shifts. Below, we calculate the waveshape diagnostic of equation (5.41) for the signal of linear arrays, by using different M_g and L (Figure 5.39).

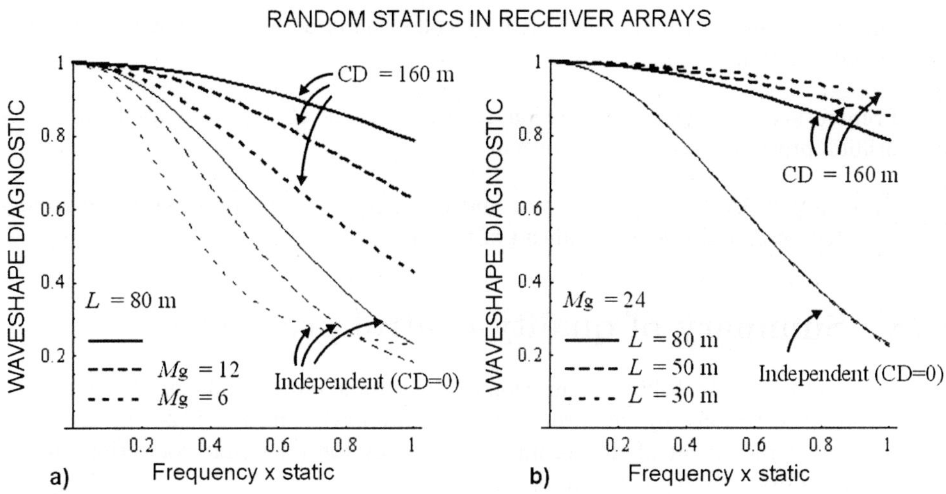

Figure 5.39: Waveshape diagnostic for linear arrays. Plot versus normalized (frequency × static shift). a) Different number of receivers (M_g) with CD = 160 m and independent statics. b) Different array lengths. After Berni and Roever (1989).

Berni and Roever (1989) suggested to obtain data as an indicator of array performance by acquiring a short (say, 200–300 m) micro spread at the beginning of a regular shooting to evaluate the distribution of static delays.

In their important work, Berni and Roever (1989) analyzed also amplitude and onset-time variations effects. It can be assumed that these effects occur also for noise.

5.17 Survey operations

The acquisition system may run in automated acquisition mode, which is driven by the drilling parameters. While-drilling survey operations, which make it possible to acquire data in automated mode, include the following activities:

◇ Control of the drilling programs, for possible problems and variations of the drilling plan.

◇ Test of acquisition procedures. Test the sensors and check the line functioning (Section 5.18). Verify any problem due to human activity along the lines, such as seasonal farming works.

◇ Set up and update the acquisition and while-drilling processing programs. Test the transmission channel and set up the transmission modalities – selection of the data to be transmitted to the office.

◇ Data acquisition, preprocessing and quality control (Section 5.18). Update drilling information used by SWD. This includes collection of detailed drill-string and drill-bit information.

◇ Preparation and daily transmission to headquarters of data sets used for while-drilling processing and interpretation.

◇ Reverse VSP (RVSP) data processing and while-drilling interpretation of the geological structure, and prediction ahead of the bit.

5.18 Summary of quality-control procedures

The quality of the while-drilling acquisition of the seismic data is influenced/determined by the drilling conditions, by the level of drilling-yard environmental noise, by the pilot-data quality, by the layout of the seismic channels, by the pilot-time correction and by the reliability of the drilling-depth information.

It is, therefore, necessary to continuously monitor the acquisition and to take operational decisions following the evaluation of seismic-while-drilling conditions and data. This is based on the measurement of the specific quality tests of the SWD data. These tests are performed periodically, according to the following specifications. A part of the quality-control specifications involves preprocessing, which is discussed in detail in Chapter 6.

Instrumental tests

These tests concern the spreading and setting of the seismic line, with the following controls:

◇ Accurate definition of the survey geometry, line orientation and azimuth from wellhead, geographic coordinates, elevation on the reference datum and field-static correction, provided by the company or measured by acquiring a refraction survey in the correspondence of the SWD geophone locations.

◇ Recording-system test. These include preliminary and daily hardware and software tests during the acquisition of the survey. The tests concern the seismic-line sensors, the rig-pilot sensors, and the telemetric equipment managed by the acquisition unit. The tests include:

- The LOOK LINE, with the control of the setting of the station units (SU);
- The NOISE and OFFSET tests for the SU, with data recording and diagnostics;
- The recording of the IMPULSE-data test of the SU with internal shunt;
- The IMPULSE test of the sensors of the seismic line.

The result of the daily controls must be checked to ensure the reliability of the while-drilling data stored on magnetic support.

Initial QC tests

Initial QC tests involve the determination of the preprocessing parameters (Section 6.18) and start up of the acquisition phase. These test are performed to optimize the acquisition parameters during regular drilling conditions. These are:

1. At the beginning of the survey, sample control of the raw (un correlated) field data, with analysis of overall signal content.

   Start of the preprocessing procedures: correlation, deconvolution, rephasing and bandpass filtering. These tests are performed on-site or, equivalently, with remote transmission of the while-drilling data to headquarters and near-real-time data analysis.

2. Selection of the pilot traces among the multi-pilot arrangement in the drilling yard. These include axial accelerometers, pressurometers, aligned optimum-stack of field traces, rig-derrick and yard-noise pilots.

3. Control of signal polarity, which must conform to the SEG standard. Rephasing of the instrumental response, for example, for a geophone signal correlated with a pilot accelerometer (Section 6.17).

4. Analysis of the reference pilots, in time and frequency, with correlation tests and stack of the common-source results for selected depth levels.

5. In agreement with the parameters of the technical specifications of Section 6.18 (for example, the required depth interval) choose the maximum bit depth interval allowed to acquire and stack the data of a selected depth level.

   The while-drilling control of the interval (number of stacks) is performed automatically to manage the progressive acquisition and preprocessing phases with different drill-string lengths.

6. Deconvolution test of the pilots, with the analysis of the long-period and short-period (BHA) reflections, with analysis of pilot correlations and synthetic modeling based on drilling program and conditions (type of pipes, BHA, bit, rig features and formation type). This step includes the automatic computation of the pilot-signal delay.

7. Analysis of signal and noise in the geophone seismograms in the common-source data. Choice of the bandpass filter to remove the low-frequency and high-frequency noise for undesired drilling vibrations after deconvolution.

The data after correlation, deconvolution, filtering, rephasing and pilot-delay correction are the preprocessed seismograms.

The acquisition phase of the survey starts with the acquisition and the preprocessing of data for the planned depth levels. The preprocessed data are sent, day by day, to headquarters via modem. From the whole multioffset dataset, good-quality traces are chosen for while-drilling quality-control VSP surveying in near-real time, with the following procedures:

◇ Preliminary correction of the pilot delay on the basis of the numerical and mechanical analysis of the drill-string features. Subsequent analysis of the pilot autocorrelations with a sufficient number of depth levels (e.g., some tens). This step provides the preliminary VSP data.

◇ Geophysical analysis of the common-shot seismograms, with support of the geophysical model and ray tracing to interpret the drill-bit arrivals and secondary noise. Identification of good-quality near-offset traces for while-drilling checkshot analysis.

QC tests with the variation of the bit depth

Usually, at the beginning of the survey, the control of the common-receiver data is performed after acquiring a sufficient number of drill-bit depth levels. These tests can be performed in the field, or equivalently with remote transmission of the data to headquarters in near-real time. The following steps have to be considered:

◇ Kinematic control of the direct arrivals without pilot signal correction in a depth interval sufficient to determine the variation of the arrival time (e.g., 500 m) for bit descent and variation of moveout at the different offsets. This is done to evidence differences with the stationary rig noise.

◇ The initial estimate of the axial pilot delay in the drill string is updated by the optimum fit of the numerical modeling with the real pilot autocorrelations. This step is necessary when a change of drill-string composition occurs and after the casing phases. Comparison of the results with the measured times by picking the pilot autocorrelations.

Periodical QC tests

The data quality and the preprocessing parameters used during the survey have to be periodically checked. The following steps are considered:

◇ Diagnostics of the drilling parameters for the preprocessed data acquired during the survey (see Section 5.9). In particular, rate of penetration (ROP), weight on bit (WOB), revolution per minute (RPM), and torque have to be observed. These parameters can vary in correspondence with changes in the drilling conditions or because calibration of the sensors of the mudlogging unit. These variations have to be taken into proper consideration while drilling.

◇ Analysis of the energy content of the deconvolved pilot and seismic signals in common-receiver gathers. This is used to determine the variation of the drill-bit source versus depth. The additional analysis of the common-shot data is useful to interpret coherent noise and signal wavefields (e.g., as in the presence of drill-bit head waves).

◇ If the signal is attenuated, tests to adjust the acquisition and preprocessing parameters, such as the total acquisition time per level, the depth interval between

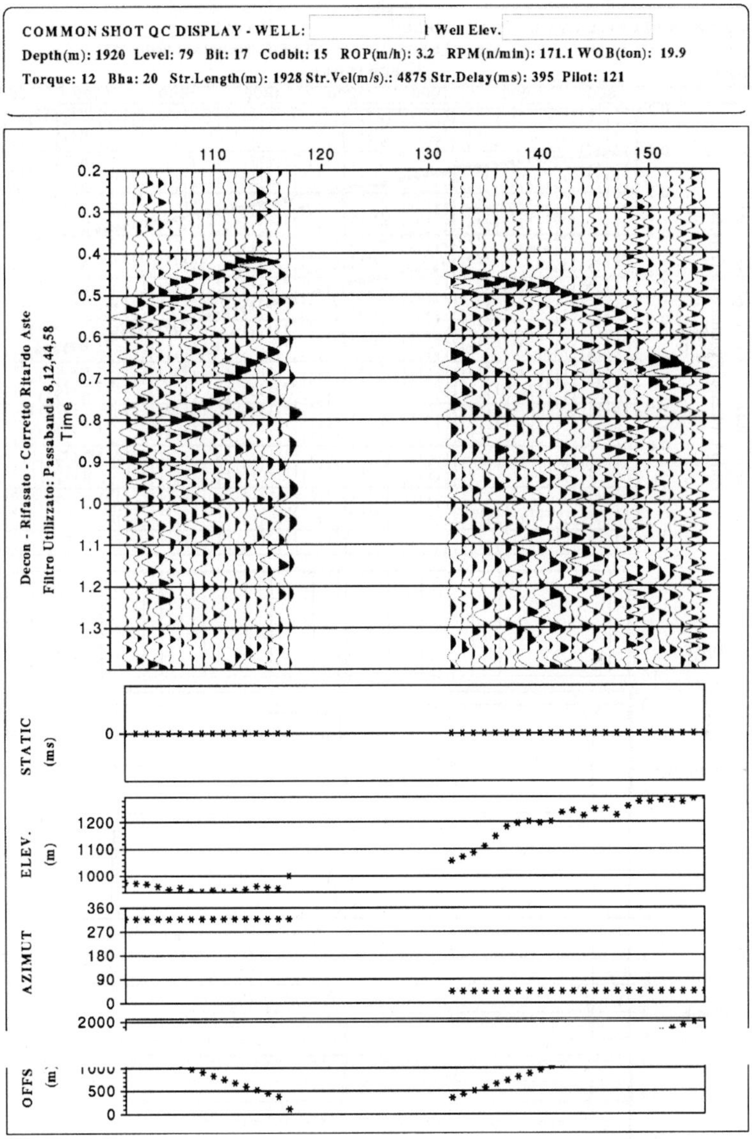

Figure 5.40: Typical common-shot field report.

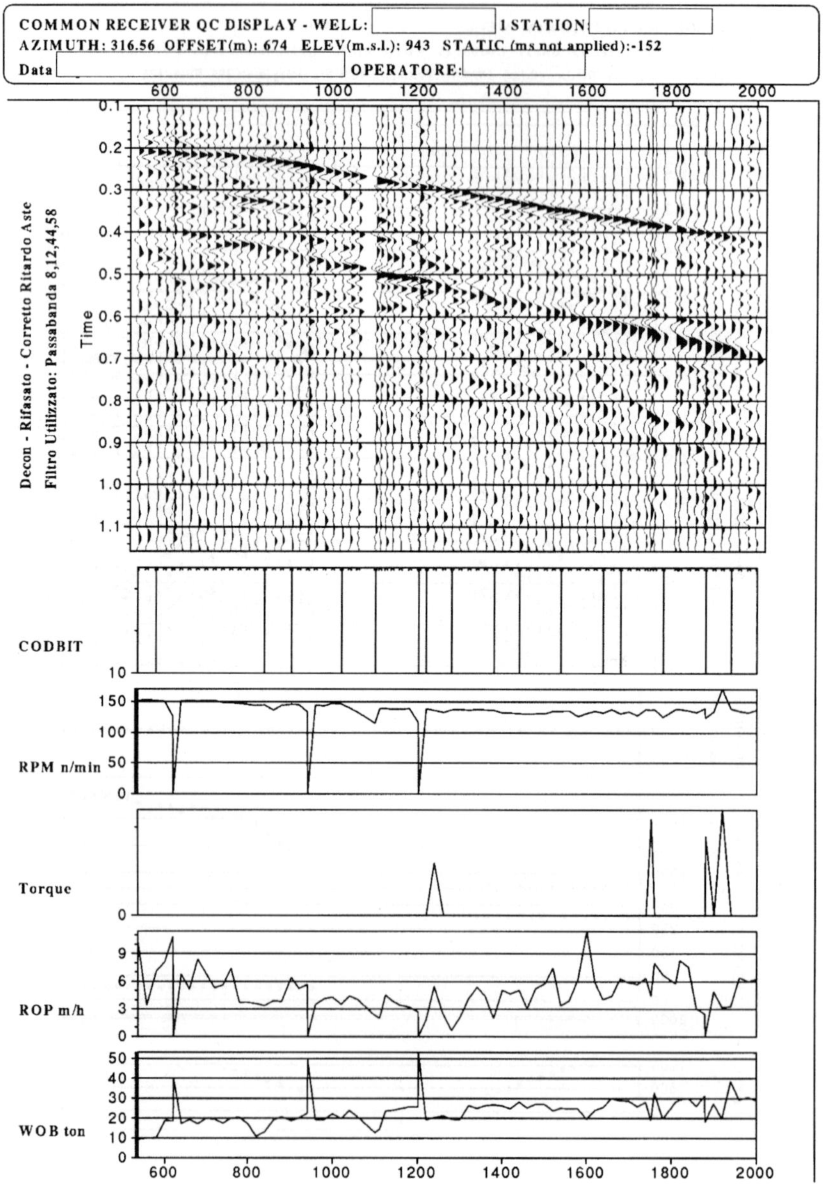

Figure 5.41: Typical VSP field report.

levels, and the use of pilots obtained by combining other pilot traces, have to be performed. The raw data are saved on disk, and the reprocessing at the headquarters can give the same results. Nevertheless, it is important to have these data in the while-drilling phase.

Figures 5.40 and 5.41 show a typical SEISBIT ® daily report with common-shot and common-receiver gathers (corresponding to a reverse VSP), respectively.

Auxiliary QC tests

The sensors and tools used in the drilling environment could suffer damage, even if properly protected. Furthermore, accidental failures are possible and require the use of substitutive sensors or immediate maintenance. This is not always possible, and the necessary interventions have to be done in agreement with the drilling operations. Furthermore, other pilot sensors are located in accessible positions to monitor the environmental noise on the derrick and drilling yard, near engines and pumps.

Noise tests are used to control the correct functioning of the pilot and noise sensors.

5.19 Onshore 3D-SWD acquisition

The aim of 3D reverse VSP (RVSP) using the drill-bit signal is to obtain detailed while-drilling 3D seismic information in the area around the well. The motivation of conventional 3D VSP is to obtain a detailed 3D seismic image in the vicinity of the well. The main limitations are: offshore applicability only when a limited number of receivers in well (say, 5–10) is used (however, recently larger borehole-receiver arrays become available for 3D VSP surveying (Input/Output.INC, 2001)); high rig stand-by costs during acquisition; risk due to receivers in well for a long time without the drill string; and no near-real-time while-drilling availability of the results. On the other hand, SWD by drill-bit signal is more suitable to obtain 3D RVSP onshore. Figure 5.43 is an acquisition layout comparison for typical configurations. The conventional 3D VSP has many source points on the surface and few receivers in the well. The 3D SWD has less-dense distribution of receiver points on the surface, and has more source positions in the well (vertical source array) . The 3D SWD results can be used either while or after drilling. We consider two cases:

◇ The location of the well is decided using 2D surface-seismic lines: in this case 3D SWD helps to improve geological structural definition.

◇ The location of the well is decided using 3D surface-seismic data: in this case 3D SWD helps to improve geological structural resolution.

For drill-bit 3D RVSP the main problems are related to cost and risk: cost for the long-time use of a large number of channels; costs associated with a more complex management of the seismic line and of the while-drilling recorded data; risks related to variability of drill-bit source versus operational conditions; and the presence of rig noise.

The 3D-RVSP data are migrated to obtain images and to update the 3D velocity model of the geological structures near the well. Another use is to increase the resolution of the

3D images, as before "side tracking". A drill-bit 3D RVSP case history is discussed in Section 8.10.

An important point for survey design, and for achieving optimal resolution, is the optimization of receiver positioning at the surface (Lavely, Gibson and Tzimeas, 1997). Let us consider the diffraction projection modeling scheme of Figure 5.42. Sources and receivers are positioned to optimize spatial resolution and diffractor reconstruction, which increases with increasing solid angle α described about the normal to the plane tangent to the ellipsoid in diffraction point. For vertical wells, the optimized layout involves circular receiver array geometry. However, this geometry is adapted to the specific surface geometry of the survey area. A saw-toothed sensor disposition (Figure 5.43, right panel) helps to synthesize array effects and discriminate signal and rig noise in the azimuth domain. In other words, what we do not want is to measure only arrivals with constant delays (horizontal) versus azimuth both for signal and noise in a common bit-depth gather.

3D acquisition system

Figure 5.14 shows the layout of the SEISBIT ® SWD-acquisition system. Its 3D version – where more acquisition systems have been put in a synchronized layout – is a multi-PC-based acquisition system able to handle also the use of the downhole instrumented tools (SUBs) to increase the quality of the cross correlated data. At present, the downhole pilot preprocessing is available as a deferred task, and this is available while drilling after the tool is pulled out of the hole.

This layout enables us to record an arbitrary large number of channels together (for example, up to 600 traces). This system is fully automated and is able to form the lines, check them, and acquire data automatically on the whole spread. Day by day, or even more often, the preprocessed data together with the QC displays are transferred, via modem or satellite, to the processing center or to the operative districts. There, the VSP data are processed and the final results are discussed with the geologists and seismic interpreters.

5.20 Offshore acquisition

Seismic while drilling can be used to optimize drilling in offshore areas and reduce the related risks. Offshore, the seismic line is limited by the features of the sea-floor recording technology, which has to be available at an acceptable cost for SWD. Moreover, care must be taken in identification of signal and removal of all unwanted waves such as the sea surface multiples, water arrival, ground roll, and shallow refractions (Hardage, 1992).

Offshore SWD wavefields

The wave types recorded during an offshore experiment are very similar to those recorded in an onshore experiment, with some exceptions which are typical of the offshore environment (Figure 5.44). A brief description is presented here below:

Direct arrival – First arrival of energy generated by the bit.

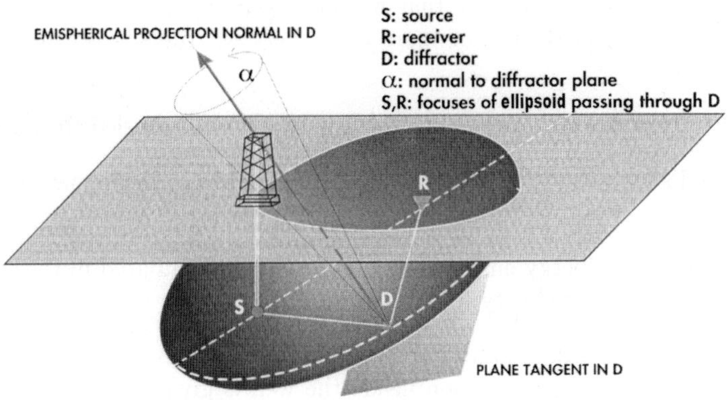

Figure 5.42: Source and receiver are positioned to optimize spatial resolution. It increases with the solid angle described by α.

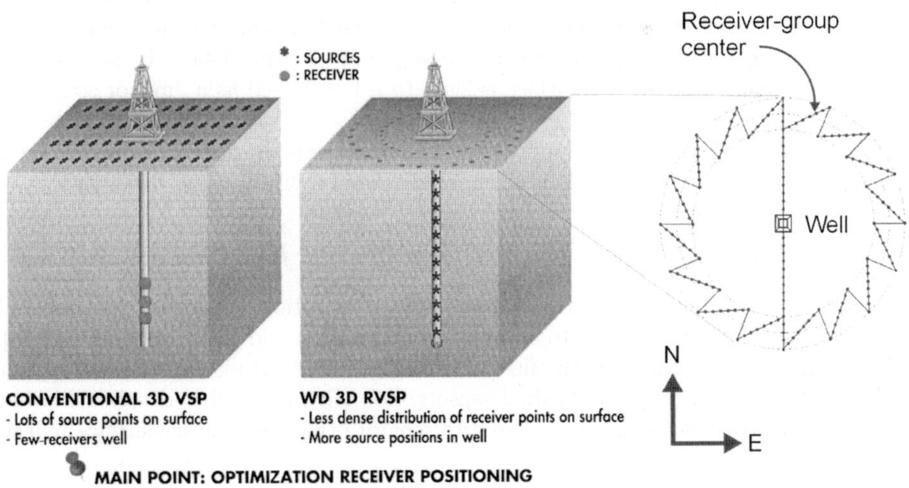

Figure 5.43: Comparison of conventional and SWD 3D VSP layouts. Conventional 3D VSP with 6000 source positions, drill-bit RVSP with 400 receiver positions. Conventional VSP with 80 m (five levels) and drill-bit RVSP with 2500 m (125 levels) of vertical-well source points. The ideal disposition of a 3D SWD seismic receiver line is a circle centered on the well. In the field, a "saw-toothed" circular geometry is adopted to discriminate signal and noise moveouts.

Reflection – Energy reflected by a reflector located below the bit.

Sea top multiple – Seismic energy that has been generated by the rig structure and that has been reflected once or more times between the sea bottom and the sea surface.

Water arrival – Energy radiated by the rig structure which traveled through the water.

Ground roll – Pseudo-Rayleigh wave involving the seafloor, analogous to ground roll on land.

Shallow refraction – Energy emitted by the rig structure radiated in the sea floor and refracted by a higher velocity layer.

Head waves – Energy that radiates into the formation whenever the velocity of source-induced seismic wave propagation inside the well is greater than the velocity of a seismic wave in the surrounding formation.

Sea-top multiples and water-arrival waves are typical of offshore acquisitions and can be regarded as noise. The last two are the direct-arrival and reflection waves that we need to isolate in the SWD acquisitions. All the other wave types described can be considered as noise that we need to eliminate in the acquisition or processing. Some sensors described above are better suited for offshore SWD acquisitions due to the fact that either the sensors are capable of filtering the unwanted waves or some sensors, due to their directivity, can better discriminate between the unwanted noise and the signal. The ghost-section – obtained from the downgoing VSP field – can be used to simulate other source positions and to increase the coverage of the reflection points (Figure 5.44). In fact, the ghost reflection coefficient of the sea surface is close to -1 and used as a "mirror source in the sky" with opposite sign.

We distinguish two cases of application. The first one is the conventional offshore SWD using fixed receivers. The second is the special case of SWD by using towed cables.

5.20.1 Offshore SWD application (using *fixed* receivers)

Shorter cables with lower hydrophone intertrace separation are typical offshore. This geometry may make it possible to apply array simulation and beam-steering techniques. With short spreads, analysis with offset of common-source data is less effective. Offshore, ocean bottom cables (OBC) with dual sensors, i.e., with coupled hydrophones and geophones in the same location, can be used to separate the one-time see-surface reflections (ghosts).

The sensors that are normally used in offshore acquisitions are:

Single hydrophones – Since these are omnidirectional, these receivers record all types of signals arriving. The more energetic part turns out to be that one of the rig; these signals are recognizable because the arrival time does not vary with the sinking of the bit. The average wave speed is representative of the velocity of the formations on the sea bottom. The data of a single hydrophone are, generally, not useful for the SWD because of their low S/N ratio.

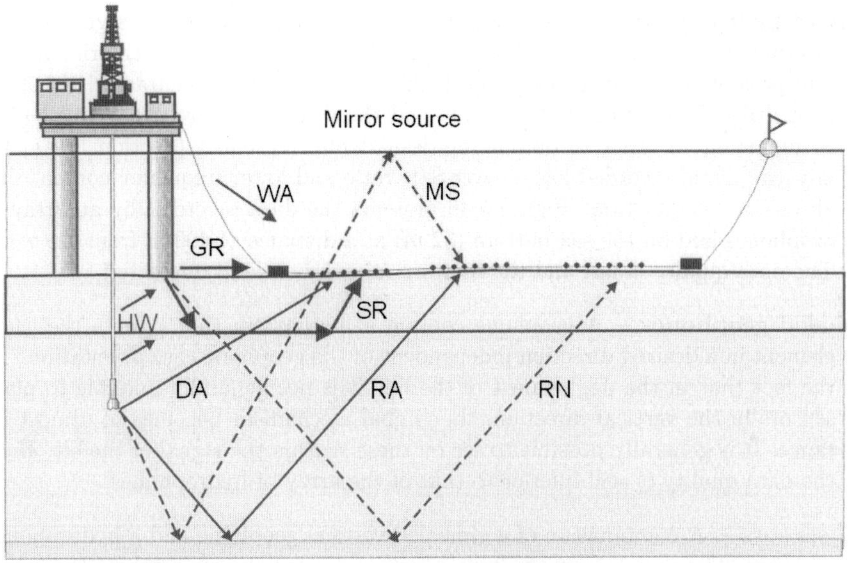

Figure 5.44: Schematic representation of offshore SWD acquisition. M: sea top multiple, WA: water arrival, GR: ground roll, SR: Shallow refraction, HW: Head wave, DA: Direct arrival, RA: Reflected signal and RN: Reflected noise (modified after Hardage, 1992). The offshore "mirror source in the sky" can be used to improve the reflection coverage.

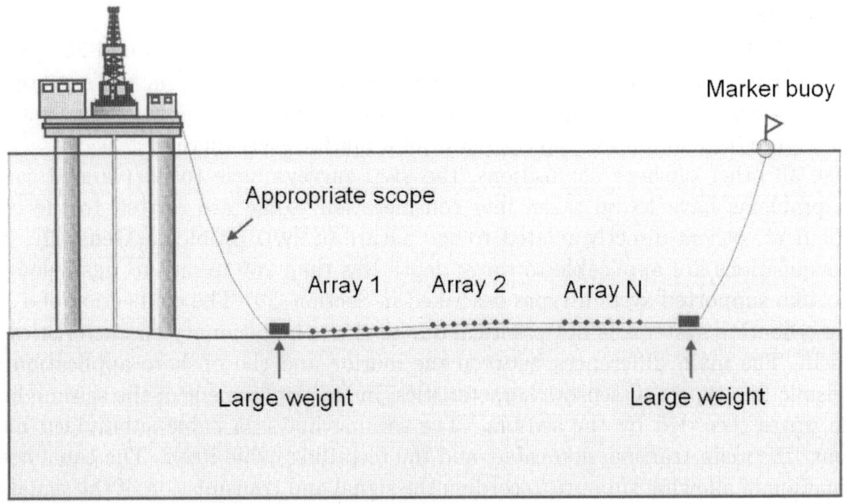

Figure 5.45: Marine SWD layout. Array of hydrophones or geophones are anchored to the sea floor (modified after Hardage, 1992; Rector and Weiss, 1989).

Array of hydrophones – These arrays record the signal usable for SWD, because the filtering effect of the array allows attenuation of the noise from the rig. It is therefore possible to identify the signal looking at its moveout characteristic of the bit as it drills. The number of elements and the distance between the single elements is important in order to obtain the desired filtering effect (Section 5.14). Generally, the signal recorded has a lower S/N ratio and lower frequency content than for an onshore acquisition. Figure 5.46 presents the data recorded by an array of hydrophones laid on the sea bottom (32 m) at a distance of 400 m from the wellhead. The data quality is fair and the first arrival can be readily detected.

Gimbaled geophones – A geophone contain a mechanism that orients the geophone element in a desired direction independent of the geophone case orientation. Due to the fact that in the deployment of the line it is not generally possible to plant the sensors in the vertical direction, the gimbal mechanism becomes of utmost importance. It is generally possible to see on these sensors the signal of the bit. However, the data quality is still inferior to that of the array of hydrophones.

Dual sensors – A combination of a gimbaled vertical geophone and a hydrophone used together. The signals from the hydrophone and geophone are combined in processing to attenuate ghost reflections and water-column reverberation. Dual sensors record the signal of the bit. However, the data quality is still inferior to that of the array of hydrophones.

Different while-drilling VSP technologies are used with seismic receivers located in fixed positions for offshore 2D investigations with short-medium offset and in limited water depth. For instance, DBSeis[3] used a system of 24 channels (hydrophones) at 12.5 m interval. This line is deployed from a work boat; after deployment, a survey is run to get its exact location.

The drill-bit source is used to acquire seismic signals with cables and 3C geophones or 4C receivers, i.e., a 3C geophone and a hydrophone, deployed on the sea bottom or with hydrophones fixed at intermediate depths by anchors and buoys (Rector and Weiss, 1989). Figure 5.45 shows the typical layout of a marine SWD survey.

Like all other offshore acquisitions, the SWD surveys have to be planned carefully; many problems have to be taken into consideration; some are related to the offshore environment, others directly related to the nature of SWD problem. Generally, marine SWD acquisitions are applicable to water depth less than 200 m and to rigs belonging to the bottom supported system types described in Section 2.4. The connection of a seismic line on a floating systems is not practical due to the non-stationary characteristics of the rig itself. The main differences between the marine and the onshore application are in the seismic line type and sensor characteristics, in the deployment of the seismic line and in the waves recorded by the sensors. The seismic line is a cable subdivided into two sections: the main transmission cable; and the recording cable itself. The cable itself has the function of allowing support, recording the signal and transmission of the signal to the recorder. The cable must, therefore, be strong enough to allow deployment and retrieval. Typical characteristics of the line can be found in Table 5.6.

[3]Mark of Schlumberger

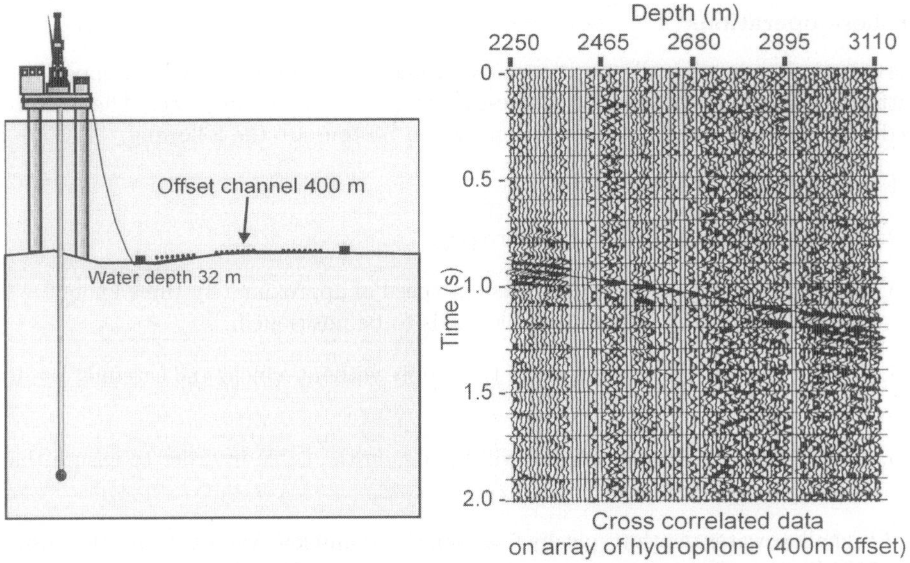

Figure 5.46: SWD data recorded by an array of hydrophones laid on the sea bottom (32 m) at an offset of 400 m from wellhead. Data are crosscorrelated, deconvolved and corrected for pilot delay (Chapter 6).

Table 5.6

	Main transmission cable
Diameter:	14 mm
Weight:	100 Kg/300 m
Length:	300 m
Conductors:	10–20 twisted pairs with protection
Cover:	Polyurethane
Load support:	Kevlar with 2500 lbs breaking load
Max. depth:	300 m
	Recording cable
Type:	Single or array of hydrophones, Gimbaled geophones and dual sensors

When we are dealing with an offshore survey, it is necessary to use the wellsite survey, recorded when the platform is put in its location, to check if any problem could occur during deployment of the seismic line. This information is needed during the planning of the SWD operation and during the line laying out and positioning. It is of utmost importance that there are no trenches or superficial gas pockets that would jeopardize the survey or the quality of the recorded signal.

Offshore operations

The seismic line is generally positioned by means of the supply vessel in support to the platform; normally 8–10 hours are necessary to put the cable in place. The operations needed to correctly position the cable on the sea bottom are the following:

1. Fix the cable to the platform.
2. Verify that the cable functions correctly.
3. Move the supply vessel with a constant speed of approximately three knots towards the direction along which the cable needs to be positioned.
4. Let out the cable from the back of the supply without winch, and carefully lay down the cable on the sea bottom.
5. At predefined positions weights will be put on the cable in order to ensure that it is firm on the sea bottom.

For shallow water depth (typically less than 50 m) and low water current, the positioning of the line is sufficiently correct and no line survey needs to follow the positioning. In more complex situations, triangulation with pingers should be performed in order to get the exact positioning of the cable on the sea floor. This triangulation is based on the time of arrival of the energy emitted by a conventional marine source fired on the sea surface (pinger). At the end of the survey, the cable is pulled in to the edge of the supply vessel using the same procedure employed to let the cable out. In this case, a winch is necessary in order to pull the cable on deck. The operations last 4–6 hours. It is important, during the deployment of the cable and during its retrieval, that the sea is relatively calm.

Throughout the survey, it is important to reduce to a minimum the circulation of ships in the zone in order to avoid undesired noises. Because the acquisition cable will have to remain in position for the duration of the survey to record an appreciable amount of RVSP data, it is indispensable to mark both extremities with floating buoys.

5.20.2 Extension of offshore SWD (using *towed streamers*)

Deep water depth and offset limitations are the main problems for offshore SWD. The conventional approach of offshore SWD discussed in the previous section has some disadvantages. Firstly, the cost of the cables and of the offshore operations makes it difficult to use a large number of channels at large offsets. Offshore maximum offsets are typically shorter than the maximum offsets normally used onshore. Secondly, sea-bottom sensors in deep sea (depth $\geq$ 1500 m) are difficult to use due to the limitations of the technology of the ocean-bottom-cables in deep water and due to their costs. These costs include the setting-up of the wireline connections and the ship-supply operations during the survey.

On the other hand, this approach has the following advantages. Firstly, the use of cables fixed at the sea-bottom partially reduces the rig noise and improves the quality of the information collected by the 4C seismic sensors (co-located hydrophones and multiple-component geophones in contact with the sea-floor to attenuate receiver ghosts and reverberations and to measure compressional and shear waves). Secondly, fixing the cables makes it possible to acquire the drill-bit signal for long listening times and to stack the correlations. In this case, the maximum listening time for a drill-bit level is related only to the rate-of-penetration (ROP) of the bit because the bit descent during the acquisition of the level is limited in order to preserve the required resolution. Several contractors have experimented with offshore SWD with fixed receivers, mainly restricted to investigations with short and medium offsets. Nevertheless, using large offsets with SWD has been demonstrated to be useful onshore, where the maximum offset is typically chosen depending on the depth of the target and can be several kilometers (say, 2 to 4 km; Miranda, Aleotti, Abramo, Poletto, Craglietto, Persoglia and Rocca, 1996). Using large offsets increases the sub-surface coverage and improves the seismic drill-bit signal (Section 8.2.4).

Recently, Poletto and Dordolo (2002) have proposed a new approach to drill-bit offshore seismic-while-drilling wherein the data are acquired close to the sea-surface by using a streamer towed by a vessel. This method is discussed as a new trend for deep-water acquisition in Section 8.12.1. The new technology simplifies the marine operations, and allows an extension of the offset investigation around the well by using compressional waves. The towing of the streamer during acquisition introduces limitations for the maximum listening time of the signal acquired at the same bit-depth position (see discussion of Section 8.12.1). However, onshore experiments showed that 5 or 10 minutes of recording time can be acceptable with signals obtained using roller-cone bits and good-quality-surface or downhole pilot signals. With poor bit-signal levels, the environmental noise, such as boat noise, is a major limitation. The measurement and removal of the independent noise (Poletto, Rocca and Bertelli, 2000) of the boat and especially the use of downhole pilot sensors near the bit (Section 7.8.1) can improve the signal-to-noise ratio and the resolution by reducing the required listening times at the same position.

To obtain reliable results while drilling, the acquisition system must be able to preprocess the whole raw dataset stored on disk, control the drilling and sailing parameters, and gather the acquired traces in common-position records. This allows a repeated stack of the correlations obtained at the same position-point through which the streamer passes during a long time-interval. This time depends on the streamer length and sailing velocity. The method uses a laboratory on the rig, and an acquisition and correlation system on the vessel. The system on the rig laboratory controls the drilling conditions and acquires the pilot traces. The communication between the rig and the vessel allows the synchronization and transmission of the pilot and VSP data. Evaluation of array and Doppler effects introduced by the continuous towing of the streamer (moving receiver line) is necessary to optimize trace gathering and time-length of the records used for correlation, respectively. Errors, introduced by sea waves, streamer buoyancy, stability and rig trim can be compensated for to improve the signal resolution.

Chapter 6

Preprocessing of SWD data

6.1 Introduction

We distinguish two main phases in seismic-while-drilling data processing. The first is the transformation of the recorded data into seismograms. It includes the data demultiplexing, pilot crosscorrelation, stacking, deconvolution, rephasing, and correction to absolute seismic time. We call this phase "*preprocessing.*" The result of the preprocessing is the raw, on-field, reverse VSP; i.e., similar to a conventional-VSP seismogram before processing. The second phase is the conventional VSP processing of the preprocessed drill-bit data. This phase typically includes picking direct arrivals, time-depth analysis, i.e., formation velocity analysis, wavefield separation, processing of reflected events for prediction ahead of the bit, and imaging. We simply call it "*processing.*"

There is no clear logical division between *preprocessing* and *processing.* Both can be complex – as when the pilot is obtained by optimum stack for signal and noise separation – and overlapping – as when the NMO correction is applied before the stack of the correlations for a depth level (Section 5.20.2). In general, because it is possible to commute the order of several steps, the whole preprocessing and processing flow could be considered as a single sequence. However, in this book we consider preprocessing and processing as distinct for the sake of clarity. In this chapter, we describe the drill-bit data preprocessing and discuss some critical steps for the subsequent reverse VSP processing. Other examples and results of the VSP processing of SWD drill-bit data are given in Chapter 8.

6.2 Preprocessing

Before the signal produced by the drill bit can be used for reciprocal VSPs, the recorded signal must be appropriately processed. Following acquisition, the signal recorded at the pilot traces, after being deconvolved, is correlated with the signals of the geophones located along the seismic line. The crosscorrelation identifies the signal emitted by the bit on the seismic traces; this is, in principle, like seismic acquisition with a Vibroseis source where crosscorrelation of the geophone traces with the reference signature – recorded generally on the base plate of the vibrator – has to be performed in order to recognize the signal. The major differences are:

◇ There is no unique start time because the bit emits noise "all the time".

◇ The exact signature of the bit is not recorded at the pilot sensor; it is filtered by the drill string.

Because of the drill-string filtering and of the source periodical patterns (Sections 3.12 and 3.16), the pilot signal does not have a "white" – or flat – spectrum. It contains repetitions of itself and energy in narrow harmonic bands ("colored"). In addition, the pilot may be noisy and the drill-bit signal may be of low amplitude, so that we need very long recording times to measure a sufficient amount of energy. These times may be tens or hundreds times longer than those typically used by vibrators (a quantitative analysis is developed in Section 3.21).

6.3 SWD data in the crosscorrelated domain

Crosscorrelation is central step of the data preprocessing. It is not the only way to achieve results with drill-bit data; however, it is the most reliable and most widely used method. The approach is similar to that used with Vibroseis, in which a pilot sweep is used for correlation. The main differences of the two methods are mentioned above. The drill-bit pilot signal used for the correlation is far from being close to an ideal pilot signal. The Vibroseis sweep comes more closer to being an ideal pilot signal.

The main steps used in the preprocessing are just those aimed at correcting the pilot-signal properties. However, the result is only an approximation of the ideal pilot signal. It is a common experience that, after preprocessing, the drill-bit data may still contain a significant amount of noise and filtering effects of the pilot signal. Hence, it is important to have a clear understanding of all the events obtained by the crosscorrelation.

6.3.1 Basic crosscorrelation properties

Crosscorrelation is widely used in seismic data processing, and its basic properties are well known to geophysicists. The crosscorrelation is a measure of the similarity between two waveforms, as a function of their relative delay. Let $p(k)$ and $g(k)$, with k integer, be two sampled sequences of signals. We assume, for simplicity, $-\infty < k < +\infty$. A standard, unnormalized definition of crosscorrelation is:

$$\Phi_{pg}(\tau) = \sum_{k=-\infty}^{+\infty} p(k)g(k+\tau), \tag{6.1}$$

where the integer τ is the relative delay. Crosscorrelation is a linear operation, as is filtering. It is equivalent to filtering the trace $g(k)$ with the trace time-reverse of $p(k)$, i.e., $p(-k)$. In fact, we have

$$\Phi_{pg}(\tau) = \sum_{k=-\infty}^{+\infty} p(k-\tau)g(k) = p(-k) \star g(k), \tag{6.2}$$

where '$\star$' denotes convolution.

Let us say that the sequences $g(k)$ and $p(k)$ are the geophone signal and the pilot signal, respectively, for which we assume a convolution model (Robinson, 1984). The trace that contains the geophysical information – the geophone signal $g(k)$ – is not reversed in time in equation (6.2), while the trace used to recognize the signal – the pilot signal $p(k)$ – is reversed in time. If we invert the order of these traces in the crosscorrelation, we have

$$\Phi_{gp}(\tau) = \Phi_{pg}(-\tau). \tag{6.3}$$

In other words, if the pilot signal is interchanged with the geophone signal, the correlation output is reversed in the delay-time axis. The pilot trace is, in many cases, the signal measured on the drill string. However, sometimes a geophone trace can be also used for the pilot signal for crosscorrelation with another geophone trace. Therefore, we must pay attention to the order of these traces in the crosscorrelation.

Because in the remainder of this book we deal with processing of crosscorrelated data, we will simply call them correlated. When it is not necessary for clarity, we will omit the use of the 'Φ' symbol to denote the result of a correlation. The correlated data will be represented in the time domain with the standard notation of a time sequence, where it is intended that the time is the relative delay of the crosscorrelation, which is shifted, when necessary, for the pilot delay.

Equivalently, the crosscorrelation can be calculated in the frequency domain. Let

$$G(\omega) = |G(\omega)|e^{\imath\phi_{\rm g}(\omega)} \tag{6.4}$$

and

$$P(\omega) = |P(\omega)|e^{\imath\phi_{\rm p}(\omega)} \tag{6.5}$$

be the Fourier transforms of $g(k)$ and $p(k)$, respectively, and ω the angular frequency. Reversal in time of a signal corresponds to changing the sign of its phase, i.e., to using its complex conjugate, denoted by superscript '*'(not to be confused with the convolution symbol). In fact, from the similarity theorem we have that $p(-k)$ and $P(-\omega)$ are Fourier transform pairs and $P(-\omega) = P^{*}(\omega)$ if $p(k)$ is real (Bracewell, 1965). Due to the convolution theorem, correlation, i.e., filtering by the pilot reversed in time, is obtained by

$$C(\omega) = P^{*}(\omega)G(\omega) = |P(\omega)G(\omega)|e^{\imath[\phi_{\rm g}(\omega)-\phi_{\rm p}(\omega)]}. \tag{6.6}$$

Equation (6.6) represents the crosscorrelation in the frequency domain, i.e., the Fourier transform of $\Phi_{pg}(\tau)$, where the relative delays of the geophone and pilot signals are expressed by the difference of the phases. The same relations hold for sequences with a finite number of samples, i.e., the digital seismic traces. From a computational point of view, it is convenient to make use of fast-Fourier transform (FFT) algorithms, which give an advantage of the order of $N/\log_2 N$, where N is the number of samples in the trace (Oppenheim and Shafer, 1975). This makes it possible to perform the task of correlating while-drilling-SWD datasets of the size of Gigabytes by using commercial power PCs with the CPU clock of MHz.

6.3.2 Energy in crosscorrelation

The crosscorrelation function (equation (6.1)) gives the cross–energy of the geophone- and pilot traces as a function of the relative delay τ. In the particular case in which a sequence $x(k)$ is correlated with itself, we obtain the autocorrelation function $\Phi_{xx}(\tau)$.

We recall here the main properties of autocorrelation. $\Phi_{xx}(\tau)$ and the power spectrum $|X(\omega)|^2$ are Fourier-transform pairs if $X(\omega)$ is the Fourier transform of $x(k)$. From equation (6.3), it follows that $\Phi_{xx}(\tau) = \Phi_{xx}(-\tau)$. $\Phi_{xx}(0) \geq 0$ is the energy of the trace $x(k)$, and we have

$$|\Phi_{xx}(\tau)| \leq \Phi_{xx}(0) \tag{6.7}$$

for every τ. Furthermore, for every τ we have (Bracewell, 1965)

$$|\Phi_{pg}(\tau)| \leq \sqrt{\Phi_{pp}(0)\Phi_{gg}(0)}. \tag{6.8}$$

We can interpret each trace as an element of a vector space and crosscorrelation as the scalar product of two traces. Equation (6.8) is the Schwartz's inequality, according to which the scalar product (Φ_{pg}) of two vectors cannot exceed the product of their absolute values ($\sqrt{\Phi_{pp}(0)}$ and $\sqrt{\Phi_{gg}(0)}$). In other words, a rule for the energy of SWD correlated data is:

Rule 1 *The cross–energy calculated by the crosscorrelation of pilot and geophone traces is always less or equal to the geometric mean of the energies of the pilot and geophone traces.*

Equality holds in equation (6.8) if and only if the "vectors" are parallel. This condition is equivalent to saying that, for some τ, $g(k+\tau)$ is proportional to $p(k)$ (see Figure 6.1).

6.3.3 Delays in crosscorrelation

Let $s(k)$, τ_0 (> 0), and Δ be a time sequence, a positive increment of the time index, and the time-sampling interval, respectively. The change of the time index k into $(k - \tau_0)$ corresponds to a time shift $(-\tau_0\Delta)$ of the sequence. In this case, the sequence is retarded (towards later times) and the shift is a delay. Contrarily, the change of k into $(k + \tau_0)$ is an anticipation of magnitude $(\tau_0\Delta)$. In this case, the sequence is advanced (towards earlier times).

Using the phase-shift theorem (Bracewell, 1965), the time delay $(-\tau_0\Delta)$ and the anticipation $(\tau_0\Delta)$ correspond in the frequency domain to phase shifts $(-\tau_0\Delta\omega)$ and $(\tau_0\Delta\omega)$, respectively.

To recall the meaning of the crosscorrelation relative delay, we write equation (6.1) as

$$\Phi_{pg}(\tau) = \sum_{k=-\infty}^{+\infty} p(k-\tau)g(k); \qquad -\infty < \tau < \infty. \tag{6.9}$$

The geophone trace is not shifted in equation (6.9). Contrarily, the pilot trace is shifted by τ. When $\tau < 0$ the pilot trace is advanced; as τ increases and $\tau > 0$, the pilot is shifted towards later times. Therefore, τ represents the position at which the pilot trace is "juxtaposed to the geophone trace", and the crosscorrelation terms of the variable delay can be regarded as the result of the shifting of the pilot trace related to the geophone trace. Due to the linear properties of the crosscorrelation operator, the events that have crosscorrelated energy in the geophone trace are then reproduced with the same time order, relative delays and amplitude ratios. If the pilot signal has no source delay, no noise and has an impulsive autocorrelation, the crosscorrelation function would be a good representation of the geophone seismogram, as if it were produced by an impulsive source. In this case, the source wavelet is the wavelet of the pilot autocorrelation.

In drill-bit data, the pilot signal may be recorded downhole and/or at the surface far from the source. The distances of the pilot sensor and of the geophone from the source are generally different. The propagation velocity of the pilot signal can be either faster or slower than the velocity of the signal in the formation. Moreover, paths are different also. Consequently, the pilot delay can be smaller or larger than the delay of the signal in the formation as measured by the geophone. Figures 6.1 and 6.2 show two examples of correlation in which the pilot wavelets have delays smaller and larger relative to the geophone signal. In the latter case, the crosscorrelated event, that is, the maximum in the crosscorrelation, occurs at negative correlation times. We have the following additional two general rules in SWD processing.

Rule 2 *The delay time of a coherent event in the crosscorrelation function is the relative delay of the corresponding events in the geophone and in the pilot recordings.*

Rule 3 *Events with larger delays (say, with slower propagation speed) in the pilot signal appear as anticipated events in the crosscorrelated data. On the contrary, anticipated events (say, with faster speed) in the pilot signal occur as delayed events in the crosscorrelated data.*

6.3.4 Crosscorrelation and filtering

Crosscorrelation is a linear operation and can be expressed by a convolution (see equation (6.2)). An important property of linear operators is that they commute. Let $f(k)$ be a filter response and $p(k)$ be the pilot signal. Then

$$h(k) = f(k) \star p(k) \tag{6.10}$$

where $h(k)$ is the pilot after filtering. (Here the same index k is used to denote the time dependence before and after convolution). The filtered pilot, reversed in time, is equal to the convolution of the time-reversed filter response and of the time-reversed pilot signal before filtering, i.e.,

$$h(-k) = f(-k) \star p(-k). \tag{6.11}$$

Therefore, the correlation with a filtered pilot signal is

$$\Phi_{hg} = h(-k) \star g(k) = [f(-k) \star p(-k)] \star g(k). \tag{6.12}$$

We use the associative property of linear operators and obtain

$$\Phi_{hg} = f(-k) \star \Phi_{pg}. \tag{6.13}$$

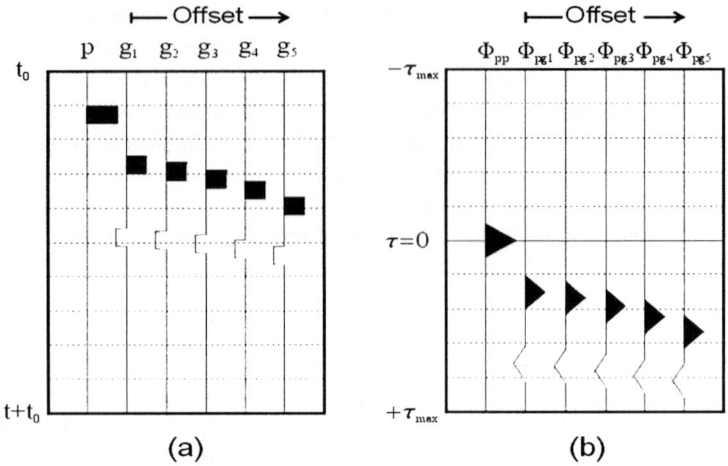

Figure 6.1: This is a simple model of a common-shot gather (i.e., at a fixed bit depth). a) Pilot p and geophone traces g_i measure a bit signal with direct arrivals and reflections versus offset. The signal has a rectangle-box shape of time duration T. b) The correlated wavelet is a triangle of time duration $2T$. If p_o and g_o are the amplitudes of the pilot and geophone signals, respectively, the pilot autocorrelation amplitude peak is the pilot energy p_o^2T and the geophone crosscorrelation amplitude peak is the cross energy p_og_oT. The initial time of the recorded data (a) starts at the arbitrary time t_0. The pilot autocorrelation and the crosscorrelation functions (b) are calculated from $-\tau_{\max}$ to $+\tau_{\max}$. The delay of the correlated geophone wavelets is relative to the delay of the pilot wavelet. The delay of the pilot wavelet is less than the delay of the geophone signals in (a) and the crosscorrelated events occur in (b) at positive delay times.

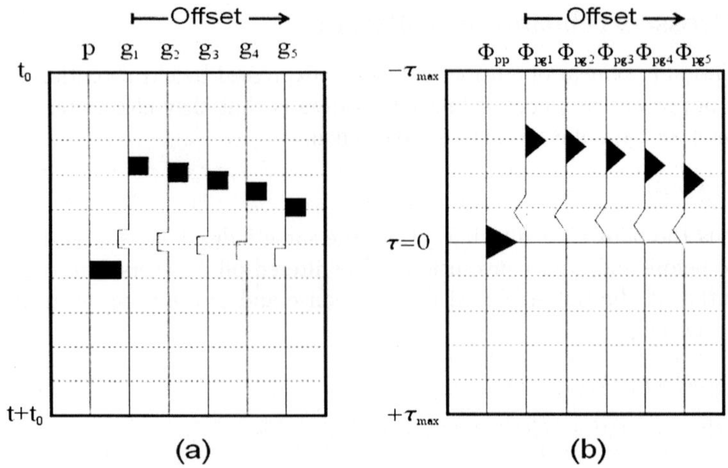

Figure 6.2: The same geophone traces of Figure 6.1 are correlated with a pilot signal with greater delay. In this case, the crosscorrelated events occur at negative delay times. If we account for the different pilot-delay shift, the information of the seismogram is substantially unchanged with respect to the result of Figure 6.1.

If $F(\omega)$ is the Fourier transform of the filter, we obtain for the frequency domain

$$[F(\omega)P(\omega)]^* G(\omega) = F^*(\omega)\,[P^*(\omega)G(\omega)]\,. \tag{6.14}$$

In other words, in equations 6.2 or 6.6 we use a time-reversed pilot signal. Using this definition for correlation, we obtain two other important rules for SWD.

Rule 4 *The filters of the pilot signal are reversed in time in the correlated data.*

Rule 5 *The filters of the geophone signal are reproduced without time reversion in the correlated data.*

Pilot filters before and after correlation are shown in Figure 6.3.

For a direct or a reversed filter we will use a direct or reversed antifilter, i.e., deconvolution, respectively. For example, the effects due to the drill-string reverberations in the pilot signal before correlation are reversed in time after crosscorrelation. Usually, we can observe long-period drill-string multiples (Section 6.10) at negative correlation times. We deconvolve them by using operators reversed in time. Contrarily, the reflectivity series, assumed as a filter in the convolutive geophone-signal model, is preserved after correlation (compare Figures 6.1 and 6.2). In summary, both correlation and filtering are linear operations, which can be distributed, commuted and associated. Under proper conditions, we can assume the following rule:

Rule 6 *It is equivalent to filter data before or after correlation.*

In general, filtering after correlation is more convenient from a computational point of view. However, rule (**6**) must be considered with care. In fact, only with correlated sequences of infinite samples rule (**6**) is true. In practice, the correlation output is calculated in a time-delay window that is much shorter than the length of the initial records. Time windowing is not a linear operation and associativity and commutativity are not preserved before and after it. Let us denote by $W\{\cdot\}$ the time windowing operator and let $d(k)$ be a pilot-deconvolution-filter operator. Untapered time windowing is equivalent to the product by a box function in time which, in turn, corresponds to convolution by a sinc function in frequency. We have

$$d(-k) \star W\{h(-k) \star g(k)\} \neq W\{[d(-k) \star h(-k)] \star g(k)\} \tag{6.15}$$

because of possible filtering edge effects. These effects can be minimized by increasing the time length of the correlation output, i.e., by "padding" the seismogram borders with more data. SWD deconvolution – like Vibroseis deconvolution (Ristow and Jurczyk, 1975) – can be performed either before or after correlation. Usually, a long operator is applied to deconvolve the pilot multiples after crosscorrelation. Hence, the operator length has to be accounted for in order to calculate the total length of the crosscorrelation output. This is obtained by increasing the length of the desired seismogram before deconvolution, that is, by adding the operator length to the ends of the data time window.

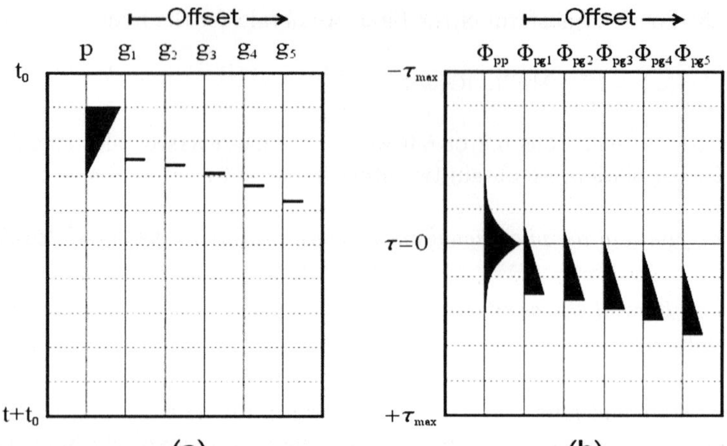

Figure 6.3: a) The geophone traces g_i contain only the direct arrivals, which are assumed to be an impulsive signal in order to better demonstrate the pilot-filtering effects. Contrarily, the pilot signal p is filtered by a filter whose impulse response is a triangle. The relative delays of pilot and geophone signals are the same delays of Figure 6.1. b) After cross-correlation, the filter shape is reversed in the delay times. The pilot autocorrelation Φ_{pp} is the inverse Fourier transform of the power spectrum of the pilot signal, i.e., $|P(\omega)|^2$. This trace can be used to design inverse filters and deconvolve the pilot signature.

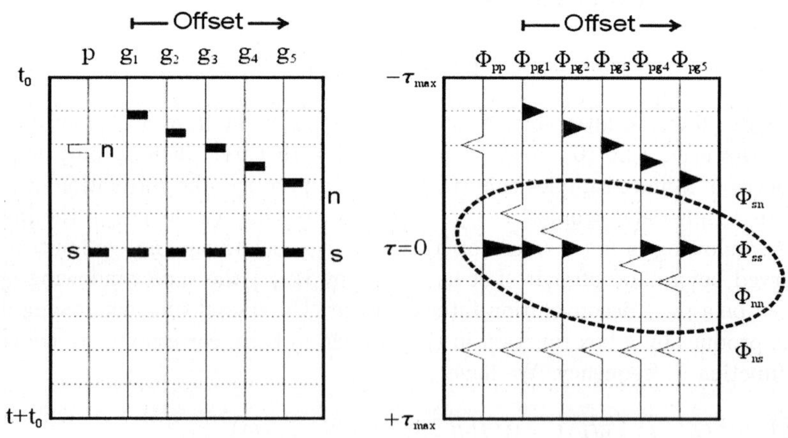

Figure 6.4: a) Pilot and geophone recordings with impulsive signal (s) to simulate a vertical plane wave and dipping noise (n) to simulate rig events. Here, the signal and noise wavelets have the same shape, i.e., that of a rectangle box. b) Crosscorrelations of signal and noise. The signal and noise are separated in recordings (a) before correlation. On the other hand, they interfere after crosscorrelation (b). The dashed circle shows the zone in which contributions of the shifted autocorrelations of signal and noise occur. The autocorrelations $\Phi_{s_p s_g}$ and $\Phi_{n_p n_g}$ are delayed for the relative delay between geophone and pilot signals as well as between geophone and pilot noise, respectively, according to rule (**2**).

6.3.5 Crosscorrelation of signal and noise

Signal and noise are summed in pilot and geophone drill-bit data. We assume

$$p(k) = s_p(k) + n_p(k) + r_p(k) \tag{6.16}$$

$$g(k) = s_g(k) + n_g(k) + r_g(k) \tag{6.17}$$

where $s(k)$, $n(k)$ and $r(k)$ denote signal, coherent noise (see Sections 4.8 and 4.9) and random noise, respectively. The crosscorrelation becomes

$$\Phi_{pg} = \Phi_{s_p s_g} + \Phi_{n_p n_g} + \Phi_{s_p n_g} + \Phi_{n_p s_g} + \underbrace{\Phi_{r_p s_g} + \Phi_{r_p n_g} + \Phi_{r_p r_g} + \Phi_{s_p r_g} + \Phi_{n_p r_g}}_{\Phi_{\text{random}}} . \tag{6.18}$$

By definition, the random-noise terms of each trace are uncorrelated with the terms of the other trace, and their crosscorrelation becomes negligible. Ideally, we can assume $\Phi_{\text{random}} \cong 0$. However, the energy of the random noise is different from zero. The signal-to-random noise ratio in drill-bit crosscorrelations is analyzed by Rector (1989) and in Section 7.3.

The coherent noise in the correlated data includes pilot noise correlated with the geophone noise (such as the rig-borehole noise) and, in general, noise correlated with the signal. This is the case of drill-string multiples, head or coupled-mud waves (Section 4.5). If we consider only the case in which signal and noise are not correlated we obtain

$$\Phi_{pg} \cong \Phi_{s_p s_g} + \Phi_{n_p n_g}, \tag{6.19}$$

plus random noise. The result of the crosscorrelation of equation (6.19) is the sum of the events produced by the correlations of signal with signal and of noise with noise. According to rule (**2**), the delay of these events is determined by the relative delay of the corresponding events in the traces before correlation: that is, between the geophone and pilot signals and between the geophone and pilot noise. When these relative delays are comparable, interference of the signal and noise events may occur in the correlated traces even if the noise and signal are separated in time in the original recordings. Figure 6.4 shows an example of impulsive signal and noise and the variation of the arrival times before and after correlation. In real drill-bit recording, both signal and noise are typically continuous and superimposed at all times (Figure 6.5a). If signal and noise are produced by uncorrelated sources, their crosscorrelation do not produce the coherent events $\Phi_{s_p n_g}$ and $\Phi_{n_p s_g}$, as in Figure 6.4b. However, if they are recorded on both the pilot and geophone traces, they always produce the crosscorrelated events $\Phi_{s_p s_g}$ and $\Phi_{n_p n_g}$. These SWD events may interfere (Figure 6.5b).

This interference may cause problems for the interpretation of SWD correlated seismograms after correction for the pilot delay time, which corrects the seismic time of the signal but changes also the apparent delays of the noise.

Noticeable cases are the following:

◇ The noise source is stationary. Here "stationary" has a "kinematic" meaning. It means that the delay of the events does not change in the recordings at different bit depths. We assume that this noise is recorded with constant time delays in the

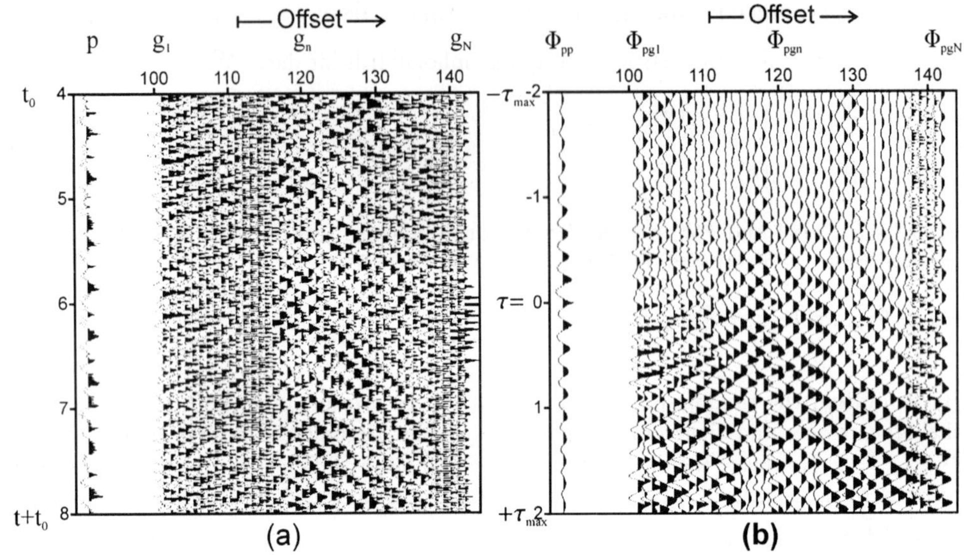

Figure 6.5: a) Time window extracted from a gather of pilot- and geophone traces before correlation. The pilot signal is an axial acceleration signal recorded at the top of the drill string. Each geophone trace is recorded with an array of 24 geophones. b) Crosscorrelated data obtained by stacking the contribution of 50 records of 45 s. The first trace is the pilot autocorrelation. Data are not filtered.

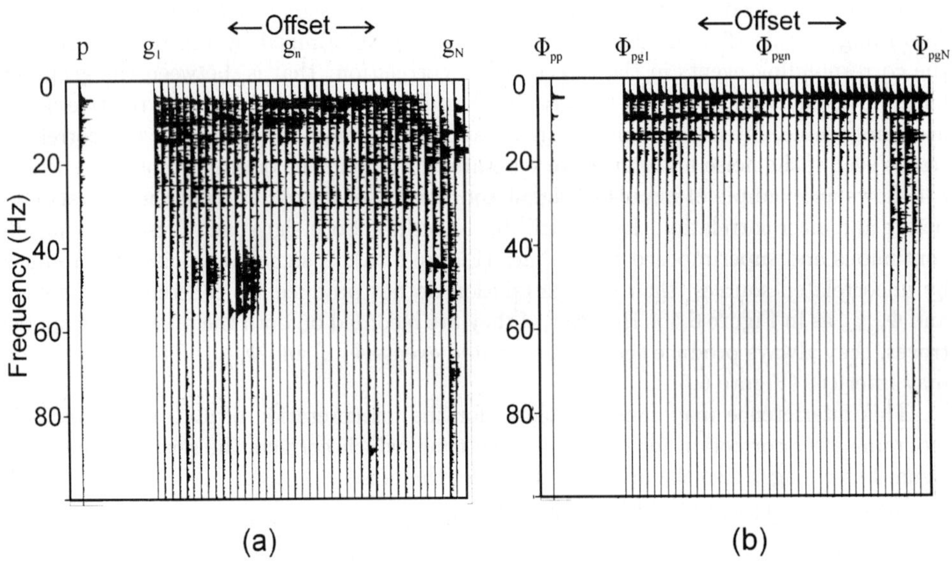

Figure 6.6: a) Frequency spectra of data of Figure 6.5 a) before pilot correlation and b) after pilot correlation.

pilot- and in the geophone traces. Therefore, their relative delays are constant and the noise event $\Phi_{n_p n_g}$ occurs at constant time in the correlations plotted at different depths. We call it "*stationary noise*".

- ◇ The source of noise is not stationary at different bit depths. For instance the bit signal is transmitted to the surface and rig, where it is recorded with variable delays by the receivers because the bit depth is changed. This signal is radiated from the rig as a secondary noise (Rector, 1991). Also in this case the difference in delay of the rig signal measured by a rig pilot and the geophones is constant at all the bit depths. This is again a "*stationary noise*" event, but produced by a non-stationary source such as the moving bit.

- ◇ At some depth interval, the delay increment in the formation for the bit signal at a geophone of a given offset may equal the pilot delay increment in the drill string. In this case, the correlated signal is stationary with constant time delay for the different bit depths of this interval.

The rig-stationary noise discussed above is sensitive to the directional filtering properties of geophone arrays (Figure 5.34). These are successfully used in SWD (Section 5.14). Conversely, filtering in the time-depth (*T-Z*) domain is not appropriate for removing stationary noise when the signal is also stationary. In other words, it is not easy to discriminate events which have, in the pilot and geophone signals, different arrival times but the same relative delays at different bit depths. All these events have a constant delay time in a crosscorrelated common-receiver gather, i.e., apparent velocity equal to zero in the data plotted versus depth. For this reason, it is important to extend the analysis of the moveouts of signal and noise to different offsets (*T-X* domain). We have the following rule:

Rule 7 *Stationary events in the correlated trace of a given geophone channel at different bit depths are produced by stationary rig noise as well as by signal arrivals with the same delay increment of the pilot waves.*

Figure 6.7 is a crosscorrelated common-receiver gather for different bit depths (VSP). Analysis of common-receiver correlated data before pilot-delay correction is an important step of SWD quality control.

6.4 Stack of while-drilling data

Let L_{T} be the total number of samples acquired while the bit is "at" a depth level. The total record of L_{T} samples can be divided into a number N_{R} of recordings of L_{R} samples so that

$$L_{\mathrm{T}} = N_{\mathrm{R}} L_{\mathrm{R}}. \tag{6.20}$$

We can subdivide the sum operator in the crosscorrelation with the total number of samples as

$$\sum_{k=1}^{L_{\mathrm{T}}} p(k) g(k+\tau) = \sum_{n=1}^{N_{\mathrm{R}}} \sum_{j=1}^{L_{\mathrm{R}}} p^{(n)}(j) g^{(n)}(j+\tau), \tag{6.21}$$

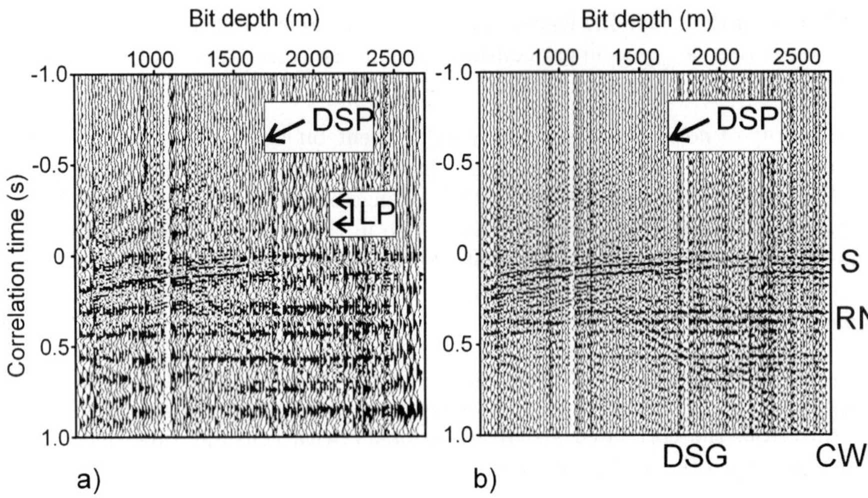

Figure 6.7: Common receiver roller-cone crosscorrelated data at different bit depths a) before and b) after bandpass (12/16/46/60 Hz) filtering. Offset is 824 m. S, RN, CW, LP, DSP, and DSG denote signal, stationary rig noise, a converted wave after reflection, roller-cone lobed pattern, drill-string waves in pilot and geophone data, respectively. The signal moveout is flat with a trend similar to that of the stationary rig noise. A change of velocity can be seen in the signal S at depth of approximately 2000 m.

where

$$p^{(n)}(j) = p(j + (n-1)L_{\mathrm{R}}), \qquad \text{with} \qquad j = 1, \ldots, L_{\mathrm{R}}; \;\; n = 1, \ldots, N_{\mathrm{R}}, \tag{6.22}$$

and a similar relation defines $g^{(n)}(j)$. Equation 6.21 holds if we break continuous recordings in records of length L_{R}. In practical implementations, acquisition for long listening times can be performed by acquiring a sequence of individual records. Equation (6.21) can be rewritten as

$$\Phi_{pg}(\tau) \approx \sum_{n=1}^{N_{\mathrm{R}}} \Phi_{pg}^{(n)}(\tau), \tag{6.23}$$

where

$$\Phi_{pg}^{(n)}(\tau) = \sum_{j=1}^{L_{\mathrm{R}}} p^{(n)}(j) g^{(n)}(j+\tau). \tag{6.24}$$

If we use recordings of length L_{R} which are acquired separately, the sign of equality holds in equation (6.21) when $\tau = 0$. When $\tau \neq 0$, the right term approximates on average the cross–energy of the left term with an attenuation equal to

$$\gamma_{\mathrm{c}} = 1 - |\tau|/L_{\mathrm{R}} \tag{6.25}$$

because of the lack of correlation at the ends of the individual records. In practice, γ_{c} is close to 1, the attenuation effect is negligible and can be compensated. For example, if $L_{\mathrm{R}} = 45$ s, at a delay $\tau = 2$ s we have $\gamma_{\mathrm{c}} = 0.96$.

Vertical stack and correlation error

The question of "how much to correlate at a depth level" becomes "how many individual records to stack in equation (6.23)". The minimum number of records to stack is determined by signal-to-incoherent noise considerations. Rector (1989) analyzed this aspect by assuming additional random noise. However, crosscorrelated drill-bit data contain both random noise and a significant amount of coherent noise. Signal-to-noise analysis of correlated data before stack is discussed in Sections 7.3 and 7.4. The maximum number of records to stack (or total length L_{T}) is determined by resolution considerations discussed in Chapter 5.

Another important aspect is the bandwidth variability of signal. The question is whether "stationarity in the wide sense" has to be assumed (or required). Stationarity in the wide sense means that a stochastic process has time-invariant mean and its autocorrelation depends only on relative delay τ (Papoulis, 1991). This stationarity condition may be assumed to be true, under the limit of resolution due to bit descent, for the bit-to-receiver transfer function. However, the drill-bit source as well as the noise waves may vary significantly in relation to drilling conditions (Section 3.12). This variability property should be regarded more as a favorable than an unfavorable condition, because it enlarges the signal bandwidth. Remember that continuous frequency variation in time is a basic property of a Vibroseis sweep, which does not "repeat itself" and produces a wideband signal.

A basic consideration for the use of "long listening times" in the presence of random noise is that they are used in SWD also to compensate "lacks in the bandwidth" of the drill-bit signal that contains more peaks in frequency spectrum than other wideband seismic sources. The importance of bandwidth to determine timing and phase in seismic crosscorrelation is discussed, for instance, by White and Simm (2003). They derive the accuracy and the standard errors with which the time shift and the phase rotation are estimated from crosscorrelation of two trace segments of duration T and bandwidth B. These standard error (S.E.) depends on the S/N ratio and are inversely proportional to the square root of B^3 and of T (White and Simm, 2003; see, also, Hamon and Hannon, 1974; and Section 7.10), i.e.,

$$S.E. \propto \frac{1}{\sqrt{B^3 T}}. \tag{6.26}$$

With the rule of thumb, for a given S/N, a halving in the bandwidth can be "compensated" by using listening times eight times as long to obtain the same time and phase errors in the correlation. Therefore, long listening times are used in SWD to reduce the time and phase errors in the correlation of the data, which are subsequently deconvolved.

6.5 Deconvolution of the drill-bit source function

In Chapter 3 we discussed the radiation properties of the drill bit as a vertical force. Let $f_{\mathrm{B}}(k)$ be the time variation of the vertical force exerted by the bit against the formation. The reaction force exerted by the formation against the bit is $-f_{\mathrm{B}}(k)$. Hence, the axial stress exerted on the positive, downward oriented, side of a cross–section A_1 inside the drill-collar attached to the bit (Section 2.3.10) is

$$\sigma_B(k) = -\frac{f_B(k)}{A_1}, \tag{6.27}$$

where the minus sign means that a positive force f_B (in the downward direction) produces a compression. The force $f_B(k)$ at the bit is the result of many effects. A stress wave produced at the bit propagates in the drill-string transmission line and is partially reflected downward so that the downhole reflections in the drill string contribute to the stress at the bit. With the notation of the drill-string model of Section 4.10, we can express f_B as

$$f_B(k) = \mathcal{H}_B^{(s)}(k) \star [f_R(k) + f_P(k)], \tag{6.28}$$

where the f_R and f_P are excitation forces generated at the bit in its rotary action: f_R is a wideband random force due to unpredictable bit behavior and f_P is a periodic-colored force due to bottomhole bit patterns (Section 3.12). Without loss of generality, we may also assume that the random force is modulated by the periodic bit pattern (Section 3.12). In equation (6.28), the filter

$$\mathcal{H}_B^{(s)}(k) = 1 + (1 - c_0)\, R_B^{(s)}(k), \tag{6.29}$$

includes the downhole reflectivity in the drill string (Figure 4.15). The superscript $^{(s)}$ denotes that the reflection coefficients in the drill string are calculated for stress waves (Section 4.12), and c_0 is the bit/rock reflection coefficient for displacement. At this stage, we assume that the reflectivity in the formation is not important for drill-string waves (reflections of the formation in the drill-string waves are interpreted and used for prediction ahead of the bit by Malusa, Poletto and Miranda, 2002).

The acceleration pilot signal at the top of the drill string is

$$p(k) = -\mathcal{H}_T(k) \star \mathcal{F}_{f,a}(k) \star [f_R(k) + f_P(k)] + n_p(k) + r_p(k), \tag{6.30}$$

where n_p and r_p are coherent and random noise, respectively. The filter $\mathcal{F}_{f,a}$ of equation (6.30) shapes force into acceleration in the drill string of characteristic axial impedance $\mathcal{Z}_1 = A_1\rho_1 c_1$ (Section 4.12). If $p(k)$ is the axial extensional signal, n_p may include coupled mud waves, rig noise and borehole noise due to lateral contacts of the string with the well bore. In equation (6.30), the filter

$$\mathcal{H}_T(k) = [(1 + c_T)\, F_{DS}(k) + (1 - c_T)\, R_T(k)], \tag{6.31}$$

represents the drill-string transfer function for acceleration and includes the reflections at the top of the drill string and the rig ghost. Here we use the notation of Figure 4.15, where F_{DS} and R_T are the upgoing field and the downgoing reflectivity at the top of the drill string, respectively (see also, Section 6.10). In general, we may express $\mathcal{H}_T$ as a composition of a minimum phase filter $\mathcal{H}_{TM}$ and of a pilot-delay shift t_p,

$$\mathcal{H}_T(k) = \mathcal{H}_{TM}(k) \star \delta(k - t_p/\Delta) = \mathcal{H}_{TM}(k - t_p/\Delta), \tag{6.32}$$

where

$$\delta(k - t_p/\Delta) \longleftrightarrow e^{-\imath\omega t_p} \tag{6.33}$$

are Fourier transform pairs and Δ is the sampling rate. For waves traveling upward in the drill string we have, from equation (4.157)

$$\mathcal{F}_{f,a}(k) = \frac{\partial_t}{\mathcal{Z}_1}, \tag{6.34}$$

where ∂_t denotes time derivative. $\mathcal{F}_{f,a}$ is assumed minimum phase so that $\mathcal{H}_{\rm TM} \star \mathcal{F}_{f,a}$ is also minimum phase, i.e., a causal stable signal with a causal stable inverse (Oppenheim and Shafer, 1975).

The filtered geophone signal is

$$g(k) = \mathcal{H}_{\rm G}(k) \star f_{\rm B}(k) + n_{\rm g}(k) + r_{\rm g}(k), \tag{6.35}$$

where $\mathcal{H}_{\rm G}$ is the formation impulse response. It includes three things: the radiation-pattern filter, which relates source force to particle velocity; the secondary rig noise and conical head waves (Section 4.6.1) generated by the bit source; and the receiver-instrumental response (Section 6.17). Terms $n_{\rm g}$ and $r_{\rm g}$ are coherent, e.g., pumps and power generators, and random noise, respectively.

Crosscorrelation of pilot and geophone traces gives

$$\begin{aligned}\Phi_{\rm pg}(k) = &-\mathcal{H}_{\rm T}(-k) \star \mathcal{F}_{f,a}(-k) \star [f_{\rm R}(-k) + f_{\rm P}(-k)] \star f_{\rm B}(k) \star \mathcal{H}_{\rm G}(k) + \\ &\Phi_{\rm n_p n_g}(k) + \Phi_{\rm r_p r_g}(k),\end{aligned} \tag{6.36}$$

where $\Phi_{\rm n_p n_g}$ and $\Phi_{\rm r_p r_g}$ are crosscorrelated- and random-noise terms. Equation (6.36) can be rewritten as

$$\begin{aligned}\Phi_{\rm pg}(k) = &-\mathcal{H}_{\rm T}(-k) \star \mathcal{F}_{f,a}(-k) \star \mathcal{H}_{\rm B}^{\rm (s)}(k) \star [\Phi_{\rm RR}(k) + \Phi_{\rm PP}(k)] \star \mathcal{H}_{\rm G}(k) + \\ &\Phi_{\rm n_p n_g}(k) + \Phi_{\rm r_p r_g}(k),\end{aligned} \tag{6.37}$$

where we assume

$$f_{\rm R}(-k) \star f_{\rm P}(k) = \Phi_{\rm RP}(k) = \Phi_{\rm PR}(k) = 0, \tag{6.38}$$

and where

$$\Phi_{\rm RR}(k) = f_{\rm R}(-k) \star f_{\rm R}(k), \tag{6.39}$$

and

$$\Phi_{\rm PP}(k) = f_{\rm P}(-k) \star f_{\rm P}(k), \tag{6.40}$$

are the source-signal autocorrelations, equivalent to the Klauder wavelet (Sheriff, 1999), for random and periodic bit excitation forces without including drill-string reflectivity.

In the crosscorrelation of equation (6.37), the formation transfer function $\mathcal{H}_{\rm G}$ is "the object of interest" to geophysicists. Therefore, at this stage we would like to focus our attention on the bit signature which filters $\mathcal{H}_{\rm G}$. In this, we ignore the problem of minimizing the noise-to-signal ratio in correlated data. The deconvolution operator $D(k)$ spiking the drill-bit signature is given by

$$D(k) = \left\{\mathcal{H}_{\rm TM}(-k) \star \mathcal{F}_{f,a}(-k) \star \mathcal{H}_{\rm B}^{\rm (s)}(k) \star [\Phi_{\rm RR}(\tau) + \Phi_{\rm PP}(\tau)]\right\}^{-1} \star \delta(k - t_{\rm p}/\Delta), \tag{6.41}$$

where the delta function represents the pilot-delay correction. If a different pilot trace is used in place of the axial drill-string acceleration, we have to substitute the filters $\mathcal{H}_{\rm T}(k)$ and/or $\mathcal{F}_{f,a}(k)$ in equations (6.32), (6.34) and equation (6.41). There are different approaches to deconvolving SWD data.

6.6 Pilot deconvolution

We can express $D(k)$ as a composition of operators

$$D(k) = D_{\mathrm{TM}}(-k) \star D_{\mathrm{B}}(k) \star D_{\Phi}(\tau) \star \delta(k - t_{\mathrm{p}}/\Delta), \tag{6.42}$$

with

$$D_{\mathrm{TM}}(k) = \mathcal{H}_{\mathrm{TM}}(k)^{-1} \star \mathcal{F}_{f,a}(k)^{-1}, \tag{6.43}$$

$$D_{\mathrm{B}}(k) = \mathcal{H}_{\mathrm{B}}^{(\mathrm{s})}(k)^{-1}, \tag{6.44}$$

$$D_{\Phi}(k) = \left[\Phi_{\mathrm{RR}}(k) + \Phi_{\mathrm{PP}}(k)\right]^{-1}, \tag{6.45}$$

where $\mathcal{H}_{\mathrm{TM}}(k)^{-1} \star \mathcal{H}_{\mathrm{TM}}(k) = \delta(k)$. The operators $\mathcal{H}_{\mathrm{TM}}(k) \star \mathcal{F}_{f,a}(k)$ and $\mathcal{H}_{\mathrm{B}}^{(\mathrm{s})}(k)$ are minimum phase. Hence, $D_{\mathrm{TM}}(k)$ and $D_{\mathrm{B}}(k)$ are minimum phase too. In analogy with Vibroseis processing (Ristow and Jurczyk, 1975), we can factorize

$$\left[\Phi_{\mathrm{RR}}(k) + \Phi_{\mathrm{PP}}(k)\right] = f_{\mathrm{FM}}(-k) \star f_{\mathrm{FM}}(k), \tag{6.46}$$

where we use equation 6.38 and assume $f_{\mathrm{FM}}(k)$ to be the minimum-phase equivalent of $[f_{\mathrm{R}}(k) + f_{\mathrm{P}}(k)]$ (Claerbout, 1976). By definition, equation 6.46 holds if the "periodic" and random-wave spectrum is such that its minimum-phase equivalent exists, i.e., has finite length and is stable with stable inverse (Berkhout, 1984; Hardage, 1987). We obtain

$$D_{\Phi}(k) = D_{\mathrm{FM}}(-k) \star D_{\mathrm{FM}}(k), \tag{6.47}$$

where

$$D_{\mathrm{FM}}(k) = f_{\mathrm{FM}}(k)^{-1} \tag{6.48}$$

is minimum phase. If we neglect the pilot-delay term, i.e., set $t_{\mathrm{p}} = 0$, equation (6.42 becomes

$$D(k) = D_{\mathrm{TM}}(-k) \star D_{\mathrm{FM}}(-k) \star D_{\mathrm{B}}(k) \star D_{\mathrm{FM}}(k). \tag{6.49}$$

An important property of minimum-phase deconvolution operators is that they preserve causality and the arrival time of the signals, which are important in SWD correlations. A part of the minimum-phase operator of equation (6.49) can be calculated from the autocorrelation (Robinson and Durrani, 1986) of the pilot. This is what we usually call "pilot deconvolution".

The pilot autocorrelation can be expressed, using equations (6.30), (6.46) and commutative properties, as

$$\begin{aligned}\Phi_{\mathrm{pp}}(k) = {} & [\mathcal{H}_{\mathrm{TM}}(-k) \star \mathcal{F}_{f,a}(-k) \star f_{\mathrm{FM}}(-k)] \star [\mathcal{H}_{\mathrm{TM}}(k) \star \mathcal{F}_{f,a}(k) \star f_{\mathrm{FM}}(k)] + \\ & \Phi_{\mathrm{n_p n_p}}(k) + \Phi_{\mathrm{r_p r_p}}(k) + \quad \cdots \quad (\text{ terms assumed } = 0).\end{aligned} \tag{6.50}$$

From equations (6.43), (6.48) and (6.50) it follows that the statistical and minimum-phase pilot deconvolution operator calculated by the pilot autocorrelation is

$$D_{\mathrm{P}}(k) = D_{\mathrm{TM}}(k) \star D_{\mathrm{FM}}(k). \tag{6.51}$$

If the pilot deconvolution is applied after correlation, we have to follow rule (**4**) and use the deconvolution operator reversed in time, $D_\mathrm{P}(-k)$. This is the part of $D(k)$ (equation (6.49)) reversed towards negative times in order to remove the apparent anti-causal components introduced by the pilot filtering.

After pilot deconvolution, the pilot autocorrelation is

$$D_\mathrm{P}(-k) \star \Phi_{\mathrm{pp}}(k) \cong \mathcal{H}_{\mathrm{TM}}(k) \star \mathcal{F}_{f,a}(k) \star f_{\mathrm{FM}}(k) \star W(-k), \tag{6.52}$$

and the deconvolved crosscorrelogram becomes

$$D_\mathrm{P}(-k) \star \Phi_{\mathrm{pg}}(k) \cong -\mathcal{H}_\mathrm{B}^{(\mathrm{s})}(k) \star f_{\mathrm{FM}}(k) \star W(-k) \star \mathcal{H}_\mathrm{G}(k), \tag{6.53}$$

where

$$W(k) = D_\mathrm{P}(k) \star \mathcal{H}_{\mathrm{TM}}(k) \star \mathcal{F}_{f,a}(k) \star f_{\mathrm{FM}}(k), \tag{6.54}$$

is the deconvolved pilot-signal signature. Equation (6.52) is the "one–side deconvolved" pilot autocorrelation. The term

$$\mathcal{H}_\mathrm{B}^{(\mathrm{s})}(k) \star f_{\mathrm{FM}}(k) \tag{6.55}$$

in equation (6.53) is the minimum-phase signature of the SWD geophone crosscorrelations deconvolved by the spiking one-sided pilot operator. Ideally, if $W(-k) \cong \delta(k)$, the deconvolved geophone signature is minimum phase. Otherwise, we can expect that it is mixed phase (see Section 6.17 on signal rephasing).

The presence of the noise terms in equation (6.50) may cause distortions in the results of deconvolution. For example, polarized shaking of the top of the drill string in the horizontal direction perpendicular to the plane of the bail connected to the swivel by pins adds noise $\Phi_{\mathrm{n_p n_p}}(k)$ in the pilot autocorrelation. Rector (1992) showed that this noise causes distortions and wrong timing of events after reference deconvolution. Furthermore, high level of additional random white noise, i.e., of additional energy $\Phi_{\mathrm{r_p r_p}}(0)$ to the peak of the pilot autocorrelation, may make inverse filtering ineffective.

6.6.1 Crosscorrelation, stack and pilot deconvolution

Ristow and Jurczyk (1975) discussed the approach and the basic steps for deconvolution of correlated seismograms; in that case, for Vibroseis – another processing method, which is alternative to correlating and deconvolving in separate steps, involves deconvolving the sweep from the recorded Vibroseis trace (Brittle, Lines and Day, 2001). The general reasoning of Ristow and Jurczyk (1975) can be extended to SWD processing based on crosscorrelation. Pilot deconvolution can be applied before correlation of the pilot data or, after, to the correlated data. Application of deconvolution after SWD correlation is discussed above and has the advantage of being more easily repeatable if some adjustment in the calculation of the operator is required.

In addition, pilot deconvolution can be applied either before or after stacking the correlations of a given bit-depth level. Before stack, i.e., when dealing with a large number of individual records acquired for a "long" total time, the deconvolution results may vary because of lack of stationarity. Here we mean stationarity in the wide sense (Papoulis,

1991). In general, this condition is not true during the entire recording of a depth level, so that suitable gates (time windows) in which the process is "piecewise stationary" (Wang, 1969) should be used. Wang (1969 and 1977) studied time-invariant and time-varying correlation functions for adaptive predictive deconvolution of seismic data with variable wavelets and reverberation periods. He defines the optimum gate length for Wiener filtering as the length for which the time-invariant and time-varying correlation functions are approximately equal. In practical SWD, recording lengths of several tens of seconds are generally assumed to be stationary for the calculation of operators of two or three seconds duration – for the stationary case, the mean square error decreases as the filter length increases (Claerbout, 1992).

Low signal-to-noise ratio is a major shortcoming in SWD data, which generally requires a robust stacking approach. After stack, the error in the deconvolution operator (Burns, 1967) is reduced because S/N increases in the pilot autocorrelation. The effect was also noted in Vibroseis deconvolution by Ristow and Jurczyk (1975), who observed that deconvolution after stack provides a better interpretation of the data compared to deconvolution before stack.

6.7 Discussion about pilot deconvolution

Advantages of pilot deconvolution compared to other SWD-processing methods (see Sections 5.2 and 7.12) are discussed herein. The pilot signal is generally a "good" representation of the drill-bit signal. Multiple reflections and other delayed components of minor amplitude such as converted modes (see Section 4.2 and Chin, 1994) are removed. Drill-bit periodic (or quasi–periodic) components like vibrations caused by bottomhole patterns are removed too and the signature is shortened with a significant improvement in the signal bandwidth (Figures 6.8 and 6.9).

Limitations are that the pilot deconvolution, if performed by Wiener- and statistical methods (Dimri, 1992), depends on the signal-to-noise ratio of the pilot data. Sometimes pilot data with poor S/N are acquired because the transmission in the drill string is weak, while the S/N in the geophone data is higher (for example in sliding mode with roller-cone bits, Section 2.5.1). In this case pilot deconvolution may not be successful in shaping the correlated data signature into a short wavelet.

Stationarity limitations and signal-to-noise improvement in the level stack have to be evaluated.

Pilot deconvolution by $D_{\mathrm{P}}(k)$ solves only a part of the problem of deconvolving the correlated bit-signal signature: it removes the pilot-filtering effect reversed in "backward times" in the correlated data (see Section 6.3.4 and Figure 6.3). Pilot data measured at the surface can not be used directly to calculate the ("forward") part of $D(k)$ remaining after deconvolution given by equation (6.55). If we assume "ideal spiking" pilot deconvolution, i.e., $W(-k) = \delta(k)$, we obtain that $\mathcal{H}_{\mathrm{B}}^{(\mathrm{s})}(k) \star f_{\mathrm{FM}}(k)$ is the signature of the SWD pilot-deconvolved seismogram. Indeed, the residual deconvolution of VSP signature can be performed by the conventional VSP processing, i.e., solved by VSP deconvolution after upgoing- and downgoing-wave separation. In addition, a beam-steering operator can be calculated by focusing analysis in the common-shot gather and used to correct the signature of the individual bit depth levels.

An important question for the wavelet-estimation methods with conventional seismograms is whether the earth response is white (Fokkema and Ziolkowski, 1987; Ziolkowski, 1991). The whiteness is invoked because it is desirable that reflectivity be preserved as "prediction error." However, in pilot deconvolution we are not interested in the prediction error but only in the drill-bit direct event. The problem becomes instead to be sure that any formation-reflectivity information contained in the pilot signal (Malusa, Poletto and Miranda, 2002) be processed so that it is not removed from the final seismograms by pilot-derived operators.

The one-side deconvolved pilot autocorrelation of equation (6.52) is the minimum-phase equivalent of the signal transmitted through the drill string if $W(-k) = \delta(k)$. Its amplitude spectrum is a good estimate of the amplitude spectrum of the transmitted drill-bit acceleration signal. Figure 6.10 shows the autocorrelations of a pilot signal measured at the top of the drill string before and after one-sided deconvolution. These autocorrelations are represented with their peaks at zero-delay and used to analyze the drill-string waves propagated in the transmission-line model (Section 6.13). In other words, from the pilot autocorrelation we get the pilot-signal spectrum (Figure 6.11), but any absolute delay information of the direct arrival is lost. However, because the drill-string multiples are detectable in the pilot autocorrelation or in its one-side deconvolved version, it is possible to draw a time-depth velocity diagram by using data of multiple drill-string lengths to calculate the pilot delay (Staron, Arens and Gros, 1988), thereby pointing out the events corresponding to the main reflection coefficients.

Least-square inversion by unit prediction distance (Peacock and Treitel, 1968) has been shown to be effective for pilot-signal processing. The $D_{\mathrm{P}}(k)$ operator length is calculated to include the BHA short- and pipe long-period multiples propagated in the 1D drill-string transmission line. A long-period multiple can have a periodicity of about 2 s in a 5000 m string. Furthermore, long operators have better resolution in frequency in removing periodic bit components. This deconvolution exploits the dynamic properties of the crosscorrelated trace. In general, spike operators with low levels of white noise, e.g., 0.001 of energy, are preferred due to the ghost–notch in the frequency of the signal recorded at the top of the drill string. This is caused by the ghost reflection at the top of the rig. Removal of the ghost by use of other pilot sensors is discussed in Section 6.14. Operator windows of 2 s or 3 s length are recommended with the caution that the second long-period multiple be included in the operator window (Figures 6.9 and 6.10).

Attenuation of correlated noise after deconvolution occurs to a certain extent if

$$D_{\mathrm{P}}(-k) \star \Phi_{\mathrm{n_p n_g}}(k) \qquad (6.56)$$

vanishes, i.e., if the deconvolution operator and correlated noise are incorrelated.

6.8 Beam-steering deconvolution

6.8.1 Focused pilot

One way to obtain a pilot signal is to use the stack of the geophone traces recorded in a given common-source gather after correction for the moveout of the direct drill-bit arrival.

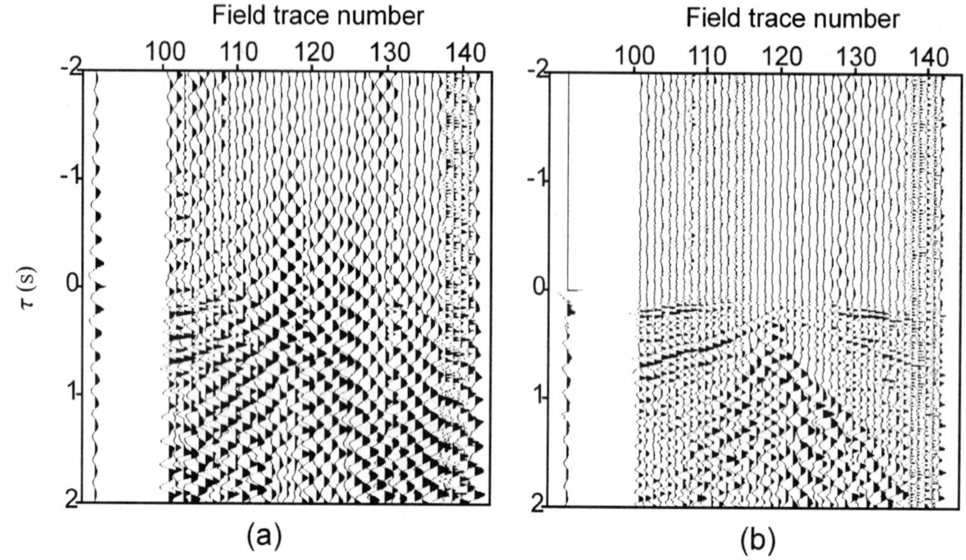

Figure 6.8: Axial pilot and geophone traces for the bit level at 2653 m. Seismograms were obtained by stacking 50 crosscorrelations of records of 45 s duration. The (axial-acceleration) pilot was measured at the top of the drill string. Data are shown a) before and b) after pilot deconvolution which removes the pilot predictable events in the negative correlation times. A 2 s spiking operator with 0.1 percent of white noise added to the autocorrelation peak was used.

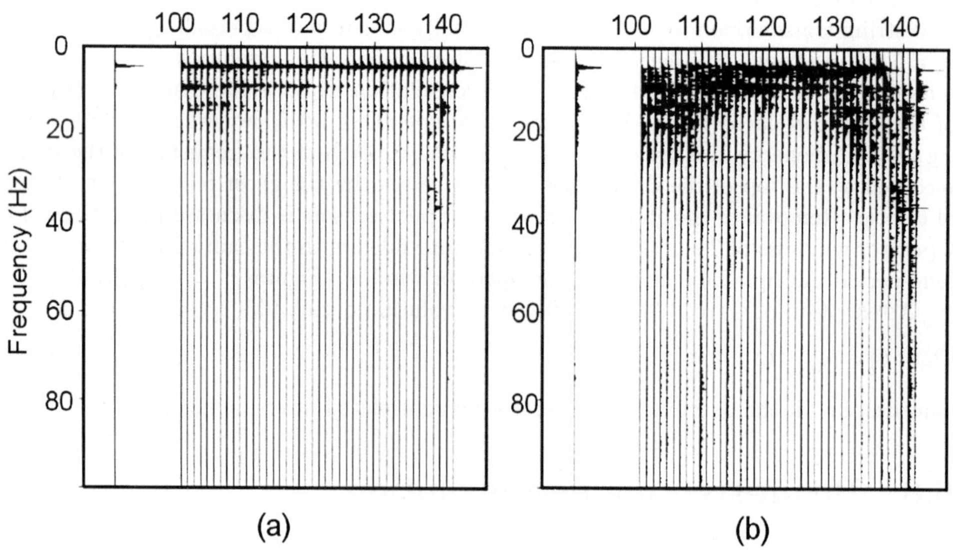

Figure 6.9: Frequency spectra of the crosscorrelations of Figure 6.8 a) before and b) after pilot deconvolution. Data are not bandpass filtered.

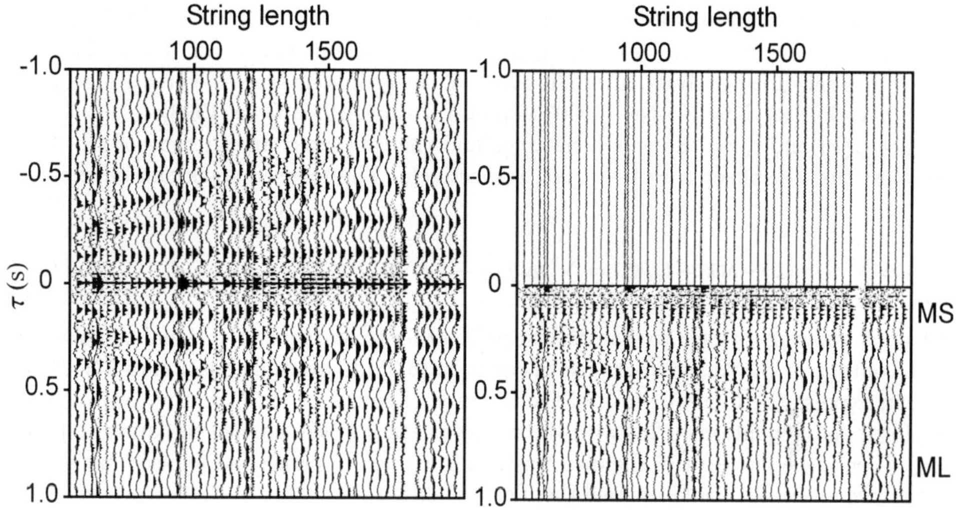

Figure 6.10: Autocorrelations of the (axial acceleration) pilot measured at the top of the drill string for different bit depths. Data are shown a) before and b) after one-sided deconvolution which removes the predictable events for the negative correlation times. A 2 s spiking operator with 0.1% of white noise added to the autocorrelation peak is used. In b) the multiples of short (MS) and long period (ML) increasing with pipe length can be seen.

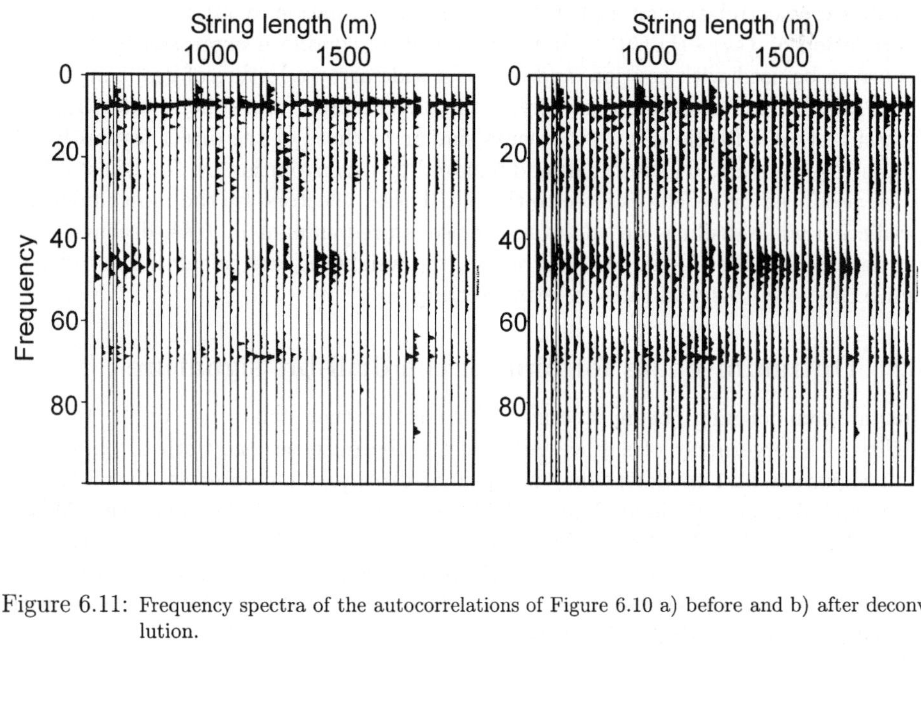

Figure 6.11: Frequency spectra of the autocorrelations of Figure 6.10 a) before and b) after deconvolution.

This is the basic concept of beam steering, or "focused stack", which emphasizes energy from a particular direction by delaying the successive channels (Sheriff, 1999). The stack of the drill-bit direct arrivals, after alignment of the geophone traces to create a geophone pilot, was discussed by Kostov (1990) (see also Kostov, 1988; Cole, 1988).

Let $g_i(k)$ and t_i be the geophone trace and the estimated arrival time of the direct-signal at the i-th offset ($i = 1, \cdots, N$), respectively. The focused stack can be expressed as

$$f(k) = \sum_{i=1}^{N} w_i g_i(k) \star \delta(k + t_i/\Delta) = \sum_{i=1}^{N} w_i g_i(k + t_i/\Delta), \tag{6.57}$$

where w_i are weights used to scale the traces in order to maximize the signal-to-noise ratio in the output $f(k)$. We can make use of equation (6.35) to expand equation (6.57). Hence, we can express the focused stack of the traces g_i of equation (6.57) as the convolution of the force $f_{\rm B}$ of equation (6.28) with the weighted stack of the aligned radiation responses $\mathcal{H}_{{\rm G}i}$. If only the moveout of the direct arrival is aligned in phase and focused in the stack of the responses, we could assume the ideal condition

$$\sum_{i=1}^{N} w_i \mathcal{H}_{{\rm G}i}(k) \star \delta(k + t_i/\Delta) \approx \delta(k) \star \mathcal{G}_{dv}(k) \equiv \mathcal{G}_{dv}(k), \tag{6.58}$$

where $\mathcal{G}_{dv}(k)$ is the filter shaping displacement into geophone velocity response. In equation (6.58), the impulse response of the formation is a delta function: it is obtained when all the reflections, wavetails that develop into spatially incoherent waves, and noise events vanish for interference effects due to different travelpaths (Claerbout, 1976). If the condition of equation (6.58) is true, the minimum-phase deconvolution operator D_f, which deconvolves crosscorrelation after pilot deconvolution of equation (6.53), is calculated from autocorrelation of $f(k) \star \mathcal{G}_{dv}(k)^{-1}$. However, the ideal condition expressed by equation (6.58) has to be used with caution, and only if a wide array aperture of geophone traces makes it possible to discriminate effectively the direct signal from other events. In fact, it is possible that reflections with moveouts similar to those of the direct signals in the responses $\mathcal{H}_{{\rm G}i}$ are stacked as residuals and removed by deconvolution by D_f. These reflections are produced by reflectors close to the bit. These reflectors are the most important for prediction while drilling. In addition, if noise is predominant in $f(k)$, deconvolution by D_f deteriorates the signal. For this reason, the weights w_i are calculated in order to minimize the noise-to-signal ratio in beam steering.

6.8.2 Optimum beam forming of the drill-bit signature

Optimum beam forming (beam steering) is a method that makes use of the focusing capability of a large array of surface geophones. The purposes are both to obtain the drill-bit signature and to determine the optimal deconvolution filter by means of the multichannel Wiener technique (Miller, Haldorsen and Kostov, 1990; Haldorsen, Miller, Walsh and Zoch, 1992; Haldorsen, Miller and Walsh, 1995). The analysis is performed using the relative moveout of the direct arrivals at progressive offsets, and is insensitive to a constant shift of the arrival times $t_i - t_0$, which can be corrected by correlation with the signal of a reference drill-string sensor.

Let the signal function be the same in all the aligned traces $g_i(k) \star \delta(k + t_i/\Delta)$, and let $\mathcal{N}_i(k)$ be the noise in the i-th geophone trace. Let $w_i = 1/N$ in equation (6.57), in which we denote $f_1(k) = f(k)$. The proposed method assumes that the aligned noise is spatially zero–mean, i.e.,

$$\frac{1}{N}\sum_{i=1}^{N}\mathcal{N}_i(k) \star \delta(k + t_i/\Delta) = \frac{1}{N}\sum_{i=1}^{N}\mathcal{N}_i(k + t_i/\Delta) = 0. \tag{6.59}$$

With these assumptions, the above authors show that the optimum filter that minimizes the average filtered noise can be expressed in frequency as (the same analysis is given by Gangi and Fairborn, 1968):

$$F(\omega) = \frac{f_1^*(\omega)}{|f_1(\omega)|^2} e^{-\imath\omega t_0} S(\omega), \tag{6.60}$$

where $e^{-\imath\omega t_0}$ is a simple time-shift operator, $S(\omega)$ is the semblance in frequency domain:

$$S(\omega) = \frac{|f_1(\omega)|^2}{E_T(\omega)}, \tag{6.61}$$

and $E_T(\omega)$ is the average total energy of the raw traces, namely

$$E_T(\omega) = \frac{1}{N}\sum_{i=1}^{N}|g_i(\omega)|^2. \tag{6.62}$$

The first factor in equation (6.60) is a spiking operator. The semblance is a band-limiting factor, and it can be shown that $F(\omega)$ is stable for construction. The filter $F(\omega)$ applied to the trace g_i produces an impulsive signal at $t = t_i - t_0$.

The authors mentioned above show applications of focusing analysis in frequency, with steps of iterative picking of the SWD arrivals. After filtering, the signal wavelet assumes a shape close to zero-phase. However, limitations may arise due the fact that direct arrivals and "close to the bit" reflections may have similar trends. In this case, the stability of the filter is not sufficient to avoid inadequate signal event removal; also, noise itself may contribute to the filter if it is not spatially zero–mean.

6.8.3 Autocorrelation bias

The crosscorrelation of the focused pilot of equation (6.57) with the trace $g_j(k)$ gives

$$\Phi_{\mathrm{fg}_j}(k) = w_j\Phi_{\mathrm{g}_j\mathrm{g}_j}(k - t_j/\Delta) + \sum_{i\neq j} w_i\Phi_{\mathrm{g}_i\mathrm{g}_j}(k - t_i/\Delta). \tag{6.63}$$

The first term of the sum in equation (6.63) is the contribution of the autocorrelation of the trace g_j delayed by t_j. This term produces a bias in the calculation of the focused-signal filter. In fact, the autocorrelation peak at t_j results from the sum of the autocorrelations of both signal and noise in g_j. This effect occurs despite the fact that the delay of the peak may – or may not – represent the delay of the bit signal. In other words, we may create a coherent signal given by the autocorrelation bias when we use for correlation the stack of the traces g_i delayed on the basis of any arbitrary time-shift curve. After all, "we find what we put in the data", at least in part. One solution adopted in the implementation of the filter in the method discussed above is to eliminate the contribution of autocorrelation by excluding the trace g_j when applying the filter $F(\omega)$ to it. The solution of excluding only the trace to be deconvolved is an improvement. However, there is no reason to say that it is sufficient, especially if the correlation of the signals in different traces is low. In Section 5.14, we discussed the coherent and random noise in arrays. A part of the random noise is characterized by a coherence distance, which may extend the effect to other traces and change for different arrays.

6.9 Deconvolution in rotation-angle domain

In Chapter 3, we showed that large variations in the downhole bit-revolution speed occur even while the surface RPM remains nearly constant. One method commonly used in the drilling literature to analyze the regularity of the bit performance is to represent the measures of the bit vibrations in cycles versus bit-revolution angle (MacPherson, 1990; MacPherson, Mason and Kingman, 1993). These measures are performed in the laboratory or in full-scale tests by using downhole tools equipped with rotation detectors, i.e., magnetic compasses, radial accelerometers and proximity switches. In SWD, the deconvolution in the bit-rotation-angle domain is a processing method that assumes the spatial periodicity of the rotary drill-bit source of excitation (Bernasconi and Vassallo, 2002). In other words, in this approach, it is assumed that the rotary bit-force signal is periodic in the angle of rotation, ϕ, rather than in time. If no assumptions are made about random rotary forces, we can use the source model

$$f_{\mathrm{B}}(k) = \mathcal{H}_{\mathrm{B}}^{(\mathrm{s})}(k) \star f_{\mathrm{P}}\left(k(\phi)\right) \tag{6.64}$$

to describe this method. Equation (6.64) is composed of two parts. The first, $\mathcal{H}_{\mathrm{B}}^{(\mathrm{s})}(k)$, represents the drill-string reflectivity and is periodic in time. The second, $f_{\mathrm{P}}\left(k(\phi)\right)$, is the bit excitation signal, which is periodic in the rotation-angle domain ϕ. The angular dependence of time is given by

$$k = k(\phi). \tag{6.65}$$

The method proposed by Bernasconi and Vassallo (2002) consists of mapping and resampling in the rotation-angle domain, the function $f_{\mathrm{B}}\left(k(\phi)\right)$. In this way, the angular periodicity of $f_{\mathrm{P}}(\phi)$ is reinforced. Narrow spatial-frequency peaks can be then removed by frequency demodulation, such as phase-locked-loop (PLL) codes (Bernasconi and Vassallo, 2001). The next step consists of inverse mapping in the time domain of the deconvolved angular signal, by using the inverse of the equation (6.65). At this stage, authors show examples of better identification of drill-string periodicity in PDC data.

In other words, this approach is aimed at recognizing the "voice" of the bit as a function of the rotation angle which is, in general, much more complex than a regular-periodic signal because it contains effects of rotation drifts, random contacts, and pattern precession, as discussed in Chapter 3. We can observe that the "voice" recognition is already realized by the fact that the bit vibrations are measured by the same tool used to detect the phase in the proximity of the bit, and that these vibrations can be used to calculate the radiated waves. An application of the proposed method is signal coding for parsimonious transmission of compressed data, by making use of downhole processing capabilities. In this case, processing by rotation angle is done to better characterize the excitation signal (Bernasconi and Vassallo, 2001).

6.9.1 Analysis of synthetic and real data in rotation-angle domain

The following example shows the analysis of PDC data in the rotation-angle domain as discussed above. The data are from a SWD gather acquired at about a 4000 m depth. Downhole accelerometers were used to measure the axial and radial acceleration in a tool near the bit. Instantaneous angular speed $\Omega(t)$ is calculated from centrifugal-radial acceleration, a_r:

$$\Omega(t) = \sqrt{\frac{a_r(t)}{r}}, \tag{6.66}$$

where r is radius. The rotation angle is obtained by integrating equation (6.66),

$$\phi(t) = \int_{t_0}^{t} \Omega(t')\mathrm{d}t', \tag{6.67}$$

where t_0 is an arbitrary initial time. Figure 6.12 shows the downhole angular speed (RPM) and the rotation angle $\phi(t)$. Figure 6.13 shows the spectra of the axial acceleration signal in a) time-frequency (cycles/second) and b) in the angle-domain frequency (cycles/radian). Synthetic traces have been calculated by assuming a periodic vibration model for the 7-blade bit and using the measured $\phi(t)$ (Malusa and Poletto, 2001). The synthetic traces are then filtered by the drill-string response calculated by the transmission-line model introduced in the next section. Figure 6.14 compares the autocorrelations of the b) synthetic to the a) real acceleration pilots.

6.10 Modeling of drill-string response

We use the Z-transform to calculate numerically the wave propagation in the drill-string transmission line (Section 4.10). We insert a unit pulse and use the notation of Figure 4.15b.

6.10.1 Propagation matrix for drill-bit signal

The drill-bit signal in the surface pilot can be written as:

$$P_{\mathrm{S}}(Z) = (1 + c_{\mathrm{T}})F_{\mathrm{DS}}(Z) + (1 - c_{\mathrm{T}})R_{\mathrm{T}}(Z). \tag{6.68}$$

We represent the propagation matrix (4.100), computed in the string, as

$$M_H^{\mathrm{String}} = \frac{Z^{H/2}}{\prod_{i=1}^{H} t_i} \begin{bmatrix} F_H(Z) & Z^{-H}G_H(Z^{-1}) \\ G_H(Z) & Z^{-H}F_H(Z^{-1}) \end{bmatrix}, \tag{6.69}$$

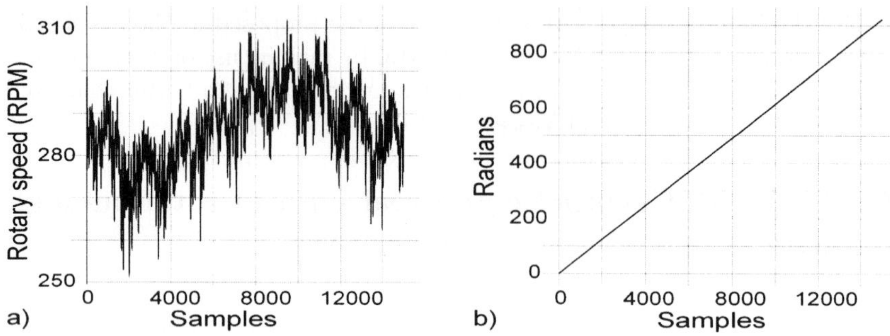

Figure 6.12: a) Angular rotation speed and b) rotation angle.

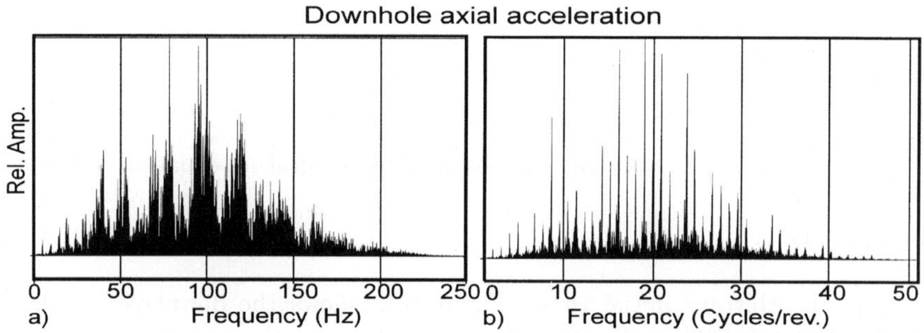

Figure 6.13: Data in frequency and angular frequency.

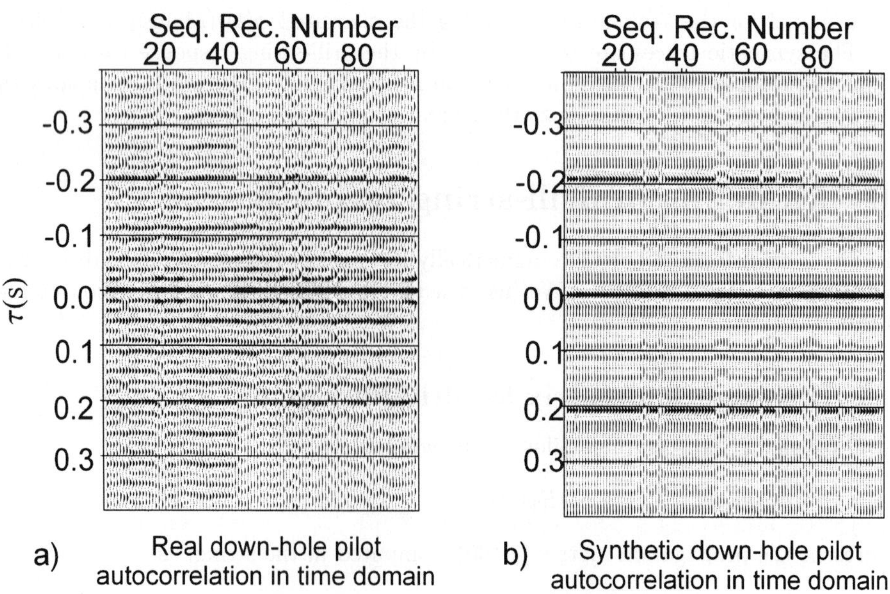

Figure 6.14: a) Real and b) synthetic autocorrelations of downhole pilot signals.

and the propagation matrix in the rig as

$$M_L^{\mathrm{Rig}} = \frac{Z^{L/2}}{\prod_{i=1}^{L} t_i} \begin{bmatrix} F_L(Z) & Z^{-L}G_L(Z^{-1}) \\ G_L(Z) & Z^{-L}F_L(Z^{-1}) \end{bmatrix}. \tag{6.70}$$

where $t_i = 1 + c_i$ is the transmission coefficient. Inserting a unit bit-signal pulse, the wave propagation in the drill-string is calculated by

$$\begin{bmatrix} F_{\mathrm{DS}}(Z) \\ c_{\mathrm{T}}F_{\mathrm{DS}}(Z) + (1-c_{\mathrm{T}})R_{\mathrm{T}}(Z) \end{bmatrix} = M_H^{\mathrm{String}} \begin{bmatrix} 1 + c_0 R_{\mathrm{B}}(Z) \\ R_{\mathrm{B}}(Z) \end{bmatrix}, \tag{6.71}$$

and in the rig by

$$\begin{bmatrix} E_{\mathrm{W}}(Z) \\ 0 \end{bmatrix} = M_L^{\mathrm{Rig}} \begin{bmatrix} (1+c_{\mathrm{T}})F_{\mathrm{DS}}(Z) - c_{\mathrm{T}}R_{\mathrm{T}}(Z) \\ R_{\mathrm{T}}(Z) \end{bmatrix}. \tag{6.72}$$

H and L are integers > 0. F_H, G_H, F_L, and G_L are polynomials of order $H-1$ and $L-1$ describing the propagation in the drill string and rig, respectively. F_{DS}, R_{B}, R_{T}, and E_{W} are the upgoing signal in the drill string at the surface pilot location, the downgoing reflected signal at the bit, the downgoing surface signal reflected in the rig system, and the escaping wavefield of the signal radiated from the rig, respectively.

Solving (6.71) and (6.72) with respect to $F_{\mathrm{DS}}(Z)$, $R_{\mathrm{B}}(Z)$, and $R_{\mathrm{T}}(Z)$, we obtain the surface pilot signal (6.68) and the downhole signal with the reflectivity at the bit for a unit bit signal given as

$$\mathcal{H}_{\mathrm{B}}(Z) = 1 + (1 + c_0)R_{\mathrm{B}}(Z). \tag{6.73}$$

If we use opposite reflection coefficients, for strain (stress, force), from equation (6.73) we get $\mathcal{H}_{\mathrm{B}}^{(\mathrm{s})}(Z)$ of equation (6.29).

6.10.2 Propagation matrix for rig noise

We assume that the rig-noise excitation point at the surface is located in the same position as the pilot sensor. Using the notation of Figure 4.15c, the rig noise in the surface pilot is given by

$$P_{\mathrm{N}}(Z) = 1 + (1 - c_{\mathrm{T}})R_{\mathrm{T}}(Z) + (1 + c_{\mathrm{T}})R_{\mathrm{DS}}(Z). \tag{6.74}$$

For a unit rig-noise input pulse, the propagation in the drill string is given by

$$\begin{bmatrix} R_{\mathrm{DS}}(Z) \\ 1 + (1-c_{\mathrm{T}})R_{\mathrm{T}}(Z) + c_{\mathrm{T}}R_{\mathrm{DS}}(Z) \end{bmatrix} = M_H^{\mathrm{String}} \begin{bmatrix} c_0 R_{\mathrm{B}}(Z) \\ R_{\mathrm{B}}(Z) \end{bmatrix}, \tag{6.75}$$

and in the rig by

$$\begin{bmatrix} E_{\mathrm{W}}(Z) \\ 0 \end{bmatrix} = M_L^{\mathrm{Rig}} \begin{bmatrix} 1 + (1+c_{\mathrm{T}})R_{\mathrm{DS}}(Z) - c_{\mathrm{T}}R_{\mathrm{T}}(Z) \\ R_{\mathrm{T}}(Z) \end{bmatrix}, \tag{6.76}$$

where, in this case, R_{DS}, R_{B}, R_{T}, and E_{W} are the upgoing reflected noise in the drill string at the surface pilot location, the downgoing reflected noise at the bit, the downgoing surface noise reflected by the rig system, and the escaping wavefield of the noise radiated from the rig, respectively.

Solving equations (6.75) and (6.76) with respect to $F_{\rm DS}(Z)$, $R_{\rm B}(Z)$, and $R_{\rm T}(Z)$, we obtain the surface rig-noise $P_{\rm N}$ (equation (6.74)) and the downhole rig-noise transmitted at the bit

$$\mathcal{H}_{\rm N}(Z) = (1 + c_0) R_{\rm TB}(Z). \tag{6.77}$$

6.10.3 Reflection coefficients in two-way travel time (TWT)

The drill-string reflection coefficients for displacement (velocity, acceleration) are calculated in Section 4.10.1. To obtain the position in time of the reflection coefficients used by the recursive algorithms of equations (4.99) and (4.100), it is necessary to know the axial and torsional propagation velocities. In the initial modeling phase, the distribution of the mass and moment of inertia in reflection two-way time is calculated using the group velocity of axial and torsional waves in a homogeneous steel rod. The group velocity can be calculated from the dispersion relation of the phase velocity $v_{\rm p} = \omega / k(\omega)$ as $v_{\rm g}(\omega) = \partial\omega / \partial k$. However, to obtain the group velocity in the drill pipes, we calculate their average elastic properties and use equations (4.73) and (4.74). In fact, the presence of tool joints, which are thicker than the pipe body, adds mass to the drill pipes while increasing their stiffness slightly, and decreases the group velocity (see Section 4.4.5). In a simplified geometry of infinite periodic pipe with tool joints, these equations assume the form for the group velocities in the low-frequency approximation of Drumheller and Knudsen (1995), namely

$$v_{\rm g(ax)} = \sqrt{\frac{Y}{\rho}} \frac{d_1 + d_2}{\sqrt{d_1^2 + (\frac{A_1}{A_2} + \frac{A_2}{A_1}) d_1 d_2 + d_2^2}}. \tag{6.78}$$

and

$$v_{\rm g(tor)} = \sqrt{\frac{\mu}{\rho}} \frac{d_1 + d_2}{\sqrt{d_1^2 + (\frac{I_1}{I_2} + \frac{I_2}{I_1}) d_1 d_2 + d_2^2}}, \tag{6.79}$$

for axial and torsional waves, respectively. We use average drill-string properties in equations (4.103) and (4.104), which are equivalent to equations (4.73) and (4.74). Examples of average values measured in drill-pipes of type S are $\bar{m}_{\rm ax} = 32.9$ kg/m, $\bar{A}_{\rm ax} = 0.00356$ m^2, $\bar{M}_{\rm tor} = 0.119$ kg m, and $\bar{I}_{\rm tor} = 12 \times 10^{-6}$ m^4. An example of velocity values computed for drill pipes of different type and for BHA is shown in column 2 of Table 6.1. Since the BHA is more massive than the drill pipes, its group velocity is closer to the theoretical steel-rod velocity.

To obtain the accurate conversion of the drill-string axial coordinate in the drill-string TWT-reflection time, we use an initial sampling interval of 10 μs. Then, we resample the distributions of mass and moment of inertia to the sampling interval of the real data, i.e., 2 ms, and calculate the reflection-coefficient time series.

Table 6.1 – Theoretical and measured group velocity

Type of pipe	Theoretical v_{ax} [m/s]	Measured $\hat{v}_{ax}$ [m/s]
Pipe S	4730	4725 ± 20
Pipe E	4850	4845 ± 15
Pipe G	4880	4890
BHA	5100	5050

Type of pipe	Theoretical v_{tor} [m/s]	Measured $\hat{v}_{tor}$ [m/s]
Pipe S	2860	2790 ± 10
BHA	3140	3100

6.11 Fitting with real data

Figure 6.15 shows the reflection-coefficient time series computed for a BHA using the mechanical features of the drill string.

Using the reflection coefficients in the recursions of equations (4.99) and (4.100), we compute the initial synthetic seismogram. To fit the deconvolved autocorrelation of a real pilot trace, we estimate c_0, c_T, and the velocity in the drill string. Changing the reflection coefficient c_0 at the bit/rock interface and the reflection coefficient c_T at the rig/drill-string rope interface in equations (6.71) and (6.72), we produce a series of synthetic traces. The optimum coefficients $\hat{c}_0$ and $\hat{c}_T$ are obtained with a least-squares minimization of the error between the real and synthetic data in a window centered on the direct drill-string arrival (Figure 6.17b). This window contains short-period BHA reverberations, rig ghost and rig noise. In this step we also estimate the optimum average-propagation velocity in the BHA, $\hat{v}_{bha}$, and the reflection coefficient at the top of the drill string, c_T.

Using the optimum $\hat{c}_0$, $\hat{c}_T$, and $\hat{v}_{bha}$, we produce the synthetic seismograms with different drill-pipe velocities, which are varied within a range of values centered on the theoretical group velocity. Then, we estimate the optimum average velocity in the pipes, $\hat{v}_{pipe}$, by the best fitting of the real data in a window centered on the long-period multiples (Figure 6.17c). Figure 6.17d shows the synthetic surface-pilot seismograms with the first arrival aligned at $t = 0$ s. Figure 6.16 shows a) the time positioning of the reflections at the bit and at the contact point between the drill pipes and the BHA, together with b) the estimated bit/rock reflection coefficients $\hat{c}_0$. The coherence of the fitting is displayed for quality control (Figure 6.16c).

The information stored in the trace headers of the real pilot seismograms is used to prepare the synthetic traces to fit with the real pilot data. This procedure is used in the field to calculate automatically the pilot delay of the bit signal through the drill string, and convert the correlation time to seismic time, for correct positioning of the events measured by the geophones (Poletto, Malusa and Miranda, 2001). Because the drill-pipe length is much larger than the BHA length, the velocity in the drill pipes is most critical.

Poletto, Malusa and Miranda (2001) obtain a good agreement between the calculated group velocities of axial and torsional components and the measured values $\hat{v}$ for different types of pipes and typical BHA. For comparison, they also calculate the velocity v_{pick} by picking the times of the long-period multiples of high-quality axial pilot data. They

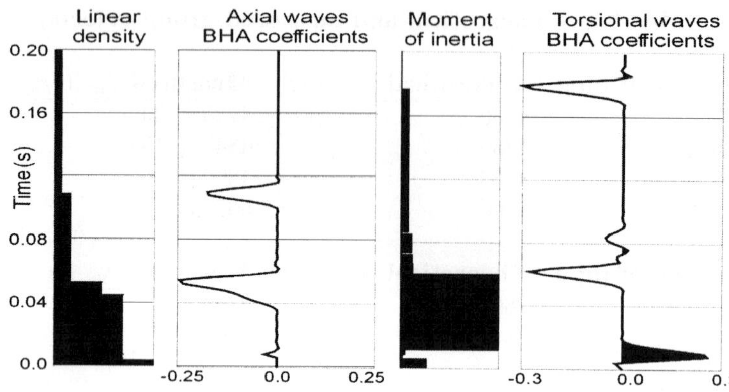

Figure 6.15: Using the mechanical description of the BHA and of the different types of drill pipes, we calculate the linear density and moment of inertia and then, assuming the properties of the steel as constant, the reflection-coefficient time series in the drill string for axial and torsional waves. The initial drill-string length-to-time conversions are obtained using the axial and torsional group velocities calculated using the low-frequency approximation.

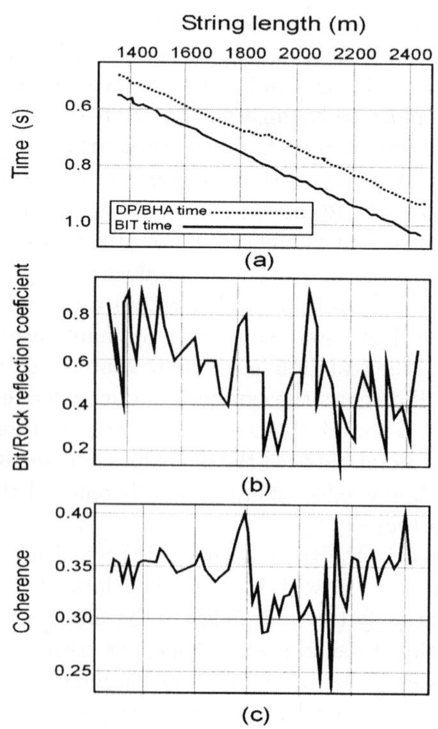

Figure 6.16: a) The reflection coefficients at the drill-pipe/BHA contact points and b) the bit/rock reflection coefficients c_0 are obtained by least-squares minimization of the difference between real and synthetic data obtained at different bit depths. c) The coherence of the fitting is displayed for quality control.

obtain, between v_{pick} and $\hat{v}$, a RMS difference equal to 12 m/s. Typical values of $\hat{v}$ are shown in column 3 of Table 6.1.

The synthetic seismograms in Figure 6.17d correspond to the real correlations in Figure 6.17a. Comparing real and synthetic data of this example allows us to interpret the change of the features of the long-period multiple at the string length of 1800 m due to a remarkable change of the BHA composition. Before the string length of 1800 m, the reflection at the drill-bit position is observable at its estimated time. After the string length of 1800 m, the reflection at the bit is masked by the reflection at the interface between drill pipes and heavy-weight pipes.

6.12 Drill-string waves in the correlated and deconvolved data

Analysis of crosscorrelated drill-string waves is a basic step in SWD processing. Among other possible options we use one-side deconvolved crosscorrelations to interpret the drill-string events of signal and noise. Let the pilot signal in the frequency domain be expressed as

$$P = a_{\text{ps}} P_{\text{S}} W_{\text{S}} + a_{\text{pn}} P_{\text{N}} W_{\text{N}}, \tag{6.80}$$

where W_{S} and W_{N} are, without loss of generality, white sources of bit-signal and surface-rig noise (Section 4.9), respectively; a_{ps} and a_{pn} are real constants; P_{S} is the response for the bit signal transmitted through the drill string, from the bit source to the surface pilot. If we separate the transfer function into a pure delay and an impulse response with zero-delay time, P_{S0}, we have $P_{\text{S}} = P_{\text{S0}} e^{\imath\phi_{\text{ps}}}$, where ϕ_{ps} is the phase delay from the bit source to the surface pilot. P_{N} represents the reflectivity response for noise at the pilot-sensor location.

Here we approximate the surface conditions by assuming that the surface noise is generated, with a delay equal to zero, at the pilot location. Let D be a one-sided deconvolution operator which spikes the pilot-trace response. Hence, in the presence of signal only, we have

$$D_{\text{S}} P \cong a_{\text{ps}} e^{\imath\phi_{\text{ps}}} W_{\text{S}}, \tag{6.81}$$

and, in presence of noise only,

$$D_{\text{N}} P \cong a_{\text{pn}} W_{\text{N}}, \tag{6.82}$$

with $D_{\text{N}} \neq D_{\text{S}}$. Moreover, let the colored seismic signal transmitted through the rocks be given by

$$X = \left[a_{\text{xs}} X_{\text{S}} W_{\text{S}} + a_{\text{xn}} X_{\text{N}} W_{\text{N}} \right] G, \tag{6.83}$$

where a_{xs} and a_{xn} are real constants; X_{S} represents the downhole signal with the drill-string reflectivity at the bit; $X_{\text{N}} = X_{\text{N0}} e^{\imath\phi_{\text{xn}}}$ is the filter for the surface noise transmitted from the source to the bit location with the phase delay ϕ_{xn}. In our approximation we assume $\phi_{\text{xn}} \cong \phi_{\text{ps}} = \phi_{\text{p}}$. Finally, G is the bit-to-seismic-receiver response.

Crosscorrelating the deconvolved pilot signal $D_S P$ with the seismic trace X and neglecting the noise term $(a_{ps}a_{xs} \gg a_{pn}a_{xn})$, we obtain for the Fourier transform

$$C_{DS} = D_S^* P^* X = a_{ps}a_{xs}|W_S|^2 e^{-\imath\phi_P} X_S G \cong a_{ps}a_{xs} e^{-\imath\phi_P} X_S G, \tag{6.84}$$

where we used the hypothesis that W_S is "white." Due to the variability of drilling conditions (Section 5.9.1) we can assume that, in general, sometimes the signal is dominant and sometimes the noise prevails. Neglecting the signal term $(a_{ps}a_{xs} \ll a_{pn}a_{xn})$ we obtain

$$C_{DN} = D_N^* P^* X = a_{pn}a_{xn}|W_N|^2 e^{\imath\phi_P} X_{N0} G \cong a_{pn}a_{xn} e^{\imath\phi_P} X_{N0} G, \tag{6.85}$$

for the rig noise. In a similar way, we obtain the (one-side) deconvolved pilot autocorrelations by the Fourier transforms

$$A_{DS} = D_S^* P^* P = a_{ps}^2 |W_S|^2 P_{S0} \cong a_{ps}^2 P_{S0}, \tag{6.86}$$

and

$$A_{DN} = D_N^* P^* P = a_{pn}^2 |W_N|^2 P_N \cong a_{pn}^2 P_N, \tag{6.87}$$

for signal only $(a_{ps}^2 \gg a_{pn}^2)$ and noise only $(a_{ps}^2 \ll a_{pn}^2)$, respectively. However, in the case of a mixture of bit-signal and surface-rig-noise with $a_{ps}^2 \sim a_{pn}^2$, deconvolution using operators derived from pilot autocorrelation may be distorted.

In the following example, we analyze the signal multiples assuming that $a_{ps}^2 \gg a_{pn}^2$. Figure 6.17a shows deconvolved pilot autocorrelations represented in time, where the traces correspond to axial pilot signals acquired at increasing string length. The data are obtained by stacking from 90 to 120 correlations of 24 s field records with a sampling interval of 2 ms. A 2 s one-sided deconvolution operator is applied to each trace. The data are filtered in the frequency band from 16 to 46 Hz. We observe the first arrival (event A) aligned at $t = 0$ s; this is interpreted as a bit signal; it corresponds to the direct arrival from the bit to the surface-pilot sensor (due to the autocorrelation, the delay disappears). Event B represents the drill-string long-period multiples of the bit signal acquired at increasing string length. The features of both these events mainly depends on the BHA composition and on the boundary conditions. To determine the optimum $\hat{c}_0$, we compare a series of synthetic traces, obtained by varying c_0, with the corresponding real trace represented on the right side of panel (b). To determine the optimum velocity, we compare a series of synthetic traces, obtained by varying the drill-pipe average velocity, with the corresponding real trace shown on the right side of panel (c). The synthetic surface pilot seismograms, with the first arrival aligned at t = 0 s (d), is the best fit to the real data.

The event B represents the long-period multiples of the drill string, reflected from the rig to the BHA and back to the pilot sensor, with the arrival time increasing according to the drill-string length. This signal is used to estimate the velocity of the pilot signal. We can observe a change in the waveform of the long-period multiple of Figure 6.17a at a string length of 1800 m. This effect is related to a significant change of the drill string, with a lighter BHA before a depth of 1800 m. This example shows how the interpretation of the phases of the reflected events can be important for pilot velocity analysis.

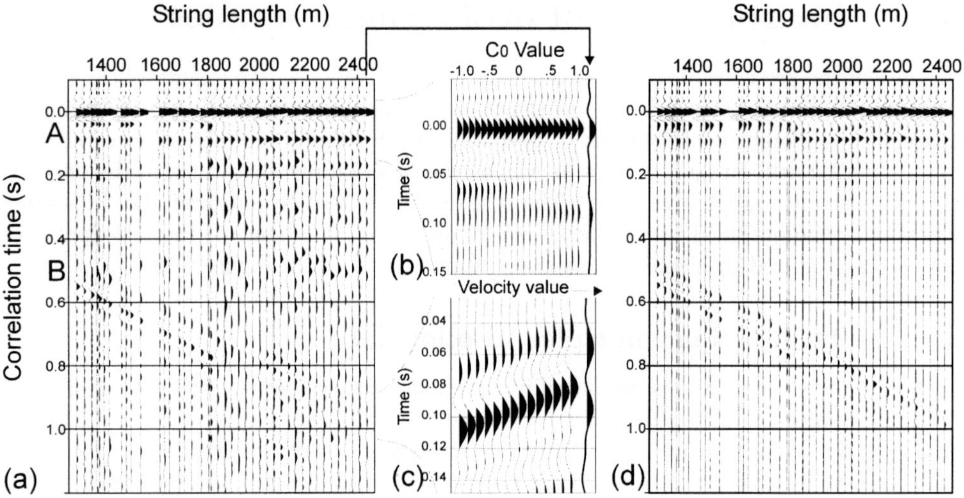

Figure 6.17: Example of one-side deconvolved autocorrelations of a) real and d) synthetic pilot data.

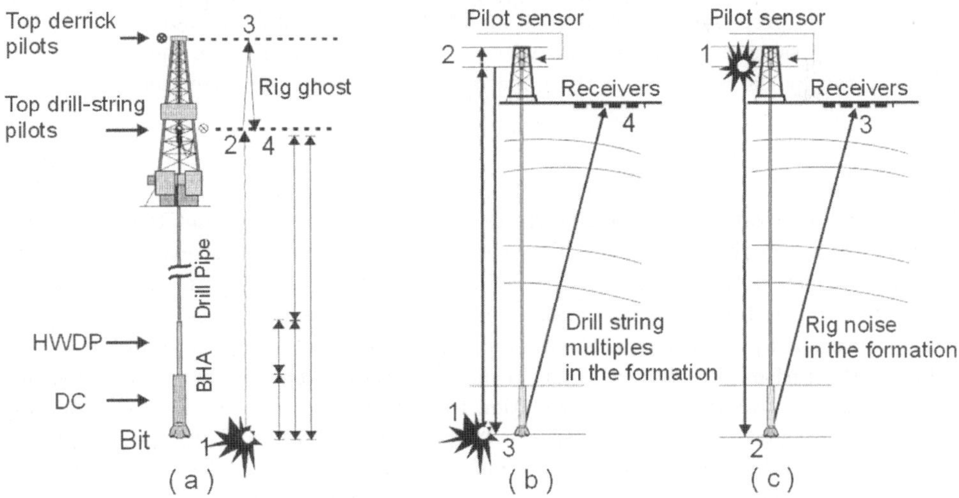

Figure 6.18: a) Long-period multiples and short-period reverberations in the BHA and rig, where the rig-ghost occurs. b) Travelpath of the bit signal and c) of the rig-noise recorded by the geophones of the surface seismic line.

6.13 Interpretation of drill-string multiples

Because of the presence of the multiples of the drill-string noise, the quality of the seismic signal is not directly shown by the presence of pilot reverberations. In the following, we analyze the different contributions of both the drill-bit signal and the surface rig noise to long-period drill-string multiples as well as short-period reverberations in the one-side deconvolved pilot autocorrelations. The real examples used to interpret the drill-string vibrations were obtained using mainly roller-cone data. A scheme of the travelpaths of the waves produced by signal and rig-noise sources is shown in Figure 6.18.

6.13.1 Drill-string multiples in pilot data

To analyze the arrival times, we simplify the model, using a single reflection coefficient c_{T} at the top of the drill pipes. Let $t_{0\mathrm{S}}$ be the arbitrary initial time of the signal, t_{pipe} and t_{BHA} the one-way travel times in the drill pipes and BHA, respectively. We observe that the first long-period multiple of the pilot signal passes through the BHA three times. However, its initial arrival time, observable at approximately 0.2 s in the first traces, corresponds to the event generated at the upper end of the BHA. Therefore it is produced by a wave passing only one time in the BHA. In the correlation time, it is given by

$$t_{\mathrm{CPS}}^{\mathrm{M1}} = t_{0\mathrm{S}} + 3t_{\mathrm{pipe}} + t_{\mathrm{BHA}} - (t_{0\mathrm{S}} + t_{\mathrm{pipe}} + t_{\mathrm{BHA}}) = 2t_{\mathrm{pipe}}. \tag{6.88}$$

Figures 6.19a and 6.19b show the real and synthetic axial pilot signals, respectively.

On the other hand, assume surface-rig noise with an arbitrary initial time $t_{0\mathrm{N}}$. The noise wavefront propagates downward through the drill string and rebounds to the surface pilot sensor passing through the BHA twice. However, also in this case the initial arrival time is given by the reflection at the upper end of the BHA. The correlated arrival time of this event is

$$t_{\mathrm{CPN}}^{\mathrm{M1}} = t_{0\mathrm{N}} + 2t_{\mathrm{pipe}} - t_{0\mathrm{N}} = 2t_{\mathrm{pipe}}. \tag{6.89}$$

Because the correlated arrival times are equal in equations (6.88) and (6.89), there is some ambiguity in the real data, as they can be caused by both these events, i.e., signal and noise which both contribute to the pilot autocorrelation. According to equations (6.86) and (6.87), they have a different content of BHA reverberations. This difference can be observed by comparing the synthetic seismograms of the drill-bit signal in Figure 6.19b to the surface-rig noise in Figure 6.19c. In realistic applications, both signal and noise contribute to the pilot autocorrelation from which the surface-pilot deconvolution operator is calculated. Surface noise can be produced by the shaking and jumping of the hoisting wireline and hook. Other types of noise sources can be the activity and the motion of the top drive on the rail and, at a lower level, other rig vibrations that are transmitted to the rig and pipes (Rector, 1992; Chapter 3). The impulse response of the surface-pilot is partially contained also in the vibrations radiated from the rig (represented by E_{W} in equation (6.72)), as surface and direct waves. For this reason, the surface-pilot deconvolution operator partially "matches" the operator for the radiated rig noise. However, equations (6.84) and (6.85) and the analysis of the geophone data show that the downhole signal needs a different operator.

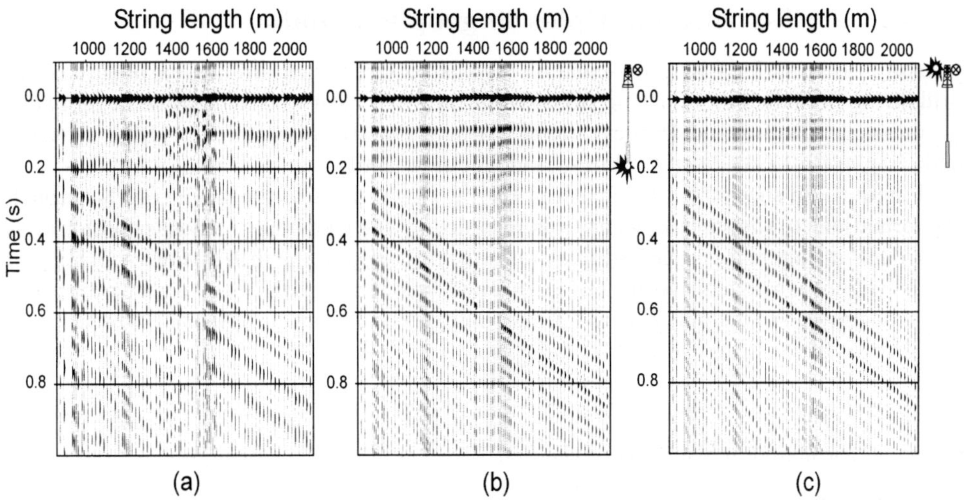

Figure 6.19: Example of real surface pilot signals (a). The corresponding synthetic drill-bit signal is shown in (b). The direct arrival is aligned at 0 s, and contains the short-period reverberations of the signal, which passed once through the BHA. The long-period drill-string multiple contains short-period reverberations of the signal, which passed three times through the BHA. The direct arrival of the synthetic noise (c) does not contain BHA reverberations, and its long-period multiple contains short-period reverberations of the rig-noise, which passed twice through the BHA.

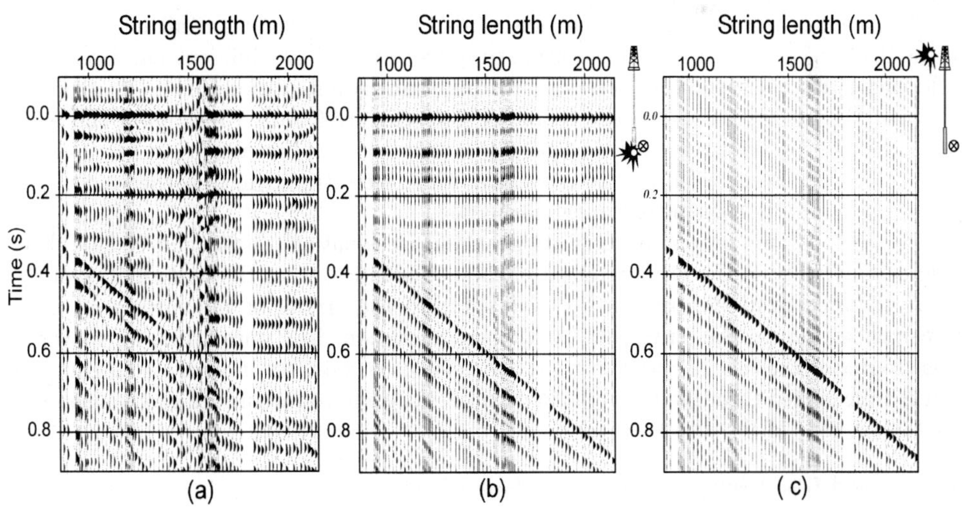

Figure 6.20: Example of real reversed VSP (RVSP) correlations of the data recorded by a geophone of the seismic line, aligned at 0 s after first-break (FB) picking (a). The events with increasing two-way time according to the drill-string length are the long period drill-string multiples. The corresponding synthetic traces are shown in (b). The synthetic direct arrival of the rig noise transmitted through the drill string and bit to the formation (c) is similar to the first long-period multiple of the bit signal.

6.13.2 Drill-string multiples in geophone data

In many cases, such as for roller-cone bit signals acquired when drilling carbonate rocks, drill-string periodicities radiating from the bit are very clear and detectable in geophone seismograms (Rector and Hardage, 1992; Aleotti, Poletto, Miranda, Corubolo, Abramo and Craglietto, 1999; Poletto, Malusa and Miranda, 2001).

Figure 6.18b shows the travelpath of the bit signal. Part of the wave travels through the formations to the surface, where it is recorded by the geophones. Another part travels in the drill string, and is reflected downward from the top of the drill string. A portion of this wavefront is transmitted into the earth and follows the path of the direct wavefront. Let t be the bit-to-geophone travel time through the formation. The first-break (FB) arrival time of the direct correlated event is

$$t_{\mathrm{CGS}}^{\mathrm{D}} = t_{\mathrm{0S}} + t - (t_{\mathrm{0S}} + t_{\mathrm{pipe}} + t_{\mathrm{BHA}}) = t - t_{\mathrm{pipe}} - t_{\mathrm{BHA}}, \tag{6.90}$$

while the correlated arrival time of the first multiple in the drill string is

$$t_{\mathrm{CGS}}^{\mathrm{M1}} = t_{\mathrm{0S}} + 2t_{\mathrm{pipe}} + 2t_{\mathrm{BHA}} + t - (t_{\mathrm{0S}} + t_{\mathrm{pipe}} + t_{\mathrm{BHA}}) = t + t_{\mathrm{pipe}} + t_{\mathrm{BHA}}. \tag{6.91}$$

With respect to the first break (FB) of the geophone signal of equation (6.90), the multiple delay is

$$t_{\mathrm{CGS}}^{\mathrm{M1}} - t_{\mathrm{CGS}}^{\mathrm{D}} = 2t_{\mathrm{pipe}} + 2t_{\mathrm{BHA}}, \tag{6.92}$$

which is different from the delay of the first long-period multiple in the pilot deconvolved autocorrelation given by equation (6.88).

We have discussed for pilot deconvolution that the pilot signal transmitted at the top of the drill string and the downhole drill-string reflectivity have different patterns. This difference appears clearly if we compare the drill-string reverberations of the bit signal in the pilot (Figure 6.19a) and in the geophone data (Figure 6.20a), which show the RVSP correlations of the data recorded by a geophone of the seismic line aligned at $t = 0$ s after FB picking. The event with increasing two-way-time with string length is the long-period drill-string multiple transmitted in the formation.

However, for the geophone correlations, given by equations (6.84) and (6.85), there is some ambiguity in the interpretation of the periodicities, since the periodicities of the signal radiated from the bit can be confused with the arrival of the surface-rig noise transmitted through the string and then to the formation. In Figure 6.18c, we show the noise source at the rig. The wavefront of this noise propagates in the drill string and part of it is transmitted through the bit/formation interface and to the surface geophones. The correlation time of the direct arrival is

$$t_{\mathrm{CGN}}^{\mathrm{D}} = t_{\mathrm{0N}} + t_{\mathrm{pipe}} + t_{\mathrm{BHA}} + t - t_{\mathrm{0N}} = t + t_{\mathrm{pipe}} + t_{\mathrm{BHA}}, \tag{6.93}$$

where we assume that, due to the fact that the noise source is placed close to the surface pilot sensor, there is not a significant pilot delay. The coherent arrival of the noise has the same time (equation (6.93)) as the signal multiples (equation (6.91)). In fact, for traces with the FB of the signal aligned to 0 s, the noise arrival time is

$$t_{\mathrm{CGN}}^{\mathrm{D}} - t_{\mathrm{CGS}}^{\mathrm{D}} = 2t_{\mathrm{pipe}} + 2t_{\mathrm{BHA}}. \tag{6.94}$$

To validate the interpretation, Poletto, Malusa and Miranda (2001) model both the bit axial signal (Figure 6.20b) and the rig noise at the geophone aligned to $t = 0$ s (Figure 6.20c). This model does not include the reflectivity in the rock layers, as well as the complex non-linear drilling conditions (such as absorption) at the bit/rock interface. A trial transmission coefficient is used at the contact between bit and rock to fit the corresponding synthetic seismogram to the multiple events transmitted into the formation. This analysis shows that the contributions of the signal's multiples and the drill-string noise can be confused in the correlations of the geophone traces, or in the aligned FB data. At the same time, with a different interpretation, the signal and the noise in the geophone correlations can give information about the pilot delay.

The signal-multiple time of equation (6.92) and the correlated-noise arrival time of equation (6.94) in the geophone seismograms (with aligned FB) differ, by a relative delay equal to $2t_{\rm BHA}$, from the multiple or noise arrival times in the pilot signal given by equations (6.88) and (6.89), respectively. This shows that the pilot surface deconvolution operators are not suited for deconvolving geophone drill-string multiples properly. In general, we can assume that the correlated pilot and geophone periodicities contain an ambiguity, because in the real data we can have the different contributions of both the signal and the noise. This may introduce limitations and distortions for drill-bit source deconvolution. Signal and noise levels could be better evaluated using downhole tools together with surface measurements.

6.14 Rig ghost

The ghost is a secondary downward reflection of upward traveling wave. A rig-derrick ghost is measured by the sensors at the top of the drill-string. The concept is shown in Figure 6.18a. The drill-bit signal travels through the drill string, the rope lines, is reflected downward by the derrick and is measured another time (ghost) by the pilot sensors at the top of the drill-string. The rig ghost is a strong reflection with a short delay, $t_{\rm g}$, of a few milliseconds, variable for different elevations of the top of the drill string above the rig floor (see also, Figure 2.7). Assume that the reflection coefficient at the top of the derrick is close to +1 for acceleration. If the signal of an upgoing plane wave is measured at the pilot position with its ghost, the pilot trace is

$$a_{\rm g}(\omega) \approx e^{\imath\omega t} + e^{\imath\omega t - \imath\omega t_{\rm g}} = 2e^{\imath\omega(t-t_{\rm g}/2)} \cos(\omega t_{\rm g}/2), \qquad (6.95)$$

which has zeros ("notches") at frequencies

$$f_{\rm g}^{(n)} = \frac{2n+1}{2t_{\rm g}}, \qquad n = 1, 2, 3, \cdots. \qquad (6.96)$$

The deconvolution of the ghost is difficult because zeros in frequency can not be inverted. This may cause distortions in crosscorrelations obtained with pilot signals measured at the top of the drill string. Long spiking operators and low levels of white noise are then used to remove short-period rig reverberations. Long operators are, in any case, required to refine frequency sampling of periodic components and detect long-delay multiples. However, low

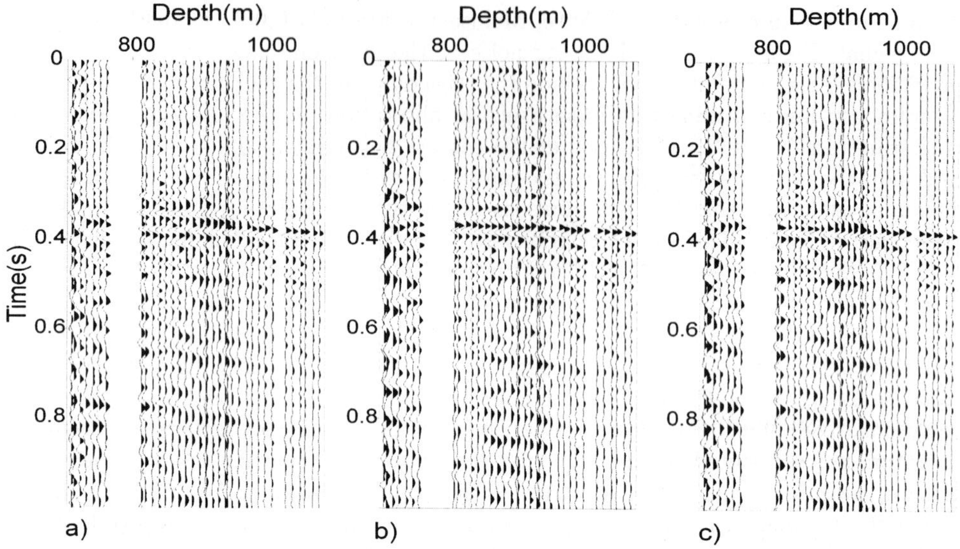

Figure 6.21: Common-receiver seismograms obtained by crosscorrelating a geophone trace at 1200 m offset from the well with a) axial acceleration pilot at the top of the drill string and b) with derrick sensors. Data are one-side deconvolved. The direct arrival of the drill-string signal is more disturbed by residual reverberations than the derrick pilot signal. The derrick signal is more disturbed by diffused noise, but the direct arrival is better deconvolved. After waveshaping and correction for the relative delay, the drill-string and derrick signals are combined with spectral averaging (c), improving the direct arrival. (After Poletto, Malusa and Miranda, 2003.)

levels of additional white noise may cause instability. If not properly removed, the short-period reflections (reversed in negative correlation times) may make it difficult to interpret the time of the direct arrival.

To make the task of removing rig-derrick ghost reverberations easier, the following solutions can be adopted. The first is to measure the signal at the derrick. A top-rig sensor is not affected by the top-rig ghost. However, this signal is usually noisier than a signal measured at the top of the drill string. Effective measurements of the rig signal were realized by Poletto, Malusa and Miranda (2003). Figure 6.21 shows a RVSP common-receiver gather obtained by a) accelerometers at the top of the drill string, b) rig-derrick sensors and c) the combination of the results.

The second approach is to use dual sensor measurements (Sections 4.12, 5.6 and 6.15). For example, a stress wave

$$s_g(\omega) \approx e^{\imath\omega t} - e^{\imath\omega t - \imath\omega t_g} = 2\imath e^{\imath\omega(t - t_g/2)} \sin(\omega t_g/2), \tag{6.97}$$

has a reflection coefficient of opposite sign, and a maximum at the frequency f_g of equation (6.96) where the spectrum of displacement of equation (6.95) has a notch. The processing of the acceleration and stress signals is discussed in the following section.

6.15 Processing of dual drill-string wavefields

Propagation and acquisition of dual drill-string fields were discussed in Sections 4.12 and 5.6. Poletto, Malusa, Miranda and Tinivella (2002) used, without any loss of generality, acceleration and strain dual wavefields. The strain waveform is shaped into the acceleration waveform by using the filter $\mathcal{F}_{s,a}$ given by equation (4.154) – alternatively one can use $\mathcal{F}_{s,a}^{-1}$ to shape acceleration into strain – and scaled to obtain direct arrivals of equal amplitude prior to sum (subtract) the dual fields. The sum (subtraction) of the dual fields removes (reinforces) the reflected events of opposite sign. If we assume that the bit-source signal is an upgoing wave from the bit, these events are produced by an odd number of reflections in the drill string, i.e., these are the downgoing reflected drill-string waves. These waves can be separated from the direct upgoing signal. Conversely, the upgoing reflections in the drill string have the same reflection coefficients and are not separated from the direct upgoing signal.

In general, the possible applications of drill-string dual-field processing are:

◇ Subtraction of the rig reflections. The downgoing rig-ghost reflection, which occurs at the derrick top above the drill string, can be subtracted from the surface-pilot signal using dual-pilot signals of acceleration and strain measured in the same location at the top of the drill string. The downgoing ghost reflection is recorded with opposite sign while the upgoing transmitted bit signal is recorded with equal sign in the dual signals. This method makes it possible to separate the rig (downgoing) and the BHA (upgoing) reflected waves.

◇ The downgoing reflections of the drill string in a signal recorded downhole in the proximity of the bit can be subtracted. The same analysis can be used to improve the reflections of the formation in the downhole dual-pilot signals and to improve prediction of the interfaces ahead of the bit.

◇ The signal and reflections measured at intermediate positions in the drill string can be separated into upgoing waves and downgoing reflections using dual measurements.

◇ After separation, the drill-string noise reflections can be analyzed with the support of the numerical modeling to characterize the noise reflections of long delay and determine the location of the noise source in the drill string.

6.15.1 Synthetic dual wavefields

Poletto, Malusa, Miranda and Tinivella (2002) investigated the acoustic transmission along the drill string by using a time-domain algorithm (Carcione and Poletto, 2000) to simulate the propagation of acceleration and strain of one-dimensional extensional waves.

The algorithm reproduces passbands and stopbands due to the presence of coupling joints, pulse distortion and delay of waves due to the non-regular geometry, wear condition, and physical parameters of a real drill string. The propagation in the time-domain is performed by using a fourth order Runge-Kutta technique. The spatial derivatives are calculated with the Chebyshev method by using the FFT. This approximation is infinitely accurate for band-limited periodic functions with cutoff spatial wavenumbers, which are smaller than the cutoff wavenumbers of the mesh. In this way, the grid points can be distributed with irregular geometry, and thus it is possible to describe in detail the structure and geometry of the drill string (Chapter 3).

The drill bit source is simulated by a pulse of axial force propagating in a drill string of steel with a cut-off frequency of 500 Hz, assuming that there is no attenuation in the drill string. Figure 6.22 shows the a) axial acceleration and b) axial strain waves recorded by receivers located at different positions along the drill string. The axial strain waves are shaped into acceleration waveform using the filter of equation (4.156). The acceleration and strain direct arrivals have the same amplitude, due to the fact that the transmission coefficients are equal for acceleration and strain. The upgoing (even number of reflections) events have positive dip and the same polarity. The downgoing (odd number of reflections) events have negative dip and opposite polarity. The sum of the dual fields, scaled to the same amplitude by a scaling factor, gives the direct arrival and the upgoing reflections (Figure 6.23a). The difference of the two fields gives the downgoing reflections (Figure 6.23b). It is important to note that, in this example, the separation of upgoing and downgoing fields is obtained without any multichannel filtering. In other words, it is sufficient to use a dual-sensor receiver in one single position in the drill string to obtain the separation.

6.15.2 Real examples with axial dual waves

A single downhole receiver location was used to acquire pilot dual fields in a SWD experiment. An instrumented downhole tool, containing an axial accelerometer and an axial strain gage, was positioned in the heavy weight drill pipes 27 m above the drill collars and 159 m above the bit. The bit was a roller-cone bit drilling with downhole motor in sliding condition, i.e., without drill-string rotation. The acquisition bit-depth levels were 1165, 1189 and 1247 m. A sequence of 15 consecutive records of 45 s, with a 2 ms sampling interval was recorded in each depth. At the surface, a conventional seismic line was used to acquire the geophone data, in split-offset configuration and with a maximum offset of 2400 m from the wellhead. Figure 6.24 shows, on the left side, an example of uncorrelated raw data with a sampling rate of 2 ms. The strain signal (a) is shaped by the deconvolution filter (b). After shaping, the dual strain (c) and acceleration (d) signals are compared. These dual fields are crosscorrelated to show the polarity of the reflections (right side of Figure 6.24). In particular, two events with opposite polarity at positive and negative times, symmetrical with respect to the peak at the zero time, can be observed. The interpretation of these events is done by calculating the axial reflection coefficients in the drill string.

Figure 6.25a shows the drill-string mass per unit length distribution versus drill-string length (measured from the bit). Figure 6.25b shows the acceleration reflection coefficients. A large reflection coefficient is at the connection between DC and HWDP, at approximately 130 m of drill-string length. Another large reflection coefficient occurs at the connection between HWDP and drill pipes, at about 265 m drill-string length above the bit. These reflections are interpreted in the autocorrelations of the acceleration and strain waves (c). The interpreted short-delay odd-reflection event is the reflection produced at the interface

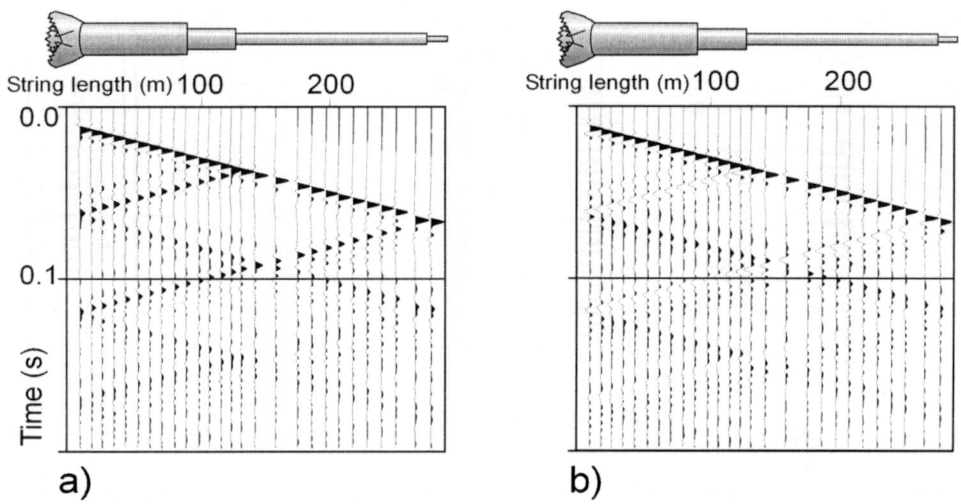

Figure 6.22: Example of synthetic data: a pulse of axial force produced at the bit is recorded by receivers located at different positions along the drill string. The drill-string length is the distance from the bit. The comparison of the a) acceleration and b) strain waveforms shows upgoing events, with increasing delays versus distance from the bit, and the same polarity; downgoing events, with decreasing delays versus distance from the bit, and opposite polarity. The direct arrivals have the same amplitude trends in acceleration and strain data.

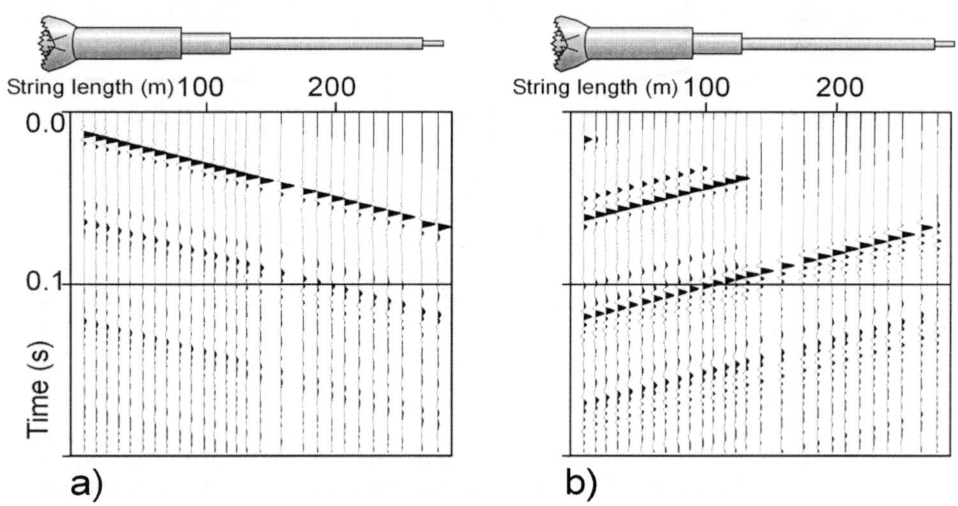

Figure 6.23: a) The sum and b) the difference of the dual fields of Figure 6.22 give the separated upgoing and downgoing events, respectively. This separation can be achieved by processing only the traces of a depth level (modified after Poletto, Malusa, Miranda and Tinivella, 2002).

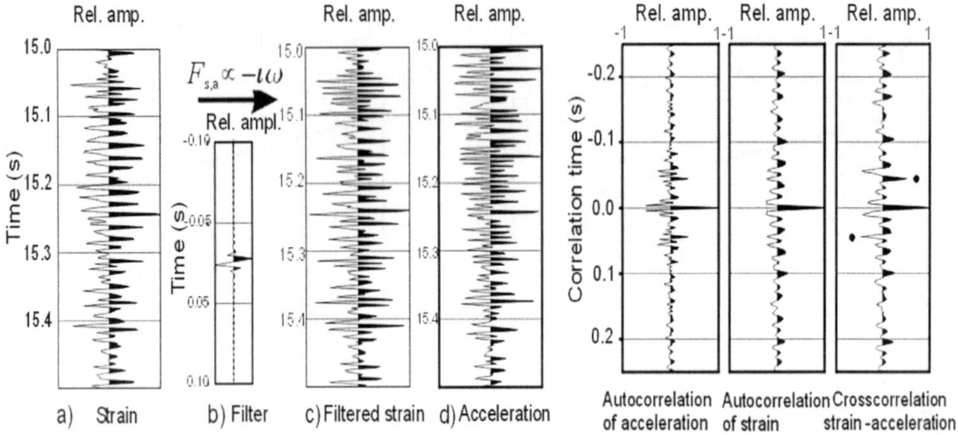

Figure 6.24: **Left:** Example of raw (uncorrelated) drill-bit data. (a) The strain trace is shaped by the (b) filter operator into (c) acceleration signal and compared to the (d) measured acceleration signal. **Right:** Autocorrelation of (a) acceleration and (b) strain signals. (c) The crosscorrelation of these two signals shows the presence of symmetric events with opposite polarity.

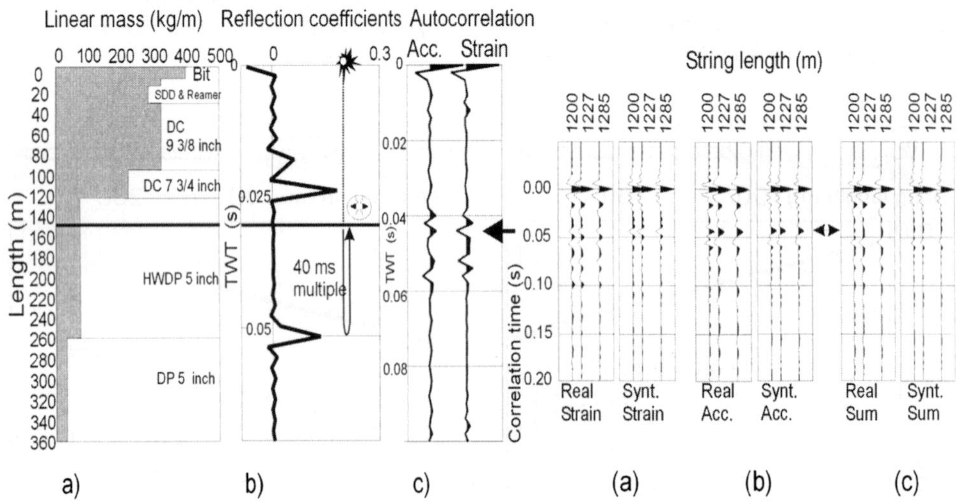

Figure 6.25: **Left:** Interpretation of reflected events in the acceleration and strain autocorrelations. The mass per unit length distribution of the drill string-components (a) is used to calculate the reflection coefficients in the drill string (b). The synthetic autocorrelations of acceleration and strain (c) show two events with opposite polarity at 40 ms, which is interpreted as the one-time reflection at the BHA interface between heavy-weight pipe (HWDP) and drill pipe (DP). **Right:** Dual signals acquired at the bit depths 1165, 1189 and 1247 m. Comparison of the real and synthetic autocorrelations of (a) acceleration and (b) strain. In the sum of these two kinds of signals, calculated using both the real and synthetic data, attenuation of the BHA reflection can be seen.

between the BHA-heavy-weight pipes HWDP and drill pipes. This event is indicated by an arrow. Other short-period events are the internal reverberations in the drill collars DC and HWDP. This interpretation is confirmed by the comparison of the real and synthetic (calculated by using a transmission-line model) autocorrelations of acceleration and strain, shown in Figures 6.25a and 6.25b, respectively. Figure 6.25c shows the autocorrelation of the real and synthetic sums of the dual fields. The periodicity of the downgoing multiple disappears, as well, about 50% of the multiple energy is removed. Not all the reflected energy was subtracted because the tool was used in a location far from the bit, where it measured both downgoing and upgoing reflected waves. Examples of separation of dual BHA reflections in walkaway SWD shots are discussed by Poletto, Malusa, Miranda and Tinivella (2002).

6.15.3 Deconvolution of pilot dual fields

Loewenthal and Robinson (2000) discussed the unified dual-field theory by assuming that the earth acts as a single-channel system. They proposed a deconvolution method which consists of two steps. First, the Einstein deconvolution (Robinson, 1984) is computed: it consists of deconvolving the linking upgoing wave by the linking downgoing wave. The output of the Einstein deconvolution is the impulse response of the layers below the transducer. Second, the output of the Einstein deconvolution is the input to a dynamic deconvolution: this is a non-linear method that separates the lower subsystems in strips to extract the unknown reflection coefficients.

We use a similar approach with drill-string dual fields. Assume a dual pilot sensor close to the bit where an upgoing bit signal is measured using acceleration and strain waves. The downgoing-acceleration wave reflected in the drill string towards the bit is $R(Z)$; the reflection coefficient of acceleration at the bit/rock interface is $c_{\mathrm{a},0} = c_0$. The upgoing- and downgoing-acceleration waves are

$$U_{\mathrm{a}} = 1 + c_0 R(Z), \tag{6.98}$$

$$D_{\mathrm{a}} = R(Z), \tag{6.99}$$

respectively. Hence, the acceleration at the bit is

$$a = 1 + (c_0 + 1)R(Z). \tag{6.100}$$

The downgoing (odd) strain wave is $-R(Z)$; the reflection coefficient of strain at the bit/rock interface is $c_{\mathrm{s},0} = -c_0$. The upgoing- and downgoing-strain waves are

$$U_{\mathrm{s}} = 1 + c_0 R(Z), \tag{6.101}$$

$$D_{\mathrm{s}} = -R(Z), \tag{6.102}$$

respectively. Hence, the strain at the bit is

$$s = 1 + (c_0 - 1)R(Z). \tag{6.103}$$

Note that equation (6.103) is similar to equation (6.29) because – if we exclude a scaling factor from strain to force – we have $R(Z) = -R_{\rm B}^{(s)}$. We can calculate the upgoing and downgoing pilot signals as

$$p_{\rm u} = (a + s)/2 = 1 + c_0 R(Z), \tag{6.104}$$

$$p_{\rm d} = (a - s)/2 = R(Z), \tag{6.105}$$

respectively. These data are deconvolved by simply estimating the reflection coefficient c_0. Let $\hat{c}_0$ be the best estimate of c_0, which is obtained by comparing $p_{\rm u}$ and $(1 + \hat{c}_0 p_{\rm d})$. If we neglect the reflectivity of the formation ahead of the bit (Malusa, Poletto and Miranda, 2002), the deconvolved pilot trace, $p_{\rm dec}$, can be expressed as

$$p_{\rm dec} = p_{\rm u} - \hat{c}_0 p_{\rm d}. \tag{6.106}$$

This method can be used to deconvolve the SWD-pilot data and estimate the bit/rock reflection coefficient. Note that the analysis shows that the optimum position for downhole dual sensors in the drill string is near the bit. When the reflectivity of the formation is not neglected and is interpreted in the drill-string pilot signal (Malusa, Poletto and Miranda, 2002), the processing of separation involves also the formation reflections ahead and behind the bit. The dual-field results may improve the analysis of the formation reflections close to the bit.

6.16 Pilot-delay correction

The numerator of the total deconvolution operator of equation (6.41) is the pilot-delay correction. This is implemented in frequency by the phase-shift filter of equation (6.33)

$$e^{-\imath\omega t_{\rm p}}. \tag{6.107}$$

The crosscorrelation of the pilot signal with the signals of the seismic line gives the travel-time difference between the drill-string path and the formation path. To find the absolute travel time from the bit to the geophone through the formation, the travel time along the drill string must be added (the drill string is the typical but not the only line that transmits the pilot signal). In other words, the pilot signal's delay must be added to the cross-correlation times to obtain the signal's travel time through the ground. This is equivalent to calculating the zero–time of the source. The unknown seismic travel time $t_{\rm s}$ through the formation (from the bit to the geophone) is related to the signal time $t_{\rm c}$ in the correlation and the travel time $t_{\rm p}$ from the bit to the pilot sensor as

$$t_{\rm s} = t_{\rm c} + t_{\rm p}. \tag{6.108}$$

The way to calculate $t_{\rm p}$ is a matter of pilot-signal analysis. The analysis can be extended to seismic waves when the geophone signals are used as a pilot signal.

A way to calculate $t_{\rm p}$ is to use the pilot-signal autocorrelation. Analysis of long-delay (long-period) multiples in the autocorrelation of the signal measured at the top of the drill string gives the two-way travel time in the drill pipes (Staron, Arens and Gros, 1988).

Figure 6.19a shows an example of one-side deconvolved autocorrelations. In this case, the drill-bit minimum-phase signature is filtered by the drill-string transfer function $\mathcal{H}_{\text{TM}}$. The event picked (at time t_{pick}) is usually the reflection between the drill pipes and the BHA, at about 0.2 s in the first trace. We obtain

$$t_{\text{p}} = \frac{t_{\text{pick}} + 2t_{\text{BHA}}}{2} \tag{6.109}$$

where t_{BHA} is the estimated time delay in the BHA.

Another approach is to analyze the long-delay multiples in the geophone crosscorrelations (Aleotti, Poletto, Miranda, Corubolo, Abramo and Craglietto, 1999), as in a VSP aligned by FB picking (Figure 6.20a). In this case, the formation response and minimum-phase bit signature is filtered by the drill-string downhole reflectivity $\mathcal{H}_{\text{B}}^{(s)}$. The picked event is the reflection at the bit, at about 0.3 s in the first trace. We have

$$t_{\text{p}} = \frac{t_{\text{pick}}}{2}. \tag{6.110}$$

In general, picked data may be noisy; also picking may be problematic at some depth intervals because of signal attenuation. Consequently, the use of an average velocity v_{ds} gives more stable results. The delay of the pilot signal can be calculated using equation (6.107), where the delay in the drill pipes is given by

$$t_{\text{p}} = \frac{z_{\text{ds}}}{v_{\text{ds}}}, \tag{6.111}$$

where v_{ds} is the average velocity calculated in pipes of the same type by time/length linear regression, and z_{ds} is the length of the drill-string travelpath. In any case, picking as well as velocity analysis by slant stack of drill-string data is made more difficult by the fact that the waveform of drill-string multiples can change because of variations of drill-string composition and of drilling conditions. Noise and signal propagation modes are superimposed and generate rapidly changing phenomena. In particular, the physical constraints at the bit/rock contact, i.e., the reflection coefficient c_0, depend on the drilling conditions (Clayer, Vandiver and Lee, 1990). Drill-string imaging (Booer and Meehan, 1993) and modeling (Poletto, Malusa and Miranda, 2001) allow a better interpretation and assessment of the velocity parameters. Moreover, in the absence of interpretable data, using the average mechanical properties of drill strings (as in the analysis of Section 4.4.3) gives a good approximation of the group velocity at low frequency (zero-frequency approximation). However, the calculated average values have to be controlled by fitting modeling and real data (Section 6.11).

Use of pilot signals with different delays

The use of different pilot signals may involve different delay corrections. This is the case of extensional-drill-string and coupled-mud waves described in Section 7.8.2. Another case may be the joint use of extensional and torsional pilot waves. The signal propagated from the bit through the formation to the geophones, i.e., compressional, shear and converted waves, has the same travel time, t_{s}, in the seismic data obtained by correlating with extensional and mud pilots

$$t_{\text{s}} = t_{\text{c(ext)}} + t_{\text{p(ext)}} = t_{\text{c(mud)}} + t_{\text{p(mud)}}, \tag{6.112}$$

so that the difference of correlation delays is

$$t_{c(ext)} - t_{c(mud)} = \frac{z_{mud}}{V_{mud}} - \frac{z_{ext}}{V_{ext}}, \tag{6.113}$$

where z_{mud} and z_{ext} are the length of the travelpaths in the mudline and drill string, respectively. Figure 6.26a shows the correlation of the axial accelerometer signal with an outer-mud pressure signal. Figure 6.26b shows the correlation of the torque with the axial accelerometer. The relative delay of the axial and torque signals is clearly measurable.

6.17 Signal rephasing

Signal rephasing is an adjustment of the phase and is used to compensate for phase distortions, for example from minimum phase, or to calculate the zero-phase wavelet with the standard-positive polarity (Figure 6.27). Wavelet processing, phase properties and resolution are discussed, for instance, by Hardage (1987) and Berkhout (1984). Signature distortions may be caused by the nature of the source (impulsive or not impulsive) and receiver properties (Landrum, Brook and Sallas, 1994). In addition, the SWD waveshape to be corrected depends also on the preprocessing (correlation and deconvolution) applied to it before rephasing.

Rephasing after pilot deconvolution

The data, crosscorrelated with the pilot signal and one-side deconvolved, are calculated using the product in the frequency-domain of the signature of the deconvolved crosscorrelogram,

$$-\mathcal{H}_B^{(s)}(\omega) f_{FM}(\omega) W^*(\omega), \tag{6.114}$$

and the radiation filter

$$\mathcal{H}_G(\omega). \tag{6.115}$$

Here we use the same notation as in equation (6.53) in frequency. The radiation filter includes the source radiation pattern, the propagation in the formation, and the receiver response. The minus sign in the signature of equation (6.114) holds if the pilot accelerometers are oriented as the field geophones, i.e., if after acceleration/velocity conversion they have the same polarity as the geophones[1] (for polarity standards see, for instance, Thigpen, Dalby and Landrum, 1975). The filter of equation (6.114) is minimum-phase if $W(\omega) = 1$.

Because we use the single vertical-force model (Section 3.8.2), the displacement radiated in the far field is proportional to the ground excitation force (Sallas, 1984). Hence, for *onshore* data, $\mathcal{H}_G(\omega)$ includes the geophone-response-to-displacement filter, namely,

$$\mathcal{G}_{dv}(\omega) = \imath\omega |\mathcal{G}_{resp}(\omega)| e^{\imath\phi_{g(resp)}}, \tag{6.116}$$

[1] The recommended standards for signal polarity are: "A negative-going output of the recording system ..." shall be produced by "... upward motion of the case" of the sensor and "... pressure increase detected by a pressure-sensitive phone."

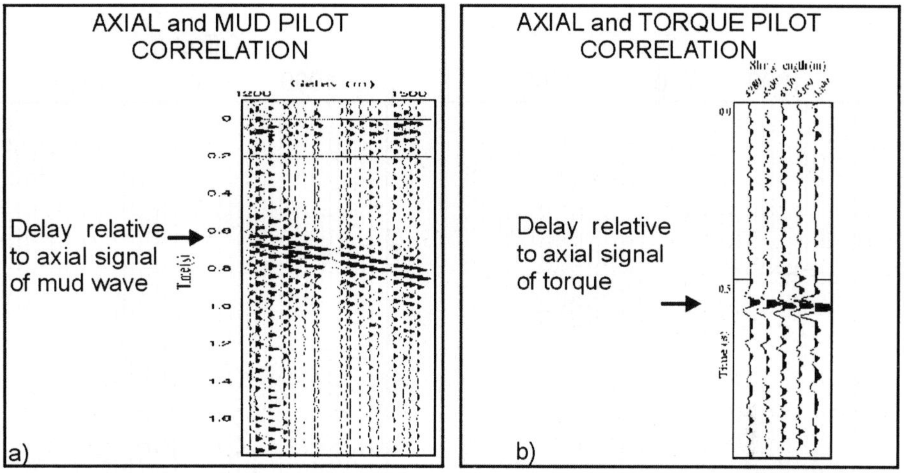

Figure 6.26: Examples of correlation with the axial accelerometer signal of (a) an outer-mud pressure signal and (b) the torque signal from amperometric pinch. Data are presented for various (a) drilling depths and (b) string lengths.

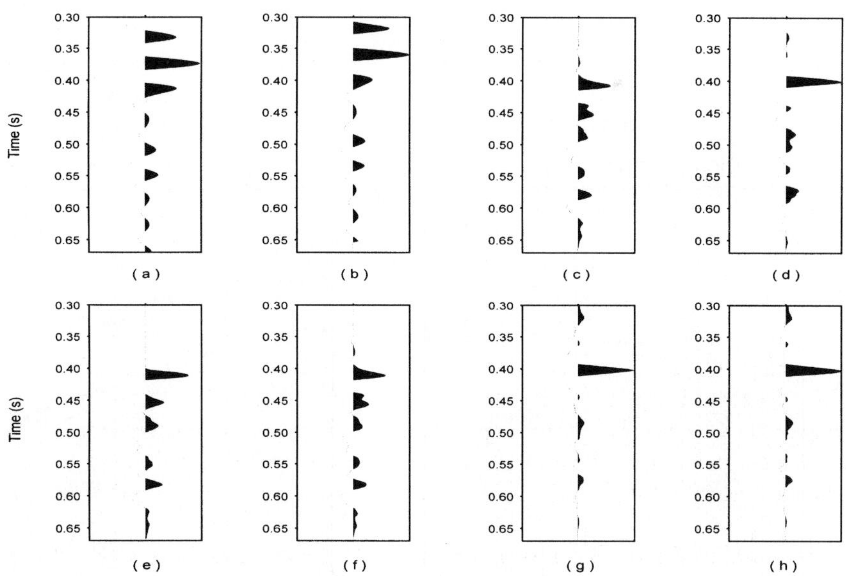

Figure 6.27: Direct arrival of real SWD data obtained using an acceleration pilot and a geophone signal. (a) Non-deconvolved correlated wavelet; (b) Non-deconvolved, rephased for geophone; (c) deconvolved wavelet (one-sided pilot deconvolution); (d) deconvolved, rephased for geophone response; (e) deconvolved, minimum-phase equivalent of the amplitude spectrum (polarity opposite to SEG standard); (f) deconvolved, shaped to minimum phase (negative SEG standard polarity); (g) deconvolved, zero-phase equivalent of the amplitude spectrum; (h) deconvolved, shaped to zero phase. The zero time is at 0.4 s.

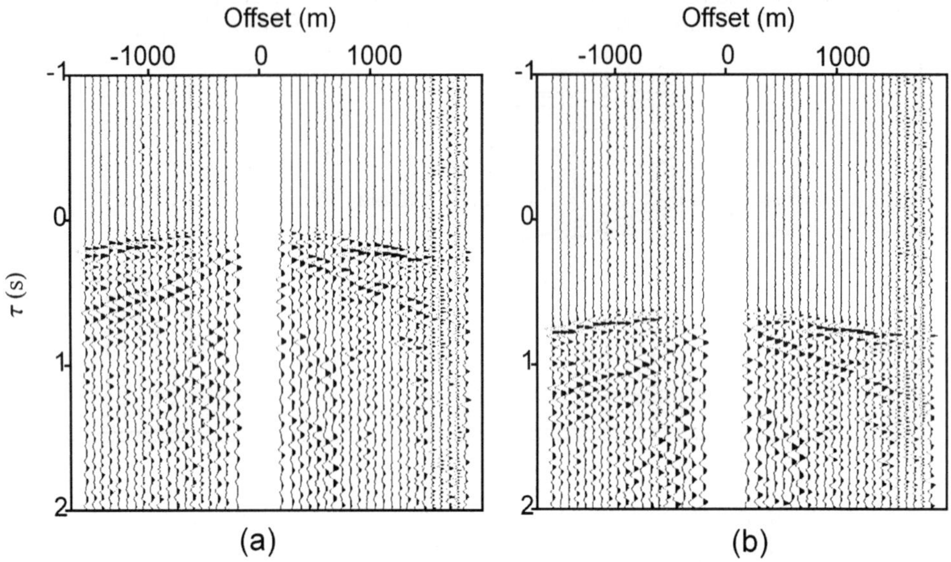

Figure 6.28: Common source data a) before and b) after pilot-delay correction and wavelet shaping.

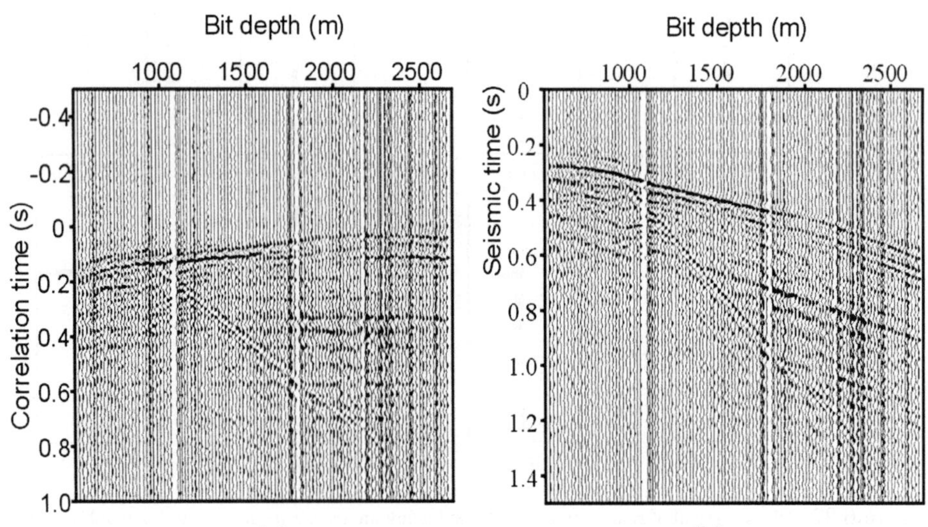

Figure 6.29: VSP data a) before and b) after pilot-delay correction and wavelet shaping. Note that Figure (b) shows, after preprocessing, the same data as Figure 6.7.

where $\mathcal{G}_{dv}$ is the geophone response to displacement and $\mathcal{G}_{\text{resp}}$ is the geophone response to particle velocity. From equations (6.114) and (6.116), we have that the phase correction for onshore correlated data after deconvolution is (Figure 6.29)

$$\Delta\phi = \frac{\pi}{2} - \phi_{\text{g(resp)}}. \tag{6.117}$$

In *offshore* data, the displacement is converted into hydrophone pressure. According to SEG polarity standard (Thigpen, Dalby and Landrum, 1975), a pressure (compression) increase is detected by a pressure-sensitive phone as a negative signal. By definition, stress is opposite to pressure. Thus, an increase in stress corresponds to a decrease in pressure which is detected as a positive signal. Assume an upgoing plane displacement wave propagating vertically from the bit to the near-surface receiver (the z axis is oriented downward), say $e^{\imath(\omega t+kz)}$. Stress is proportional to strain which is calculated from displacement by using $\partial/\partial z = \imath k$ for the upgoing wave. Therefore, as stress is proportional to particle velocity, the phase that corrects to standard polarity the correlated and deconvolved upgoing-wave is

$$\Delta\phi = \frac{\pi}{2}. \tag{6.118}$$

Rephasing before pilot deconvolution

In the above calculations, we have assumed that the pilot deconvolution inverts the filter of equation (6.34) that shapes force into acceleration,

$$\mathcal{F}_{f,a}(\omega) = \frac{\imath\omega}{\mathcal{Z}_1}, \tag{6.119}$$

where the numerator is the Fourier transform of the time derivative ∂_t. In fact, derivation of sampled signals of bandwidth below Nyquist frequency is equivalent to the application of a minimum-phase operator (for example the bilinear transform, Claerbout, 1976). This means that a spiking deconvolution performed by a one-sample delay Wiener-prediction filtering flattens the amplitude spectrum and corrects also the phase of the filter that shapes force into acceleration. Before deconvolution or if deconvolution is not a spiking deconvolution, the application of the inverse filter of equation (6.119) could be necessary to remove the $\pi/2$ shift for the propagation of the pilot signal in the drill string.

Shaping to minimum phase

Let $W(\omega)$ be the Fourier transform of $W(k)$ (Section 6.54). We have shown that $W(\omega)$ of equation (6.54) is minimum phase after minimum-phase pilot deconvolution by $D_{\text{P}}(\omega)$. Therefore, the filter of the crosscorrelated and deconvolved drill-string transfer function of equation (6.114) is mixed phase because of $W^*(\omega)$, and is ideally minimum phase if $W(\omega) = 1$. From equation (6.52) we have that the two-side deconvolved pilot autocorrelation is

$$D_{\text{P}}(\omega)D_{\text{P}}^*(\omega)\Phi_{\text{pp}}(\omega) \cong W(\omega)W^*(\omega), \tag{6.120}$$

and in time

$$D_{\text{P}}(k) \star D_{\text{P}}(-k) \star \Phi_{\text{pp}}(k) \cong W(k) \star W(-k). \tag{6.121}$$

Equation (6.121) shows that spiking deconvolution is applied two times in opposite directions in order to spike the pilot autocorrelation. The equivalent minimum phase

$$\phi_{\mathrm{WM}}(\omega) = F_{\mathrm{Hi}} \{\log [W(\omega)W^*(\omega)]\}, \tag{6.122}$$

where F_{Hi} is the Hilbert transform, shapes the filter W^* of equation (6.114) to minimum phase. Hence, the phase correction that shapes the correlated and deconvolved wavelet to minimum-phase is

$$\phi_{\mathrm{M}}(\omega) = \phi_{\mathrm{WM}}(\omega) + \phi_{\mathrm{GM}}(\omega), \tag{6.123}$$

where ϕ_{GM} is the equivalent minimum-phase correction for the geophone and recording instrumental response. The phase compensation by equation (6.122) is similar to the procedure used by Ristow and Jurczyk (1975) in order to calculate the noise-free Vibroseis twofold deconvolution operator after correlation. This, in turn, is equivalent to the phase compensation used for bandlimited Vibroseis signals (Bickel, 1982). Bickel (1982) observed that Vibroseis signals can be assumed to be bandlimited. As for narrow-band signals, he discussed distortions because of additional noise, showing that the envelope and phase functions are weak functions of the S/N ratio. We observe that in pilot-deconvolved SWD data the minimum-phase conditions depend on the results of pilot deconvolution applied on the base of the factorization of equation (6.46).

6.18 Example of preprocessing parameters

Figure 6.30 shows a basic standard SWD correlation processing flow, in which the choice of the pilot signal in not limited to rig pilots but can be extended to combinations of seismic traces.

The following parameters are usually set during a drill-bit survey. These parameters are utilized by the automatic acquisition and preprocessing system.

Sampling time interval – A standard value is 2 ms. The seismic signal has significant energy content below 100 Hz. However, the pilot signal contains higher frequencies limited by the attenuation and the stopband effects for signals transmitted in the drill string (Section 4.4.2). A higher sampling rate (say, sampling interval 1 ms) can be used to preserve a downhole pilot signal for deconvolution. This higher rate is not typically used in current SWD surface measurements.

Recording time – The minimum time length T_{R} of a single raw uncorrelated field-seismic record is calculated as follows

$$T_{\mathrm{R}} = 3\,(T_{\mathrm{S}} + 2 \cdot T_{\mathrm{D}}), \tag{6.124}$$

where T_{S}, and T_{D} are the time lengths of the seismic data recordings and of the deconvolution operator, respectively. The factor 3 is used in equation (6.124) to limit the attenuation effects due to the lack of overlapping with the crosscorrelation

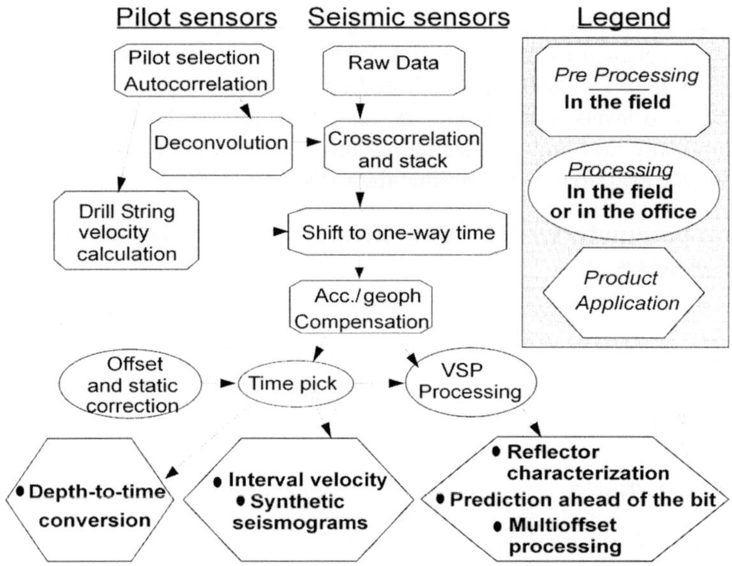

Figure 6.30: Flow of seismic-while-drilling basic correlative preprocessing of the signals from pilot sensors and seismic sensors, processing and product application. All this processing can be performed in the field.

shift. Equation (6.124) also accounts for the tapering at the ends of the output trace for deconvolution purposes. We can estimate the seismic time as

$$T_S \leq \frac{D_{max}}{V_{min}}, \tag{6.125}$$

where V_{min} is the minimum propagation velocity, D_{max} is the maximum travelpath related to the maximum investigated depth z_{max} and maximum offset x_{max} by

$$D_{max} = \left(x_{max}^2 + 4z_{max}^2\right)^{1/2}, \tag{6.126}$$

where the factor four is included to account for the reflected travelpaths, i.e., the two-way time for records at shallow bit depths. If, for instance, we assume P waves with $V_{min} = 2500$ m/s, and $D_{max} = 10\,000$ m. For example, if we desire to obtain a seismogram of 4 s, and use a 2 s deconvolution operator, we have

$$T_R = 3\,(4 + 2 \cdot 2) = 24 \text{ s}. \tag{6.127}$$

Number of records per level – The maximum number of raw records per level is given by

$$N_{R(max)} = \frac{\Delta_{LEV}}{(T_R + T_P) \cdot \text{ROP}}, \tag{6.128}$$

where $\Delta_{\rm LEV}$ is the maximum allowed descent interval of the level (e.g., 10 m), $T_{\rm P}$ is the pause between two subsequent records (e.g., 6 s), and ROP is rate of penetration (e.g., 5 m/h or 5×3600^{-1} m/s). Note that $0 \leq \Delta_{\rm LEV} \leq \Delta_{\rm z}$, where $\Delta_{\rm z}$ is the interval between two levels (e.g., 20 m). $\Delta_{\rm z}$ and $\Delta_{\rm LEV}$ are set by the operator, controlled automatically and limit the signal resolution when the intervals are as large as the wavelength available in the seismic data (Section 5.12).

Correlation parameters – The total correlation length is the sum of the positive and negative delays

$$\tau_{\rm c} = \tau_{\rm pos} + \tau_{\rm neg}, \tag{6.129}$$

where $\tau_{\rm neg}$ and $\tau_{\rm pos}$ are the maximum absolute values of negative and positive correlation shifts. It is calculated as follows

$$\tau_{\rm c} = T_{\rm S} + 2 \cdot T_{\rm D}, \tag{6.130}$$

i.e.,

$$T_{\rm R} = 3\tau_{\rm c}. \tag{6.131}$$

For instance, if we consider P waves with $V_{\rm min} = 2500$ m/s, $D_{\rm max} = 10\,000$ m, and $T_{\rm D} = 2$ s, we obtain $\tau_{\rm c} = 8$ s. In this calculation, as a rule-of-thumb, we may neglect the effect of the pilot delay because, in general, pilot correlation tends to shift the seismic arrivals towards the zero of correlation.

Pilot deconvolution parameters – Statistical operators can be calculated from pilot autocorrelation. The length of the pilot-deconvolution operator $T_{\rm D}$ is usually set to contain at least the first long-period multiple of the drill string. Moreover, long operators perform better with bit signals with narrow-band periodic patterns (Section 6.4; see, also, equation (6.26)). A typical operator length is $T_{\rm D} = 2$ s. A spiking-deconvolution operator with low levels of white noise, e.g., 0.1 percent, is often used.

Rephasing – The correlations between geophone traces and pilot accelerometers are rephased to correct the geophone response (depending on natural frequency and damping) and the $\pi/2$ phase shift of equation (6.117).

Pilot-delay correction – The pilot-delay correction is calculated by using the pilot autocorrelations and the modeling of the drill-string waves. This delay is calculated by averaging the drill-pipe elastic properties and dimensions to obtain an estimate of the group velocity in the low-frequency approximation (see Section 4.4.3). Different velocities are used for pipes and BHA. The pilot correction time $t_{\rm p}$ can be calculated automatically by using the drilling information (string-length) and by monitoring the addition of the new pipe elements in the drill string. The seismic time $t_{\rm s}$ is given by equation (6.108).

Filtering – Filtering is typically used for the following purposes:

- ◇ remove the low-frequency noise radiated from the rig (low-cut typically 12–16 Hz).
- ◇ remove the drilling high-frequency noise from near-surface data, when the source is shallow (high-cut typically 60–100 Hz).
- ◇ remove the drilling high-frequency noise from deep data, when the source is deep (high-cut typically 48–60 Hz).

Chapter 7

Processing of signal and noise RVSP fields

7.1 Introduction

The seismograms produced by preprocessing are used for seismic-while-drilling analysis. This step involves the separation of signal and noise wavefields as well as the processing of the wavefields acquired in the well-surface geometry. Several techniques can be used to separate signal and noise. Some of these are applied after crosscorrelation; other methods can be used before crosscorrelation or before stack of crosscorrelations. That is, some of the techniques discussed in this chapter are used with preprocessed data, other methods are applied using field data before complete preprocessing. Moreover, the differences between processing of conventional normal VSP and RVSP wavefields are pointed out. Other SWD methods not based on pilot-geophone crosscorrelation are also discussed.

7.2 Entropy and repeatability of the drill-bit source

In thermodynamics, entropy is a measure of the unavailable energy, and high entropy represents increased disorder (Sheriff, 1999). The entropy analogy, for which the drill-bit cross-correlation result is assumed to be a signal from a stochastic source, is a matter for discussion. Let us use correlated data from the same receiver station and assume that $P(s_i)$ is the probability of obtaining the signature s_i for the direct arrival of the trace at the i-th bit depth. The (Shannon) entropy H_s (Walden, 1985) is given by

$$H_s = -\sum_i P(s_i) \log_2 P(s_i). \tag{7.1}$$

Zero entropy means that there is no stochastic uncertainty, i.e., $P(s_i) = 1$. If

$$I_i = -\log_2 P(s_i) \tag{7.2}$$

is the "information" of the i-th event, then entropy H_s simply gives the average information. Before calculating equation (7.1), we have to calculate $P(s_i)$, which is, in turn, related to the expected correlated waveform s_i. This waveform is subjected to changes in the drilling condition, source performance and propagation in formation. Thus the entropy of dynamic drilling conditions can be high. These conditions can be monitored while drilling. The analysis of drilling vibrations of Chapter 3 shows that the average drill-bit source performance can be characterized by performing drilling-condition monitoring. In this way, we can compensate for measurement variations, uncertainty (and entropy in the analogy) can be significantly reduced, as well as repeatability controlled by drilling information. This can be done, for example, by processing separately data obtained with different drilling modes, such as sliding and rotary.

Hardage (2000) observed that a poor repeatability of the conventional VSP source is one of the factors that adversely affect the quality of the borehole seismic stratigraphic sections. Thus, repeatability – at least local in bit-descent intervals – of the direct-arrival signature versus depth (Section 1.2.3) is a basic requirement that has to be satisfied in order to effectively process the VSP data.

The repeatability of the crosscorrelated drill-bit signature is mainly limited by the variability of the source performance in relation to the drilling conditions. Factors that may limit repeatability in the SWD direct arrivals are:

◇ Changes of the drilling plant, BHA composition, type of bit.

◇ Changes of the drilling parameters (such as ROP, WOB, RPM, rotary mode).

◇ Changes of well verticality or deviation from verticality layout.

◇ Changes of the properties of the drilled formation.

◇ Changes of surface recording conditions, such as those due to seasonal changes and human works in the field.

In most cases, these changes arise because of operational variations essentially due to the drilling needs. The analysis of correlated SWD data collected during the drilling of wells has shown that the repeatability of the drill-bit source is a condition only partially met. In fact, the analysis showed that the drill-bit SWD source repeatability is met – at least – over large depth intervals. These may correspond to the main drilling phases (such as the different casing sections, between the main formation changes, or within intervals drilled with the same drilling mode).

The control of the source variability in relation to drilling conditions is an important task in acquisition, which has to be accomplished while drilling to assess the quality of the preprocessed SWD data and improve signal repeatability. The analysis of the signal waveform repeatability is important both in the preprocessing and in the processing phases. Kragh and Christie (2002) discussed different "metrics" that can be used to measure repeatability. One is the normalized RMS error (NRMS) defined as the RMS value of the difference between two time sequences in a given time window, normalized by the sum of the RMS values of the two sequences. Another is the "predictability", equivalent to coherence, and defined as the square of the crosscorrelation between two traces divided by the product of their autocorrelations. These metrics have different sensitivities. The second one is more sensitive to noise and is not sensitive to amplitude and polarity changes. Conversely, the first one is extremely sensitive to the smallest changes of phase in the data.

7.3 Common-level stack of correlations with noise

"*A large class of seismic processing operations involve the linear combination of several traces in such a way that certain 'signal' events are emphasized over other 'noise' events. If the noise events are randomly distributed between traces, then simple addition of the traces represents the best possible technique for suppressing the noise. However, in many cases a noise event on one trace will also occur at relatively well-determined points on other nearby traces. For these cases, multichannel linear filters offer better discrimination against organized noise than the straight addition of the traces*" (Galbraith and Wiggins, 1968).

In this section, we discuss stacking of crosscorrelations of signals from the same depth level. This is the "vertical stack" of common-receiver data – in which signal and coherent-noise moveouts are stationary and their moveouts are not used to improve the stack of the data. Conversely, focusing is the "horizontal stack" of traces of common-source records – in which signal and coherent noise are defined by their relative moveout. In both these applications of multichannel stacking filters, we are only concerned with one output channel.

In Section 5.12 we have analyzed the variation of the radiation pattern with drill-bit descent and have shown that the maximum thickness of the acquired interval (level) is limited by resolution consideration. Rector (1992) analyzed the random noise in drill-bit data by assuming that the stochastic drill-bit signal and the incoherent noise are Gaussian. From this assumption it follows that the signal-to-noise ratio achieved by crosscorrelation improves as the square root of the total recording time. With the Gaussian-noise assumption, the stack of $N_{\rm R}$ crosscorrelated records of the same recording-time length and statistical properties should give

$$\frac{S}{N} \propto \sqrt{N_{\rm R}}. \tag{7.3}$$

However, coherent noise may reinforce as $N_{\rm R}$ in the stack of the level, so that equation (7.3) is not realistic for actual drill-bit crosscorrelations.

In this discussion, it is important to understand the distinction between random and coherent noise. Organized events (propagating waves) measured by the receivers can produce random noise if they are not correlated in the pilot and seismic signals. On the other hand, random time sequences measured both by the pilot and seismic signals occur as coherent events in the crosscorrelated data. A significant part of coherent noise is produced by secondary bit waves radiated from the rig and borehole.

We can use the stack of the correlations for the same level to reduce the noise components. Improvement from stacking depends on noise conditions. If a trace were noise free, there would be no reason to stack other traces with it. We may use only this perfect record. If two traces have equal signal and random-noise levels, i.e., have the same S/N ratio, the best we can do is to sum them. If two traces contain equal signal and the same coherent noise with different amplitudes, we can subtract them after proper weighting in order to remove noise completely and extract only the signal (orthogonalization). In general, the seismic drill-bit data are affected by both random and coherent noise.

Optimum vertical stacking to suppress noise of common-level correlated data requires evaluation of the signal and noise amplitudes. In particular, a separate identification of random and coherent noise is necessary to calculate the correct weights for the individual records.

7.3.1 Vertical-stack model

Optimum-weighted stacking to attenuate noise of common-level drill-bit data is applied to crosscorrelated traces of the same receiver in the level gather. A theory for weighting seismic records in the stacking process of traces signals (with similar shapes after dynamic and static corrections) and statistically independent noise was developed by Robinson (1970). His theoretical analysis shows how optimum frequency-independent weighting factors are related to S/N. Optimum weighted stack of offset marine data is treated by Cassano and Rocca (1973) and Schoenberger (1996) for suppression of multiple reflections. Because of moveout variation with offset in their data, the weighted stack acts as a frequency filter. The solution is achieved in frequency by a Wiener filter calculated using a Lagrange multiplier, which minimizes coherent multiple and random noise.

Conversely, in the following model for the optimum common-level stack of correlated drill-bit data, stationarity conditions and no frequency-filtering effects due to moveout variation are assumed for stacking of signal and noise. We only assume that signal and noise amplitudes may vary in the records. In other words, we use the approximation that in the data of the same level the transfer-function filters related to signal generation and propagation do not vary, but only the amplitude in their input excitation varies. Let

$$w_i = \Phi_{\rm pg}^{(i)}, \qquad i = 1, \ldots, N_R, \tag{7.4}$$

be the crosscorrelated trace of a fixed receiver repeated $N_{\rm R}$ times for the same depth level. The weighted stack is

$$w = \sum_{i=1}^{N_{\rm R}} p_i w_i, \tag{7.5}$$

where p_i are weights subject, without loss of generality, to the constraint

$$\sum_{i=1}^{N_{\rm R}} p_i = 1. \tag{7.6}$$

Let x, y_i and v_h ($h = 1, \ldots, H$) represent the signal, random and coherent-noise components in the i-th correlated trace, respectively. Each coherent noise component is the same in each trace of the common-receiver gather, except for an amplitude change. The crosscorrelated trace model is

$$w_i = \alpha_i x + \beta_i y_i + \sum_{h=1}^{H} \gamma_{ih} v_h, \tag{7.7}$$

where α_i, β_i and γ_{ih} are constants. In addition, we assume that x, v_h and y_i have unit energy, and that – in a first approximation – the drill-bit seismic signal and noise (like the coherent noise radiated from the rig) are orthogonal (have negligible crosscorrelation) because of their different travelpaths and waveforms after correlation. So we have

$$\begin{aligned} &\mathrm{E}[x^2] = 1, \\ &\mathrm{E}[v_h v_l] = \delta_{hl}, \\ &\mathrm{E}[y_i y_j] = \delta_{ij}, \\ &\mathrm{E}[x v_h] = \mathrm{E}[x y_i] = \mathrm{E}[v_h y_i] = 0 \end{aligned} \tag{7.8}$$

where δ_{ij} is the Kronecker delta. The symbol E[.] is the sample-mean operator (or expectation value, Papoulis, 1991). That is, $E[x] = \int p(x) x \, dx$ where $p(x)$ is the probability distribution of x. With sampled data $x(n)$ with $n = 1, 2, \cdots, N$, we can approximate the expectation value by the time average $E[x] = \sum x(n)/N$. The total energy and the signal-to-noise ratio of thc i-th trace are

$$\sigma_i^2 = \mathrm{E}[w_i^2] = \alpha_i^2 + \beta_i^2 + \sum_{h=1}^{H} \gamma_{ih}^2, \tag{7.9}$$

and

$$R_i^2 = \frac{\mathrm{E}[s_i^2]}{\mathrm{E}[n_i^2]} = \frac{\alpha_i^2}{\beta_i^2 + \sum_{h=1}^{H} \gamma_{ih}^2} \tag{7.10}$$

We express the sum of equation (7.5) as

$$w = s + n_{\mathrm{ran}} + n_{\mathrm{coe}}, \tag{7.11}$$

where, using (7.5) and (7.7),

$$\begin{aligned} &s = x \textstyle\sum_{i=1}^{N_{\mathrm{R}}} p_i \alpha_i, \\ &n_{\mathrm{ran}} = \textstyle\sum_{i=1}^{N_{\mathrm{R}}} p_i \beta_i y_i, \\ &n_{\mathrm{coe}} = \textstyle\sum_{h=1}^{H} v_h \sum_{i=1}^{N_{\mathrm{R}}} p_i \gamma_{ih}, \end{aligned} \tag{7.12}$$

with $i = 1, \ldots, N_{\mathrm{R}}$ and $h = 1, \ldots, H$. Using equation (7.12) and the condition of equation (7.8), the signal-to-noise ratio of the weighted sum becomes

$$R_w^2 = \frac{\mathrm{E}[s^2]}{\mathrm{E}[n^2]} = \frac{\left(\sum_{i=1}^{N_{\mathrm{R}}} p_i \alpha_i\right)^2}{\sum_{i=1}^{N_{\mathrm{R}}} p_i^2 \beta_i^2 + \sum_{h=1}^{H} \left(\sum_{i=1}^{N_{\mathrm{R}}} p_i \gamma_{ih}\right)^2}. \tag{7.13}$$

The problem is to determine the optimum weights p_i that maximize the signal-to-noise ratio (7.13) of the summed trace.

7.3.2 Optimum weights (general case)

In the general case, the coefficients of random and coherent noise are $\beta_i \neq 0$ and $\gamma_{ih} \neq 0$. The noise energy is different from zero for any choice of the weights p_i not identically null. The maximum value of the signal-to-noise ratio (7.13) of the summed trace is obtained by the N_R conditions

$$\frac{\partial R_w^2}{\partial p_i} = 0. \tag{7.14}$$

We can calculate the solution of equation (7.14) minimizing the error energy

$$\frac{\partial \mathrm{E}[e^2]}{\partial p_i} = 0, \tag{7.15}$$

where

$$e^2 = (x - w)^2. \tag{7.16}$$

The solution for the optimum weights is

$$p_i = \frac{\alpha_i - \alpha_i m_\alpha - \sum_{h=1}^{H} \gamma_{ih} m_{\gamma h}}{\beta_i^2}. \tag{7.17}$$

where m_α and $m_{\gamma h}$ are functions of α_i, β_i and γ_{ih} (Poletto, 2000b). The terms m_α and $m_{\gamma h}$ are the weighted sums of the coefficients of the signal and coherent noise,

$$\begin{cases} m_\alpha = \sum_{k=1}^{N_\mathrm{R}} p_k \alpha_k \\ m_{\gamma h} = \sum_{k=1}^{N_\mathrm{R}} p_k \gamma_{kh} \end{cases} \tag{7.18}$$

and contain the unknown weights. The coefficients m_α and $m_{\gamma h}$ can be calculated by solving the $H + 1$ equations (Poletto, 2000b).

$$\begin{cases} (1 + R_{\alpha\alpha}) m_\alpha + \sum_{k=1}^{H} R_{\alpha\gamma_k} m_{\gamma k} = R_{\alpha\alpha} \\ R_{\alpha\gamma_h} m_\alpha + \sum_{k=1}^{H} (\delta_{hk} + R_{\gamma\gamma_{hk}}) m_{\gamma k} = R_{\alpha\gamma_h} \end{cases} \tag{7.19}$$

where

$$\begin{cases} R_{\alpha\alpha} = \sum_{i=1}^{N_\mathrm{R}} \frac{\alpha_i^2}{\beta_i^2} \\ R_{\gamma\gamma_{hk}} = \sum_{i=1}^{N_\mathrm{R}} \frac{\gamma_{ih}\gamma_{ik}}{\beta_i^2} \\ R_{\alpha\gamma_h} = \sum_{i=1}^{N_\mathrm{R}} \frac{\alpha_i \gamma_{ih}}{\beta_i^2} \end{cases} \tag{7.20}$$

Solving equation (7.19) we obtain $(H + 1)$ values of m_α and $m_{\gamma h}$ as function of the terms in equation (7.20), if the amplitudes of signal and noise components are known. Finally, using m_α and $m_{\gamma h}$ in equation (7.17), we obtain the optimum weights. In summary, once α_i, β_i and γ_{ih} are known, that is, measured in suitable windows, equation (7.17) allows us to compute the maximum signal-to-noise ratio of equation (7.13) for the optimum summed trace of equation (7.5). Optimum weights of equation (7.17) are frequency independent.

7.3.3 Special cases of level stack

We analyze the following special cases of level stack.

Only one coherent noise – We assume the limit case of one single coherent noise v (the index h is not necessary) and $\beta_i = 0$. We exclude the trivial case of traces w_i all proportional, with $\alpha_i \propto \gamma_i$. The weights of the optimum sum are obtained from the following "orthogonality" condition, which makes the weighted sum of the organized noise zero,

$$\sum_{i=1}^{N_R} p_i \gamma_i = 0. \tag{7.21}$$

Equation (7.21) means that – ideally – in the absence of random noise, we could obtain the total separation of signal and noise if we know the noise vector γ_i. We simply take a vector of weights p_i orthogonal to the vector of components γ_i to separate x from v. In this case, the number of measurements w_i is not critical, because, for example, two different measurements (if w_1 is *not* proportional to w_2) are sufficient to separate the signal x from a single noise v.

Only random noise – In this case $\beta \neq 0$ and $\gamma_{ih} = 0$. From equation (7.17) we obtain

$$p_i \propto \frac{\alpha_i}{\beta_i^2}, \tag{7.22}$$

and, neglecting the overall gain term $(1 - m_\alpha)$, the optimum sum is given by

$$w_{\text{opt}} = \sum_{i=1}^{N_R} \frac{\alpha_i}{\beta_i^2}(\alpha_i x + \beta_i y_i). \tag{7.23}$$

Because $R_i^2 = (\alpha_i/\beta_i)^2$, equation (7.23) becomes

$$w_{\text{opt}} = \sum_{i=1}^{N_R} R_i(R_i x + y_i). \tag{7.24}$$

A noticeable case occurs when the signal-to-noise ratio is constant, $R_i = R$. From equation (7.24) we obtain, for $R_i = R$, the well-known rule that the signal-to-noise ratio of the sum, R_w, increases as the square root of the number of stacked traces. In this case the number of summed traces is important.

Signal-to-noise ratio of the simple sum – If we set $p_i = 1$ for every $i = 1, \cdots, N_R$, from equation (7.13) we obtain the signal-to-noise ratio of the simple sum. This can be compared to the signal-to-noise ratio of the optimum stack.

Analysis of signal-to-random-noise ratio requires the separate estimation of the coefficient of the random noise β_i, which gives only a part of the noise. When coherent noise is also included, as calculated by using equation (7.17), the stack is performed with positive and negative coefficients. Because the signal waveform can be subject to fluctuations, this step requires caution, as it may lead to instable waveform after stacking of opposite-sign wavelets. The analysis of the separated records (before sum) aligned at the same arrival time ("diagnostic" of the common-level correlated wavelet before stack) is performed to test the waveform before and after optimum stack. Real-data diagnostic S/N and correlation results obtained with simple and weighted stack and filtering are shown in Figures 7.1 and 7.2, respectively. The efficacy of weighted stack is appreciable.

7.4 Selective stack by drilling parameters

A method proposed by Miranda, Abramo, Poletto and Comelli (2000) improves the signal-to-noise ratio in the processing of drill-bit vertical seismic profiles (VSP) on the basis of the almost simultaneous collection of:

◇ seismic and pilot signals (A), thus obtaining seismic traces in which the signal is disturbed by noise;

◇ drilling parameters (B) relating to the signals (A) (see Section 5.9).

These signals, obtained with repeated measurements for the same depth range, are sorted and analyzed according to one or more drilling parameters or their relative combinations (see also, Section 3.6.1). Then they are partially or totally summed up by weight, on the basis of one or more drilling parameters (B) or their relative combinations, thus obtaining an improvement in the signal-to-noise ratio and a more distinct separation of signal and noise.

The parameters (B), which describe the instantaneous drilling state during the measurements (A), are collected at regular intervals during drilling itself, for example every second. They can also be sub-sampled to supply average information in the time interval relating to the single registration of the group of seismic recordings (A) collected for the same depth level. Parameters used in the process of selective processing are typically those which describe the dynamic drilling conditions, for example torque, weight on bit (WOB), rate of penetration (ROP), rotation speed (RPM) and mud flow (Section 5.9). Combinations of the above parameters can be used to calculate dimensionless parameters of drilling rotary equations (Section 3.6.3). The criteria with which the most representative signals can be identified are based on the cross-analysis of the reordered (sorted) data against the drilling parameters. In this way, it is possible to more effectively analyze the correspondences between the drilling parameters and the signal and noise levels in the seismic data obtained with the bit in the various operating contexts. From these correspondences, the criteria for improving the sum of the data collected for the same drilling level are determined, for example by summing up the correlations of different groups of ten of 24 s recordings obtained in a penetration interval of the bit of a few meters. The process of "selective stacking" comprises the following steps:

1. Picking the arrival time of the direct wave in the correlated and deconvolved traces of one or more channels of the seismic line, or using a representative channel focused by moveout correction from various channels of the seismic line.

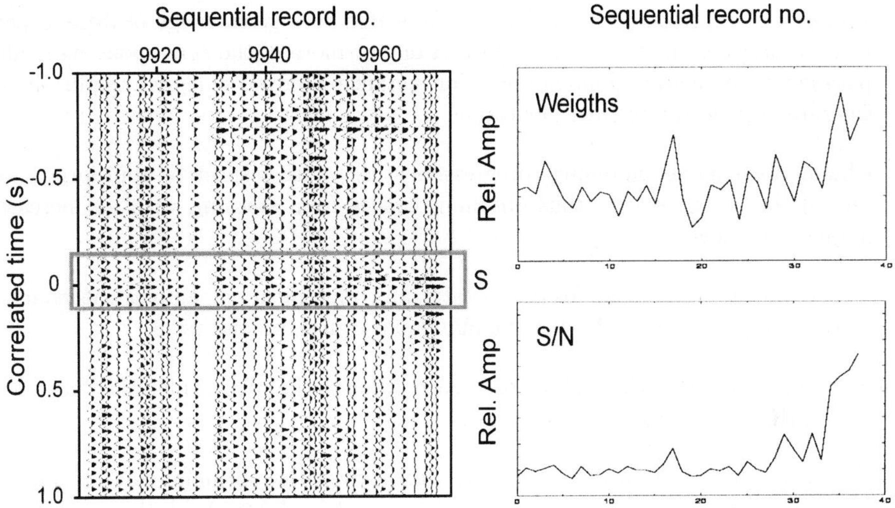

Figure 7.1: Correlated and deconvolved geophone traces are used as (left) a diagnostic before stack, with signal and noise windows and (right bottom) S/N ratio measured on each singular record; (right top) Calculated weights.

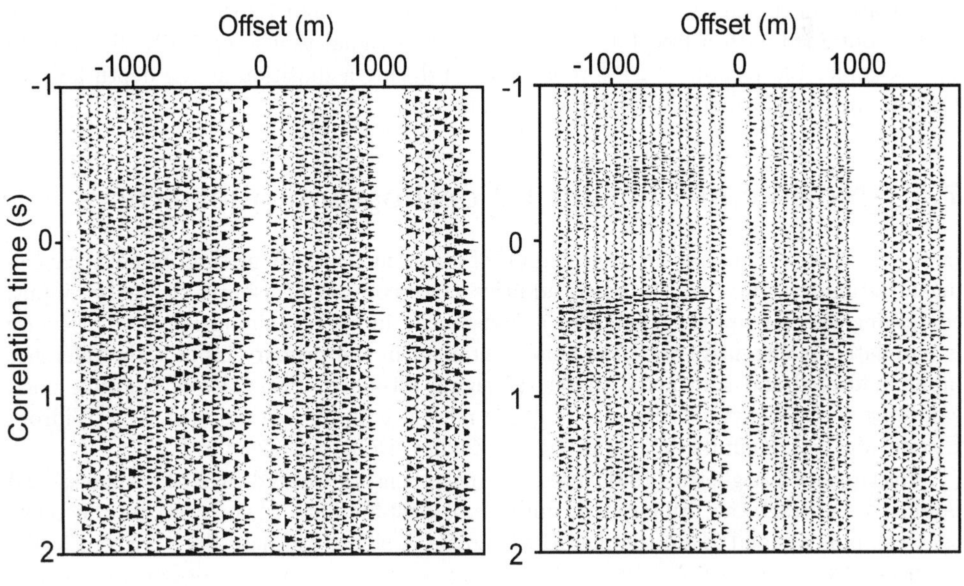

Figure 7.2: Correlated and deconvolved records (left) before and (right) after optimum stack and signal filtering. The direct arrival is observable at about 0.5 s.

2. Sorting and representation of the traces, related to a given range of drilling parameters, in accordance with the values of one or more of the corresponding drilling parameters or relative combinations: for instance, analysis of signal amplitude versus rate of penetration (ROP) or torque.

3. Channel-by-channel summing (progressive stack) with respect to appropriate subsets of the sorted traces, thus obtaining traces from stacking with an increasing number of traces.

4. Measurement of the signal-to-noise ratio and identification, among the analyzed subsets, of the subset(s) having the maximum signal-to-noise ratio.

5. Summing channel by channel the traces according to the subsets identified in step #4 above, thus obtaining the optimum sum.

6. Following identification of the subset, which gives the optimum-stack result, the complementary stack is calculated to obtain the "noise" wavefield; this noise can be used to subtract the residual noise in the signal wavefield.

The selective stack has the following advantages. It can be applied using a single pilot channel with a single seismic trace. In addition, it uses only the part of the data having higher signal-to-noise ratio, which is selected and summed in the final data. This process is advantageous with data in which the direct identification of the signal-to-noise ratio is not possible, but where it can be associated with drilling conditions, like ROP, torque, WOB, RPM and their combinations. This allows the final data to be improved with a consequent optimization of the geophysical monitoring. In Figures 7.3 and 7.4 the method is applied in data obtained with a PDC bit, which generates a signal that is typically difficult. In this example, 100 records of 24 s were obtained during acquisition at the depth level at 1380 m, during which the bit penetrated about 1 m.

7.5 Noise cancellation by orthogonal pilot traces

In this section, we discuss the cancellation of coherent noise. A method of noise canceling to estimate signals corrupted by additive noise consists of using a "primary" input, containing the corrupted signal, and a "reference" input containing noise. This noise is correlated (coherently) with the noise contained in the primary input. The reference input is filtered and subtracted to get the signal estimate. Principles of adaptive noise canceling with Wiener solutions for multiple reference inputs were discussed by Widrow, Glover, McCool, Kaunitz, Williams, Hearn, Zeidler, Dong and Goodlin (1975).

Poletto and Craglietto (1991) used reference sensors located near the engines and pumps to subtract the drilling-yard noise from drill-bit data. Let $N_j(\omega)$ be the reference components and $X(\omega)$ the disturbed signal. A signal estimate $\mathbf{S}(\omega)$ is given in the frequency by

$$\mathbf{S} = X - (F_1, \ \dots \ , F_N)(Y_1, \ \dots \ , Y_N)^T, \tag{7.25}$$

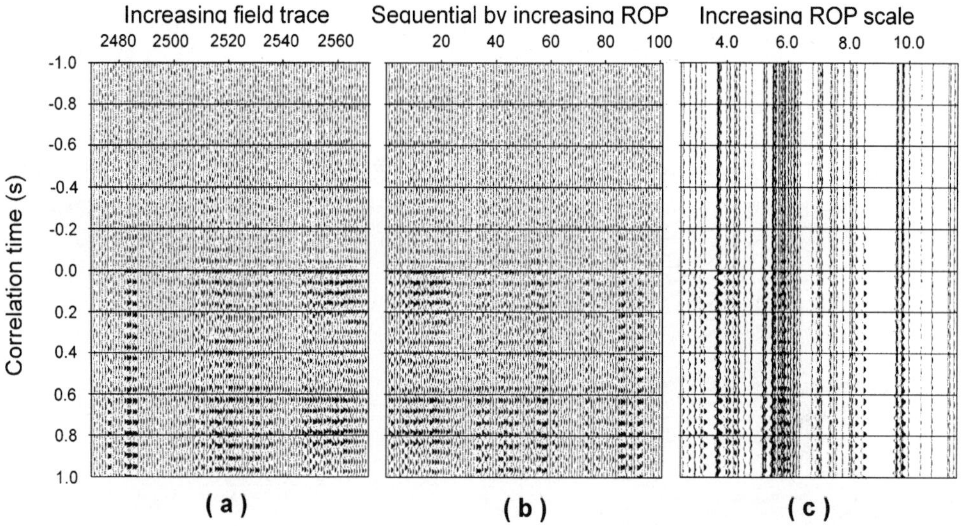

Figure 7.3: a) Horizontal stack after moveout correction gives the focused and aligned traces of correlated and deconvolved records provided by 100 measurements sequentially repeated for the same depth (1380 m). b) and c) The same 100 focused traces are re-arranged with increasing ROP [m/h] drilling parameter.

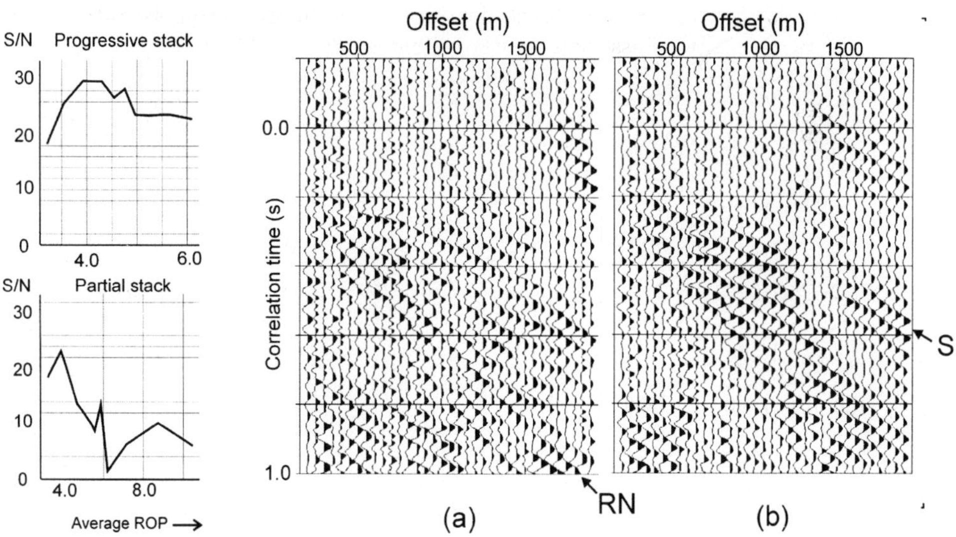

Figure 7.4: The selected stack at S/N equal 30 is obtained by adding the traces up to ROP = 4.2 m/h. This result is applied to the original deconvolved, correlated traces. The selected stack provides a clearer identification of the signal. In panels (a) and (b) the seismic traces of the level 1380 m are represented with respect to the well distance (offset). Panel (a) is the result of the sum of records with ROP higher than 4.2 m/h. Here the rig noise (RN) event is stronger. Figure (b) shows the result of the sum with ROP lower than 4.2 m/h, in which the signal (S) is separated from rig-noise events.

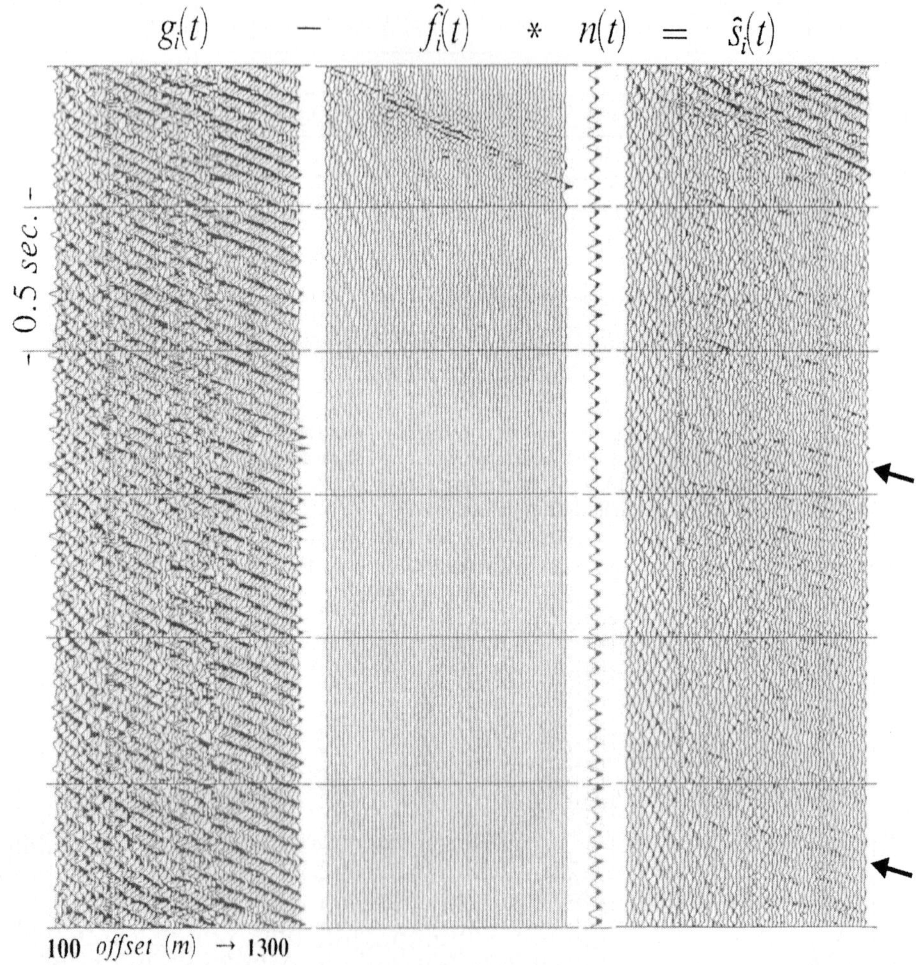

Figure 7.5: Data before correlation. (g) Disturbed geophone signal with noise of linear moveout in offset, (n) reference yard noise, (f) filters and (s) estimated drill-bit signal after noise subtraction in drill-bit geophone traces before pilot crosscorrelation. After noise canceling, the hyperbolic signal moveout, different from the moveout of noise, can be appreciated (arrows) (after Poletto and Craglietto, 1991).

where the superscript '^{T}'denotes transpose and

$$(Y_1, \ \dots \ , Y_N)^T = \mathbf{S}_N (N_1, \ \dots \ , N_N)^T, \tag{7.26}$$

is the orthonormalized noise vector, i.e., of unit covariance. In other words, each noise component Y_j is not correlated with (it is orthogonal to) the others. $\mathbf{S}_N$ is a square root of the inverse of the noise covariance matrix $R_{N(ij)} = N_i N_j^*$, such that

$$\mathbf{R}_N^{-1} = \mathbf{S}_N \mathbf{S}_N^\dagger, \tag{7.27}$$

where † denotes transpose conjugate. The filters $F_j(\omega)$ calculate the part of orthogonalized correlated noise to be subtracted by

$$F_j = \frac{Y_j^* X}{Y_j^* Y_j}. \tag{7.28}$$

The hypothesis of measuring reference recordings dominated by noise is required for this method. If signal is contained in reference noise traces, distortions may occur in the output. Results of noise canceling in geophone data are shown in Figures 7.5, where the reference-noise trace is estimated by stacking the geophone recordings corrected with the noise moveout. Other applications are shown by Rector (1992), who subtracts, after orthogonalization, the noise measured by the horizontal-pilot accelerometers from the signal of the vertical pilot accelerometer at the top of the drill string. This noise is polarized and is due to the vibrations perpendicular to the plane containing the bail in a rotary-kelly system (Section 4.9).

7.6 Noise separation by independent pilot traces

Statistical independence was used by Poletto, Rocca and Bertelli (2000) to separate signal and environmental noise in drill-bit reverse-VSP data. This approach is based on measurements of signal and noise with two or more reference (pilot) channels, in which signal and noise have different amplitudes. This method uses statistical independence to determine the combination coefficients, and makes it possible to estimate the optimum pilot signal from a set of reference traces that does not contain drill-string measurements (Angeleri, Persoglia, Poletto and Rocca, 1996).

The concept of statistical independence of stochastic processes is the following. Let x and y be two stochastic variables and $p(x, y)$ their joint associated probability density. The processes are defined as statistically independent when

$$p(x, y) = p_x(x) p_y(y), \tag{7.29}$$

where $p_x(x)$ and $p_y(y)$ are the probability densities of x and y, respectively. Let $f(x)$ and $g(y)$ be two functions of the stochastic variables x and y. The expectation value of $f(x)g(y)$ is defined as

$$\mathrm{E}\left[f(x)g(y)\right] = \int f(x)g(y)p(x, y)\, dx dy. \tag{7.30}$$

If the stochastic variables x and y are statistically independent, we have (Papoulis, 1991) from equation (7.29)

$$\int f(x)g(y)p(x,y)\,dxdy = \int f(x)p_x(x)\,dx \int g(y)p_y(y)\,dy, \tag{7.31}$$

therefore

$$\mathrm{E}\left[f(x)g(y)\right] = \mathrm{E}\left[f(x)\right]\mathrm{E}\left[g(y)\right]. \tag{7.32}$$

In fact, for independent variables, the expectation values of $f(x)$ and $g(y)$ are defined as

$$\mathrm{E}\left[f(x)\right] = \int f(x)p_x(x)\,dx, \tag{7.33}$$

$$\mathrm{E}\left[g(y)\right] = \int g(y)p_y(y)\,dy, \tag{7.34}$$

respectively. Equation (7.32) holds, for instance, for the moments of order n and m of x and y, respectively, when x and y are independent, i.e.,

$$\mathrm{E}\left[x^n y^m\right] = \mathrm{E}\left[x^n\right]\mathrm{E}\left[y^m\right]. \tag{7.35}$$

In particular, when $n = m = 1$,

$$\mathrm{E}\left[xy\right] = \mathrm{E}\left[x\right]\mathrm{E}\left[y\right]. \tag{7.36}$$

$\mathrm{E}\left[xy\right]$ is the statistical form used to express crosscorrelation, and equation (7.36) may be the definition of *incorrelation* (the property of not being correlated or being linear independent). In general, we deal with zero-mean processes, for which $\mathrm{E}[x] = \mathrm{E}[y] = 0$, so that incorrelation of signals x and y becomes "orthogonality" (of vectors x and y), namely,

$$\mathrm{E}\left[xy\right] = 0. \tag{7.37}$$

Incorrelation of x and y can be verified by equation (7.36) or (7.37) for zero-mean processes. Independence can be verified by equations (7.32) or (7.35) with arbitrary n and m. If these equations hold for arbitrary functions or powers of the stochastic variables, statistical independence (say, independence) of the variables holds too. It is important to understand that incorrelation (say, linear independence) is a necessary consequence of (statistical) independence. But incorrelation does not necessarily imply independence. Therefore, equation (7.32) (or (7.35)) is not necessarily a consequence of equation (7.36).

Gaussian and non-Gaussian processes

Assume a zero-mean statistic variable. For normal Gaussian variables, i.e., with zero mean, incorrelation implies independence. The higher order moments of a random variable x with Gaussian normal distribution

$$N(x) = \frac{1}{\sigma_x\sqrt{2\pi}} e^{-\frac{x^2}{2\sigma_x^2}}, \tag{7.38}$$

are unambiguously defined by its standard deviations σ_x given by

$$\sigma_x^2 = \mathrm{E}\left[x^2\right]. \tag{7.39}$$

We have, with k arbitrary positive integer,

$$\mathrm{E}\left[x^n\right] = \begin{cases} 0 & n = 2k+1 \\ 1 \cdot 3 \cdots (n-1)\sigma_x^n & n = 2k \end{cases}. \tag{7.40}$$

In other words, with variables of normal distribution with zero mean, the products of the higher moments of equation (7.35) are unambiguously defined by the energy of the traces. Contrarily, if the processes are not all Gaussian, (statistical) independence conditions may give additional information after incorrelation (linear independence) condition. This information can be successfully used for "blind" separation of signals. Figure 7.6 is a graphical representation of incorrelation and independence.

7.6.1 Uncorrelated drill-bit signal and RVSP's

The implementation of the independence technique is to record the bit signal and noise with pilot sensors located at the surface, say, on the drill string or on the drilling tower and yard. These recorded signals should then be used as a replica of the source, to be crosscorrelated with the geophone outputs, to recover the RVSP. Naturally, we could also combine the techniques of using both the near-the-bit memory recordings and the surface sensors so that the intermediate results, albeit imperfectly, are available in real time.

Many types of pilot-sensors can be used: accelerometers or strain gages on the drill-string, mud pressure sensors, torque measurements like the rotary-motor current, etc. As we will see later, the independent-signal estimation will be an optimal combination of pilot signals taken at different points on the drill tower, on the drill string or in the yard, some of them measuring the signal and others measuring the disturbances to be subtracted from the pilot signal (Angeleri, Persoglia, Poletto and Rocca, 1996). A signal related to the bit signal can be recorded at the surface, but several other disturbances will be added to it: noise from the mud pumps, the engines, and in general the activity in the fields surrounding the well. The techniques for achieving an optimum separation of these "noise" signals exploits the multiplicity of recording points that we can use. At any frequency, the different signals will have different delays and amplitudes at different locations.

7.6.2 Incorrelation and independence

Poletto, Rocca and Bertelli (2000) use a combination technique based on (non-linear) statistical analysis. This technique was first proposed by Jutten and Herault (1988), Comon (1989) and Cardoso (1990) to carry out the decomposition by exploiting the non-Gaussianity of the signals involved.

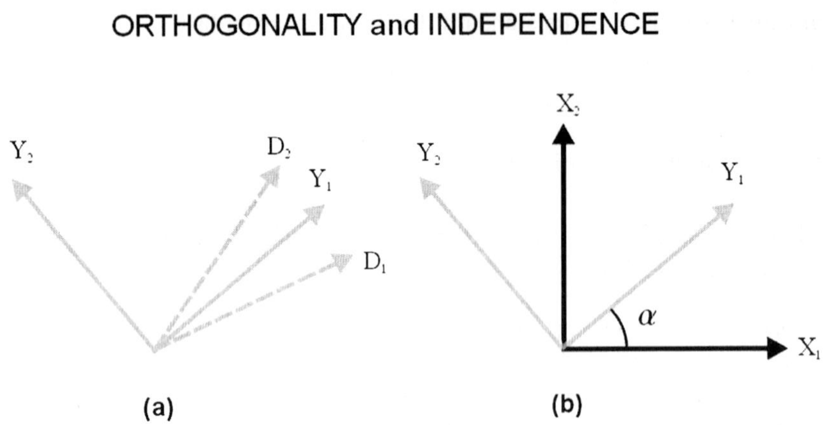

Figure 7.6: a) The measured data D_1 and D_2 are zero mean and not orthogonal in the space of the independent-signal vectors X_1 and X_2. Both have non-null components in these directions, i.e., they are a mixture of the independent signals X_1 and X_2. Y_1 and Y_2 are obtained by one of the infinite possible expressions of D_1 and D_2 in terms of orthogonal vectors: for example, using the orthonormal vectors $Y_1 = (D_2 + D_1)/|D_2 + D_1|$ and $Y_2 = (D_2 - D_1)/|D_2 - D_1|$. Y_1 and Y_2 are orthogonal, i.e., zero mean and uncorrelated, but do not give the independent solution X_1 and X_2 (b). Assuming that X_1 and X_2 are not Gaussian, the unknown orthogonal-rotation-angle α can be determined by "blind" analysis of Y_1 and Y_2, for instance, by equation (7.35). Note that rotations $\alpha + k\pi/2$, k integer, are also solutions (i.e., we have two solutions with two polarities each).

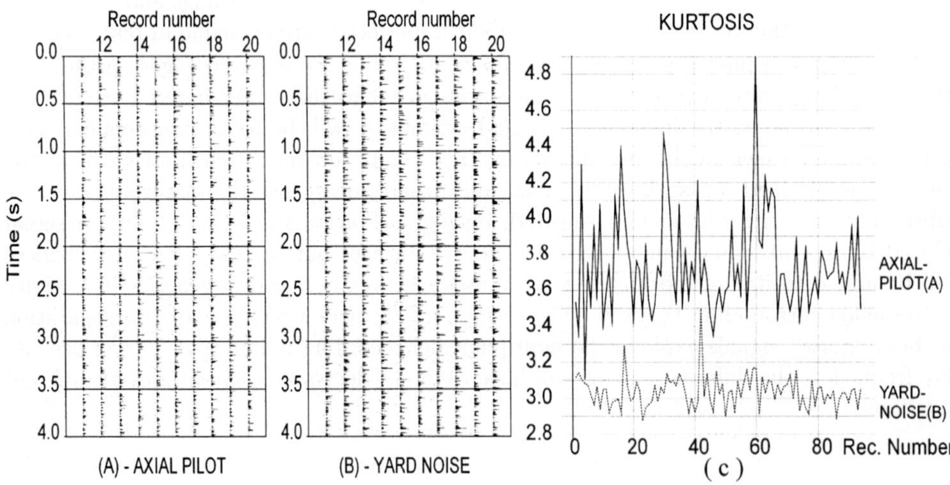

Figure 7.7: Example of axial pilot traces measured at the top of the drill string (a), and of yard-noise traces (b). The traces have been whitened. The source-level is 3560 m. c) Kurtosis of the whitened axial-traces (Figure 7.7a) and yard-noise traces (Figure 7.7b), computed using a 18 s analysis window length. The figure shows the values computed for 94 records all belonging to the same source-depth level (3560 m).

General approach

Let us consider an N-dimensional random vector $\vec{\mathbf{x}}$; its N components are assumed to be independent random processes with the same variance. This does not entail any loss of generality since the observation vector, that has the same dimension N, is

$$\vec{\mathbf{d}} = \mathbf{A}\,\vec{\mathbf{x}}. \tag{7.41}$$

The matrix $\mathbf{A}$ is totally unknown, but not random. We suppose we can measure the covariance matrix of $\vec{\mathbf{d}}$. Let it be

$$\mathbf{R} = E[\vec{\mathbf{d}}\vec{\mathbf{d}}^{\dagger}] = \mathbf{A}\,E[\vec{\mathbf{x}}\vec{\mathbf{x}}^{\dagger}]\,\mathbf{A}^{\dagger} = \mathbf{A}\;\mathbf{A}^{\dagger}, \tag{7.42}$$

where $\dagger$ denotes transpose conjugate, and the vectors' juxtaposition denotes row-by-column product. We wish now to estimate $\vec{\mathbf{x}}$ from $\vec{\mathbf{d}}$, and let the unknown estimation matrix be $\mathbf{P}$,

$$\vec{\hat{\mathbf{x}}} = \mathbf{P}\,\vec{\mathbf{d}}. \tag{7.43}$$

A necessary but insufficient condition for $\vec{\hat{\mathbf{x}}}$ is that its covariance matrix is orthogonal like that of $\vec{\mathbf{x}}$, i.e., its components are incorrelated. Let us call $\vec{\hat{\mathbf{x}}}_0$ the vector obtained as

$$\vec{\hat{\mathbf{x}}}_0 = \mathbf{S}\,\vec{\mathbf{d}}, \tag{7.44}$$

where $\mathbf{S}$ is the square root of the matrix $\mathbf{R}^{-1}$, i.e., has its same eigenvectors and square rooted eigenvalues,

$$\mathbf{S}\;\mathbf{S}^{\dagger} = \mathbf{R}^{-1} \tag{7.45}$$

Then we have

$$E[\vec{\hat{\mathbf{x}}}_0\vec{\hat{\mathbf{x}}}_0^{\dagger}] = \mathbf{S}\;\mathbf{A}\;\;E[\vec{\mathbf{x}}\vec{\mathbf{x}}^{\dagger}]\,\mathbf{A}^{\dagger}\,\mathbf{S}^{\dagger} = \mathbf{I}. \tag{7.46}$$

Now, to get the best linear estimate $\vec{\hat{\mathbf{x}}}$ of $\vec{\mathbf{x}}$ from $\vec{\hat{\mathbf{x}}}_0$ we still have to multiply by an unknown matrix $\mathbf{Q}$, that is, however, orthogonal (unitary). Let

$$\begin{aligned} \vec{\hat{\mathbf{x}}} &= \mathbf{Q}\,\vec{\hat{\mathbf{x}}}_0, \\ E[\vec{\hat{\mathbf{x}}}\vec{\hat{\mathbf{x}}}^{\dagger}] &= \mathbf{Q}\,E[\vec{\hat{\mathbf{x}}}_0\vec{\hat{\mathbf{x}}}_0^{\dagger}]\,\mathbf{Q}^{\dagger} = \mathbf{Q}\;\mathbf{Q}^{\dagger} = \mathbf{I}. \end{aligned} \tag{7.47}$$

Herein, $\mathbf{Q}$ is the inverse of its transpose and therefore is an orthogonal (unitary) matrix, characterized by $N(N-1)/2$ degrees of freedom. In other words, any choice for the orthogonal matrix $\mathbf{Q}$ will yield a vector $\vec{\hat{\mathbf{x}}}$ with identity covariance matrix. Incorrelation – that is, linear independence or "orthogonality" – alone is not able to give us the inverse of the matrix $\mathbf{A}$ that we need. Therefore, we have to use the (statistical) independence hypothesis to determine $\mathbf{Q}$ and thus the optimum-combination matrix $\mathbf{P} = \mathbf{Q}\;\mathbf{S}$.

To do so, it is enough to ascertain that the average of the products of higher order moments is also null, thus obtaining the $N(N-1)/2$ remaining unknowns (the $N(N-1)/2$ angles that characterize the unknown rotation of the coordinate set in the N dimensional space). Poletto, Rocca and Bertelli (2000) discussed the nature of the vector $\vec{\mathbf{x}}$; it cannot be a single frequency component of a vector time series, because then it would be Gaussian due to the central limit theorem (Papoulis, 1991). To analyze higher order moments, they use a vector time series. For the method to work, the unitary matrix $\mathbf{Q}$ should be either frequency independent, or slowly changing with frequency. In that case, we can combine several frequency components to have a bandpass-vector time series and then estimate its higher order moments. Thus, we could establish bandpass-vector independence separately from incorrelation. In other words, long time series and stationary conditions – we mean stationarity in strict sense, i.e., the stochastic process has time-invariant statistical properties (Papoulis, 1991) – are needed. To obtain independence in practical seismic drill-bit applications, traces of tens of seconds are successfully used.

It is important to stress that the solution cannot be unique, since all permutations of the indexes of the vector $\vec{\mathbf{x}}$ correspond to viable solutions; for $N = 2$, we have two solutions; six for $N = 3$, etc. This method is non linear, and we can expect severe numerical difficulties, especially when random Gaussian noise is added to the data as well. It is true that we could choose moments that are systematically zero for Gaussian variates; however, the problem is difficult to solve in a case of some generality. In general, a reasonable procedure might be navigating the $N(N-1)/2$ dimensional space looking for minima of some figure of merit that quantifies the independence of the variables identified.

7.6.3 Statistical independence of drill-bit data

The necessary assumption to recover the drill-bit signal using the statistical-independence method is the non-Gaussianity of signal and noise components. Transmission effects, through complicated transfer functions, strongly affect the bit signal. Nevertheless, reference (pilot) downhole and surface measurements show that the characteristics of the signal from the working drill-bit are different from those of the surrounding noise sources. The kurtosis K, the fourth order moment normalized by the square energy, can be used for the description of the non-Gaussianity, namely,

$$K = \frac{E[x^4]}{E[x^2]^2}. \tag{7.48}$$

The kurtosis of axial acceleration traces is far from the Gaussian value $K = 3$: it ranges prevalently between 2 and 7 in the case of tricone data. To exploit statistical differences, Figure 7.7a shows a sub-set of whitened axial pilot traces. These were acquired at the top of the drill string, when using a roller-cone bit. Figure 7.7b shows the corresponding noise traces, acquired with a pilot geophone in the drilling yard. These are whitened too. Figure 7.7c is the kurtosis of all 94 traces belonging to the same depth level, and only partially shown in Figure 7.7a and Figure 7.7b.

Below, we show single-record applications.

Independence of real data

The following analysis is performed for $N = 2$. Experimental data were obtained with a roller-cone bit at the depth of 826 m, using continuously repeated acquisition of records of 24 s and 2 ms sampling rate. Two pilot signals were acquired from two sensors, one located on the derrick and the other in the drilling yard. We call $\vec{\mathbf{D}}$ the vector that has

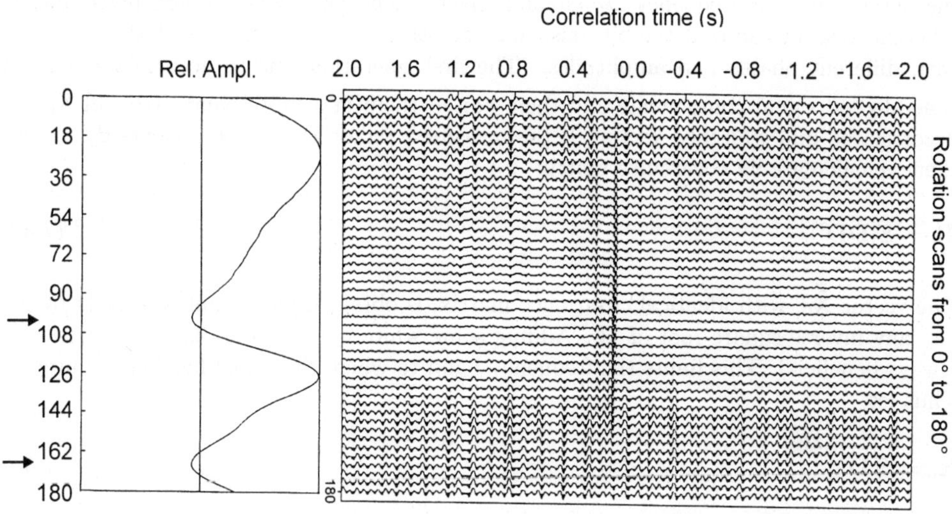

Figure 7.8: Rotation analysis with the trial function of equation (7.52) (left), when combining a derrick-pilot and a drilling-yard-pilot (aligned with the derrick-signal). This result is compared with the correlations of the rotated-pilot with a near-offset geophone trace, which gives a correlation peak at about 0.1 s. The correlations are plotted versus the rotation angle (right).

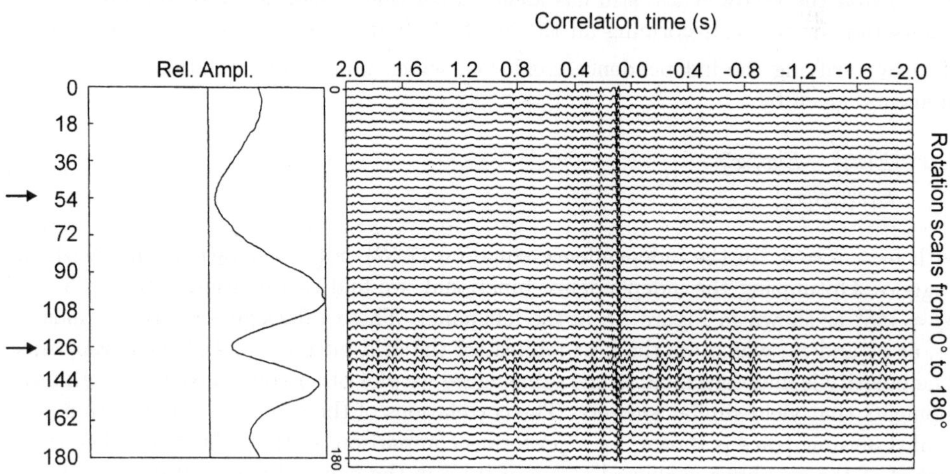

Figure 7.9: Rotation analysis with the trial function of equation (7.52) (left), when combining a derrick-pilot and a mud-pressure-pilot (aligned with the derrick-signal). This result is compared with the correlations of the rotated-pilot with a near-offset geophone trace, which gives a correlation peak at about 0.1 s. The correlations are plotted versus the rotation angle (right). Modified after Poletto, Rocca and Bertelli (2000).

the observations at frequency ω_0 as components. The pilot sensors were positioned in different locations so that the bit noise and the other disturbances arrived at the sensors with different phases and amplitudes. The first operation was to decorrelate the two time-aligned signal components and then whiten them in a finite band. We call $\vec{\hat{\mathbf{X}}}_0$ the decorrelated and whitened vector. If the measured vector $\vec{\mathbf{D}}$ has components d_0 and d_1, then we calculate its covariance matrix

$$\mathbf{R} = E[\vec{\mathbf{D}}\vec{\mathbf{D}}^{\dagger}]. \tag{7.49}$$

The vector $\vec{\hat{\mathbf{X}}}_0$, with identical covariance matrix, is obtained by premultiplying $\vec{\mathbf{D}}$ by $\mathbf{S}$, i.e., any inverse square root of $\mathbf{R}$. Adding some white noise to stabilize the inversion is always beneficial. This limits the frequency band to where the signal has been whitened. Thus let

$$\vec{\hat{\mathbf{X}}}_0 = \mathbf{S}\,\vec{\mathbf{D}}. \tag{7.50}$$

Then any orthogonal combination $\vec{\hat{\mathbf{X}}}$ of the two components of $\vec{\hat{\mathbf{X}}}_0$ at a single frequency ω_0, such as

$$\begin{aligned} x_1 &= \hat{x}_{01}\sin\alpha + \hat{x}_{02}\cos\alpha, \\ x_2 &= \hat{x}_{01}\cos\alpha - \hat{x}_{02}\sin\alpha, \end{aligned} \tag{7.51}$$

is such that the vector $\vec{\mathbf{X}}(\omega_0)$ still has identical covariance matrix. Now, we revert to the time series, inverse transforming all the available frequency components. Then, we get $\vec{\hat{\mathbf{X}}}(t)$ by verifying the independence condition, which implies incorrelation of the higher moments, given by

$$\frac{E[x_1^m]E[x_2^n] - E[x_1^m x_2^n]}{E[x_1^m]E[x_2^n]} = 0, \tag{7.52}$$

with $m = n = 4$. In the independence equation (7.52), $x_1(t)$ and $x_2(t)$ change with the angle α. Optimizing the independence functional, we find two solutions: $\alpha_1 = 105^o$ and $\alpha_2 = 165^o$ as shown in Figure 7.8 in the left panel. These values give the two signals x_1, x_2, or x_2 and x_1, which are the best replicas of the drill-bit noise and of the disturbance. The polarity is undetermined; in the same figure we plot other combinations of high order moments that yield similar answers. The correlations of the combined pilots with a good quality seismic trace is shown, for a-posteriori comparison, in the right panel. The direct arrival at the geophone is aligned at about 0.1 s in the figure. Figure 7.9 shows a similar analysis using the independence-trial-functions of equation (7.52), with a derrick-pilot and a mud-pressure-pilot. Here, the data are shown in the central signal bandwidth, from 20 to 50 Hz. The separation angles are 126 and 54 degrees for independent noise and optimum signal. The correlations of noise and optimum signal with all the geophones of the surface seismic line are shown in Figure 7.10, in the left and right panels respectively. The plots agree substantially and indicate the correctness of the approach.

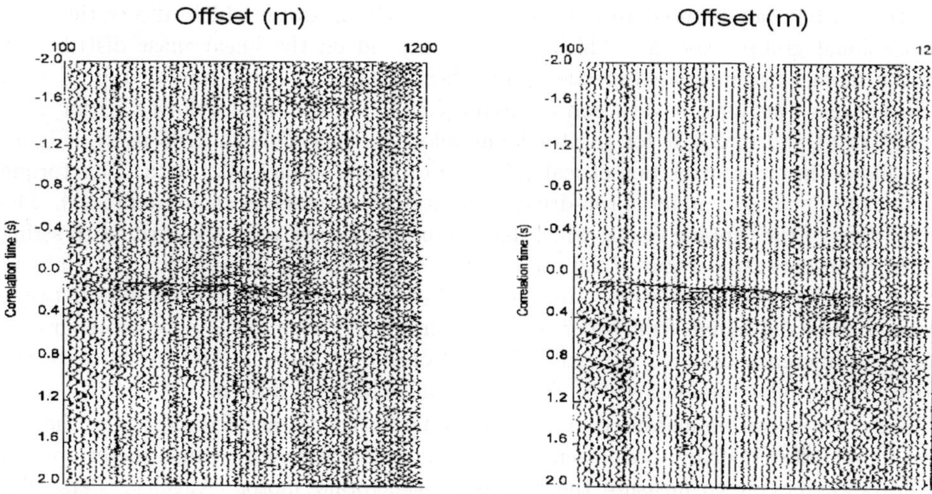

Figure 7.10: Correlations of the traces of the seismic line with the noise (left) and the optimum signal (right), obtained using the rig-pilot and the mud-pressure-pilot.

7.7 Analysis of torsional pilot waves

The usual pilot signal is the axial pilot acceleration in the drill string. However, other components such as torsional drill-string waves are detected by the pilot sensors and may produce coherent events. Torsional and mud waves are slower than the extensional ones. Figure 7.11 shows an example of crosscorrelation of an axial-top-drive pilot with the torque signal obtained measuring the intensity of the electric current of the rotary motor (equation (2.5)), the signal of a geophone of the seismic line, the signal of a horizontal pilot accelerometer on the top drive and the autocorrelation of the axial signal itself. Cross-sensitivity of axial and torsional components can be observed. This interpretation involves the analysis of the torsional pilot waves.

Moreover, detection of axial drill-bit multiples can, in some cases, be difficult because of noise and amplitude decay. Axial information about propagation velocity of elastic waves in the drill string can be complemented by using the torsional components. By crosscorrelating axial and torsional signals from surface pilot components, we measure their relative delay. This gives information about the drill-string mechanical properties and the propagation velocity of waves. Let $\Delta t_{at} = (t_{tor} - t_{ax})$ be the time delay difference between torsional and axial waves. The signals transmitted from the bit through a drill string with length L_S have $t_{ax} = L_S/v_{ax}$ and $t_{tor} = L_S/v_{tor}$. The time difference is obtained from $v_{ax} = (Y\bar{A}_{ax}/\bar{m}_{ax})^{1/2}$ and $v_{tor} = (\mu\bar{I}_{tor}/\bar{m}_{tor})^{1/2}$, to give

$$\frac{\Delta t_{at}}{L_S} = \sqrt{\frac{\bar{m}_{tor}}{\mu\bar{I}_{tor}}} - \sqrt{\frac{\bar{m}_{ax}}{Y\bar{A}_{ax}}}. \tag{7.53}$$

Equation (7.53) can be used to determine the pipe-wear state, which affects the axial and torsional group velocities, which, in turn, depend on the linear mass distribution and the moment of inertia, respectively (see Section 4.4.3). For a pipe of type S with 80 % of wear, we have a variation of about 2.1 % for the axial velocity (from 4727 to 4626 m/s) and of 2.3 % for the torsional velocity (from 2860 to 2795 m/s). Figure 7.12a shows the crosscorrelation of real axial and torsional surface-pilot data. This torque signal is obtained by combining top-drive horizontal accelerometers (equation (5.9)). The arrival time of the dipping event, which has increasing two-way-time according to string length, represents Δt_{at}. Figure 7.12b shows the crosscorrelation of the synthetic axial and torsional surface pilot data, which are calculated to correctly separate the delays between torsional and axial signals from other coherent periodical events in the pilot seismograms. The time difference between the axial and torsional pilot data is used to calculate the velocity of the pilot torsional waves, once the axial velocity is known.

These waves can be measured at the surface by horizontal accelerometers on the top drive (Section 5.4.1) or by torsion sensors, such as an amperometric pinch meter applied to the power circuit to measure the current of the rotary motor (Angeleri, Persoglia, Poletto, and Rocca, 1996).

7.8 SWD with downhole-motor drilling

One of the main trends for SWD has been to enlarge the applicability of the method under difficult operational conditions (Poletto and Savini, 1997). A particular case, is the use of a downhole motor (DHM) at the lower end of the drill string for directional drilling (Section 2.3.7). Directional drilling is done using a downhole mud driven hydraulic motor with the rotating tool oriented in the desired direction (Section 2.5). This requires the absence of surface drill-pipe rotation. Any interruption of the downhole motor rotation is not possible because it requires stopping the mud-flow in the drill string. Interruption of directional deviation is achieved with the tool in the well by "distributing the orientation" of the rotating downhole motor. This effect is obtained by adding pipe rotation from surface, with the result to average the tool-direction angle to zero around the drill string axis, thus drilling straight.

To drive drilling during the use of a DHM for deviation, typically a series of phases with and without surface rotation, which are defined as *rotary* and *sliding* drilling, respectively, is used (Section 2.5.1). It is common SWD experience that the quality of the axial rig pilot signal obtained in the presence of additional surface rotation is "acceptable" and comparable to the quality of axial signals obtained without downhole motor tools. On the other hand, it is also common experience that very poor results are obtained with the conventional use of axial pilot signals during drilling without surface rotation. In this case, the detectability of the compressional drill-bit arrivals is a problem, and the while-drilling information for RVSP surveying is often interrupted. This effect is attributed to the increased friction of the non-rotating drill string in contact with the wall of the well. In this case, the higher static friction and buckling may cause a decrease of the SWD axial-pilot signal (Poletto and Miranda, 1998). Transmission of axial loads (axial-pilot signal) in pipes during sliding appears to be characterized by a threshold after which the signal is

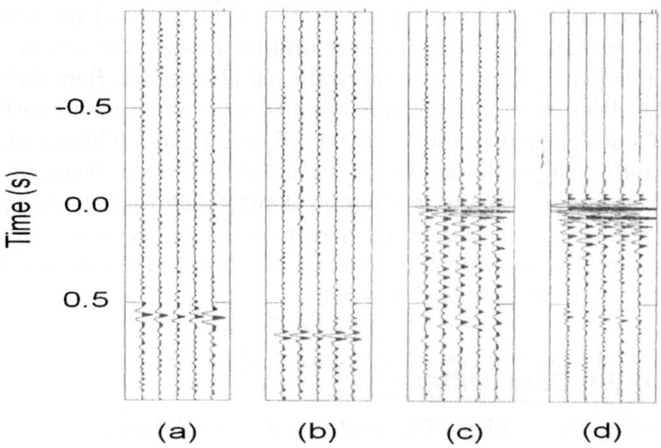

Figure 7.11: Crosscorrelation of axial pilot signal with a) torsion, b) geophone and c) horizontal pilot signals. d) Autocorrelation of the axial pilot. Data are measured at depth of about 4200 m.

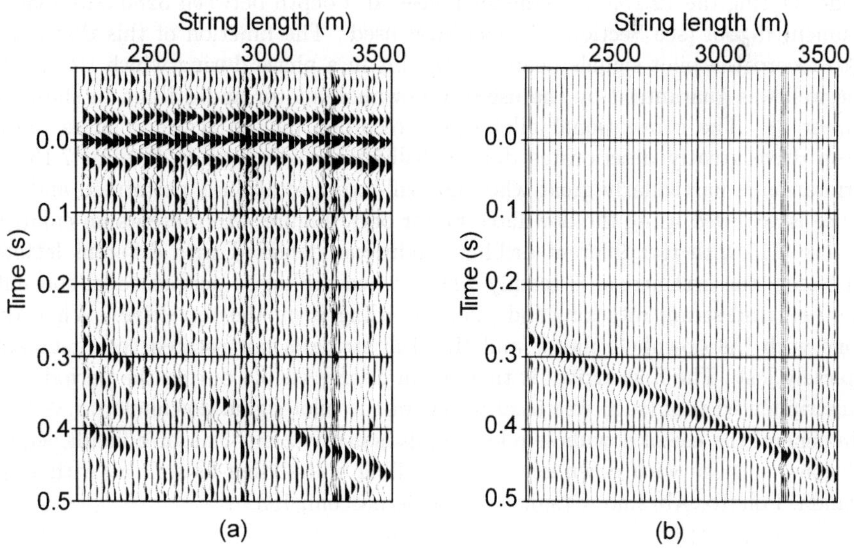

Figure 7.12: a) Real and b) synthetic crosscorrelations of axial and torsional surface pilot data. The time of the dipping event represents the delay between axial and torsional waves. This auxiliary (to axial) information about signal propagation in the drill string can be used to verify the drill-string parameters.

very low (see Chapter 2). A possible explanation for this effect may be the passage from pre- to post-critical buckling conditions (see Section 2.6.6).

Typically, sliding drilling leads to the apparent absence of signal when using conventional pilot preprocessing. Figure 7.13 shows drill-bit reverse VSP represented in true amplitude, for a roller cone, using a conventional axial pilot signal. Here the data are not corrected for pilot delay. This dataset was acquired with and without surface rotation during the use of a downhole motor for deviation (Figure 7.13b). It shows that the signal energy is significantly reduced during the directional control of the deviation phase. This undesired effect causes discontinuity in the VSP. A practical expedient to avoid any survey interruption and to allow better data acquisition is to ask the drillers to give surface rotation for limited drilling intervals, but in many cases the application of this solution is difficult due to drilling constraints and may involve risks.

7.8.1 Downhole pilot signals

A key point in solving the problem of loosing the pilot signal during sliding is in the fact that the axial signal is still generated at the bit, even if it is not properly transmitted through the drill string to the surface. This is demonstrated in the example shown in Figure 7.15, in which a common-receiver SWD VSP gather is shown. The signal of a geophone trace is correlated with the signal of surface and downhole pilots during sliding and rotary drilling. The dramatic change between surface and downhole measurements in sliding condition can be appreciated, with improved signal-to-noise ratio and frequency content. During the 12 1/4 in. diameter phase, at a depth between 3280 and 3336 m, an instrumented tool (see Section 5.5) has been used. The function of this downhole tool was to record the pilot signals nearer to the bit in a phase during which the verticality of the well was guaranteed by the use of a downhole motor and drilling in sliding mode (Chapter 2). The conventional pilot signal recorded by the accelerometers placed on the top drive greatly deteriorates during drilling in sliding mode (Figure 7.15). This deterioration is particularly great when no drill-string rotation is performed and drilling advance is achieved using the downhole motor only. Figure 7.16 presents an enlargement of Figure 7.15 with the different drilling conditions identified with different letters. As can be seen, the first arrival is readily visible when drilling is performed in conventional mode, i.e., drill-string rotation, and pilot recording is on the top drive. When drilling is done with the downhole motor and the drill string is not rotated the first arrival is completely lost. The first arrival is then again clearly visible during the period in which the instrumented tool (sub) was used to record the pilot signal downhole.

In general the use of downhole pilot signals is highly beneficial for difficult acquisition conditions, such as with downhole motor drilling and sliding as well as with PDC bits (Miranda, Poletto, Abramo, Craglietto and Bernasconi, 1999).

7.8.2 Mud guided-waves pilot signals

Examples obtained with surface-rotary and sliding conditions show how the processing of SWD data after the synthesis of different pilot signals is able to recover the signal even when it is apparently not present in conventional preprocessed axial seismograms. The

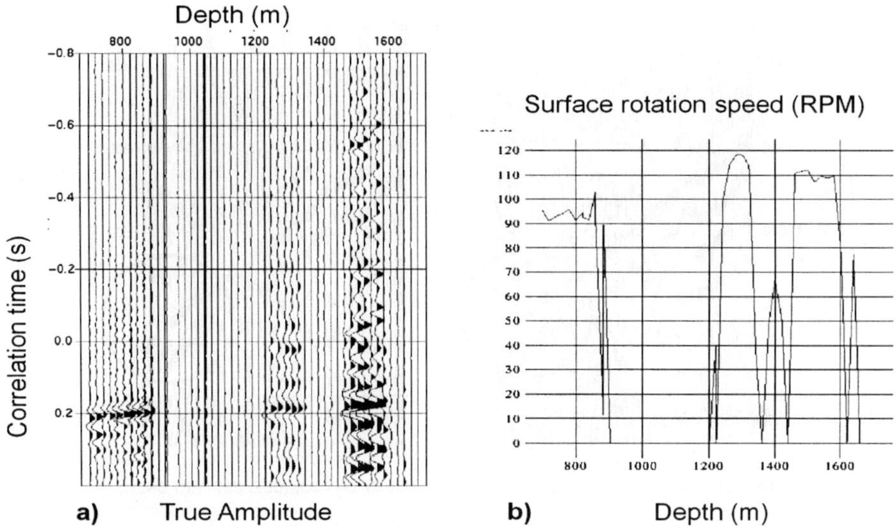

Figure 7.13: a) Common-receiver correlations vs depth. b) Surface RPM vs depth.

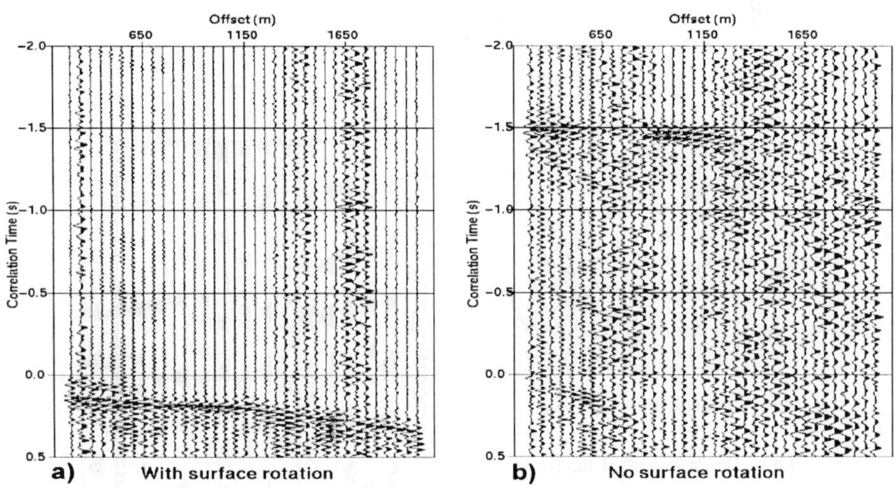

Figure 7.14: Data correlated with an axial pilot signal measured at the top of the drill string. a) Deconvolved common-source data acquired during rotary mode (depth 2863 m). b) Deconvolved common-source data acquired in a drilling position close to that of panel (a) during sliding mode (depth 2860 m).

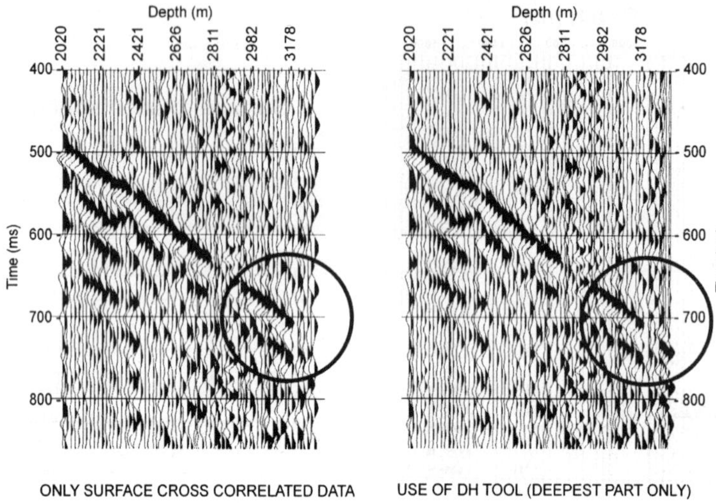

Figure 7.15: A SWD VSP is acquired at different depths with a geophone trace correlated with surface axial pilot signal (left) with and without drill-pipe rotation. In comparison the last SWD VSP five traces of the right panel are crosscorrelated with an axial pilot of a downhole instrumented tool (sub) without drill-pipe rotation.

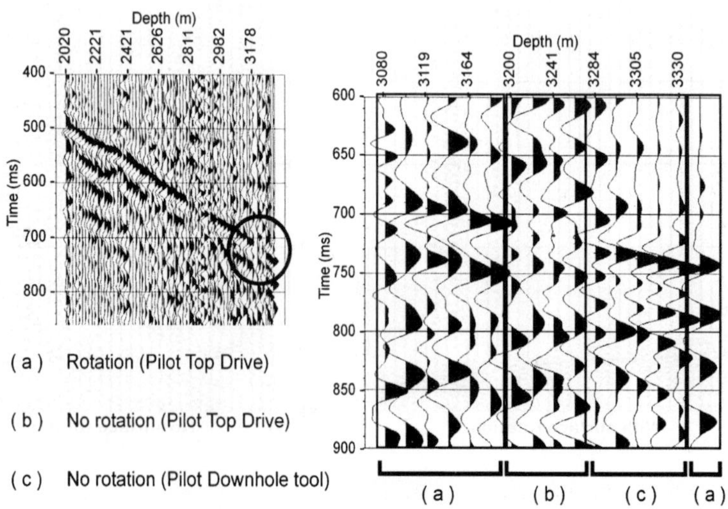

Figure 7.16: The same SWD data as in Figure 7.15 are shown with the enlarged display of the traces with the signal improvement.

presence of non-negligible levels of signal can be demonstrated by processing the geophone traces to get a reference pilot. Above we discussed the solution of using a downhole pilot signal. Another solution is to use only surface pilot sensors and guided waves that are processed usually as noise (Poletto and Miranda, 1998). These are the guided waves in the mud coupled to the drill string (Section 4.6). This alternative approach requires determination of the proper time delay.

In fact, a non-conventional processing of the traces of the same rig axial pilot sensors used to measure axial waves can give while-drilling seismic information during sliding. This allows us to fill the loss of the reverse VSP data during no rotation of the drill string. In the absence of surface rotation, all the drilling power is supplied through the mud flow, the torque is strongly reduced, and the level of coupled drill-string noise (due to string contacts, see Section 4.9) is quite low. In this case, coupled mud-drill string guided waves (interpreted as tube waves) are more clearly detectable during the SWD survey, and represent an alternative pilot signal. These dispersive components have a velocity which is about three and a half times lower than the drill string extensional velocity. Poletto and Miranda (1998) analyzed data from several wells, with and without surface rotation when using a downhole motor. They consider examples with roller-cone data.

The first example involves common-source data. Here, the deconvolved correlations are obtained with a vertical (axial) pilot sensor fixed at the top of the drill string (not corrected with the pilot delay time). The common shot in Figure 7.14a is obtained during surface rotation at a depth of 2863 m. The drill bit direct arrival appears at about 0.2 s. Using standard pilot correction (with an average drill-string correction velocity of 4750 m/s), we obtain the seismic signals corrected in time. On the other hand, in Figure 7.14b, which shows a common-shot obtained at the adjacent depth position (2860 m) in the absence of surface drill string rotation, the signal is obtained by mud-guided pilot waves and it is quite different in frequency content, S/N ratio and horizontal continuity. The direct arrival appears clearly at about −1.5 s correlation time, while the arrival at 0.2 s, given by the extensional pilot wave, disappears. This negative delay means that the pilot (mud wave) velocity is very low with respect to the propagation velocity in the rock – this data time-window could not be observed in the usual applications because of the conventional windowing of the compressional SWD data during processing and interpretation. The closeness of the two levels allows the calculation of the relative delay and, using the known axial-wave propagation velocity, the calculation of the total delay and of the velocity of the guided-wave component.

During acquisition of both levels (2860 and 2863 m), the same drill string, with a total length $L_S = 2884$ m, was used. Thus, by picking, on the same field trace, the arrival times of the pilot extensional waves in Figure 7.14a, and of the arrival times of the guided waves in Figure 7.14b, we obtain, respectively, the first break times: $\mathrm{FB_{EXT}}$ (with surface rotation) = 0.177 s and $\mathrm{FB_{GW}}$ (no surface rotation) = −1.449 s, and a relative delay

$$t_{\mathrm{FB}} = \mathrm{FB_{EXT}} - \mathrm{FB_{GW}} = 1.626 \qquad [\mathrm{s}]. \tag{7.54}$$

The guided-wave velocity v_{GW} is related to the axial velocity v_{EXT} (equation (6.113)), by

$$t_{\mathrm{FB}} = 1.626 = \frac{2884}{v_{\mathrm{GW}}} - \frac{2884}{v_{\mathrm{EXT}}} \qquad [\mathrm{s}]. \tag{7.55}$$

Measuring $v_{\mathrm{EXT}} = 4750$ m/s from the pilot signals, we get $v_{\mathrm{GW}} = 1291$ m/s. Poletto and Miranda (1998) showed examples of VSP with corrected seismic times, obtained using different pilot-signal corrections for levels acquired with and without surface rotation. They observed dispersive effects for mud-guided waves. Integration of the axial and guided correlated signals requires calibration for every depth.

7.8.3 Interpretation of mud-guided waves

The method proposed above was to use an alternative signal to provide drill-bit SWD correlations when using downhole-motor drilling without surface rotation. The nature of the signal propagated in the mud coupled with the drill string was discussed by Poletto, Carcione, Lovo and Miranda (2002). Moreover, to integrate drill-bit seismograms obtained in the presence and in absence of surface rotation, polarization analysis and guided-wave dispersion-curve recovery are important steps in validating and processing this dispersive pilot-signal component.

A SWD experiment is shown in Figure 7.17, where the dipping events are interpreted as guided waves and used as pilot signals (Poletto and Miranda, 1998). The seismic traces belong to a correlated common-receiver gather, obtained (a) with (rotary mode) and without (b) drill-string rotation (sliding mode). The zero correlation time is 4 s, and data are displayed without pilot-delay correction. Figure 7.17a shows the correlation with the extensional wave recorded by an accelerometer located above the drill string, and Figure 7.17b shows the correlation with a slower signal (interpreted later as a pipe wave) recorded by the same accelerometer. The well under investigation is located in southern Italy (Basilicata province), where the reservoir rock is the Apula formation, consisting of limestone. The average mud density versus drill-bit depth is shown in Figure 7.19a. Acoustic properties of drilling mud and of the different materials are discussed in Section 4.5. We assume, for simplicity, an average value of the volume fraction of cuttings versus depth, with ROP = 8 m/h, $\mathrm{FLOW_m} = 4000$ lit/min and the bit radius $r_b = 0.3$ m. This approximation should be modified to consider varying rates of penetration and bit radii versus depth. However, the influence of the fraction of cuttings is not very important for low and moderate rates of penetration.

Because the average mud weight varies as a function of drill-bit depth, the sound speed down to each depth varies as a function of drill-bit depth. At any particular depth, the sound speed of any mode will then be a function of the average density and the drill-string hardware at that depth. Both will change as the drilling progresses. Figure 7.19b shows the velocities of the different wave modes as a function of depth. They have been calculated by using the drill-pipe, tool-joint and BHA composition, dimension and length given below and considering worn and new pipes (see Section 4.5.4). The calculations do not take into account pressure and temperature effects. The drill-string radii are $r_{\mathrm{O}} = 0.063$ m and $r_{\mathrm{I}} = 0.054$ m for the pipes, $r_{\mathrm{O}} = 0.084$ m and $r_{\mathrm{I}} = 0.033$ for the tool joints, and their lengths are 9.2 m and 0.5 m, respectively. For the worn pipes, we assumed a 20 % reduction of the pipe thickness. The outer and inner radii of the casing are $r'_{\mathrm{O}} = 0.178$ m and $r'_{\mathrm{I}} = 0.163$ m, and the formation shear modulus has been calculated as $\mu = 800\ V_{\mathrm{P}}^2$, where V_{P} is the P-wave velocity obtained from the sonic log, given in m/s.

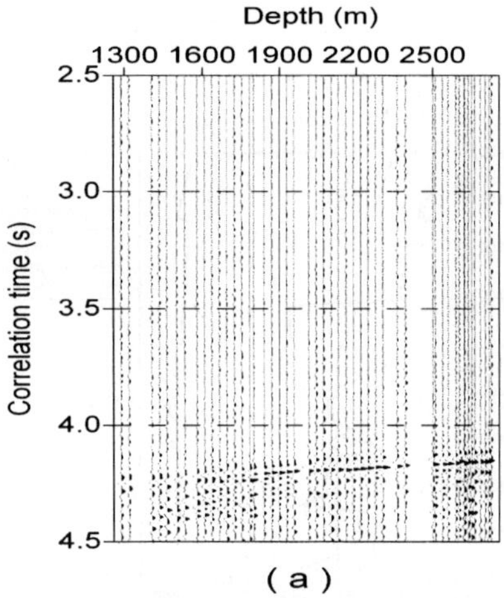

Figure 7.17: Correlated SWD seismograms, where the guided waves can be recognized. Seismic-while-drilling experiment showing the correlation with the extensional wave recorded by an accelerometer located above the drill string (a), and the correlation with a slower signal recorded by the same accelerometer (b). Note that in the correlated data, a slower pilot signal appears at lower arrival times.

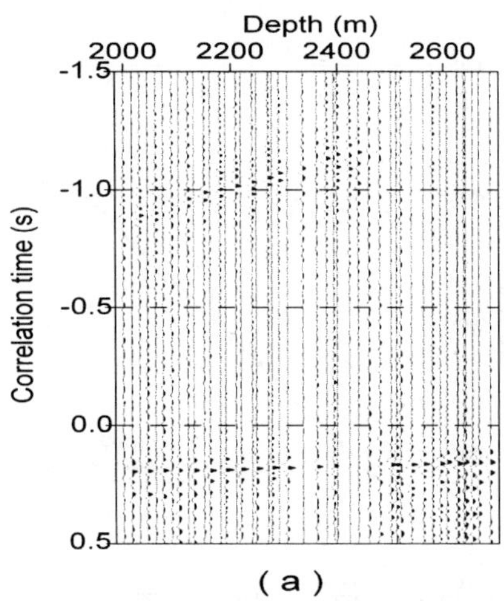

Figure 7.18: Complete seismogram (b) (referred to as the extensional pilot wave), using both pilot signals, the extensional wave – with drill-string rotation –, and the pipe wave – without drill-string rotation. In panel (a), we can recognize the same event with different travel times.

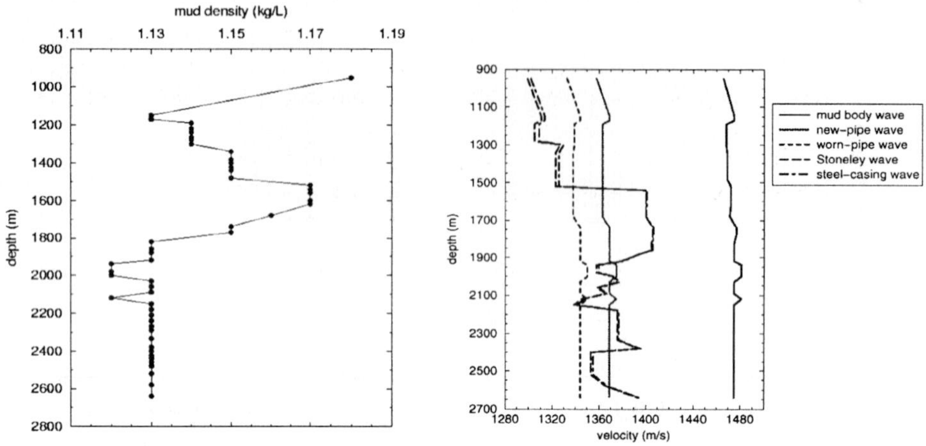

(a) (b)

Figure 7.19: a) Average mud density versus drill-bit depth. The proportion of fluids (water) and bentonite versus depth, are provided by the mudlogging company. b) Velocities of the different wave modes as a function of depth: mud body wave (equation (4.80)), worn-pipe and new-pipe waves (equation (4.84)), Stoneley wave (equation (4.86)) and tube wave in cased hole (equation (4.87)).

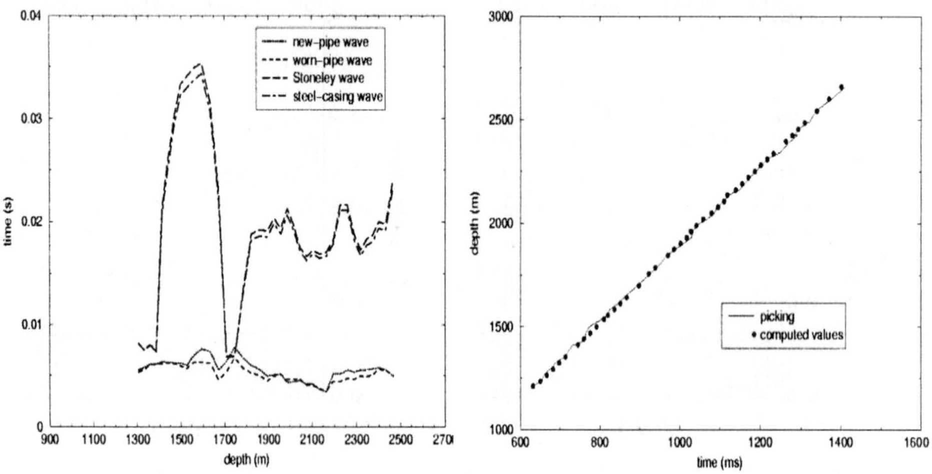

Figure 7.20: a) Mean squared time errors between the calculated and measured velocities. b) Comparison between the picked travel times (dots) and the travel times calculated with the computed pipe-wave velocity (solid line).

The highest velocity in Figure 7.19b is the sound velocity in the drilling mud. The variations with depth are mainly due to the changes in drilling-mud density. The mean squared time errors between the modeled and measured velocities are shown in Figure 7.20a. These results indicate that the slower pilot signals recorded by the accelerometer are, with good probability, pipe waves. The minimum error is obtained for worn pipes (Premium class and 20 % thickness reduction). The comparison between the picked travel times (dots) and the travel times calculated with the theoretical pipe-wave velocity (solid line) is shown in Figure 7.20b. The agreement is good.

Sensitivity analysis for mud velocity

We use the sensitivity analysis of Chapter 3. Equation (4.91) allows us to perform a sensitivity analysis of the pipe-wave velocity (equation (4.84)) for solid fraction and worn conditions. Assuming $\phi_s = 0.09$ and $\rho_m = 1148$ kg/m^3 (see Figure 7.19), $r_O = 0.063$ m and $r_I = 0.054$ m, we obtain $c_m = 1463$ m/s and $c_p = 1356$ m/s (the tool-joints have been neglected in this calculation). The bulk-modulus and density terms in the right side of equation (4.91) have opposite signs. If $\delta\phi_s = 0.01$, we get $\delta c_p = -3.8$ m/s, and $\delta c_p/c_p = -0.0028$, of which 0.0044 corresponds to the bulk-modulus term and -0.0072 is due to the density term. Let us consider now the third term in the right side of equation (4.91). A relative wear $\delta_w = -0.05$ implies $\delta c_p = -4.8$ m/s, and $\delta c_p/c_p = -0.0035$. Hence, the total decrease in the pipe-wave velocity is 8.6 m/s, in agreement with the variations observed in Figure 7.19b.

7.9 RVSP processing of SWD-VSP seismograms

Applications of VSP-SWD processing are shown in Chapter 8. Critical steps for conventional-VSP processing of drill-bit data are discussed in this section.

7.9.1 Direct-arrival, First-Break picking (FB)

Here we use the terms "*direct arrival*" and "*first break*" as synonymous even if it has to be remembered that in correlated SWD data events can appear before the direct arrival from the bit. Picking is a basic step of SWD data processing. Picking is used while-drilling for: time/depth checkshot analysis and geological model update; VSP-wavefield separation and processing; focused-data processing and S/N improvement.

The quality and repeatability of the direct drill-bit arrival is a basic aspect of picking. In the following we will evaluate the potential of methods to analyze the impulsive feature of the drill-bit signal, because this signal is obtained with different conditions. It is recommended that the SWD picking be performed in the following two gather domains.

Picking arrivals of common-receiver data

The data of a single geophone's traces are picked at different bit depths. Signals may be continuous only in intervals because of partial losses of signal energy. Interpolation is sometimes required. Interference with coherent-stationary noise may be a major shortcoming. This makes also the interpretation of waveforms difficult.

Picking of common-source data

This is the picking in the common-shot gathers. Signal and coherent noise are interpreted in the traces at different offsets. Interpolation between traces measured at different locations is not easy because local surface conditions may change the response of adjacent traces. Receiver array patterns may change the waveform. In general, the identification of suitable traces is a straightforward task in common-bit-depth (common-source) gathers, even if the picking of the whole dataset is generally better performed in common-receiver gathers.

7.9.2 Gain recovery

Gain recovery corrects for the geometrical spreading of the reflected waves (Hardage, 2000; Balch and Lee, 1984). Due to the presence of coherent noise, this step is sometimes critical. The geometrical-spreading correction is applied to SWD seismograms with a control for the maximum gain after the first breaks. In fact, organized noise is typically over amplified by the time-varying gain function used to correct the amplitude of the reflections ahead-of-the-bit. Therefore, the reflections in the seismograms are often processed without gain recovery or with gain recovery after improvement of downgoing reflections with dip filters.

7.9.3 Wavefield separation

Note: In the normal VSP geometry the *downgoing* wavefield contains the direct arrivals, and the *upgoing* wavefield contains the primary reflections. In the reciprocal geometry the terms should be interchanged, so that we should compare as equivalent the *downgoing-reciprocal* with the *upgoing-normal* VSP reflected field. This concept is shown in Figure 7.21. Due to the fact that conventional commercial VSP processing packages can be used to process also the SWD RVSP seismograms, in the practice sometimes these terms are interchanged. However, we prefer, for sake of clarity, to use the names with the physical meaning.

SWD wavefield separation into upgoing and downgoing fields, as well as into signal and noise fields can be performed by several approaches. We mention the properties of some of them for sake of clarity and completeness. SWD-wavefield separation methods include:

Separation of downgoing and upgoing waves for reflection processing

Effectiveness of wavefield separation is related to repeatability of the source and regularity of acquisition. Median filters usually are robust and successfully used for the separation of upgoing and downgoing drill-bit SWD waves. Another approach is using velocity filters, like fk. This approach requires data adequately sampled in space to avoid aliasing. τ-p slant stack (Moon, Carswell, Tang and Dilliston, 1986; Carswell and Moon, 1989) is also used, in which case coarse sampling, i.e., aliasing, may introduce high-frequency noise.

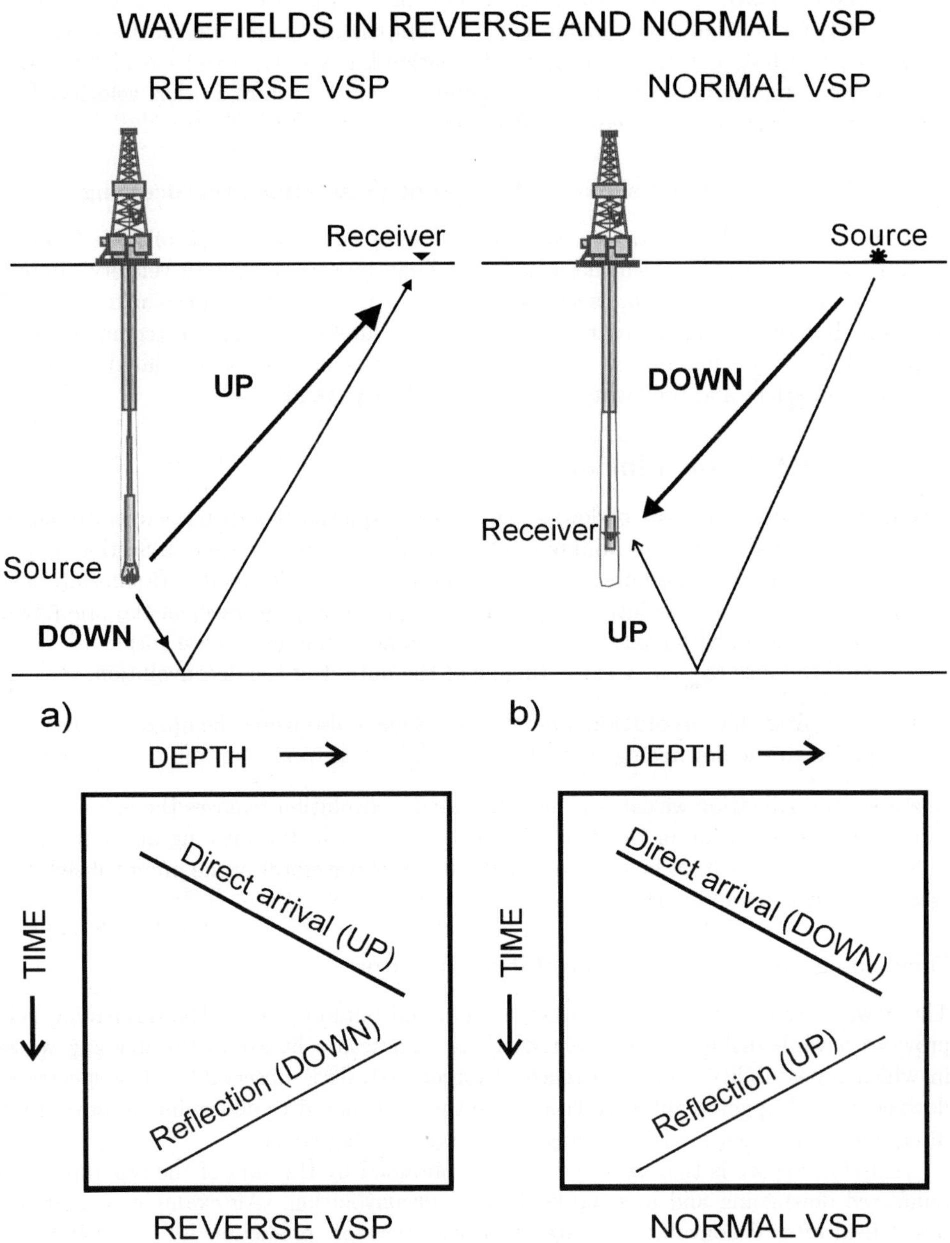

Figure 7.21: Upgoing and downgoing waves in a) reverse (or reciprocal) and b) normal VSP.

Singular-value decomposition of flattened VSP wavefields require that the amplitude of the wave to be extracted be larger than noise. This method is discussed by Glangeaud and Mari (1994) for conventional wireline VSP's in comparison to spectral matrix separation methods – which do not require a priori information but are sensitive to local stability of the wavelet – and parametric separation methods – which require a priori velocity/delay information and are robust if noise-to-signal ratio is low.

Separation of correlated signal and coherent noise after preprocessing

Depending on available geometry, SWD data are sampled in depth (Z), offsets (X and Y in 3D) and time (T). Coherent noise and signal can be separated with velocity filters in the in the $X-Z-T$ domain, as well as in the $X-Y-Z-T$ or in the azimuth-$Z-T$ domains. Removal of signal and noise using velocity filters in the $X-T$ (common source only) and $Z-T$ (common receiver only) domains can be critical for correlated signal and noise that interfere and propagate with similar velocity trends.

7.9.4 VSP deconvolution

Predictive deconvolution of downgoing waves is typically based on the operator calculated on upgoing waves. Minimum-phase conditions are discussed in Section 6.17 on signal rephasing. Assuming a non-minimum delay wavelet model, Robinson (1998) discussed wavelet resolution, optimum gap (prediction distance) in the prediction-error operator, with analysis of minimum and non-minimum phase "orphans". Poor signal-to-noise ratio may cause ringing of the output of the deconvolution.

A waveshaping deconvolution operator is designed also using the upgoing waves and applied to the downgoing waves to shape the signal to zero or minimum phase.

In a first approximation we can assume that VSP deconvolution removes the upgoing filter (due to the propagation of the signal in the formation) in the upgoing and downgoing wavefields, as well as it deconvolves the drill-bit signal waveform in the preprocessed data (see equations 6.55 and 6.54).

Downgoing and upgoing deconvolution operator

The downgoing field is extracted by separation and reinforcement. The downgoing field provides a while-drilling section of reflections. These are linked to the upgoing waves in which repeatability and consistence of direct arrivals is observable. Log-spectra of downgoing and upgoing autocorrelations are then calculated to determine deconvolution filters for signals measured in the upgoing and downgoing traces.

A useful display is the "*recomposed field*" obtained by the sum of the separated and reinforced downgoing and upgoing fields after deconvolution. An example of a recomposed field after deconvolution using a 500 ms operator, prediction lag to second zero crossing of the autocorrelation and cepstral sum (see next section) of autocorrelations of the downgoing an upgoing waves with combination angle $\theta = 45°$ is shown in Figure 7.22. In this display, the variation of the reflections related to changes of direct arrivals can be readily seen.

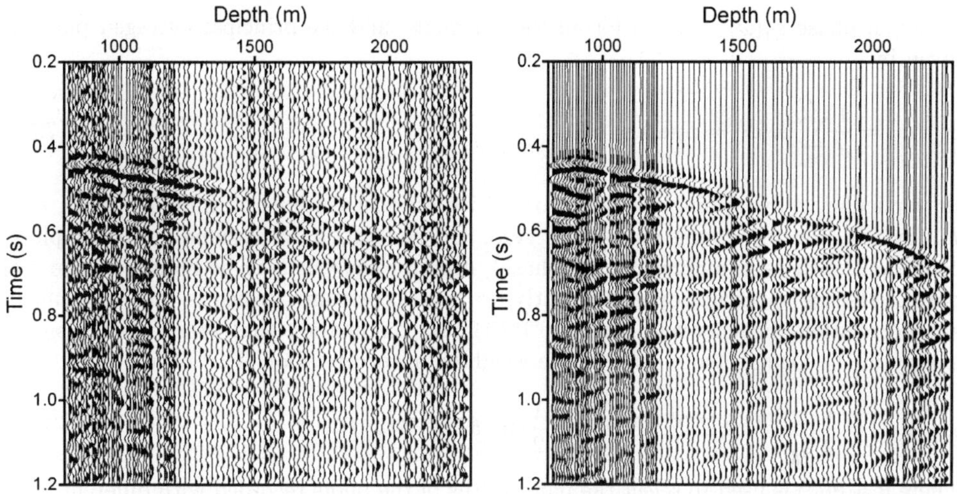

Figure 7.22: Common-receiver SWD data before (left) and after (right) wavefield processing and re-composition.

Log-spectrum (cepstral) analysis of drill-bit correlations

Convolutive noise can be removed by inverse filtering of an individual pilot trace, i.e., such as deconvolution of drill-string reflections. However, when using more than one pilot signal some different local filtering effects may occur – besides additional noise – which may make identification difficult. Removal of common-filtering effects modified by local effects can be achieved by log-spectral or "cepstral" averaging of drill-bit data (Ulrych, 1971). This homomorphic technique transforms data from the time domain, where composition corresponds to convolution, into an equivalent domain, i.e., log spectrum or "cepstrum", where composition corresponds to linear summing (Oppenheim and Shafer, 1975). A shortcoming of this method is that averaging the phase, i.e., imaginary part of log spectrum, wrapped in intervals of 2π in the log-spectrum domain causes noise (say, averaging the wrapped phase is equivalent to compressing random phase towards zero phase). In fact, let $\phi_j(\omega)$ $\left(0 \leq \phi_j(\omega) \leq 2\pi\right)$ be the phase in the principal domain and $\phi_j(\omega) + 2\pi k_j(\omega)$, with k_j an integer, be the continuous phase in function of ω. We obtain the averaged phase from J signals ($j = 1, \ldots, J$) by averaging their continuous phases, namely,

$$\phi_{\mathrm{av}}^{(c)}(\omega) = \frac{1}{J}\sum_{j=1}^{J}\phi_j(\omega) + \frac{2\pi}{J}\sum_{j=1}^{J}k_j(\omega). \tag{7.56}$$

In the presence of noise, phase unwrapping, i.e., determination of $k_j(\omega)$, can be unstable (Cambois and Stoffa, 1993; Spagnolini, 1993). However, combination of SWD data can be achieved without wrapping distortions if signals of the same shape are aligned in time in the correlations obtained with different pilots. In this case, we can assume the same principal phase $\phi_j(\omega) = \phi_s(\omega)$ for all the J signals, and the principal averaged phase is given by

$$\phi_{\rm av}^{(p)}(\omega) = \frac{1}{J}\sum_{j=1}^{J} \phi_j(\omega) = \phi_s(\omega). \tag{7.57}$$

In particular, no-wrapping problems arise with autocorrelations of different pilots, e.g., P_1 and P_2, for which the phases are zero. In general, if we assume a known phase condition (such as minimum or zero phase), cepstral analysis may be limited to the real part of the log spectrum or, equivalently, to "autocepstrum", which is the cepstrum of autocorrelation (Scheuer and Wagner, 1985). A combination that preserves the signal energy in the log-spectrum space is the weighted autocorrelation $A(\omega)$ obtained as

$$\log A(\omega) = \frac{1}{2}\cos^2\theta \log\left[P_1(\omega)P_1^*(\omega)\right] + \frac{1}{2}\sin^2\theta \log\left[P_2(\omega)P_2^*(\omega)\right], \tag{7.58}$$

where $\theta = \theta(\omega)$ is used to select the bandwidths of the pilots recorded with different local transfer functions, assuming that both the pilot channels measure the same signal and drill-string filter. Equation (7.58) is used by Bellezza and Poletto (2002) for deconvolution of downgoing and upgoing waves.

7.9.5 Velocity analysis

Velocity analysis of SWD data is discussed in Chapter 8. Velocity analysis is based on picking direct arrivals in near-offset seismograms and coherence analysis in the beam forming of multioffset data. The interval velocity is a basic product calculated by the checkshot of the seismograms corrected to the vertical travelpath (Sections 8.2 and 8.8.2). Analysis of multioffset CDP (or migrated) data gives average and RMS (migration) velocity. Tomographic inversion of first arrivals is also performed to update the geological model while drilling (Section 8.8.6).

7.10 Analysis of impulsive drill-bit signal

Drill-bit and reverse VSP data are characterized by a strong direct arrival, unlike surface seismic data, where the energy is distributed along the trace. Equivalent SWD results can be obtained using different pilot signals. However, they have different quality of the impulsive source signal. The role of the kurtosis in evaluating the quality of correlated VSP drill-bit data is investigated by Poletto (1998). This analysis was applied to synthetic and real common-offset and common-shot drill-bit seismograms in order to evaluate the prominence and the quality of the first arrival and other coherent events. High values of the kurtosis correspond to an isolated first arrival or to a compressed large-amplitude coherent noise event, while low values are typical of low signal-to-distributed-noise ratio traces.

SWD interpretation problems arise from the presence of noise (Sections 4.8 and 4.9) and the complex nature of the drill-bit source. In addition, the pilot signal is affected by reverberations of the drill-string system (extensional, torsional and flexural vibrations),

vibrations of the rig and periodic bit patterns. For these reasons, the signal obtained by pure crosscorrelation is not reduced to an effective impulse, but requires further processing, such as deconvolution (Section 6.5). Different operational conditions may lead to significant variations in the data quality. Hence, while-drilling quality control of the signal is required. One possible approach among those in current practice is based on an evaluation of the direct wave traveling from the bit to the receivers. The quality control can be achieved by computing the energy of the signal compared to the energy of the noise, or by analyzing the stacking of the aligned events in a common-shot gather (Haldorsen, Miller and Walsh, 1995). Another proposed method is to evaluate the quality of the data by computing the kurtosis of the crosscorrelated signal. The kurtosis can be used in an automatic quality control technique, which does not require time picking. The use of this objective function is justified by the presence of prevalent, typically strong direct arrivals. In conventional VSP and drill-bit data, the reflection events are significantly weaker than the direct wave (a factor of 0.2 for the reflections to direct-wave ratio is quite realistic).

The "impulsive" VSP drill-bit data

The drill-bit SWD signal depends on several factors. The frequency-domain raw seismic trace, recorded at the ith receiver, can be expressed as

$$G_i = (S_{\mathrm{B}i}E_{\mathrm{B}i} + S_{\mathrm{H}i}E_{\mathrm{H}i} + S_{\mathrm{Sw}i}E_{\mathrm{Sw}i})B + N_i, \tag{7.59}$$

where B is the wide-band drill-bit signal; $E_{\mathrm{B}i}$, $E_{\mathrm{H}i}$ and $E_{\mathrm{Sw}i}$ are the frequency responses for waves traveling (through the earth) from bit to receiver, the head-wave response and the secondary rig-wave response, respectively; $S_{\mathrm{B}i}$, $S_{\mathrm{H}i}$ and $S_{\mathrm{Sw}i}$ are the respective instrument frequency responses (they differ, since they include the receiver array), and N_i is a term containing coherent and incoherent noise. For simplicity, the bit radiation pattern is omitted. On the other hand, the pilot signal can be written as

$$P = S_{\mathrm{P}}HB + N_{\mathrm{P}}, \tag{7.60}$$

where S_{P} is the pilot sensor response, H is the transfer function of the rig/drill-string system, and N_{P} is noise from the rig system. Crosscorrelating the signals of equations (7.59) and (7.60) gives $C_i = P^* G_i$. After correlation, deconvolution by using the operator D, and filtering by the filter F, we obtain the VSP traces

$$X_i = FDC_i. \tag{7.61}$$

In addition, the deconvolution operator D may be used to remove S_{P} and $S_{\mathrm{B}i}$ and to correct the travel time delay from the bit to the pilot sensor, t_{p}, multiplying the spectrum by $e^{-\imath\omega t_{\mathrm{P}}}$. The aim of this preprocessing step is to find the earth's response $E_{\mathrm{B}i}$ at each trace by obtaining an impulsive filter, namely,

$$I_{\mathrm{B}i} = FD(S_{\mathrm{P}}H)^* S_{\mathrm{B}i}|B|^2. \tag{7.62}$$

Here we define the noise as a sum of filtered/deconvolved coherent and random noise terms, $\mathcal{N}_i \equiv FD(N_{\mathrm{C}i} + N_{\mathrm{R}i})$. We obtain from the preceding equations

$$X_i = I_{\mathrm{B}i}E_{\mathrm{B}i} + I_{\mathrm{H}i}E_{\mathrm{H}i} + I_{\mathrm{Sw}i}E_{\mathrm{Sw}i} + \mathcal{N}_i, \tag{7.63}$$

where

$$I_{\mathrm{H}i} = FD(S_{\mathrm{P}}H)^{*}S_{\mathrm{H}i}|B|^{2},$$
$$I_{\mathrm{Sw}i} = FD(S_{\mathrm{P}}H)^{*}S_{\mathrm{Sw}i}|B|^{2}. \qquad (7.64)$$

In VSP configurations, the earth response is dominated by the direct arrival, and the seismic trace is mainly impulsive, depending on the subsurface conditions and the reflectivity profile. In good-quality drill-bit SWD data this feature is generally preserved. Notwithstanding the presence of the head and secondary rig waves $I_{\mathrm{H}i}E_{\mathrm{H}i}$ and $I_{\mathrm{Sw}i}E_{\mathrm{Sw}i}$, respectively, for a given earth response $E_{\mathrm{B}i}$, the result will depend on the impulsive character of $I_{\mathrm{B}i}$ and the magnitude of $\mathcal{N}_i$, containing random and "colored" components.

A statistical method of evaluating the "spikiness" of the data is to estimate the kurtosis of equation (7.48) or

$$K = \frac{N\sum_{j=1}^{N} x_j^4}{\left(\sum_{j=1}^{N} x_j^2\right)^2}, \qquad (7.65)$$

where x_j is the trace-time inverse Fourier transform of X_i, N is the number of samples, and the expectation value is expressed by the sample mean average. Equation (7.65) also means that for an isolated wavelet in the absence of noise, the kurtosis is proportional to the window length $N\Delta t$, Δt being the sampling rate. The kurtosis is N times the varimax norm, an objective function widely used in signal processing (Wiggins, 1978; Wiggins, 1985; Levy and Oldenburg, 1987). It is well known (Papoulis, 1991) that for sequences with a Gaussian distribution, the fourth-order moment equals three times the square of the second-order moment of the time series; thus $K = 3$ (Section 7.6). The maximum value of the kurtosis is N, and this occurs when the trace is a single spike at a given sample. The kurtosis is minimum when the amplitude distribution is uniform along the trace ($K = 1$). Note also that, in general, from equation (7.65) the kurtosis of N samples with M ($1 \leq M \leq N$) unit-amplitude events is given by $K = N/M$.

Due to the variability of the drilling conditions, drill-bit data often contain periodic patterns and also distributed noise. In some cases, restricted-band low- and high-pass filters are used to remove noise. In general, the kurtosis depends on the bandwidth and phase of the seismogram. Zero- or minimum-phase wavelets are considered as a reference, although mixed phase conditions could also be realistic for correlated real data. Knowledge of the kurtosis and bandwidth of the signal helps to evaluate the quality of the preprocessing in terms of an optimal phase. Figure 7.23 shows nine synthetic zero-phase and minimum-phase wavelets generated with Butterworth filters (Kanasewich, 1981). The linear time-shifts are not relevant. The wavelets in Figure 7.23 have the same ratio of high cut-off frequency to low-cutoff frequency, i.e., the bandwidth is 2.5 octaves. The central frequency increases linearly from 9 to 81 Hz. The kurtosis, represented in Figure 7.24b, is computed with a window of 800 ms (the sampling rate is in this case 1 ms). The horizontal axis indicates the central frequency of the traces with increasing linear bandwidth (see Figure 7.24). As can be seen, the kurtosis increases linearly with the central frequency, and the kurtosis of the minimum-phase wavelets is about 68 % of the kurtosis of the zero-phase wavelets. As shown by Levy and Oldenburg (1986), zero-phase wavelets have the maximum kurtosis for a given amplitude spectrum. For zero-phase wavelets, each with the same bandwidth shifted linearly in frequency, the kurtosis is constant when the

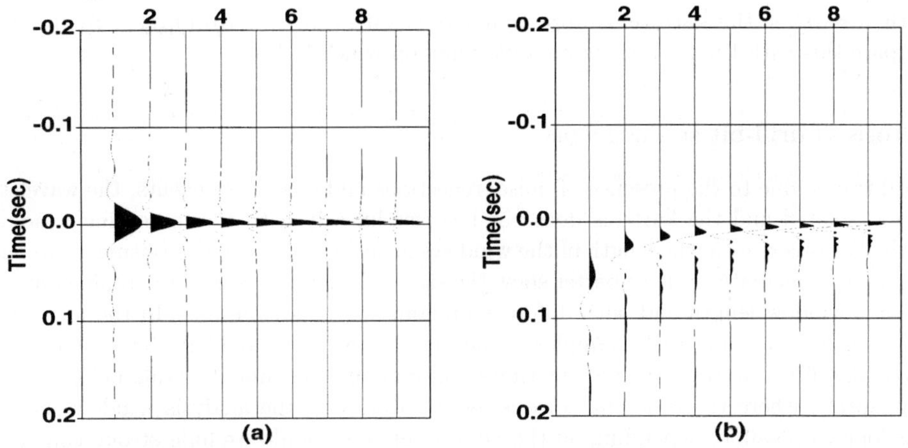

Figure 7.23: Nine zero-phase (a) wavelets generated with Butterworth filters at a 1 ms sampling rate. They have the same bandwidth in octaves, with the central frequency increasing linearly from about 9 to 81 Hz. b) Minimum-phase equivalent wavelets of the wavelets in (a).

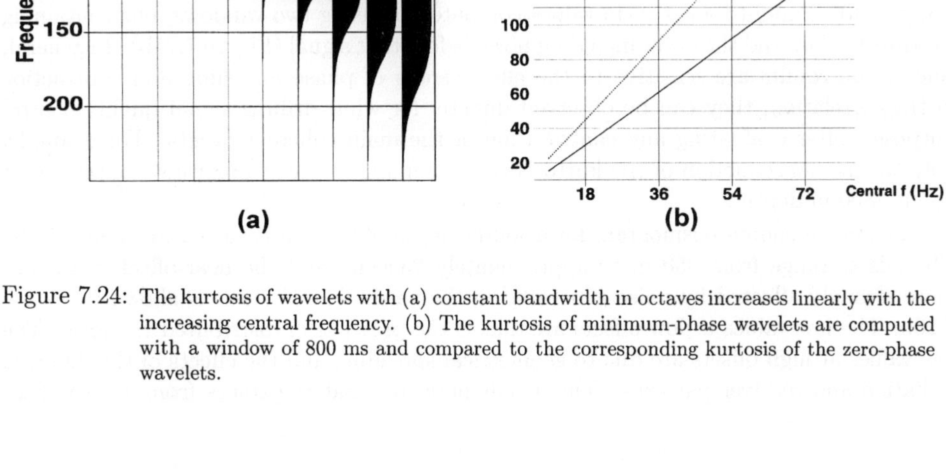

Figure 7.24: The kurtosis of wavelets with (a) constant bandwidth in octaves increases linearly with the increasing central frequency. (b) The kurtosis of minimum-phase wavelets are computed with a window of 800 ms and compared to the corresponding kurtosis of the zero-phase wavelets.

low-cut frequency is greater than half bandwidth. As a rule of thumb, for zero-phase wavelets of linear shifts in frequency of a $\Delta/(2\pi) \cong 34$ Hz bandwidth, using the window length $L = 0.8$ s, the kurtosis is approximately equal to $(3/4)(4/3)L(f_{\max} - f_{\min}) = 27$ (compare this with the kurtosis of a cosine function which is 1.5).

Kurtosis of drill-bit seismograms

In real traces, due to the presence of noise, reflections and secondary events, the wavelets are not isolated and the kurtosis does not increase linearly with the increasing window length. For this reason, the length of the windows is not set on the entire seismogram, but set to an intermediate value to better show the signal. Different results are obtained using the same window length and with different random noise distributions. In the following discussion we see some real examples. The analysis window could be long enough to enclose the direct arrivals at different times and may include time intervals before $t = 0$ (anti-causal), where the coherent noise is lower. Moreover, the analysis window should be as long as possible, depending on the noise level, and should exclude strong coherent events, such as rig waves, multiples and shear waves. The onshore seismograms shown in Figures 7.25 and 7.26 are obtained by crosscorrelating an axial pilot signal (recorded at the rig) with the traces of a seismic line acquired using groups of 24 vertical geophones of 10 Hz natural frequency. The uncorrelated record length is 24 s, and the sampling rate is 2 ms. Each depth level is the result of stacking from 60 and 90 correlations. The correlated data were deconvolved using an operator of length 2 s, spiking the pilot signal, rephased for the instrumental effects and corrected to seismic time by adding the pilot delay $t_{\rm p}$. No spherical divergence was applied to correct the amplitude decay caused by the geometrical spreading. Figure 7.25a shows a correlated and deconvolved common near-offset (550 m) drill-bit dataset $x(t)$, consisting of 102 depth levels recorded with roller-cone bits between 1607 and 3290 m. The data are normalized for display purposes.

The traces have a similar frequency bandwidth. They are filtered with a bandpass filter of corners 10, 14, 48, and 56 Hz, and the analysis window length goes from 0.2 s to 1.2 s. The kurtosis is shown in Figure 7.25b. High kurtosis values are approximately half the maximum value computed for the filtered zero-phase wavelet (Poletto, 2000). These values correspond to a roller-cone bit signal of good quality. This behavior can be compared to signal-to-noise ratio values computed by using two windows, one containing the direct signal and the other including noise before that signal (Figure 7.25b). In general, since these results are sensitive to the effectiveness of phase matching and subtraction in the correlation, they can be observed directly for while-drilling signal quality-control purposes, after evaluating any contribution of the main coherent events. For example, note the partial reduction of the kurtosis corresponding to the strong wave conversion at about 2800 m depth.

A common-source seismogram for a source depth of 2440 m is shown in Figure 7.26a. The offsets range from 350 m to approximately 2350 m, and the near-offset traces are more strongly affected by noise generated in the yard. The kurtosis is shown in Figure 7.26b. The low values at low offsets are due to the presence of distributed rig noise. The low values at high offsets are due to geometrical spreading and the effects of the different radiation and receiver patterns. This result indicates that the traces from 475 m offset

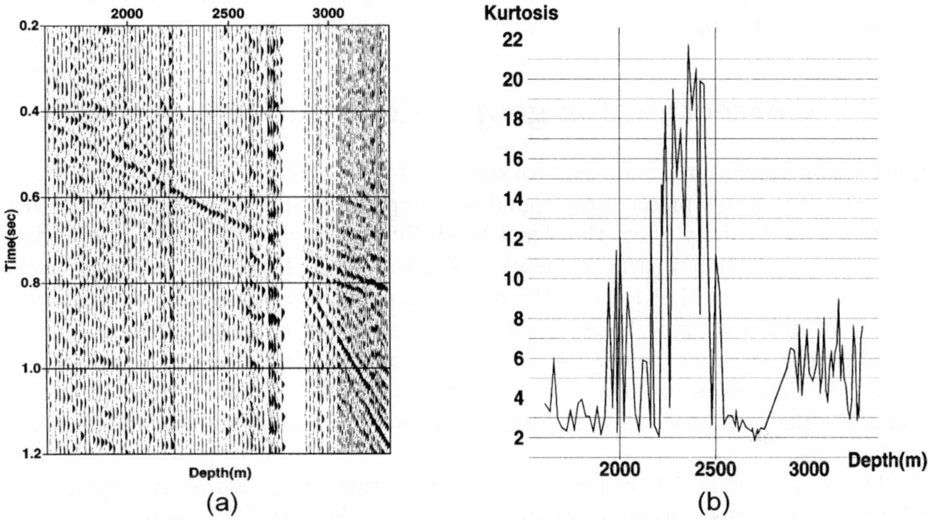

Figure 7.25: a) Real reverse common offset VSP drill-bit data showing variable signal-to-noise ratio and coherent noise presence. b) The kurtosis of the real drill-bit seismograms computed with a window of 1.0 s plotted versus depth.

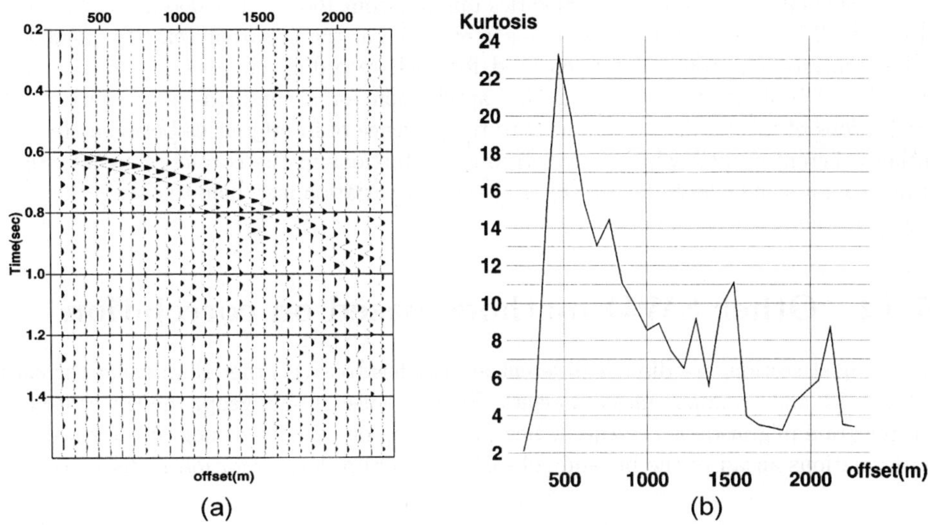

Figure 7.26: a) Real common shot drill-bit data showing variable signal-to-noise ratio versus offset b) The kurtosis of the common-shot seismogram computed with a window of 1.0 s and plotted versus offset (after Poletto, 2000).

to 1350 m offset (in particular that at approximately 475 m offset), with a kurtosis value higher than 8–10, are suitable for common-receiver monitoring.

7.11 Correction for geophone group-array filters

Behavior of receiver-group arrays with signal and noise is discussed in Section 5.14. Long receiver-group arrays preserve the signal but suppress low-frequency noise radiated from the rig. However, they affect the signal bandwidth because of NMO effects, especially at large offsets and shallow drill-bit depths. The loss of the high-frequency components is an undesirable effect that reduces the resolution of the drill-bit signature. Furthermore, the use of different array lengths along the seismic line is typically made to increase noise-filtering effects in the near-offset traces, where noise is higher and signal moveout is lower. The differences in the array configurations introduce inhomogeneity in the signal, and affect the multioffset RVSP processing and waveform analysis, e.g., for tomographic applications.

The signal bandwidth can be recovered by means of inverse array-response filters, calculated using the optimum apparent velocity of the signal. To compensate for the group-array-filtering effect, inverse filters are computed using different apparent velocities. The linear group-array configuration, for length and number of receivers, is the same as that used for the field traces. Before inversion, white noise is added to the frequency response (to fill the zero-amplitude array notches) of the filters. The kurtosis is calculated in a time window centered on the drill-bit direct arrivals, after inverse filtering. The optimum group-array-inverse filter is obtained when the kurtosis is maximum. Figure 7.27a shows a preprocessed trace (300 m offset, 1242 m source depth) corrected by inverse filters calculated with apparent velocities ranging from 1000 and 10 000 m/s with steps of 150 m/s. The maximum of the kurtosis is obtained at the apparent velocity of 4000 m/s. This analysis is automatically repeated for each recording channel and source depth. Figure 7.27b shows the results of this application for a common-source gather. The limitations of this method are related to the presence of interferences between signal and other coherent events, which may introduce distortions in the corrected signal. Shorter time windows and constraints in the trial-velocity variation may help to obtain the correct result in the presence of noise.

7.12 Other SWD methods based on correlation

In the next sections we discuss different approaches to SWD other than using correlation of rig-pilot and geophone measurements. First, the use of only geophone traces for cross-correlogram migration is presented. Secondly, using only pilot seismograms for prediction of formations ahead of the bit and subsurface-structure interpretation is discussed.

7.12.1 Crosscorrelogram migration

This method is based on migration of autocorrelograms (Yu, Followill, Katz and Schuster, 2001) and crosscorrelograms (Yu and Schuster, 2001) of drill-bit VSP data measured only by geophones located at the surface. This inverse VSP while-drilling (IVSPWD) method is based on Katz's patent (1990). It uses the ghost reflections – the multiple reflections given by the free-surface – for "*daylight*" imaging of subsurface reflectivity with unknown source locations (Claerbout, 1968; Rickett and Claerbout, 1999). In conventional VSP processing, the "surface-ghost" reflection is part of the downgoing field. The "primary reflection" migration image is obtained (Schuster, Followill, Katz and Yu, 2000) by

$$m(x) = \sum_{g,s} \ddot{\varphi}\left(\mathbf{r}_g, \mathbf{r}_s, \tau = \tau_{sx} + \tau_{xg} - \tau_{sg}\right), \tag{7.66}$$

where the summation is extended over all source and receiver positions (r_s and r_g, respectively) in the well and at the surface. Herein, $\ddot{\varphi}$ is the second order time-derivative of the autocorrelation function. τ_{sx}, τ_{xg} and τ_{sg} are the delays of raypaths of Figure 7.28a. Moreover, the surface-related "ghost reflection" imaging condition for migration is (Schuster, Followill, Katz and Yu, 2000)

$$m(x) = \sum_{g,s} \ddot{\varphi}\left(\mathbf{r}_g, \mathbf{r}_s, \tau = \tau_{sx_0} + \tau_{xx_0} + \tau_{xg} - \tau_{sg}\right), \tag{7.67}$$

(see raypaths of Figure 7.28b). Figure 7.28b shows the basic concept of migration using ghost and primary reflection crosscorrelograms. Because the method does not require a pilot signal, it was successfully applied to horizontal-well drilling, where the axial pilot transmission is poor (Yu and Schuster, 2001). The data need some processing before migration, such as noise removal and velocity analysis by beam forming.

Three aspects of crosscorrelogram migration deserve attention. First, the application of drill-bit seismic in horizontal wells is an important target. Using the ghost reflection makes it possible to image reflection patterns different from those obtained by using only primary reflections, i.e., not only below the bit. Secondly, in the absence of reliable pilot signals, the autocorrelation migration, based on imaging travel times, can be used to image the reflections. However, autocorrelation collapses both signal and noise. The latter has to be removed as much as possible before autocorrelation. Several papers show that the focusing of a signal trace by stacking the geophone traces (corrected of travel time) before autocorrelation is a good way – not equivalent to stacking autocorrelations – to separate signal and noise (see Section 6.8). Thirdly, a time-depth model of this method is obtained by imaging which, in turn, gives an estimate of the signal travel times. One important product of the drill-bit seismic-while-drilling method is the checkshot analysis, i.e., the identification of the seismic time of the bit while drilling (Section 8.2). This information is obtained using pilot signals for correlation without the need of data migration and can be used to correct the seismic model while drilling.

7.12.2 Formation analysis by pilot seismograms

Drill-string axial pilot signals can be used to determine the reflections in the formation ahead of the bit. In fact, the part of the drill-bit signal propagating downward in the formation and reflected upward by a seismic interface is transmitted also to the drill-string and to the surface pilot sensors. These reflections are interpreted in drill-bit pilot signals by means of the numerical modeling of the drill string coupled to the formation at the

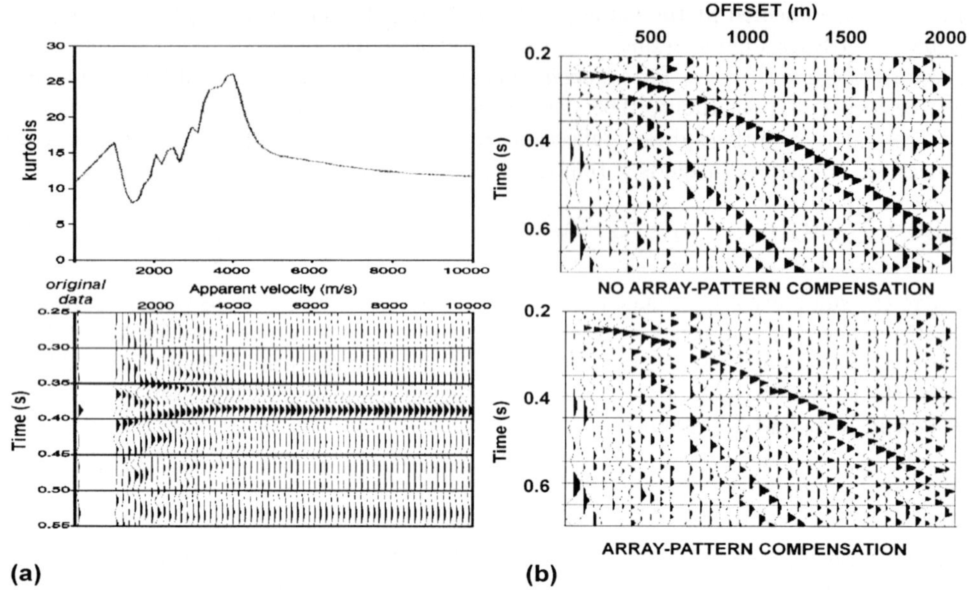

Figure 7.27: a) Kurtosis analysis on direct arrival (1150 m offset, 1242 m): the trace corrected with an apparent velocity of 4000 m/s shows the maximum "spikiness". b) Common-source-gather (1242 m), before (top) and after (bottom) array-pattern compensation.

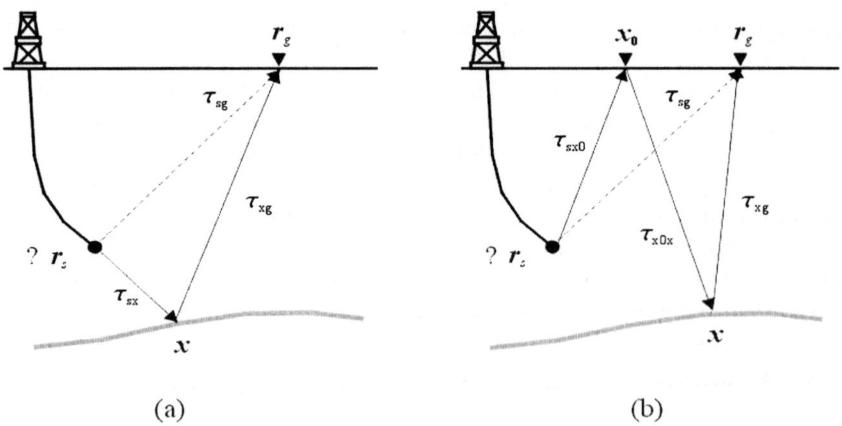

Figure 7.28: a) Travelpaths of the waves used by autocorrelogram migration. b) Travelpaths of the waves used by crosscorrelogram migration. The delay due to the unknown source position disappears in the direct-arrival/ghost-arrival correlation term. Modified after Schuster, Followill, Katz and Yu (2000).

bit/rock contact, and processed as a sort of additional low-cost reverse VSP in the zero-offset approximation. Malusa, Poletto and Miranda (2002) integrated the while-drilling results obtained with the pilot seismogram with the conventional drill-bit reverse VSP measurements, and compared it with other geophysical and well results.

The inclination of reflectors below the bit means that the pilot signal also contains events produced by the interaction of the drill string with the surrounding formation. For instance, Dubinsky, Henneuse and Kirkman (1992) analyzed the drill-string vibrations and interpreted the events produced by a stabilizer (see Section 2.3.10) entering a new formation. The concept of using the surface pilot signal as auxiliary information for prediction of formation to be drilled ahead of the bit is shown in Figure 7.29. As can be seen in particular datasets, when the downward wavefront emitted by the drill bit is reflected upward by a seismic discontinuity, part of the reflection of the formation is transmitted through the bit/formation interface to the drill string and to the pilot sensor. These reflections in the pilot data can be processed as a reverse VSP, and the result matches the measurement from a conventional drill-bit RVSP (Malusa, Poletto and Miranda, 2002).

Reflections in pilot and geophone seismograms

The vibrations induced in the drill string and the quality of the pilot signal are influenced by the type of bit and by many other factors such as the drill-string composition, the drilled rocks, and the operational conditions, like frequency of rotation (RPM), weight on bit (WOB), rate of penetration (RPM), and the verticality of drilling. In addition to the reflections in the drill string, the interaction of the drill string with the formation produces a complex distribution of reflections. The following main types of events are detectable in SWD RVSP data: drill-string reflections in pilot data; reflections of rock layers in geophone data; drill-string reflections in geophone data; reflections of rock layers in pilot data. Malusa, Poletto and Miranda (2002) modeled the waves in the drill-string and the interaction between drill string and formation at the bit/rock contact including the compressional waves reflected in the formation for pilot-VSP interpretation.

Pilot VSP interpretation

The reflections of the formation ahead of the bit are interpreted in pilot seismograms as follows. Figure 7.29 shows the pilot-VSP signal acquired for increasing drill-string lengths. The event marked by A is the direct arrival, while B is the long-delay drill-string multiple. Part of the drill-bit signal travels along the drill string, and is detected by the pilot sensor on the rig and part of the signal propagates downwards into the formation. When the bit approaches a seismic discontinuity (R), part of this wavefront is reflected and reintroduced into the drill string, with a travelpath from the bit to the reflector and back to the bit. Then, the wavefront travels through the drill string and is recorded by the surface-pilot sensor with a delay with respect to the direct arrival owing to the distance from the bit to the reflector and to the formation velocity. This event is marked by C.

As the drilling proceeds, this delay decreases until the bit reaches the reflector. Then, after the bit enters the new formation, the upward wavefront emitted by the bit is reflected downwards from the other side of the seismic discontinuity and again a part of it is

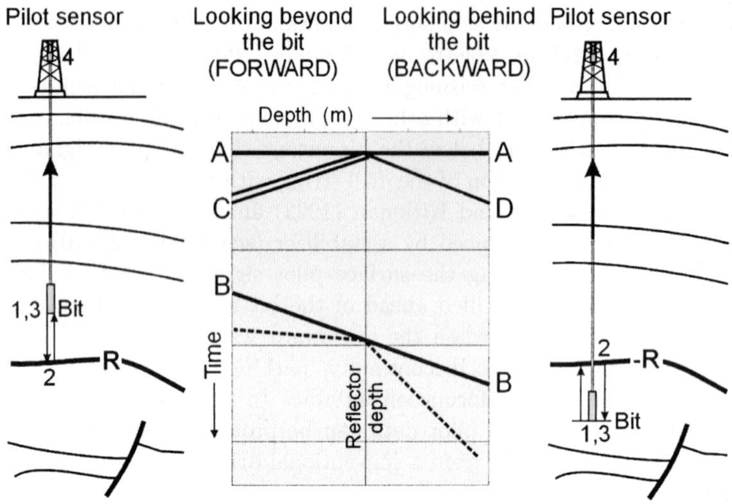

Figure 7.29: Pilot-VSP and interpretation of reflections in pilot auto-correlations. The event marked by A is the direct (FB) arrival; due to the auto-correlation, the time of the direct arrival is zero. The event marked by B is the long-period drill-string multiple. If the bit approaches a seismic discontinuity, the wavefront emitted is partially reflected and reintroduced in the drill string (event C), and is recorded by the pilot sensor with a delay, with respect to the FB, owing to the bit distance from the reflector. When the bit goes through the reflector, the wavefront emitted by the bit is reflected with opposite polarity by the other side of the seismic discontinuity, and again affects the drill string. In this case the delay, with respect to the FB, increases with the depth (event D).

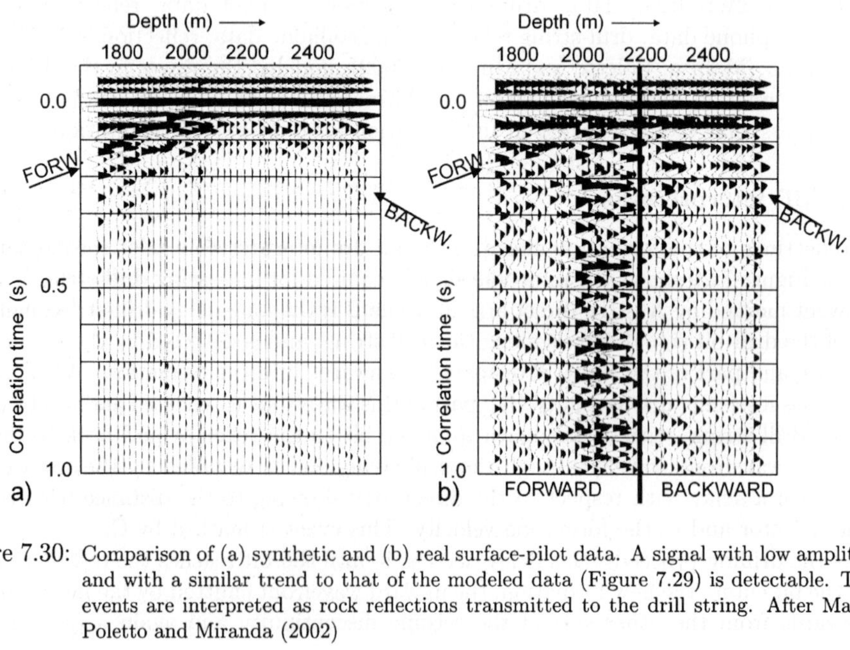

Figure 7.30: Comparison of (a) synthetic and (b) real surface-pilot data. A signal with low amplitude, and with a similar trend to that of the modeled data (Figure 7.29) is detectable. These events are interpreted as rock reflections transmitted to the drill string. After Malusa, Poletto and Miranda (2002)

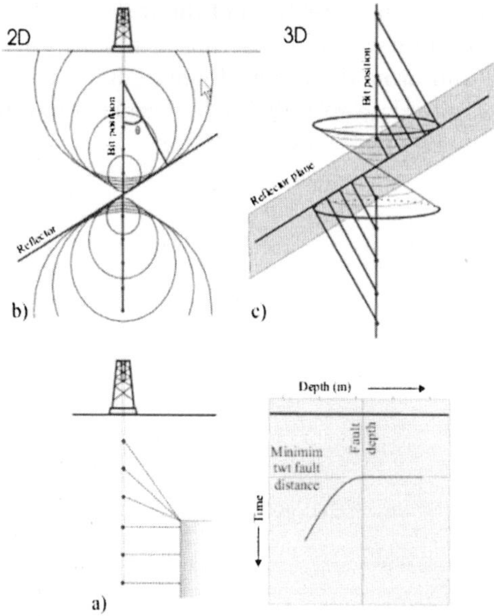

Figure 7.31: Lateral effects in the presence of a fault, and a dipping reflector, which is predicted by the pilot VSP in 2D and 3D dimension. After Malusa, Poletto and Miranda (2002).

transmitted into the drill string through the rock/bit interface. In this case, the delay, with respect to the drill-bit direct arrival, increases with increased drilling depths. The events related to the same reflector have an opposite polarity before and after the bit has passed through it. We mark this event with a D in Figure 7.29, and call its relative delay relative to the direct arrival time.

Figure 7.30b shows an example of an axial pilot signal obtained with a roller-cone bit during a conventional SWD survey, by stacking from 90 to 120 correlations of 24 s field records with a sampling rate of 2 ms. The data are one-side deconvolved and bandpass filtered (12–42 Hz). Low-amplitude upward and downward sloping events are detectable near the direct arrival, and indicated by arrows. The interpretation is related to the presence of a reflector at about 2200 m depth. The time positioning of the rock reflector was confirmed while drilling and after the well completion, by means of checkshot and sonic-log information. Figure 7.30a shows the numerical modeling of the corresponding surface-pilot signal. It was used to support the interpretation of the low-amplitude events in the real data, which show a trend similar to that of the synthetic data. Malusa, Poletto and Miranda (2002) processed surface-pilot VSP seismograms to detect reflectors ahead of the bit ("forward"), and behind the bit after they have been drilled ("backward"). They showed examples of real pilot data compared with conventional SWD results to detect changes of the formation, from limestone to shales, at a depth of about 2000 m. The drill-bit pilot VSP can be used together with the surface RVSP data to locate objects, like faults, ahead and near to the well.

For correct processing and interpretation of the rock-reflections detected by the pilot-signal VSP, the lateral effects have to be also taken into account to correctly investigate the stratigraphy ahead and around the bit (Figure 7.31). The analysis of these events would be more accurate with data recorded by a downhole tool (Section 5.5).

Chapter 8

Applications

8.1 Introduction

This chapter shows applications of drill-bit SWD for geophysical well monitoring. This method is successfully used in oil and gas exploration, in particular in areas characterized by a complex geology, or land where the poor surface seismic response does not allow a reliable prognosis of the depth of the target and a confident definition of the geological structures (Bertelli, Abramo and Gatti, 1998). The main problem in these situations is the uncertainty of the subsurface depth-model. This uncertainty does not allow one to forecast the position and depth of the geological structures, or to predict the conditions of instability of the drilled formation and overpressure zones. With a well plan based on an uncertain geological model, drilling may be seriously affected by the variation encountered during drilling itself. This fact explains how critical certain decisions (such as setting the casing points) are for drilling safely and efficiently. Conventional geophysical techniques, such as wire-line logging and VSP, give information to solve some of these uncertainties, but the results of these techniques are not available in real time.

While-drilling measurements are used to support drilling and geophysical exploration. The MWD technology, introduced to monitor directional drilling (Section 2.3.18), has allowed for the development of the logging-while-drilling (LWD) technology that acquires borehole data such as formation resistivity, rock density, porosity, and others in real time. These data, that are all grouped under the term "formation-evaluation-while-drilling" (FEWD), give high-resolution information in a few meter range of investigation around the well.

Seismic-while-drilling (SWD) is a geophysical tool which has more potential in terms of structural investigation at larger distances from the well. Using the drill bit as a downhole seismic source, we are able to obtain reverse vertical seismic profiles (RVSP) (Rector and Marion, 1990). SWD surveys are run without any drilling breaks and risk, and produce significant results which can be immediately used with an impact on drilling operations (Miranda, Aleotti, Abramo, Poletto, Craglietto, Persoglia and Rocca, 1996). Using this technology, it is possible to do the following: to revise while drilling the initial geological model; to update the drilling program by reprocessing the surface seismic data; to set the casing points in the most appropriate position; to revise the hole trajectory; and to predict anomalous drilling conditions. An overview of drill-bit SWD application for drilling is also

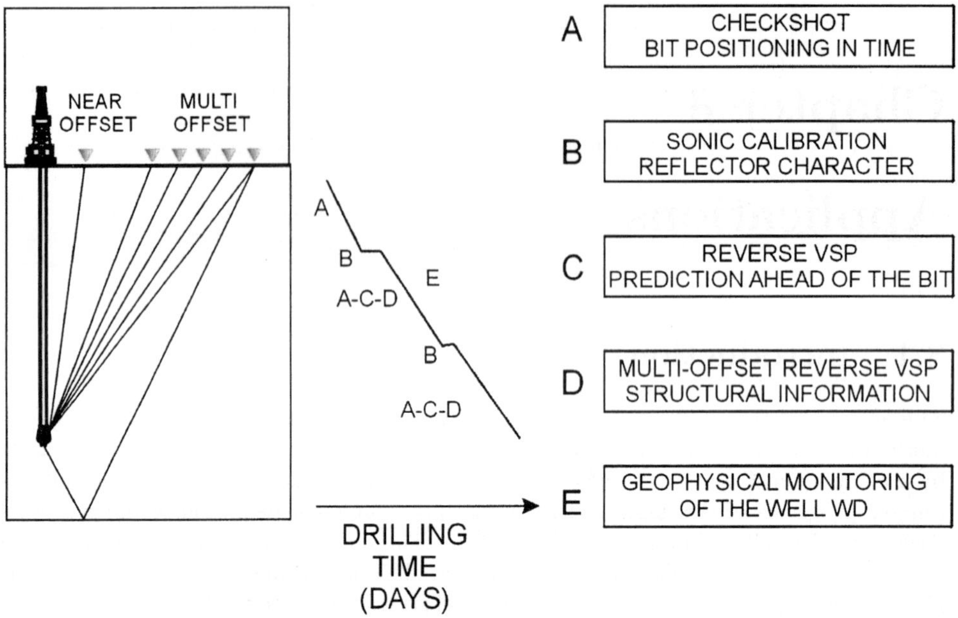

Figure 8.1: Different drill-bit-seismic products and applications obtainable during drilling.

given by Meehan, Nutt, Dutta and Menzies (1998).

8.2 SWD products

The SWD products are useful if they are available in real time during the drilling operations. Five main SWD products are identified. These products are summarized in Figure 8.1, where they are denoted by the letters A, B, C, D and E. These different products are associated with the drilling phase during which they can be available and with their relevant geophysical applications, where these products can be of valid assistance in assessing the drilling results.

Generally these applications are used while drilling, together with surface-reflection-seismic data. Because the surface-seismic sections are, in general, obtained using P waves, the most common application is to exploit these products with compressional drill-bit signals.

8.2.1 Checkshot

The first product is the "checkshot". It requires the identification of the travel times of the direct compressional arrival from the source point (the bit) to a receiver close to the wellhead (near offset). This basic information gives the bit position in time on a reference seismic section crossing the well location (Figure 8.2a). As shown in Figure

8.1, this information and its application may be obtained during all the stages of drilling. Successive data points provide the seismic-velocity function on the well location, since the times of the first arrivals of the bit signal can be picked on the RVSP seismograms and because the depth of the drill bit is known (Section 5.10).

The quality of the available data is crucial for while-drilling checkshot monitoring and putting this application into practice. In general, the "*verticalized*" (corrected to vertical travelpath) direct-arrival times t_{FB} of near-offset traces are used for this purpose. In a constant-velocity model, the verticalized time is calculated by (Figure 8.2b)

$$t_0 = t_{FB} \cos \vartheta \tag{8.1}$$

where

$$\vartheta = \arctan \frac{X}{Z}, \tag{8.2}$$

and where X and Z are the receiver offset and bit depth, respectively. Equation (8.1) is an approximation only for small offsets and large depths (as a rule of thumb, $X/Z < 0.2$). The measured time is then corrected for the seismic-datum time t_s, i.e., surface static correction, see Section 5.16.1, to obtain the two-way seismic time at the datum plane, namely,

$$t_{\text{seismic}} = 2\,(t_0 + t_s)\,. \tag{8.3}$$

In marine applications water-depth corrections are also calculated and corrected for variation of sea-water acoustic velocity with seasonal temperature, water density and salinity (Meehan, Nutt, Dutta and Menzies, 1998). In addition to zero-offset checkshots, offset checkshots are also a while-drilling product that may provide timetables that can be used as additional information for accurate migration of surface-reflection seismic data. A comparison of the results obtained with the conventional-VSP vertical checkshot and the travel time tables based on conventional-VSP offset-checkshot data of a salt-proximity survey is reported by Krebs, Fara, and Berlin (1995).

8.2.2 Reflectivity characterization

The second product obtainable with the use of SWD data, is the "sonic-log calibration". It is indicated by the letter "B" in Figure 8.1. Its relevant application is the computation of a calibrated synthetic seismogram using well-log data and SWD calibration times during intermediate drilling phases. This product allows for the while-drilling seismic characterization of the horizons reached by the bit.

Using the integrated travel times of equation (1.2) and the intermediate-log data (velocity and density), it is possible to process the intermediate synthetic (acoustic) seismogram to accurately identify the drilled horizons and to characterize them seismically (Section 1.2.5).

From an operational viewpoint, this application is generally carried out when the hole diameter changes and intermediate wireline logs are recorded. The computation of a synthetic seismogram may take place while the casing is run in and cemented. In this way, it is possible to identify all the main seismic markers drilled through in the previous

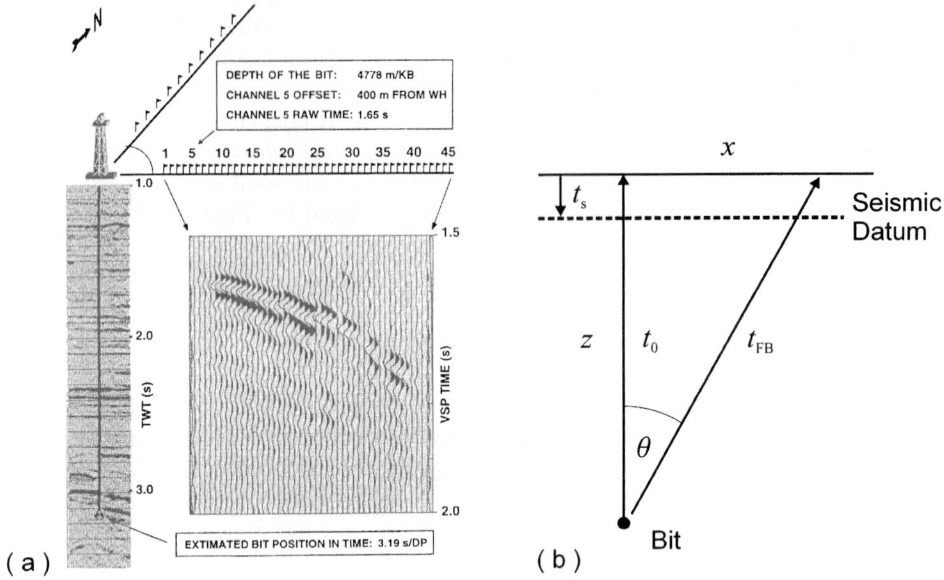

Figure 8.2: Checkshot, synthetic and prediction. Modified after Miranda, Aleotti, Abramo, Poletto, Craglietto, Persoglia and Rocca (1996).

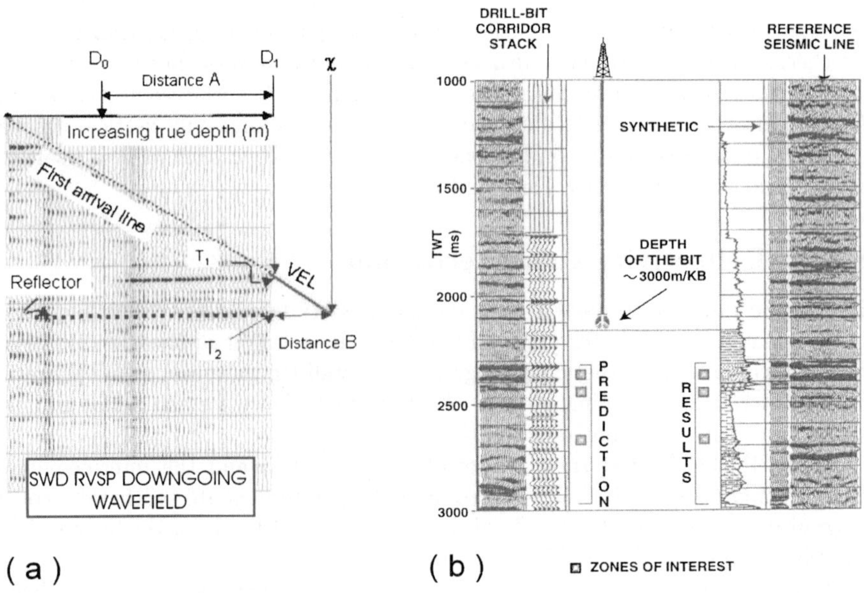

Figure 8.3: a) Prediction ahead of the bit using SWD data. b) Corridor stack and synthetic seismogram calculated by using sonic-log data are tied to surface seismic.

phase, and to assess any additional static shift required to correct the seismic time of the well data, i.e., the time shifts between the seismic and the well data can be determined by comparison of the seismic-section and synthetic traces. Consequently, the bit position in time can also be more accurately determined. When a sonic log is recorded while drilling (LWD) using a downhole tool and MWD telemetry, the use of the drill-bit-SWD arrival times for calibration makes it possible to produce a synthetic seismogram while actually drilling.

8.2.3 Prediction ahead-of-the-bit

The third product is the determination of the formation ahead of the bit using the reverse VSP drill-bit data (letter "C" in Figure 8.1). It can be obtained when the number of the recorded depth levels is sufficient (say, at least a few tens in a 200 m or 300 m depth interval) for wavefield separation (Section 7.9.3). Moreover, it is an important requirement that the SWD data have repeatability (Section 7.2) and a sufficient signal-to-noise ratio. These data may be successfully processed using the entire seismic trace to extract the upgoing wavefields (direct arrival and multiples) and the downgoing wavefields (primary reflections). The latter, free from multiple reflections after VSP deconvolution, provide the real seismic trace at the well location.

Among the various applications of a VSP, its use for predicting the horizons below the bit seems to be the most interesting during drilling. Figure 8.3a shows how to obtain the prediction of a reflection in two-way time in a single-offset SWD RVSP.

Graphical approach – The direct-arrival-time line is projected in depth with the average slope of the last drilled interval. In Figure 8.3a, D_1 and D_0 are the bit total depth and an arbitrary depth, respectively; χ is the predicted distance below the total depth. If A and B are the distances measured in the scale of the graphic, we simply have

$$\chi = \frac{(D_1 - D_0) \cdot B}{A}. \tag{8.4}$$

The calculated depth of the reflector is

$$D_P = D_1 + \chi. \tag{8.5}$$

Analytical approach – Velocity V_{pred} is extrapolated below the total depth by, as well as measured by an analysis of seismic-reflection data or from sonic-log data of nearby wells. T_2 and T_1 are the measured times of the reflector and the total depth, respectively. The predicted depth of the reflector is

$$D_P = (T_2 - T_1)V_{pred} + \chi. \tag{8.6}$$

Figure 8.3b shows the prediction and the reflectivity characterization by SWD data.

8.2.4 Multioffset VSP

The fourth product is the "multioffset VSP" (letter "D" in Figure 8.1). This product can be processed and used essentially at the same time as the product "C". As the recording seismic line extends several kilometers from the well, several tens of receiver groups are typically recorded with different offsets (Section 5.13.1). These traces can be processed while drilling to improve the results obtained with the zero-offset RVSP. The multioffset processing consists either of migration of 2D RVSP data or of common-depth-point (CDP) stacking of the reflections, which gives the familiar VSP section of "bell" shape at the sides of the well (see, for example, Figure 8.46).

Multioffset processing of SWD seismograms has the following advantages.

Lateral images and resolution

Multioffset processing gives, besides the information close to the well, while-drilling structural information at a certain lateral distance from the well. The lateral investigation distance extends up to half the maximum offset if the geological model is flat. Images are obtained from migrating RVSP data or processing the VSP-CDP transform (see Section 1.2.3). The maximum lateral resolution is determined by the Fresnel zone of the VSP signal. The Fresnel radius $R_{\rm F}$ can be approximately expressed by (Hardage, 2000)

$$R_{\rm F}^2 = \lambda \frac{z_{\rm R}\,(z_{\rm R} - z)}{2z_{\rm R} - z}, \tag{8.7}$$

where λ, z and $z_{\rm R}$ (with $z_{\rm R} > z$) are the signal wavelength, the bit and reflector depths, respectively (Figure 8.4). We can verify that, with the bit source at a distance from the reflector of 0.1 times the reflector depth and with a signal frequency of 20–30 Hz, one obtain a The lateral resolution intermediate between that obtained with a typical conventional VSP (having shorter λ) and a typical exploration surface seismics (of comparable λ and $z = 0$).

Improved velocity analysis

Multioffset processing improves velocity analysis by providing interval and RMS velocities in the proximity of the well. These data are compared with the tomographic inversion maps (Section 8.8.6).

Improved reflections for prediction

Multioffset processing improves significantly reflections and velocity analysis, also when single-offset processing (Dillon and Thomson, 1984) of SWD drill-bit data may be poor. The concept is shown in Figure 8.5a, where a seismic line of $N_{\rm g}$ geophone traces with offset ranging from Δx to $N_{\rm g}\Delta x$ and intertrace distance Δx is assumed. The depth of the reflector to be drilled is $z_{\rm R}$. The reflector prediction in a given point $x_{\rm D}$ is done initially, say, when the bit is at the position z_1. The bit depth is chosen in such a way that the ray from the bit to the geophone is reflected in the fixed point $x_{\rm L}$. As the bit approaches the reflector, the coverage COV, i.e., the number of traces stacked, in a point at a distance

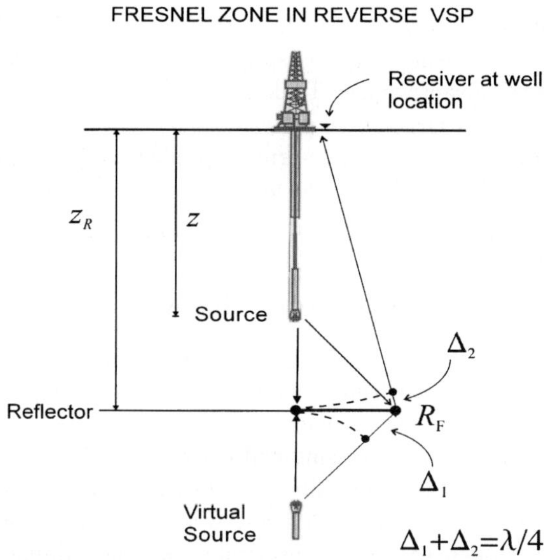

Figure 8.4: First Fresnel zone in reverse VSP. At the radius R_F, the length of the diffracted travelpath is longer by $\lambda/4$ than the length of the vertical reflected travelpath.

x_L in the CDP stack (or migration) increases. That is, coverage is given by the number of offset traces measuring the reflections in x_L and, in any case, COV $\leq N_g$. If, for simplicity, we make the assumption that we start the survey from the surface (bit depth $z = 0$), the coverage is approximated by

$$\mathrm{COV} \sim \frac{(z_R + z)x_L}{(z_R - z)\Delta x}, \qquad \text{(in any case } \leq N_g). \tag{8.8}$$

The coverage trend of equation (8.8) is shown in Figure 8.5b, where also

$$S/N \sim \sqrt{\mathrm{COV}} = \sqrt{\frac{(z_R + z)x_L}{(z_R - z)\Delta x}}, \qquad (\leq \sqrt{N_g}) \tag{8.9}$$

is plotted versus increasing depth. In other words, if we assume unit S/N for each VSP-reflection trace, we have approximately

$$S/N_{\text{stack}} \to \sqrt{N_g}, \tag{8.10}$$

where N_g is the total number of the offset traces. Figure 8.5b shows this trend as the bit approaches the reflector. This geometrical ray calculation does not include geometrical spreading, model complexity, absorption losses (Toksöz and Stewart, 1984), AVO effects (Leaney, 1994; MacBeth, 2002) and radiation properties of the bit source, which may be not negligible. Indicatively, in real applications with some tens of traces we could expect an average improvement in S/N of the SWD reflections from about three to five times in the approach to a reflector. Because the S/N ratio for the direct drill-bit arrivals can be of the order of some units, using large 2D spreads with tens of geophone traces in SWD surveying is an effective advantage (see Table 1.1).

Although in many cases large-offset pre-critical reflections show a better S/N ratio, near-offset traces are important for while-drilling prediction in the proximity of the well (Figure 8.6). Products "C" and "D" are the most important for taking operative decisions such as setting casing points.

8.2.5 Geophysical monitoring of the well

This product (denoted by the letter "E" in Figure 8.1) is a summary of all the previous ones; essentially during the whole drilling of a well, the well-drilling evolution is monitored from a geophysical point of view through the use of checkshot, sonic-log calibration, reverse-VSP and multioffset-VSP processing and interpretation.

This application is important especially in geologically complex areas, where the definition of the correct geological model is difficult. If the geological model is wrong, actions must be taken early enough to correct the model and the drilling plan. As wells are normally drilled on the basis of information coming form other wells in the same area, correlation between the wells is done using seismic lines. However, in the area between the reference well and the newly drilled well, some uncertainties may be present that can lead to mispositioning. Moreover, the geological model for the well has to be given in depth. This model is usually obtained by performing the time-to-depth conversion of the interpretation done on the time seismic line. Therefore, a correct definition of the velocity model is one of the most critical issues, especially in geologically complex areas. This is one of the reasons why it is important to be able to follow the drilling of the well and to monitor its evolution from a geological and geophysical point of view.

8.3 Drilling and real-time migration

Another application is the while-drilling (or "near-real time") reprocessing of the surface reflection seismic. The approach consists in starting from an initial, or reference, depth model (built before drilling) and continuously refining it, using all the information derived from SWD or LWD measurements.

Bertelli and di Cesare (1999) proposed that we utilize the available velocity information acquired during the drilling of a well to adjust, in "near-real time", the initial geophysical model. Where the initial seismic data quality is poor, the iteration of the depth migration using an updated velocity model can produce a sensible improvement in the geological model itself. The new seismic depth section is migrated while drilling with the updated velocity model and the integration of all the geological/geophysical and drilling data available while drilling. This procedure makes it possible to estimate the drilling results in real time. For this purpose, the well-velocity information is interpreted and integrated with the seismic velocity to define a "new" velocity function. The adjusted function is used to update and calibrate the initial seismic macro model. The new model is then checked through seismic modeling and used to run a new depth migration. In this way, it

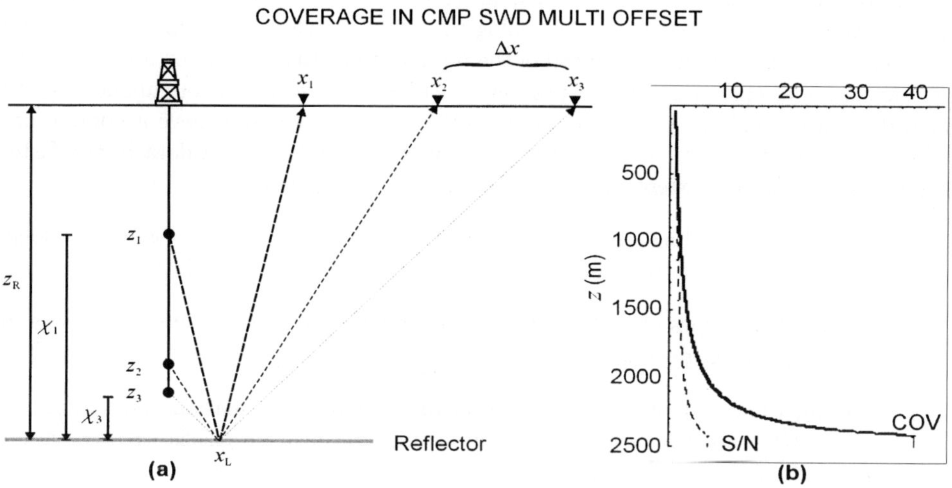

Figure 8.5: a) Prediction and b) coverage (continuous line) by multioffset SWD. In (b) S/N is the dotted line and 40 receiver traces are assumed.

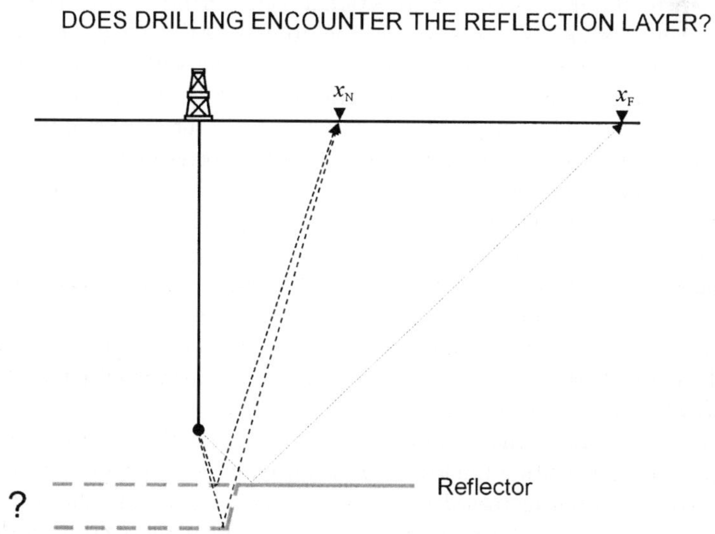

Figure 8.6: Prediction by near- and far-offset traces may lead to different results in the well proximity.

is possible to calibrate the depth section up to the drill-bit position by using the while-drilling information velocity. As the geological model is refined and updated, it is also possible to better image what is ahead of the bit and, thus, to make important operational decisions such as changing the well program, picking the casing position, abandoning the well, drilling beyond the prognosed total depth, etc. In order to achieve an operational impact it is necessary to collect, process and interpret while drilling data in the fastest possible way. The basic steps of this methodology are the following:

1. Define, before drilling, the depth reference model, using all the available velocity information derived from seismic, well data, regional geological data, etc.

2. Collect all the types of while-drilling information derived from SWD, LWD data, as well as logs, geological and drilling data.

3. Perform, while drilling, evaluation of all the available information and verify, on a continuous basis, the coherency of the initial geological prognosis with the actual situation.

4. Improve the initial geological prognosis coherently with the while-drilling information, refining subsurface depth image via real-time reprocessing of seismic data.

8.4 Deviated-well monitoring

SWD data can be used to monitor directional drilling (Section 2.5). Figure 8.7 shows a real case where the deviation of a directional well can be seen in the signal of the different geophones of an SWD line laid on the ground. At shallow bit depths the well was not deviated, or only very little (say 50 m) from the vertical axis, and the signal direct arrivals appear with a nearly symmetric moveout in the geophone traces of the common-source gather. The left panel of Figure 8.7 shows the synthetic traces calculated with a ray-tracing code, while the right panel shows the corresponding shot-gather traces acquired in the field. As the well became deeper, at a depth of 3300 m, the wellpath was deviated laterally around 500 m from its vertical axis. The panels of Figure 8.7 show both the synthetic and real data after deviation. A strong asymmetry can be observed in the arrival times of the drill-bit signal. This is a simplistic approach to measure the deviation in directional drilling using SWD data, since the deviation from a symmetric arrival can depend also on the geological structure. However, in this example, we can appreciate how a SWD survey can be used to give an approximated measure about well deviation. A refined tomographic analysis of the velocity/geological model is necessary to obtain information about the deviation of a well from its planned vertical trajectory.

In general, during deviation the loss of transmitted pilot signal, extra radiation points at the drill string, and an increased presence of coherent noise are the factors affecting the signal-to-noise ratio (see Section 4.9). Figure 8.8 shows the SWD data measured after deviation which, although less energetic, have features in good agreement with those of the data recorded during the vertical drilling. In this case deviation was around 45°. Commercial applications of drill-bit seismic with deviation up to 65° are reported by Meehan, Nutt, Dutta and Menzies (1998). Some experiments in horizontal drilling geometry

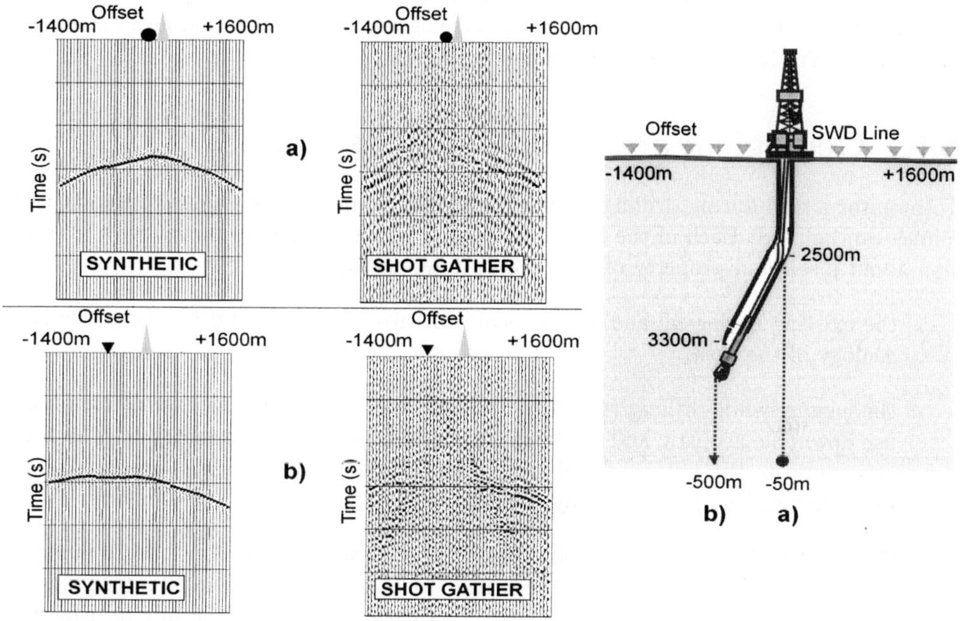

Figure 8.7: SWD monitoring of deviation.

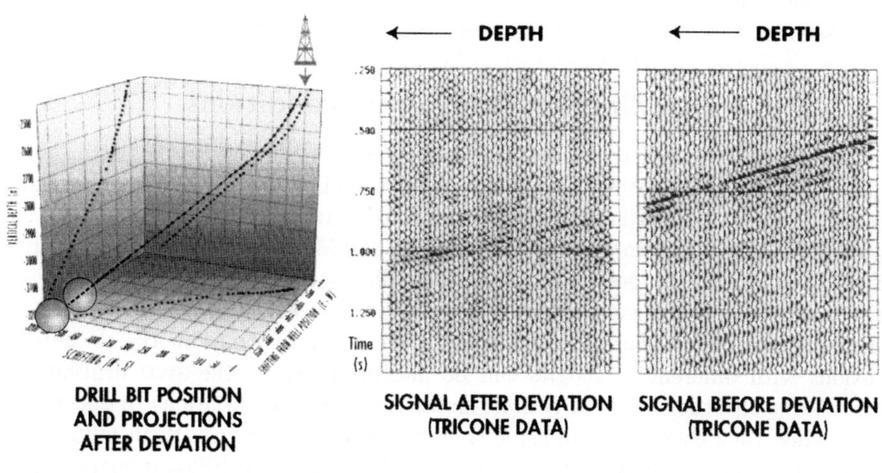

Figure 8.8: P-wave radiation pattern and SWD data acquired after and before deviation.

were performed by using shear waves to monitor the drill-bit trajectory (Chameau, Batini and Omnes, 1991) as well as by using crosscorrelogram migration techniques for imaging purposes by Yu, Katz, Followill and Schuster (2001).

8.5 Geological and lithological aspects

To monitor a well during drilling, from a geological point of view, several different techniques can be used. Each of the following tools provides the explorationist with information about a selected property of the rock at different scales:

- ◇ the existing geological and geophysical data give prior information about the local geology of the area;
- ◇ the logging-while-drilling (LWD) and measurement-while-drilling (MWD) techniques use downhole-acoustic and electromagnetic tools with mud telemetry to detect and communicate the petrophysical parameters of the rocks measured only in the immediate vicinity of the well, say, at a distance of the order of a few meters;
- ◇ the mudlogging service, generally present on well site (see Section 2.3.17), gives the lithological information about the drilled rocks analyzing the rock cuttings which come out of the well. Moreover, the drilling parameters, also generally measured by the mudlogging company, are used for quality control of the drilling activity;
- ◇ finally, the seismic-while-drilling (SWD) technique gives information about the acoustic and elastic properties of the drilled and to-be-drilled rocks, both in the vicinity of the drilling bit and in the area around the well.

With the joint use of all these different types of information, it is theoretically possible to successfully monitor while drilling a well from the geological point of view.

8.5.1 SWD with different lithological conditions

The SWD response in different lithologic conditions was analyzed by Poletto, Miranda, Corubolo and Abramo (1997). They analyze the signal collected in several vertical wells drilled in soft (sand and shale) and hard (limestone) rocks. Analysis of drill-bit signals propagated in salt domes are reported by Casserly and Marion (1992). Because in the salt-proximity surveys the raypaths are often perpendicular to the borehole axis, the S-wave energy is often the dominant energy.

According to the bit-source-radiation pattern and acquisition geometry, the subsurface conditions with different lithologies can be interpreted using the SWD drill-bit-seismic waves. As in all other types of seismic data, the lithology variations of the rocks can be seen when they correspond to some acoustic/elastic variations. With SWD data it is important to keep in mind that the lithology of the rock being drilled influences also the way the rock is drilled in terms of type of selected bit, drilling parameters (Section 7.4) and bit/rock interactions (Section 4.11.2). It is, therefore, of utmost importance to look at the drill-bit seismic data keeping in mind the bit type used and monitoring the drilling parameters such as torque, RPM, WOB and ROP. These parameters are stored in the seismic-trace headers for quality-control purposes (Section 5.9). Different case histories are presented below. The characteristics of the different lithologies are presented together with the drilling parameters used during the recording of the SWD data.

The first case history concerns a well drilled in a shale and alternating shale-sand sequence (Figure 8.9). Here, and in the next two figures, we can see: in the middle panel the total field of the near-offset reverse VSP SWD recorded data; in the right panel, some typical drilling parameters (torque, RPM, WOB and ROP) and the signal-to-noise ratio (S/N) of the SWD data. The signal-to-noise ratio is calculated measuring the RMS trace amplitude on two windows, one before the direct arrival (noise) and the other on the direct arrival (signal). To the left of these figures, the lithological column of the drilled sequence is shown.

We can see in Figure 8.9 the strong decrease in the signal-to-noise ratio when the drilling gets into the alternating clay/sand formation drilled by a PDC bit. Here the overall data appear noisier even if the PDC-bit signal is detectable in continuity with the roller-cone-bit signal. The unfavorable conditions, from a SWD point of view, are mainly tied to the decrease of the WOB drilling parameter and increase in ROP, which may limit the total listening time per single depth level of a prefixed depth interval. To some extent, the total source energy for depth level should be related to the energy expended to drill a unit volume of rock – i.e., the specific energy E_S (Section 3.6.1), which is inversely proportional to ROP – and to the formation strength, which is in turn related to rock density, ρ. However, a correct calculation of E_S requires the measurement of the downhole torque since the surface value is different due to friction losses (see discussion of Section 3.5.1). Unfortunately, while-drilling downhole measurements are not frequently available. Even if a lower ROP may produce favorable S/N conditions (Miranda, Abramo, Poletto and Comelli, 2000; see also Section 7.4), in some cases an increase in ROP can be related to an increase of signal amplitude (see Sections 3.12.3 and 8.12.3). In general, it is important to consider the drilling parameters that are essential for quality control of the data and for possible source normalization. Any analysis of amplitude effects on the recorded SWD data must be done keeping in mind the different drilling parameters and bit types.

The second case history is a well drilled in a sequence which begins with clay and, at a depth of 1240 m, gets into a limestone formation (Figure 8.10). The slope of the first arrival (middle panel) shows the strong velocity variation at the top of the limestone sequence. In this case, we can see a strong SWD signal up to 3500 m. Furthermore, on this dataset strong signal variations have been recognized at the lithological change.

The third case history is important too because a part of the well has been drilled with a PDC bit (from 1300 to 1350 m depth) and shows a good signal (Figure 8.11). Here we can see the raw acquired/preprocessed SWD-RVSP data showing a strong reflection. This event is shown here in one-way-time (OWT). As can be seen, even if all the four drilling parameters reported on the right have rather constant values, at the beginning of the anhydrite/dolomite sequence, the S/N ratio decreases quite sharply. S-wave arrivals can also be seen readily.

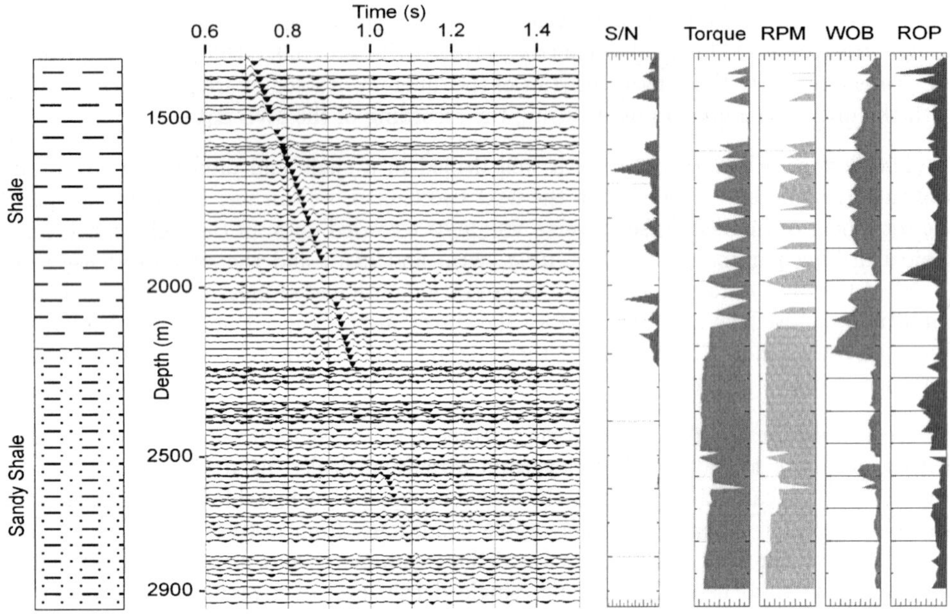

Figure 8.9: Well drilled in a shale and sandy-shale sequence.

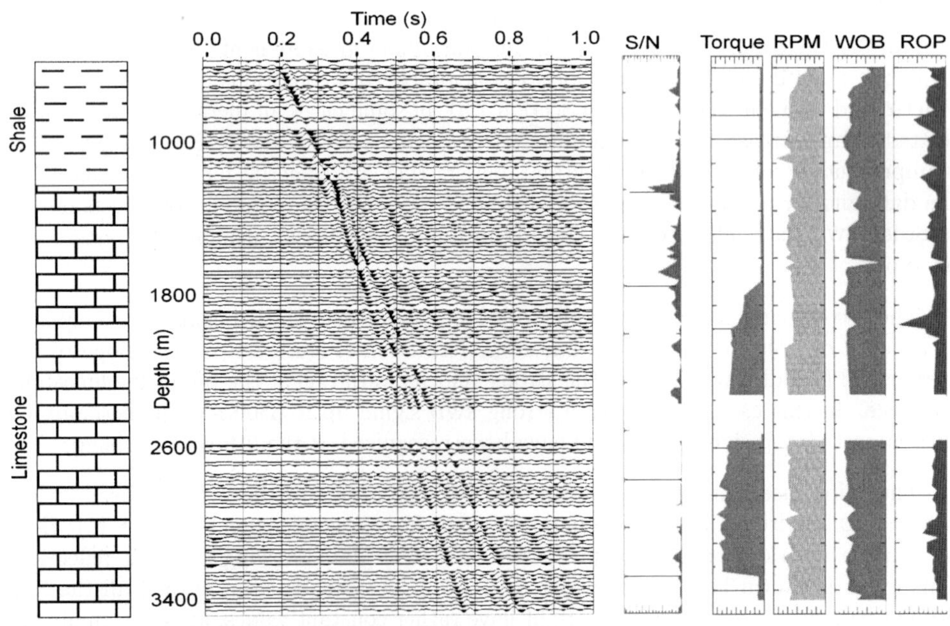

Figure 8.10: Well drilled in a shale and limestone sequence.

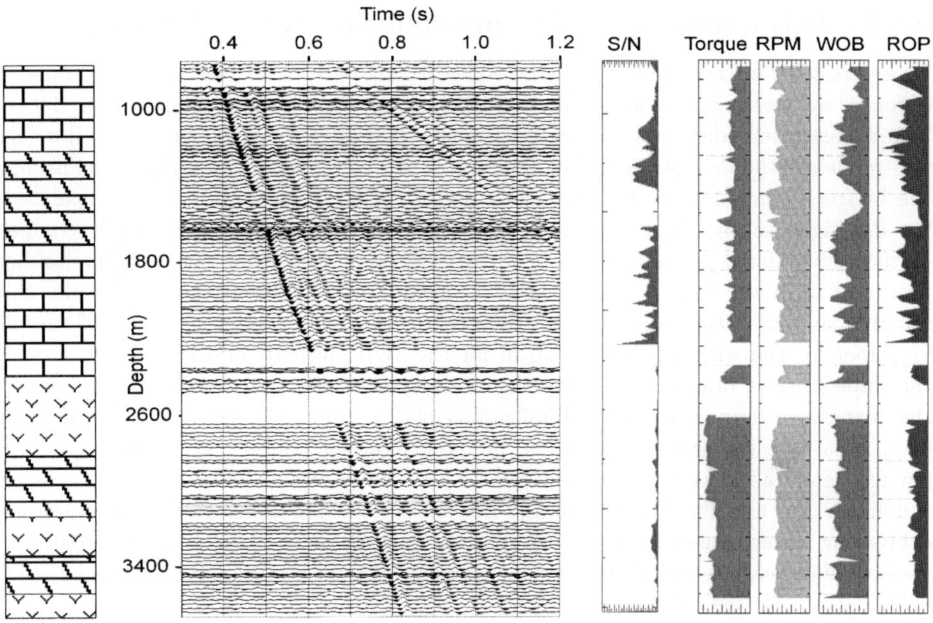

Figure 8.11: Well drilled in limestone, dolomite and anhydrite alternances.

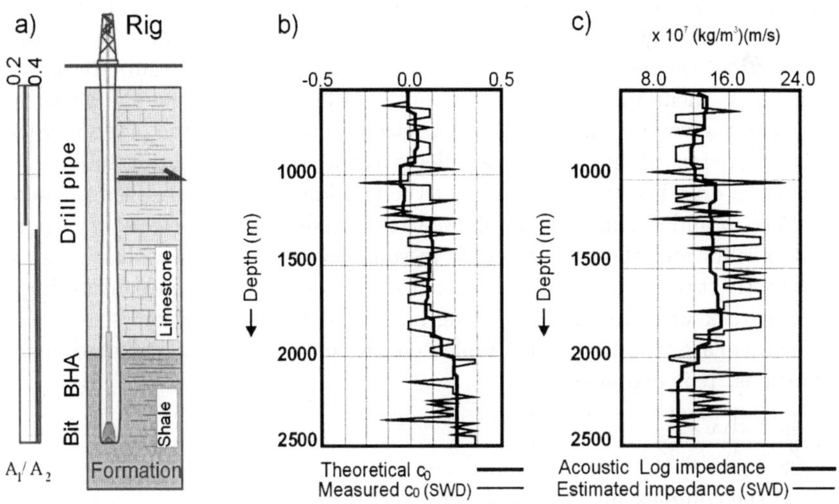

Figure 8.12: Analysis of bit/rock reflection coefficient c_0 with real examples. Panel (a) shows the well with the lithostratigraphic profile and the variation of the ratio A_1/A_2 of the drill-collar cross-section (a). Panel (b) shows the measured reflection coefficient (thin curve), compared with the theoretical reflection coefficient (thick curve) calculated by using conventional sonic-log data. Figure (c) shows the acoustic impedance calculated using drill-bit data (thin curve) and the acoustic impedance measured by conventional sonic-log data (thick curve).

8.5.2 Estimating acoustic impedance from SWD data

As a particular further application, the while-drilling detection of the acoustic impedance of rocks have been carried out in different lithological conditions for selected SWD datasets. The acoustic impedance is calculated from SWD data after analyzing the pilot signal and modeling the drill-string waves (Section 6.10), using proper boundary conditions at the bit/rock contact (see Sections 4.11.1 and 4.11.2). These data can be calculated as the drilling goes on. In Figure 8.12, the drill-bit acoustic-impedance log is compared to the conventional acoustic impedance (AI) log calculated using the velocity log being produced after the well was finished. The comparison between the two results shows remarkable similarities; some of the differences are due to different types of sampling (1/2 foot for the wireline log and 20 m for the SWD method) and due to the fact that the impedance is obtained by inverting the bit/rock reflection coefficient calculated in the plane-wave approximation. Limitations of this approach are discussed in Section 4.11. The reflection coefficient at the bit–rock interface contact, c_0, is analyzed in the low-frequency approximation in Section 4.11.1. By assuming ideal conditions, as with a non-working and flat bit, a laterally-free bar in the borehole, and using extensional drill-string waves transmitted to the formation in the form of plane P-waves, Poletto and Malusa (2000) calculated the reflection coefficient between bit and formation as be written as

$$c_0 = \frac{A_1\sqrt{\rho_1 Y_1} - A_2\rho_2 V_{P2}}{A_1\sqrt{\rho_1 Y_1} + A_2\rho_2 V_{P2}}, \tag{8.11}$$

where A_1, ρ_1, and Y_1 are the cross-section of the drill-collars in the BHA, the mass-density, and the Young modulus of the drill string, respectively. A_2, ρ_2, and V_{P2} are the borehole cross-section, the density and the P-wave velocity of the formation. Actually, at the bit-rock contact, shear waves in the formation and flexural waves in the string are also produced, and c_0 given by equation (7.55) has to be considered as the plane-wave approximation calculated assuming only linear effects at the drill-bit reflection interface. Moreover, the model does not take into account the non-linear effects involved in the drilling process. Poletto and Malusa (2002) showed that, under this low-frequency approximation, these results and those obtained from sonic-log data match with a good average trend. Recently, the use of the bit-rock reflection coefficient for measuring impedance of the drilled rock was discussed including near-field effects (equation (4.145)) (Poletto, 2002a). In this case, the bit/rock complex coefficient accounts for radiated energy and local vibrations. It is expressed as

$$c_0(\omega) = \frac{\mathcal{Z}_b - A_1\sqrt{\rho_1 Y_1}}{\mathcal{Z}_b + A_1\sqrt{\rho_1 Y_1}}, \tag{8.12}$$

where $\mathcal{Z}_b$ is the complex bit impedance given by equation (3.100) and c_0 is the complex bit/rock reflection coefficient variable with the frequency of the bit signal.

8.6 Comparison of SWD and wireline VSP results

Although results of seismic while drilling are obtained with reciprocal geometry (Section 3.2) and with typically different receiver-offset configurations, the comparison with conventional VSP is worthwhile for evaluation of the quality of the single-offset traces at different recording depths.

Figure 8.13b shows the field crosscorrelated seismograms of a wireline VSP acquired by using a Vibroseis source and a SWD VSP acquired in the same well. Some minor differences are due to the fact that the conventional wireline VSP was acquired at zero offset while the SWD-VSP traces were acquired at an offset position of 400 m, where the weathering-correction are different. The data, recorded between 3200 and 3500 m, show some differences in the signal-to-noise ratio. The signal-to-noise ratio was calculated with a time window centered on the direct arrivals and a window in times lower than the first break for measuring signal and noise, respectively, in both SWD (Figure 8.13a) and wireline-VSP data. Figure 8.14a shows the S/N curves for Vibroseis and SWD VSPs. In the trend of the S/N curve of the wireline VSP, geometrical spreading and attenuation effects with distance are more evident. The trend of the S/N curve of the SWD VSP is flatter, also at surface recording levels. These data show some large amplitude variations between close bit depths due to different drilling conditions (such as casing, drilling mode, type and status of bit and drilling-dynamic parameters). Obviously, the SWD data contain a higher level of noise introduced by the correlation and preprocessing steps, in which the borehole-radiated waves and noise recorded in the pilot signal make a coherent contribution. However, the difference between the wireline and SWD results is more marked for shallow-surface recording levels, which are also more sensitive to receiver-array spatial filtering. For traces at intermediate and higher recording depths, the signal-to-noise ratio of wireline and SWD data is comparable. In this example, S/N is calculated by using direct arrivals. The lower S/N value of the drill-bit data indicates that the reflections in SWD data may have a S/N of the order of unit before the processing of the SWD reflected wavefields.

Figure 8.14b shows the results of the two-way-time reflections obtained by processing the wireline-conventional and the SWD VSPs. The carbonate target of the well is clearly identified by the troughs at 3.3 s reflection time. This event is marked by the bullets in the figure and is clearly recognizable in both the drill-bit (while-drilling) and wireline (after-drilling) results. These single-offset data show some difference in frequency content. The target depth was at about 4800 m depth. It was predicted by SWD about 300 m before it was reached by drilling.

8.7 Prediction by SWD in favorable conditions

The SWD case history presented below has been acquired in an area where the first 600 m were drilled in a sand and shale sequence. The rest of the well crossed a limestone/dolomite sequence which, from a SWD point of view, can be considered among the best types of lithology for signal response. In addition, the well was vertical, drilled by rotary, mainly with roller-cone bits. These particularly favorable conditions allowed geophysicists to investigate the achievable prediction distance and resolution.

Figure 8.15 shows the comparison of the SWD-VSP reflections recorded using the bit as a source between 708 m and 3522 m and the reflections of a conventional VSP acquired using an air gun (200 c.i. at 110 bar pressure) positioned at 59 m offset from the wellhead.

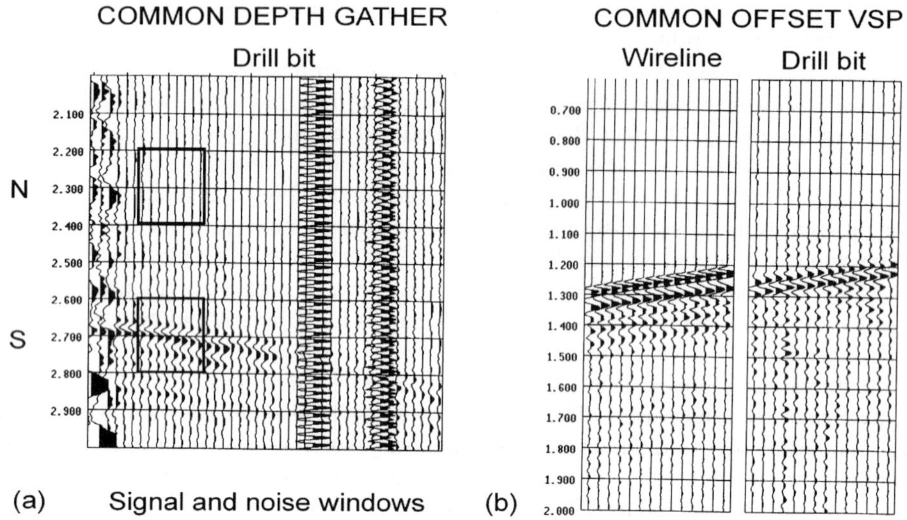

Figure 8.13: a) Time windows of signal and noise analysis (horizontal scale is offset, vertical scale is correlation time). b) Wireline and drill-bit SWD VSP for the same depth levels (approximately between 3200 and 3500 m).

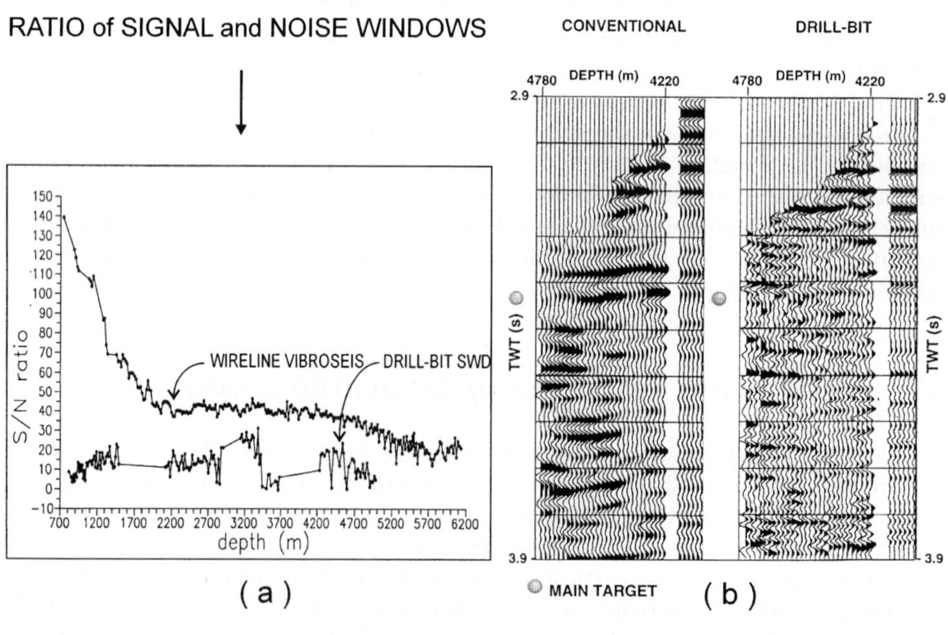

Figure 8.14: a) S/N (energy) based on analysis of direct arrivals of conventional wireline and SWD VSPs versus drilling depth. b) Reflections in two way time of conventional and SWD single offset VSPs. The bullets mark the predicted target.

The conventional VSP logged the interval between 148 m and 2270 m with a spacing interval of 20 m in depth. The correct comparison between the two VSPs should take in account the differences in the acquisition geometries and the possible residual-static differences. However, a very good matching between the SWD and the conventional VSP data can be recognized. The data quality is comparable in terms of frequency content, and the fact that the reflections are recognized in the different panels demonstrates how good the while-drilling VSP can be. In these situations, where the frequency content of the SWD data is as high as 60–80 Hz (Figure 8.16), a high precision, in terms of direct-arrival picking and velocity calculation, as well as prediction ahead of the bit are achieved. In this case, $V_P \sim 5000$ m/s at the bit and the vertical-resolution distance is of the order of 20–30 m. For this particularly good SWD dataset, prediction ahead of the bit was possible with high precision already very early during drilling.

Figure 8.17 shows three different panels of the processed while-drilling VSP, with increasing depths from right to left: these single-offset reflection data have been processed at three stages during the drilling of the well. The target/marked horizon corresponds to the top of an anhydritic layer and was a position in which to set the casing point. This reflector is marked in each of the three panels by an arrow. As the drilling was getting deeper, from 1383 to 1783 m and then to 2181 m, this marker was getting more and more evident. The depth of this marker was estimated at 2244 m using the time information from the SWD data. The prediction of the marker, given as early as 1783 m, was compared with the measured depth of the target reached during the drilling. The error in the prediction was of the order of 10 m, in a measure obtained more than 450 m before the target. Generally, the prediction distance from which the SWD data can produce reliable results is of the order of a few hundred milliseconds, which can correspond to approximately 200–300 m of distance from the reflector, depending on the interval velocity of the to-be-drilled section.

Prediction by multioffset-SWD data is shown in Figure 8.18. In this particularly favorable condition, both roller-cone and PDC data have been processed and used; a comparison between multioffset VSP processed using roller-cone-bit and only PDC-bit data is shown in Figure 8.18. Despite the large variability of the data quality in the only 13 depth levels of the PDC dataset, the multioffset processing of the PDC dataset allowed a good detection of the reflections, with a prediction distance of about 650 m ahead of the bit. This result is confirmed by the comparison with the tricone walkaway VSP and by later drilling data. The lateral structural reconstruction obtained with these VSPs is around 400 m for the roller-cone result and 300 m for the PDC result shown in the figure.

8.8 SWD in geologically-complex and poor-seismic-response area

We will now present a case history from an area (Val d'Agri) located in the Southern Apennines in Italy. In this area with complex geology and poor seismic response, the potentials of SWD were recognized. Several exploration wells drilled in the last years have been monitored using the SWD methodology. The mountain range in this area is characterized from a geological point of view by a series of overthrusted nappes faults

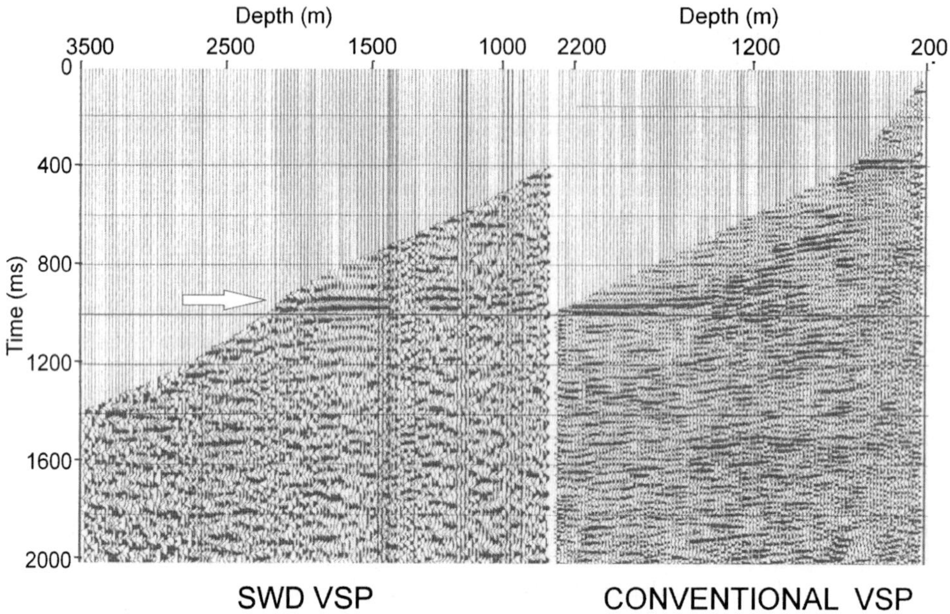

Figure 8.15: SWD VSP compared with a conventional wireline VSP. The logged interval is different in the two datasets, but the strong reflection at 980 ms is clearly visible on both panels.

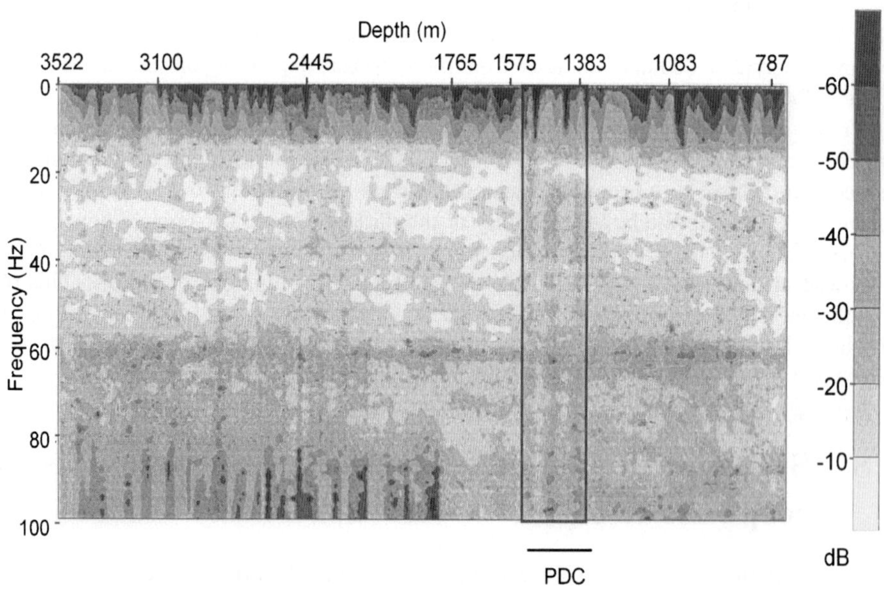

Figure 8.16: Plot of the frequency content of SWD data between 708 m and up to 3522 m. A PDC portion, around 1383–1575 m, is clearly visible with lower frequency content and a less organized signal.

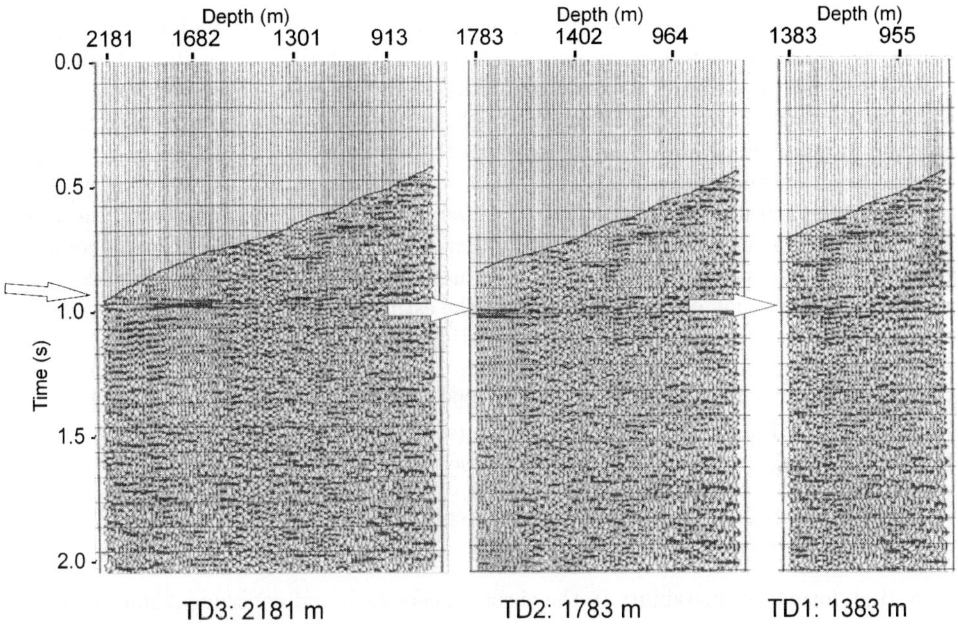

Figure 8.17: Prediction by a single-offset (400 m) trace. Three SWD-VSP-reflection fields processed as the depth of drilling was increasing. The marked horizon becomes more and more evident as the bit gets closer to the target.

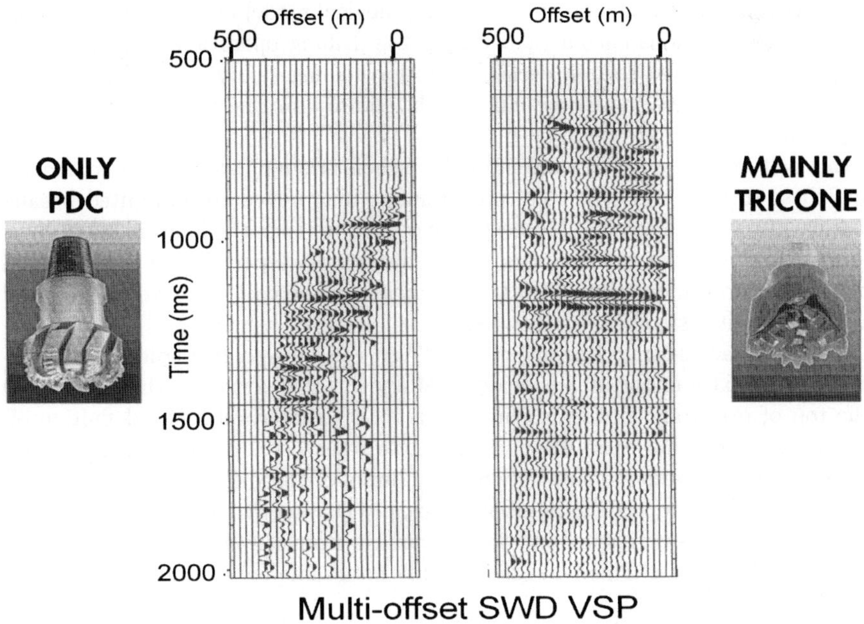

Figure 8.18: Multioffset SWD-VSP reflections processed using roller-cone (right) and PDC data (left).

which, from the tectonically higher sequences, from west to east include the Liguride sequences (typically open sea deposits), the Apennine platform (carbonate and calciclastic sediments) and the Lagonegresi sequences (shales and sands). These nappes overlap on the Apula Interna carbonates Platform which is affected by compressive tectonic. The reservoir is made up of Mesozoic carbonate of the Apula Interna Platform.

The poor seismic data quality is a direct result of the structural complexity and of the outcropping fractured carbonates. The very poor seismic response over the target area means that neither a reliable "a priori" definition of the target depth nor a confident geological prognosis can be made and – even more important – that unpredictable lateral velocity variations may often seriously affect the estimate of the depth of the target. This target is often interpreted on the seismic section as a "phantom reflection", i.e., drawn where one can not follow the event far enough to develop a map on that event alone but following nearby reflection events (Sheriff, 1999). These conditions make this belt area very challenging both in terms of exploration (seismic surveying) and drilling operations. The main drilling problems to be faced are due to:

- ◇ Great uncertainty of the formation top depth forecasts due to unpredictable velocity variations in the overburden.
- ◇ High formation instability in the thrust sheets zone and consequent mud losses.
- ◇ Overpressure zones close to the main thrust areas.
- ◇ Overpressure of the formation above the top fractured carbonate reservoir.

Therefore, the well plan based on an uncertain geological model may of course be seriously affected by the variations encountered during the drilling operations. All these factors may explain how critical certain decisions are for drilling safety and efficiency, such as setting the casing points, increasing mud weight and minimizing rig stand-by to mitigate hole instability in shaly areas etc.

In this belt area, the use of conventional borehole seismic (VSP) was fundamental for tackling these drilling and structural uncertainties but proved to be limited because of the absence of real-time results. Moreover the operative risks, the long rig stand-by time, the poor surface coupling which causes "fair" data quality especially for offset VSPs, shot to reduce structural uncertainties led to an increase in the cost of conventional borehole operations without a significant added value.

A typical surface seismic line of the Val d'Agri can be seen in Figure 8.19. Here the structural complexity of the area is evident. Many thrust faults cross the seismic section and the top of reservoir, shown in white, is difficult to interpret. The definition of the allochthonous layers above the Apulian carbonate reservoir is quite problematic mainly due to the lack of correspondence between the limits of the different units and acoustic variation whereas the top of the reservoir is usually characterized by an abrupt increase of acoustic impedance corresponding to the passage from the more marly sequences to the carbonatic limestone. This passage is readily noticeable on seismic sections, VSP and on sonic logs.

The average frequency response of the seismic sections in this area can be seen in Figure 8.20. It varies from 15–45 Hz at shallower, between −0.4 s to 0.2 s two-way-time,

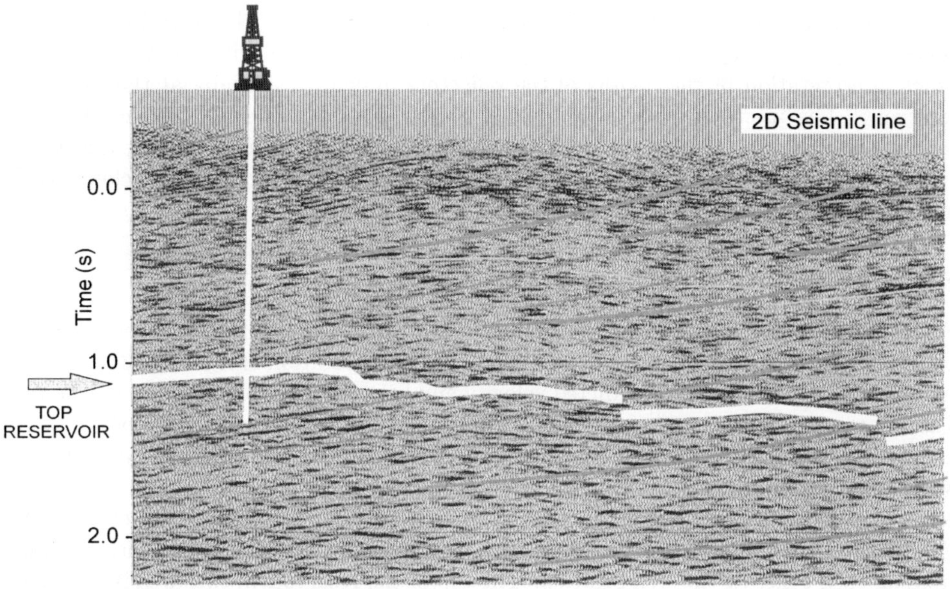

Figure 8.19: Typical 2D-seismic section in the Val d'Agri area. The top of the carbonatic reservoir is marked in white.

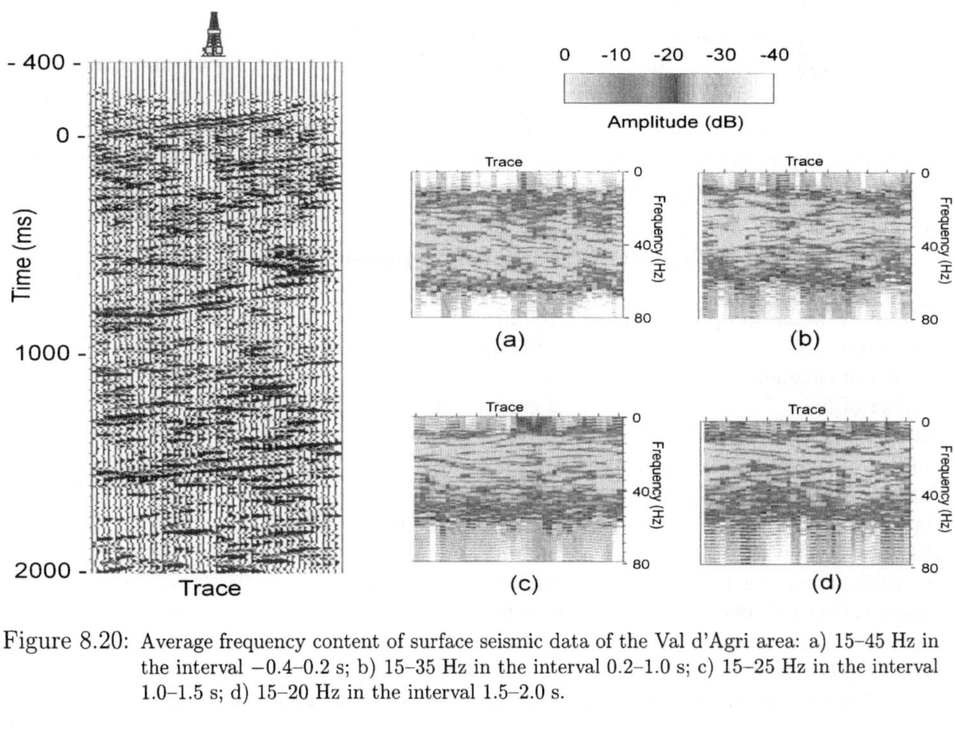

Figure 8.20: Average frequency content of surface seismic data of the Val d'Agri area: a) 15–45 Hz in the interval −0.4–0.2 s; b) 15–35 Hz in the interval 0.2–1.0 s; c) 15–25 Hz in the interval 1.0–1.5 s; d) 15–20 Hz in the interval 1.5–2.0 s.

to 15–20 Hz at deeper, between 1.5 s and 2.0 s two-way-time. This very low frequency content, together with the difficulty of performing a correct velocity analysis, makes an exact interpretation of the depths of the reflectors difficult, especially at the top of the reservoir. The interpretation is done with the support of borehole seismic. Borehole seismic data as well as SWD data have a slightly higher frequency content due to the type of acquisition with geophones in one case or the bit in the other case, closer to the investigated formation.

8.8.1 SWD applications in the Val d'Agri

For all the SWD acquisitions performed in this area, the layout of the SWD lines is selected after ray tracing performed on models based on the interpretation of the 2D or 3D reference surface seismic lines and the predicted velocity and litho-stratigraphic column for the well. The ray tracing allows us to see the reflection point distribution on the top of the target and to optimize the receiver-line layout to collect the data especially in the event the well will have a horizontal-hole section for production purposes. The ray-tracing simulation is very important, especially in an area with high structural complexity such as in the Val d'Agri. Generally, one or two 2-km long SWD lines are used, typically with two lines crossing each other at the well location. The top of the reservoir in this area is at around 3.5–4.0 km depth from the ground level, and has a maximum dip of 30°. Drill-bit data are typically recorded on the surface seismic lines with receiver-group intervals of 75 m. The depth sampling in the well is kept at around 20 m intertrace to avoid spatial aliasing with a maximum frequency of 60 Hz.

Typical acquisition parameters are listed in the Table 8.1.

After the theoretical modeling, due to the particularly complex topographic and logistic problems of this area, a scouting in the field was carried out together with a topographic survey, and a refraction survey is performed to calculate weathering velocities and static corrections. The main products of SWD in the Val d'Agri area are: continuous checkshot information, sonic-log calibration, reverse VSPs and multioffset VSPs while drilling.

Table 8.1 – Typical SWD acquisition parameters in Val d'Agri

Offset from wellhead of first channel	250 m
Line length	1800 m
Number of channels	40
Number of lines	2 crossing at well location
Distance between channels	75 m
Number of geophones per group	24
Natural frequency of geophones	10 Hz
Type of geophone array	Linear over 75 m
VSP interval depth	20 m
Acquisition time per record	24–50 s
Number of records per level[1]	30–70
Beginning of SWD survey	$\sim$ 400 m
End of SWD survey	Top of reservoir

[1] Also related to ROP and use drilling in sliding mode (in that case available time may be much less)

8.8.2 Comparison of SWD and seismic velocities

In the Val d'Agri area, with high lateral velocity variations due to the heterogeneties present in the allochtonous zone above the carbonatic reservoir, the value of results obtained by continuous-checkshot while drilling is of utmost importance. The comparison of the checkshot results with the velocities before drilling (derived from seismic-stacking velocities) allows us to improve and update the velocity prognosis and the expected target depth (Figure 8.21).

This result has been possible thanks to the general good-quality of the direct-arrivals in the near-offset channels. The checkshots acquired while drilling have been used to calibrate the sonic log and to process the synthetic seismogram (Figure 8.22). The high correlation between the reverse VSP-reflected field (downgoing), the VSP corridor stack and the synthetic seismogram makes it possible to validate, from an interpretative point of view, the quality of the SWD data. This analysis is performed also with the determination of the drift between the integrated-sonic-log times and the checkshot times (see Section 8.2.2). In Figure 8.22, the top of the reservoir is readily noticeable on all the data. The SWD-downgoing VSP, plotted at the left side of the section, shows the high-velocity carbonates overlaid by the lower-velocity, more shaly, sequence. The top of the carbonates is evident also on the acoustic-impedance log (AI) and on the derived synthetic seismogram.

8.8.3 Prediction of acoustic interfaces ahead of the bit

Reverse VSPs are used to predict acoustic (also elastic if S or converted waves are used) interfaces ahead of the bit; in the Val d'Agri area this study is done in order to monitor anomalous pressure ramps. The prediction distance depends on the drilling conditions and characteristics of the acoustic interfaces. Generally the top of the reservoir in the area was predicted 150–200 m before being drilled, with a precision of the order of 5–20 m (Figure 8.23). During the monitoring of the wells, several drill-bit RVSP (more than 10) are processed at increasing depth during drilling. This analysis is done for each SWD-acquisition offset channel.

8.8.4 Structural reconstruction near the well by multioffset

Efficient "while-drilling" structural reconstruction may be achieved in the well neighborhood using several receiver channels of the SWD lines (Figure 8.24). Generally, out of the 40 receiver channels on the SWD line, 10–15 channels are selected for the while-drilling multioffset RVSP processing. The choice of the channels is based on data quality, which is closely tied to the radiation pattern and depth of the bit but also to the particular surface geological conditions as well as drilling conditions. In general, the quality of the SWD-multioffset VSP is quite high and the comparison with the surface-seismic lines crossing the well allows us to obtain a more precise interpretation of the reflectors and faults. This type of information allows us to support drilling decisions such as well deviation and changing casing profiles. In this case, multioffset RVSPs are used to mitigate structural uncertainties around the well position in this area where the number of faults and lateral lithology variations is very high.

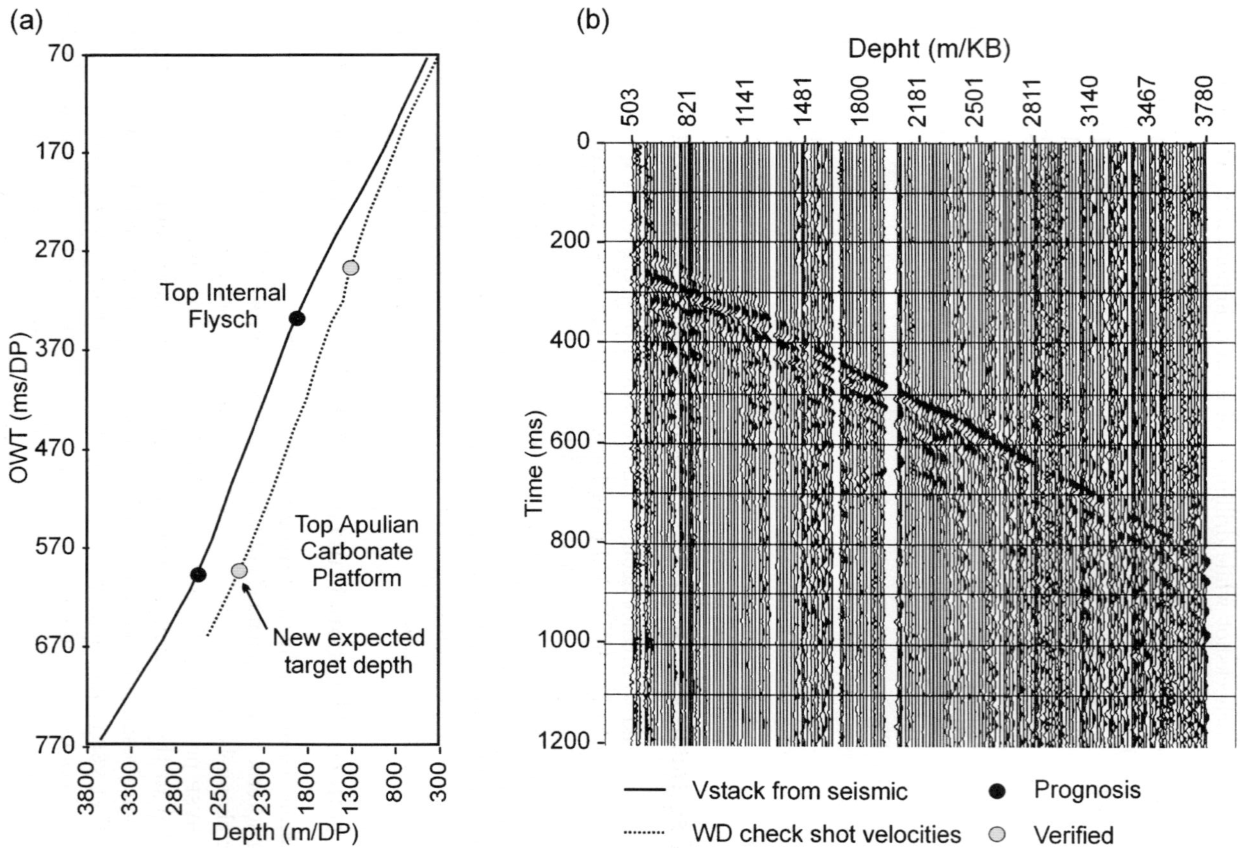

Figure 8.21: a) Comparison between travel-time versus depth derived from information available before drilling and checkshot data while drilling. b) Near-offset while-drilling VSP channel. Good-quality first arrivals are readily noticed through the section.

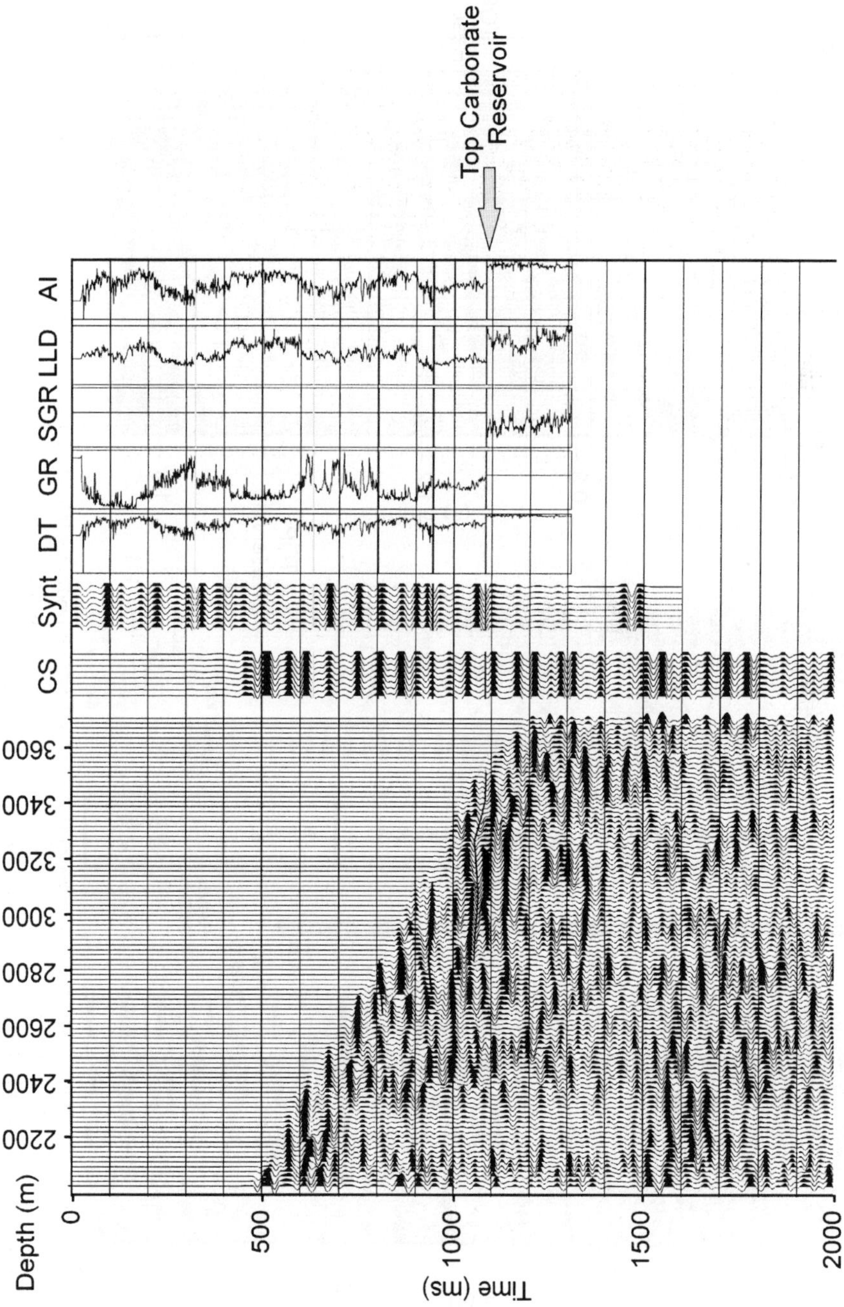

Figure 8.22: Composite display containing the processed SWD-VSP data together with the calibrated sonic log and synthetic seismogram. From left to right: interpreted downgoing while-drilling reverse VSP, corridor stack, synthetic seismogram, sonic log, gamma-ray logs, resistivity log and acoustic-impedance log.

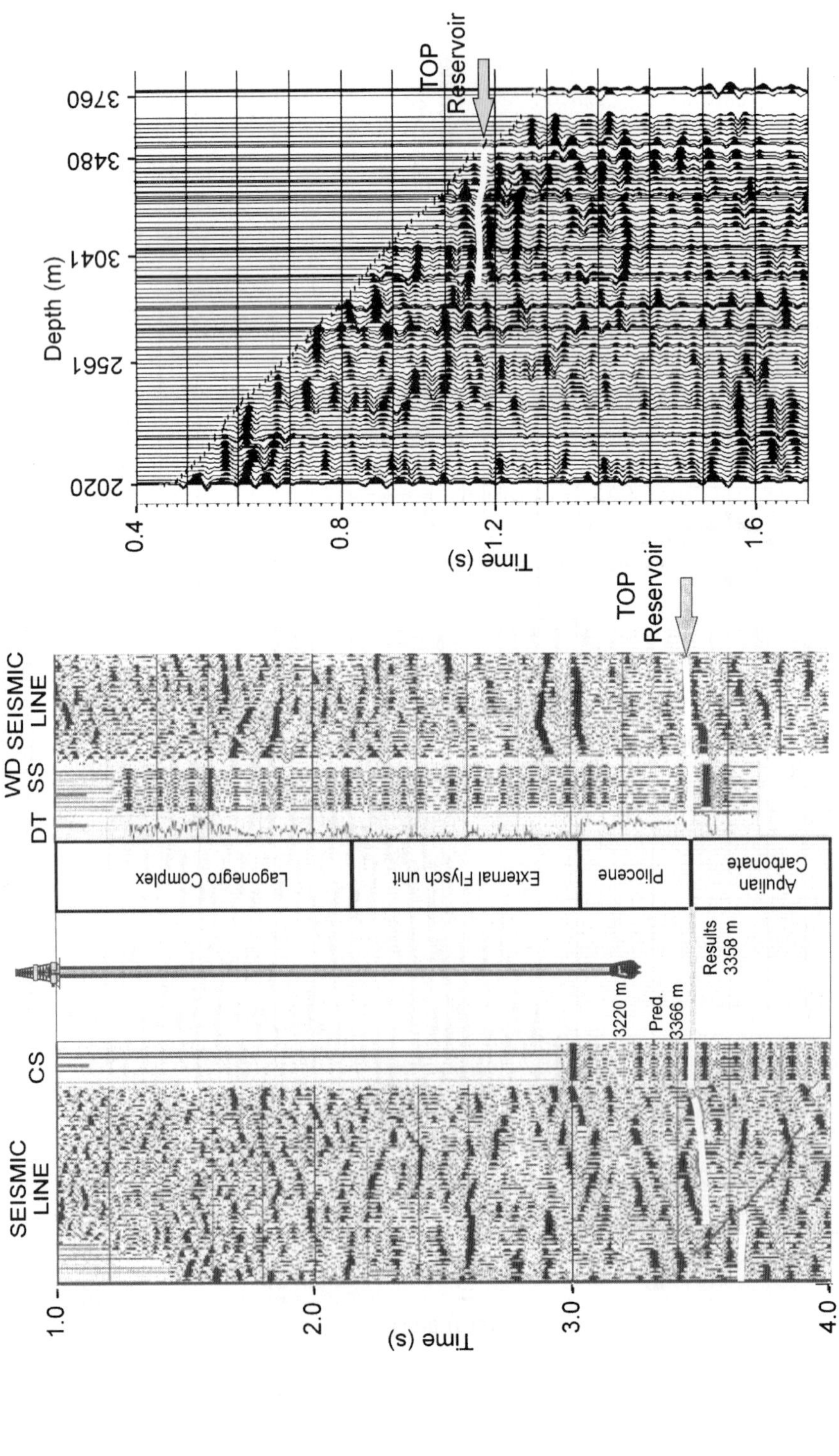

Figure 8.23: Near-offset processed VSP data (right) with marked the top of the reservoir.

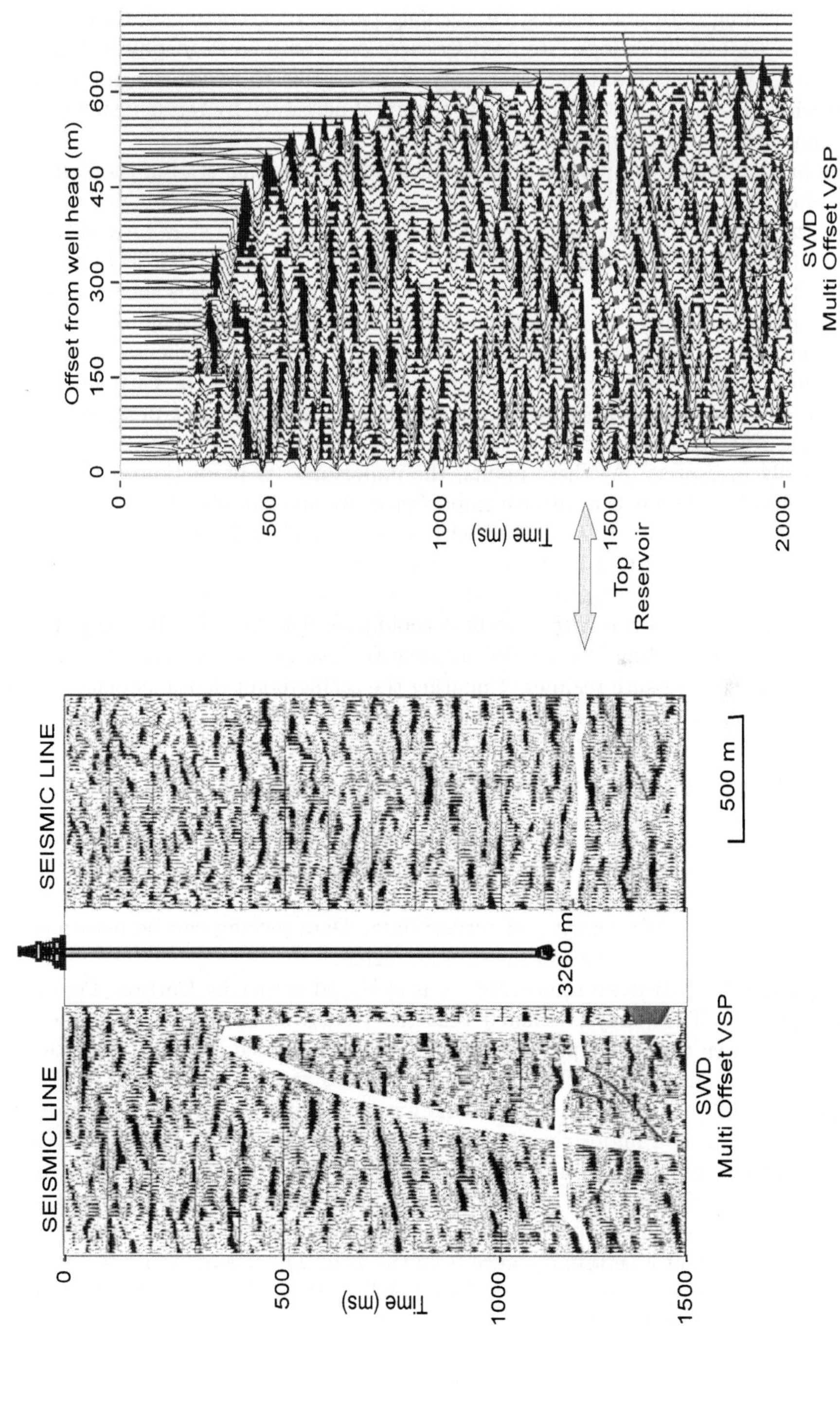

Figure 8.24: Multioffset RVSP SWD processed data.

8.8.5 Isolation of zones with different pressure gradients

All while-drilling data and results, i.e., drilling parameters, cuttings information, intermediate or while-drilling logs, are used to carry out continuous geophysical monitoring of the well. With this real-time technology, it is possible to revise the initial geological model while drilling, update the drilling program if necessary by reprocessing the seismic data, thus optimizing operations and costs.

Important results, supported by this methodology, such as setting the casing points in the due positions, revising the hole trajectory and predicting anomalous drilling conditions, have been achieved in the Val d'Agri area. Figure 8.25 shows a geological cross section of one of the wells drilled in the Val d'Agri region. On this cross section, beside the predicted formation top, the formation top reached during drilling is shown. The critical decision points are when the drilling mud has to be changed and casing has to be lowered in the well in order to isolate a formation with different pressure gradients. Here the zones in the main thrust areas and in the formation above the top of the carbonate reservoir are generally in overpressure with respect to the normal hydrostatic pressure, whilst the carbonatic reservoir is generally associated with fractures, and therefore is usually hydrostatically pressured. At these points, the importance of prediction of the formation changes ahead of the bit is of utmost importance. As an example, the formations on the top of the reservoir have to be drilled with a heavy mud (1.9–2.0 g/cm^3). The drilling mud has to be changed to a light mud (1.06 g/cm^3) at the fractured reservoir level. The concept is explained by Figure 8.26, which shows the difference between the predicted pressure gradients and the actual results encountered by the well. By using the while-drilling data for prediction, it is possible to identify the different formation tops associated with the different pressure regimes. Updating the casing-point depth associated with the different pressure regimes ensures an efficient and safe drilling.

8.8.6 Prediction by SWD-RVSP tomography

The advantage of tomography with respect to conventional seismic velocity analysis (stack, interval, RMS) is to provide a velocity macro model in depth. The tomographic inversion of SWD data gives a result which can be integrated with the velocity model obtained from the conventional velocity analysis of surface data. Data picking can be performed while drilling with the direct arrivals of the drill-bit signal.

Resolution in SWD-RVSP tomography was analyzed earlier by Carrion, Persoglia and Poletto (1992). The analysis showed that the results are constrained to the interval velocities between the well-depth intervals. If not compensated by constraints, the results obtained by the inversion of the direct arrivals differ at a great distance from the well and in the vicinity of the well. In fact, the limited angular aperture in the coverage by direct-arrival rays is a major shortcoming in the RVSP geometry. This limited aperture causes large smearing effects in the direction of the rays connecting far-offset traces to the bit-source positions. Use of constraints or joint inversion of direct, refracted and reflected arrivals, either of SWD or surface-seismic data is, therefore, recommended to improve resolution at a certain distance from the well. The use of SWD direct waves only in the inversions gives results with a lower reliability (Rossi, Corubolo, Bohm, Ceraggioli, Dell'Aversana, Morandi, Poletto and Vesnaver, 2001).

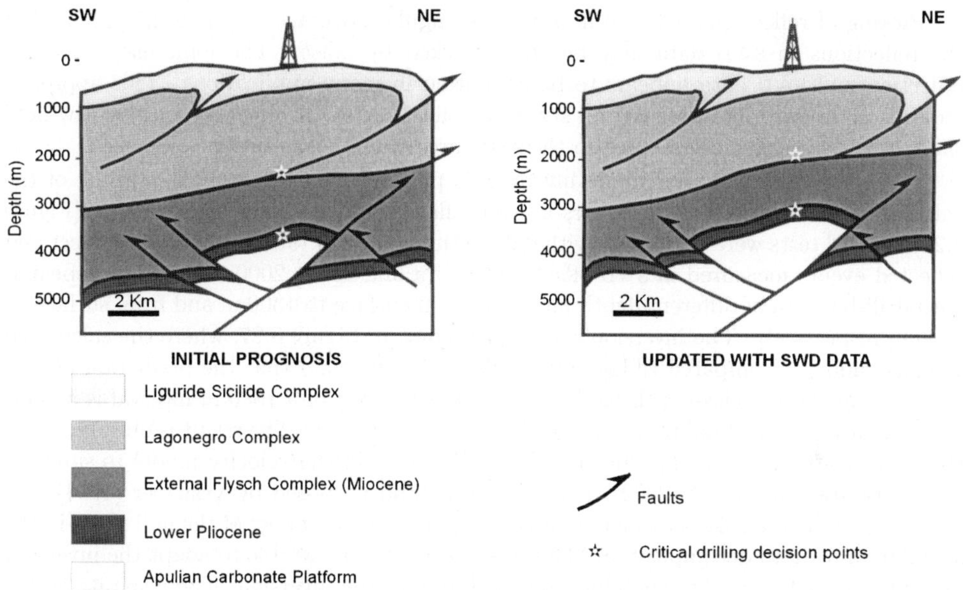

Figure 8.25: Geological section of one of the wells drilled in the Val d'Agri region. Both the prognosed and the actual formation boundary are marked. (left) Initial prognosis, (right) section updated using SWD results.

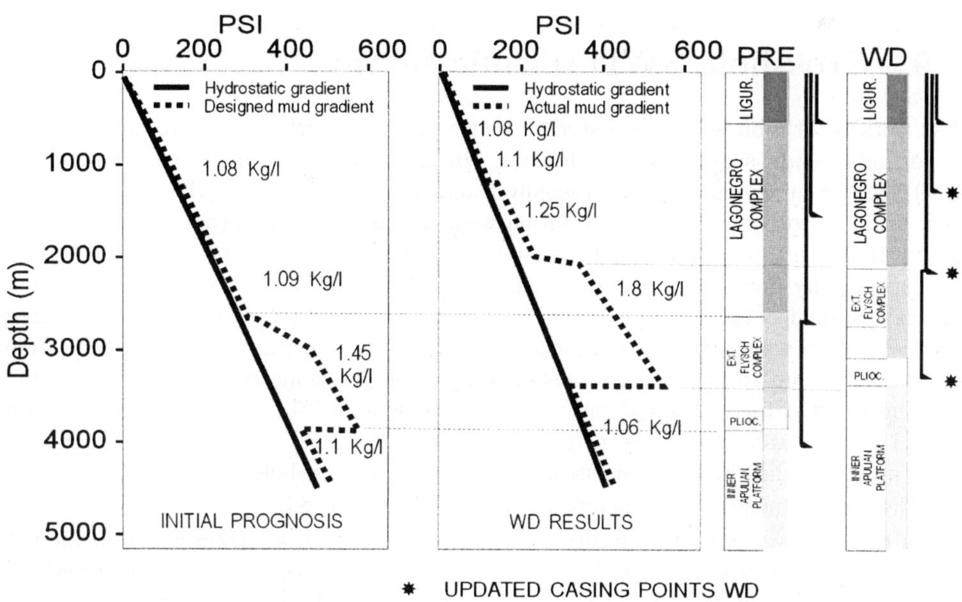

Figure 8.26: Comparison between prognosed pressure gradients and while-drilling results with the associated casing points.

Picking of reflections helps to improve the angular coverage. A main difficult is that the reflections in SWD data may be often masked by noise. The joint use of surface reflection and SWD data appears to be a solution in many applications. This approach requires an appropriate preparation study to collect before drilling and analyze raw-field shots of surface seismic measured in the same area where the SWD lines are used.

The while-drilling tomographic inversion is performed by subsequent refining of the velocity model as the bit approaches a deep reflector. In the following example (Figure 8.27), several tests were performed with data of the same well. Joint inversion of direct and reflected events measured in SWD-offset traces of a line about 2000 m long was repeated with drill-bit data of different depth intervals, from surface to 1300 m and from surface to 2600 m respectively. The inversion results are shown in Figure 8.27, where the calculated velocity model is compared to log curves. We can appreciate that the prediction ahead-of-the-bit of the velocity-depth model obtained using data of a 1300 m interval is a good approximation of the final result obtained with all depth data. The result is obtained with the identification of some intermediate layers, inserted in the velocity model to simulate a velocity gradient (Rossi, 2002). Following a method proposed by Vesnaver (1994), the reliability of the inversion can be measured by analyzing the energy of the null space in the singular-value data decomposition. This analysis makes it possible to adapt the inversion grid so as to optimize the reliability and resolution of the inversion. The accurate while-drilling map of the velocity model with the resolution given by tomographic inversion is an important product for prediction ahead of the bit of the properties of the rock to be drilled (Rossi, Corubolo, Bohm, Ceraggioli, Dell'Aversana, Morandi, Poletto and Vesnaver, 2001).

8.9 Crosshole SWD seismic survey

The working drill-bit source was studied under particularly high signal-to-noise ratio in a crosshole seismic survey (Aleotti, Poletto, Miranda, Corubolo, Abramo and Craglietto, 1999). In this survey, in addition to geophysical monitoring of the well while drilling, other post-drilling processing and interpretation were carried out. The layout of the crosshole survey is shown in Figure 8.28a. Data were acquired at one of ENI E&P's special testing site, specifically prepared for seismic monitoring of gas storage in a sandy reservoir and for conventional crosshole experiments. At this site, ENI E&P set up an instrumented vertical well in which 50 three-component geophones were cemented in, at depths from about 1600 m up to 900 m. The installation of the geophone sensors was done using the cementing string to lower them into the well (Figure 8.28b). During the drilling of a second vertical well, about 500 meters away, the SWD system was employed using both surface sensors and the geophone string cemented in the borehole.

The main acquisition parameters are described in Table 8.2.

In this survey, pilot signal measurements by sensors such as strain gages on the rig, and a high-frequency sampling of the WOB were measured. For pilot correlation, a vertical (axial) component recorded by sensors at the top of the drill string was used. Continuous control of the drilling parameters was automatically performed by the SWD system during data acquisition. In addition to the depth of the bit, other important drilling parameters

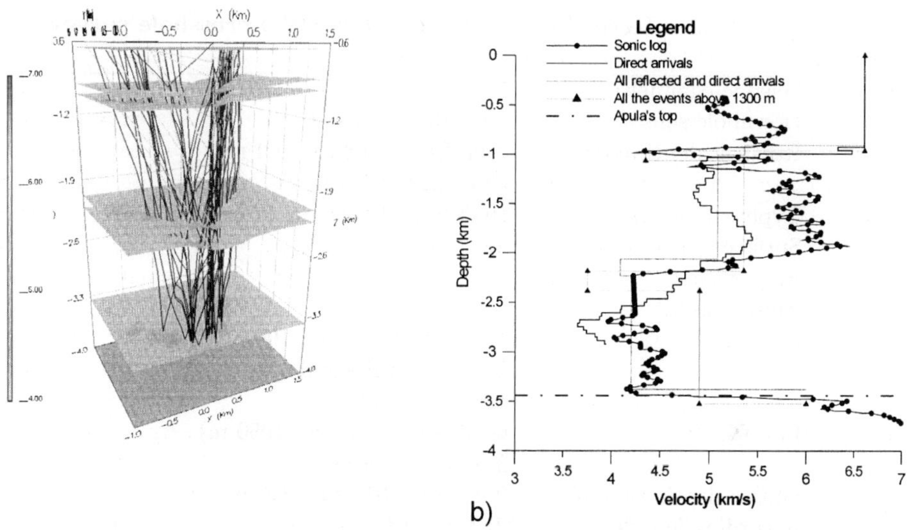

a) b)

Figure 8.27: a) Transmitted and reflected rays calculated in the 3D-tomographic inversion model. b) Comparison of the sonic-log curve (dotted) with the curves obtained from the tomographic inversion of the direct arrivals (continuous black line), of the direct and reflected (all data) (gray continuous line), and of the direct and reflected arrivals from surface to 1330 m depth (gray line and triangles).

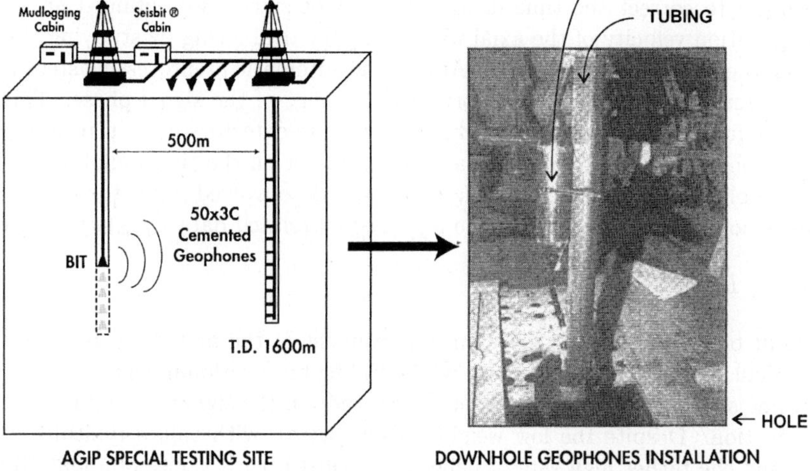

Figure 8.28: a) Test site (ENI E&P Division) for crosshole seismic while drilling. b) Downhole geophones are fixed on the production tubing and installed in the receiver well.

Table 8.2 – Recording parameters for SWD crosshole survey

Seismic channels:	192
Downhole sensor:	50 × 3C (Z, H_1, H_2)
Receiver depth interval:	9 m, from 1598 to 1364 m 18 m from 1346 to 950 m
Type:	HGS SM-14 (natural frequency 14 Hz)
Surface sensors:	42
Rig sensors:	15
Noise sensors:	2
Line Z:	19 (offset from well 100 m, 50 m interval, total spread 1050 m) 12 + 12 geophones
Line X, Y:	6 (at 500 m, 800 m, 1050 m) 3 + 3 geophones
Source depth interval:	9 m from 1012 to 1463 m
Recording length:	21 s/record
Number of records/level:	15–30
Sampling rate:	1–2 ms

driving the acquisition were the length of the drill string, the weight on bit WOB [tons], the rotation speed RPM [rotations/minute], and rate of penetration ROP [m/h]. Data acquisition and preprocessing are automatically steered by the system as a function of the above drilling parameters. A quality-control display of acquisition intervals versus drilling parameters is shown in Chapter 5 (Figure 5.16).

In order to correct the time delays of the pilot signal, we obtained an estimate of the propagation velocity of the axial wave along the drill string by studying the multiple events on the pilot's autocorrelation. Among the automatic preprocessing steps performed by the system, one of the most important is recovery of the signal phase. This includes a phase shift to compensate for the different seismic detectors used in recording (in this case a geophone on the ground, and an accelerometer on the rig as pilot sensor). Let ϕ_g be the geophone phase and $D(\omega)$ the correlated deconvolved data, taking into account the correction of equation (6.117), the rephased correlated data $D_{\rm reph}(\omega)$ are given by

$$D_{\rm reph}(\omega) = D(\omega) e^{\imath(\pi/2 - \phi_g)}. \tag{8.13}$$

The result of equation (8.13) gives an improvement in the first-break picking, useful, in this particular case, both for a correct while-drilling bit positioning in time with respect to the seismic section, and later for an accurate study of the signature and for tomographic reconstruction. Despite the low weight on bit (WOB), with values typically lower than 4 tons, and the rather high rate of penetration (ROP), with values of about 10 m/h, the quality of the crosshole dataset was quite good. An example is shown in Figure 8.29, where the seismic data are presented in common-source depth for the bit at 1089 m; the traces of all the downhole cemented geophones are plotted. Z is the vertical geophone component, while H_1 and H_2 are the horizontal components, whose orientations are variable in depth. Taking into account the special layout of the survey, the very clear event in the Z component has been interpreted as the direct shear wave, while some compressional energy is only partially shown (at about 0.25 s) by the deepest geophones of the H_1 and H_2 components. This energy distribution is in agreement with the roller-cone drill-bit radiation pattern proposed and measured at the surface by Rector and Hardage (1992) and discussed in Chapter 3.

Given the good data quality, an alternative method for determining the pilot-correction time shift was used, based on picking the pilot periodicity versus drill-string length in the correlated seismic signals, after focusing them by stacking the aligned S-wave first arrivals (Figure 8.30). From a comparison of the two results, a correction velocity of 4940 m/s was used throughout the survey.

Data reorientation and wavefield analysis

While a continuous geophysical monitoring of the drilling results was guaranteed in the field, some additional processing for a better understanding of the drill-bit seismic source was performed at the end of the drilling operations. This was particularly aimed at studying the drill-bit radiation pattern and, secondarily, at a shear-velocity tomographic reconstruction of the area between source and receiver wells. Figure 8.31 shows the reorientation of the horizontal components of the geophones cemented into the receiver well: this was done using a rotation angle derived from several standard offset VSP and walk-away VSPs run around the receiver well instrumented for gas-storage monitoring. These results were then checked by analysis of the energy distribution of the compressional first arrivals after reorientation. The radial-horizontal component (HR) and the corresponding transverse-horizontal component (HT) were recovered. As forecast, the compressional direct arrivals in the HR component at a time of 0.25 s for the deepest geophones are clearer. Furthermore, it can be shown that after data reorientation, P and S components of the coherent surface yard noise are clearer.

As a comparison, Figure 8.32 shows a dataset acquired at the same position with a borehole air gun of 140 cu.in. from Bolt Ltd. In order to make the two datasets comparable, the air-gun signature was used to deconvolve the signals recorded by the downhole geophones. After deconvolution, the air gun signature was converted to a zero-phase signal of frequency similar to that of the while-drilling data. In the Z component, the dataset acquired with the air gun shows very poor lateral direct SV arrivals, in particular at the depth corresponding to the source level, whereas it shows clearly the reflection from the top of the sandy reservoir and a reflection below bit total depth (TD) of the receiving well. The same reflections are weak (top of the sandy reservoir) or totally absent (reflection below TD) in the dataset acquired with the drill-bit source, but there are energetic lateral SV arrivals. The two horizontal components show very similar results. Again, the drill-bit-source data show a marked presence of multiples from the rig site totally absent in the air-gun data as expected. Lastly, the fact that the shear arrivals have the same times denotes a good choice of the drill-string velocity used for correction and gives assurance on the quality of the drill-bit data.

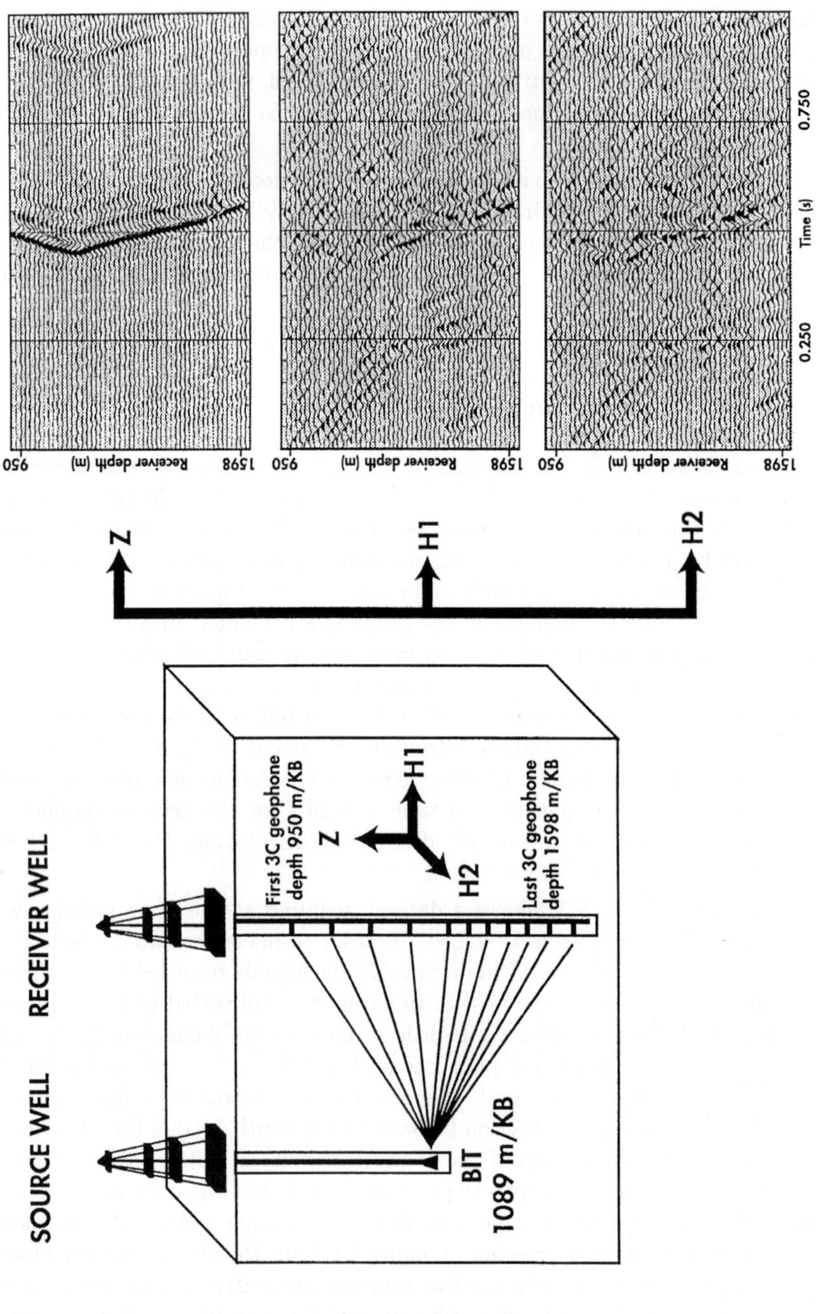

Figure 8.29: Example of crosshole data after automatic acquisition and preprocessing. 3C data are shown in common source, with the drill bit at a depth of 1089 m.

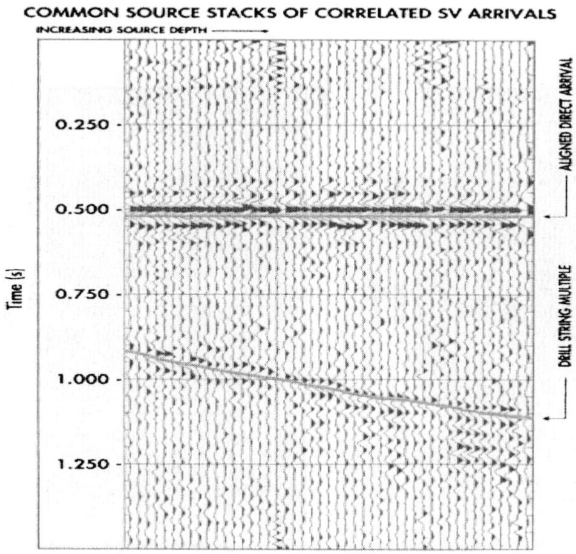

Figure 8.30: Example of drill-string multiple periodicity versus depth in the stack of the correlated seismic signals. Focused stacking is done after alignment of the first arrivals at 0.5 s.

Polarization analysis and vertical reorientation

After horizontal reorientation, a vertical, depth-variable rotation was applied to the HR and Z components to get the vertical, radial R and vertical-transverse T components of the direct arrivals. The T and R components were derived from a detailed analysis of the direct SV energy traveling in the vertical plane – which includes source and receivers – and vibrating orthogonally to the wave-propagation direction. For this analysis, data in a 100 ms window centered on the SV direct arrivals in the Z and HR components were used. The hodograms of these two components, shown in Figure 8.33 for three different receiver positions and the same bit position, illustrate not only the change in orientation of SV polarization with respect to the relative position of source and receiver, but also the strong influence of the geological stratification. In the depth-versus-angle diagram we can see the difference between the result obtained with a simple geometrical assumption (straight rays) and that obtained with the actual angles of incidence at the geophone of the SV energy coming from the bit.

If V_S/V_P = constant, we have, of course,

$$\frac{V_S(i)}{V_S(j)} = \frac{V_P(i)}{V_P(j)}, \tag{8.14}$$

and P and S waves propagate with the same geometry. In this case, the radial component derived from SV arrivals is the same as the radial component derived from the direct compressional P arrivals. Figure 8.34 shows a common-source reorientation example of the R and T components, showing direct SV-energy separation in the T panel and also

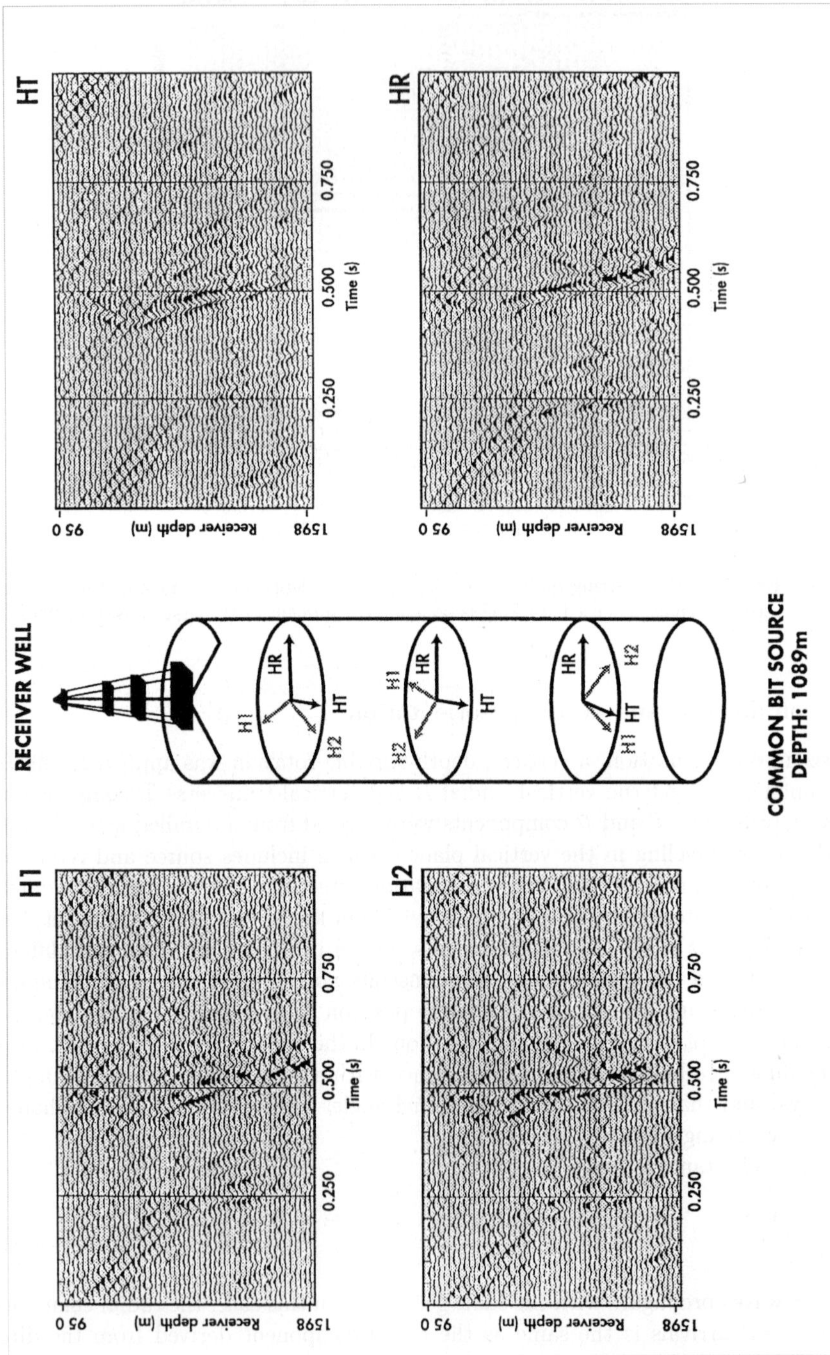

Figure 8.31: Reorientation of the horizontal components of the geophones cemented into the receiver well.

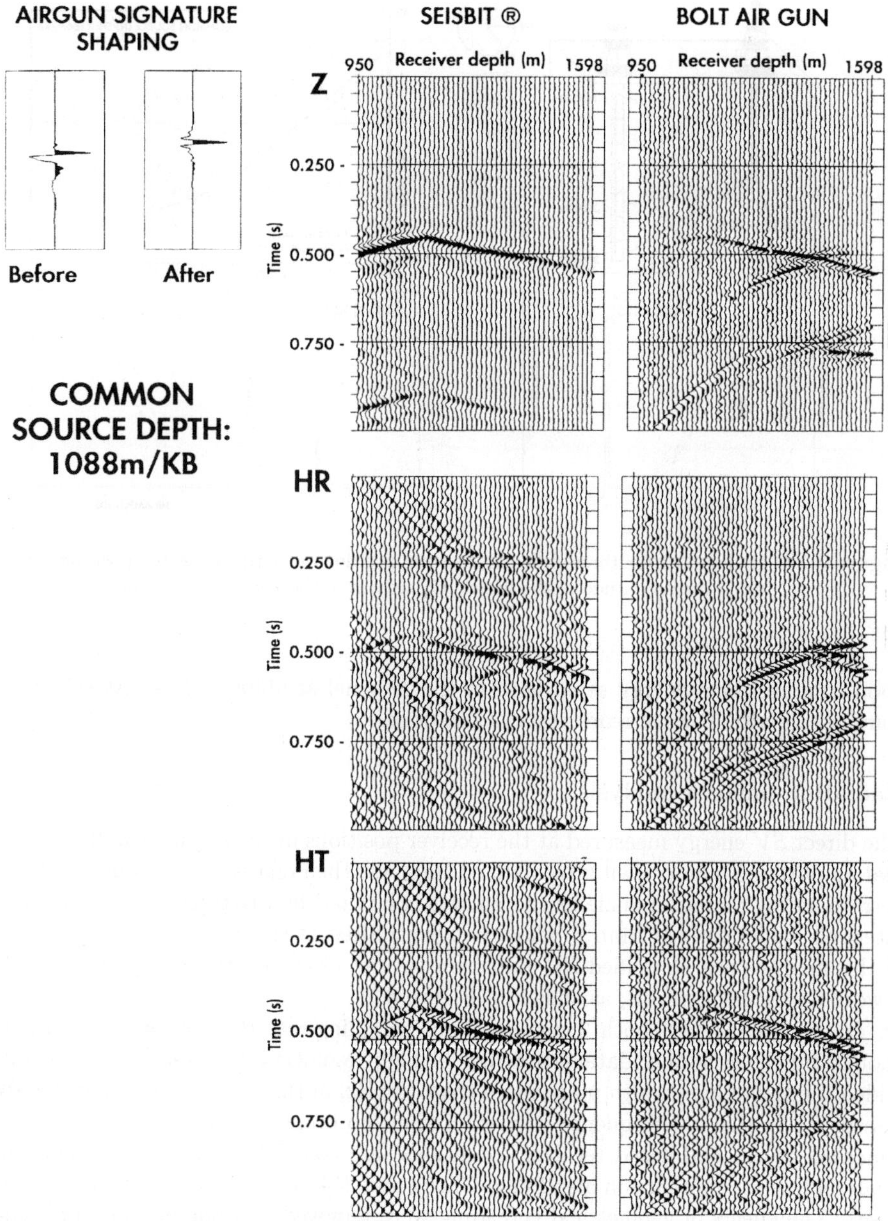

Figure 8.32: Comparison between datasets acquired with the SWD system and a borehole air gun of 140 cu.in at the same position. In order to make the two datasets comparable, the air gun signature was used to deconvolve the signals recorded by the downhole geophones.

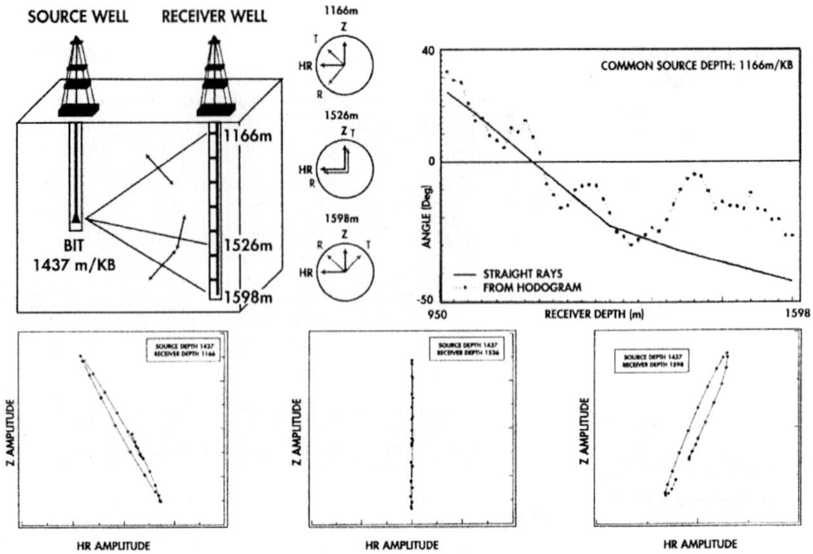

Figure 8.33: Hodograms for three different receiver positions and the same bit position, computed using the amplitude of the direct SV arrivals in the Z and HR components.

good compressional-energy separation in the R panel at about 0.25 s. At this stage, no comparison with P-wave reorientation was done.

Radiation-pattern analysis

The direct SV energy measured at the receiver positions in the listening well, given by the real-amplitude T component after data editing, was then related to the source points. The corresponding radiation angle at the bit was computed by ray-tracing programs using an interwell model derived from S velocity logs acquired at the source well. The first result of this analysis is a simplified SV radiation pattern of the seismic energy measured from a working tricone drill bit, as shown in Figure 8.35.

During the survey, both 2 ms and 1 ms sampling-rate datasets were acquired. Although by multichannel data processing of high-resolution 1 ms data, it is possible to widen the signal bandwidth up to more than 100 Hz, in this analysis a smaller bandwidth was used. In this low-frequency approximation, no borehole effects for the receiver well (uncased source and cased receiver conditions) are considered in the radiation patterns analysis (see for comparison also Peng, Cheng and Toksöz, 1994 for a fluid-filled borehole case). The effects of geometrical spreading were removed by applying an amplitude loss factor proportional to d_{ij}^{-1}, where d_{ij} is the distance between the i-th source point and the j-th receiver, and by compensating for the transmission effects. Acoustic-impedance parameters were derived from wireline logs measured both in the receiver and source wells. No assumptions were made about absorption effects and anisotropy, which are not completely negligible.

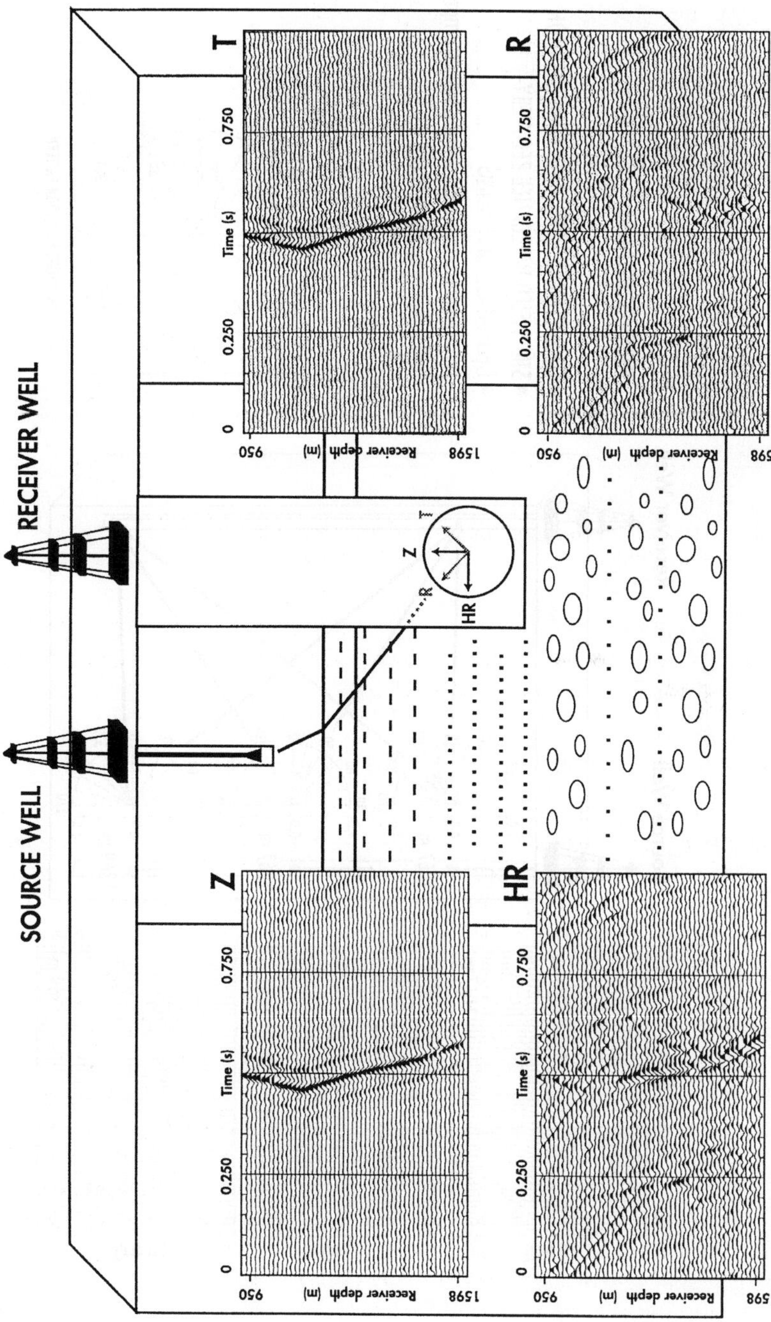

Figure 8.34: Common source example of vertical reorientation. The direct SV energy separation in the T panel is seen, and also good compressional energy separation in the R panel at about 0.25 s.

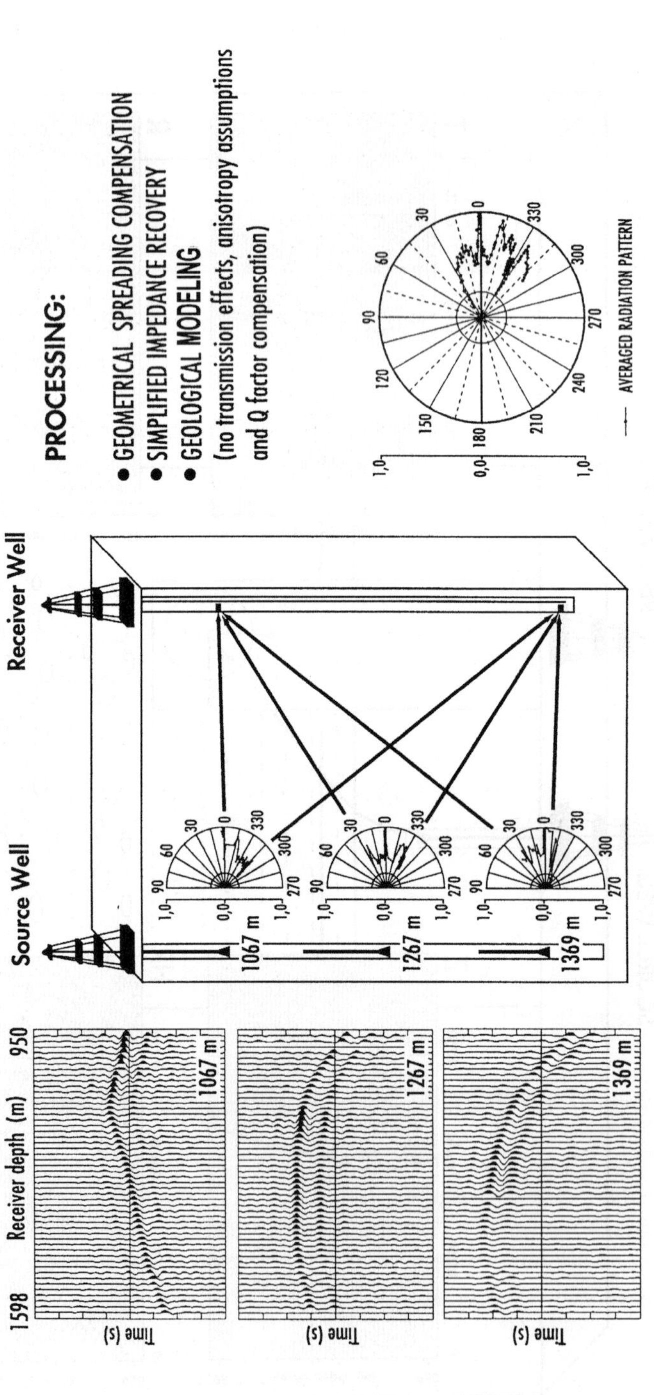

Figure 8.35: Radiation pattern of the direct SV amplitude measured for three different source positions and averaged.

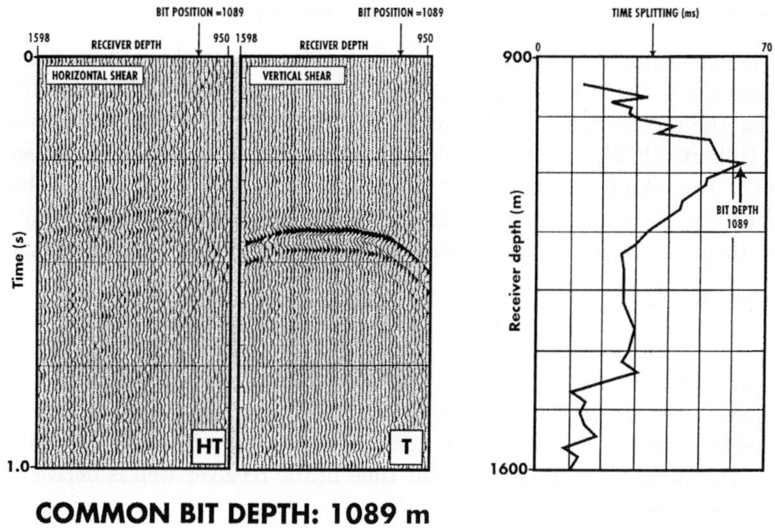

Figure 8.36: Comparison between *HT* and *T* components, in which the horizontal shear waves (left) and the vertical shear waves (central) are represented. This example shows a clear shear-wave splitting, with a maximum at the receiver at the same bit position depth.

Three different bit positions are shown in Figure 8.35, together with the corresponding radiation patterns and, on the left, with the relevant *T*-components of the seismic data. At the bottom right, an average *SV* radiation pattern is given: analogies with the theoretical tricone *SV* radiation pattern can be seen (Figure 5.20).

Anisotropy

The good quality data allows a direct comparison between the *HT* and *T* components, which show, respectively, the horizontal shear wave and the vertical shear-wave direct arrivals (Figure 8.36).

The data are organized into common-bit depth (equal to 1089 m) and a strong time splitting is visible in the diagram on the right, with maximum at a geophone depth corresponding almost perfectly to the depth of the bit. In this case, the horizontally polarized component of the shear waves *SH* radiated by the bit (collected on the *HT* dataset) is traveling almost parallel to the fine layering of the shaly formation penetrated by the well. At the same time, with identical source and geophone depth, the vertically polarized component *SV* (collected in the *T* dataset) is practically perpendicular to the stratification. So the probable explanation for the shear-wave splitting is the strong geometrical anisotropy (see also Thomsen, 1986; MacBeth, 2002) due to thin-layer stratification ("thin-layer anisotropy"), which diversely affects the vertical *S* component (slower) and the horizontal *S* component (faster) for wavelengths that are appreciably larger than the layer thickness. The computed anisotropy coefficient ranges from 5 to 15 %.

Tomographic reconstruction of the SV velocity field

Figure 8.37 is the results of an SV tomographic reconstruction of the velocity field between source and receiver wells. In total, about 2000 rays and a square grid of 10-m × 10-m pixels were used for the reconstruction. The result was obtained after 10 iterations of a SIRT algorithm (Gilbert, 1972) with the straight-ray approximation and without anisotropy assumptions. The signal bandwidth of the picked dataset is limited to below 80 Hz and, consequently, the expected resolution limits for transmission tomography of the S component (by assuming $V_S = 1300$ m/s) is related to the radius (Williamson, 1991)

$$R = \frac{L}{2} = \frac{1300/80 \times 475}{2} = 88 \qquad [\text{m}], \tag{8.15}$$

where the interwell distance is 475 m. In this very unconventional image, the top of the sandy reservoir, used for gas storage, is interpretable, with the low velocity zone also matching the increase in the shear-wave interval transit time from logs recorded in the source well. The shear-wave interval transit time in the receiver well is derived from the compressional sonic log.

8.10 3D-RVSP application

In Section 5.19 we have presented the peculiarities and characteristics of a typical 3D-SWD data acquisition. Here we present a case history of a 3D-SWD acquisition, processing and interpretation (Poletto, Miranda, Petronio, Bertelli, Malusa, Luca and Schleifer, 2001; Poletto, Malusa, Petronio, Lovo and Miranda, 2002).

Geological setting of the 3D-SWD case history

The well is a vertical well situated in southern-eastern Sicily (Italy) on the Ibleo-Maltese plateau. The overlaying sediments in the well area are quaternary marine marly and shaly limestone deposits. The drilled well then runs across a complete lithostratigraphic sequence reaching the Upper Triassic. The feasibility phase was prepared by modeling the 3D area to optimize the acquisition layout (Petronio, Poletto, Carcione, Seriani, Luca and Miranda, 2001).

8.10.1 Modeling of 3D-SWD survey

The main reason for modeling the area around the well is that the analysis of the synthetic data allows us to better choose the layout and the acquisition parameters (offset, receiver group-array length, etc.), and to adjust the position of the lines. The geological model was calculated to simulate the seismic signal and noise waves by synthetic seismograms before the acquisition. The preparation of the model uses 2D-seismic sections, the isobathic map of the top reservoir, the formation velocities obtained from nearby wells and the weathering and sub-weathering velocity. Every line was interpreted, recognizing 14 interfaces in the time domain. The conversion in depth was performed on 64 points utilizing the velocity function obtained in correspondence with one of the reference wells.

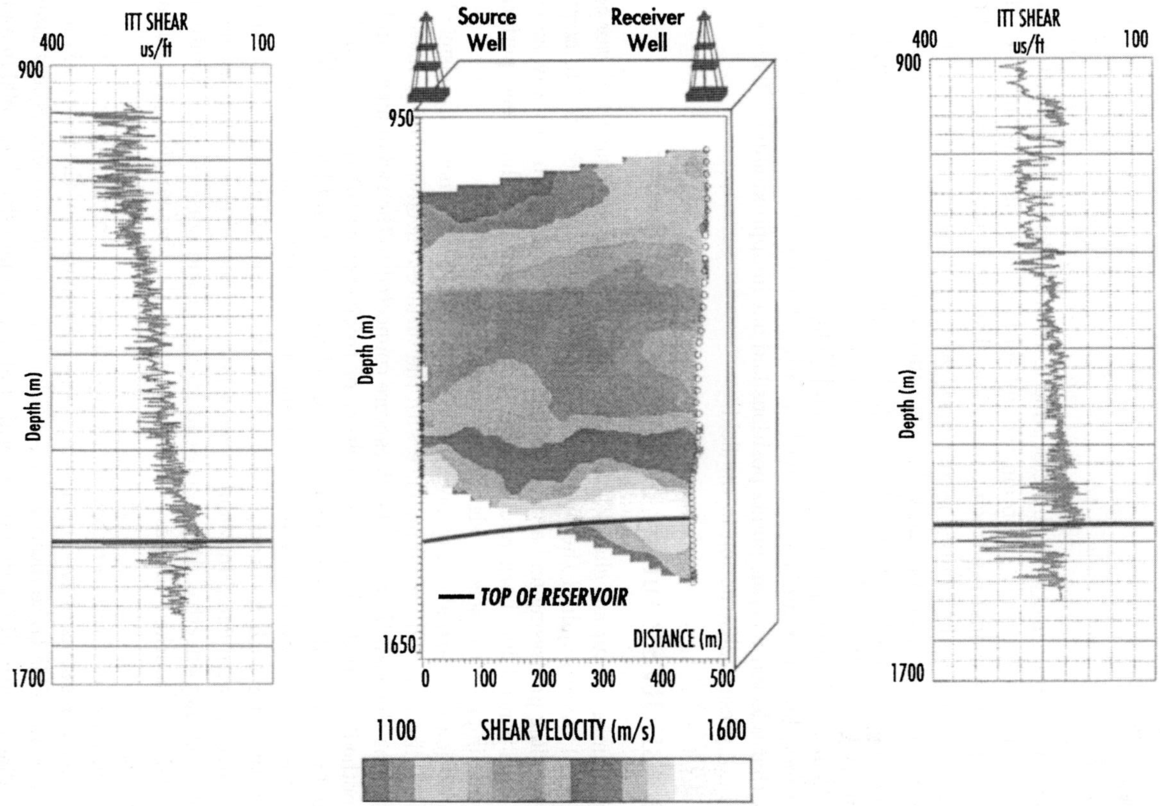

Figure 8.37: Result of a tomographic reconstruction using the *SV* velocity field between source and receiver wells, compared with the shear interval transit times from sonic logs.

The P-wave velocities were derived from sonic logs of the reference wells and interval velocities obtained from seismic reflection data. The approximation by Poisson relationship $V_S = V_P/\sqrt{3}$ was utilized to determine the S interval velocities, while the densities were obtained by using the modified empirical Gardner's relationship,

$$\frac{\rho}{\rho_0} = \left(\frac{V_P}{V_{P0}}\right)^{0.25}, \tag{8.16}$$

where ρ is formation density. In this case, due to the complexity of the survey, the modeling was performed in order to obtain the full seismograms and not only the reflection location as in conventional ray-tracing modeling. The algorithm used for the 3D forward modeling is based on the Fourier pseudo-spectral methods and on the elastic full-wave equation for heterogeneous media, which performs without compromising material parameter assumptions. Moreover, for improving the numerical accuracy a staggered scheme has been used, and efficiency has been obtained by a parallel implementation (Carcione, Kosloff, Behle, and Seriani, 1992; Carcione and Helle, 1999).

After modeling, a preliminary survey ("scouting") was performed in the field in order to check that selected receiver points positions laid out in approximately 30 km of seismic lines did not present obstruction in the field.

8.10.2 3D-RVSP survey organization and layout

The high number of surface channels and the utilization of a downhole tool (see Sections 5.5 and 7.8.1) have required the presence of a reinforced crew, dedicated to acquisition, to while-drilling processing and analysis of drill-bit seismic data. In this way, an accurate quality data control has been realized, and data pre-processing has been performed to control and improve the seismic signal during the drilling phase.

Acquisition was carried out by laying out four radial (two diametrical) seismic lines, in orthogonal directions, in correspondence with the surface-reflection-seismic lines acquired before drilling the well. Two circular lines were connected to the radial lines. An irregular ("saw-toothed" shaped) circular disposition of receivers was used to discriminate the signal and noise arrivals in the azimuth domain (Section 5.19 and Figure 5.43). A circular layout is the most appropriate geometry for imaging the formations surrounding the well. However, since for such a constant-offset layout the shot gathers record bit-signal and rig-noise arrivals with equal moveout, wave separation techniques – such as median and fk filters – are not effective. Hence, to separate signal and noise in the azimuth domain, an irregular "saw-toothed" disposition of receivers in the circular geometry was adopted (Figure 8.38).

The inner circle of receivers has an average radius of approximately 1100 m and the outer circle has a radius of 2200 m (Figure 8.39). The spacing of the traces of the circular lines was 75 m. This trace interval was calculated to record frequencies less than 80 Hz without aliasing for a diffraction point located at a minimum depth of 1500 m and 400 m offset away from the well. Geophone arrays of variable lengths (75 m, 50 m, 40 m, and 30 m) were aligned in the radial direction with respect to the well's central point. During the survey, measurements of the apparent velocity of signal and coherent noise were executed, to evaluate efficiency of array configurations calculated before the survey using the 3D model. Horizontal geophones were used in correspondence with the junctions between the radial lines and the inner and outer circles.

The SWD yard system consisted of two cabin units with recording quality-control instruments, local processing capabilities and 15 pilot channels (Table 8.3). These pilot sensors allowed us to pick up the signal for crosscorrelation and the unwanted noise from the surface-rig assembly. The survey started at a depth of 428 m and was completed at a depth of 3060 m. The configuration with the two circles was used from 428 m to 2700 m drilling depth. After this depth, the configuration of the seismic lines was modified because it was necessary to remove part of the line because of seasonal harvesting works. The new configuration was studied to maintain the continuity of the radial lines and a consistent coverage of the 3D recordings in the different quadrants of the investigated area. The well was mainly drilled using roller-cone bits, with a downhole motor and, after 2700 m, by only rotary drilling. A short interval (from 2101 m to 2395 m) was drilled with a PDC bit. During the survey, 159 depth levels were acquired. The while-drilling recordings consist of about 159 Gigabytes of while-drilling raw data and 10 Gigabytes of preprocessed and processed RVSP seismic data. All the raw data were re-processable (for pilot synthesis and correlation) in the field with access on line. For each acquisition-depth level, 30 records of 45 s have been acquired. Initially, a level every 10 m and later every 20 m were acquired. The average rate of penetration was 10 m/h. The recording parameters for the well section are shown in Table 8.4.

During the test, the acquisition system was thoroughly checked, together with the performances of the synchronization and quality control procedures. Auxiliary pilot sensors in an instrumented downhole tool were also used.

Table 8.3 – SEISBIT ® **pilots of 3D-RVSP survey**

Pilot sensor	**Position**	**No.**
Vertical accelerometers	Top drive	4
Horizontal accelerometers	Top drive	4
Pressure meters	Mud line (in, out)	2
Geophones z	Top and base rig	2
Geophone z	Shallow hole	1

Table 8.4 – SEISBIT ® **recording parameters (3D RVSP survey)**

Sample rate:	2 ms
Record length:	45 s
Recording time per level:	up to 30 min.
Maximum allowed penetration per level:	10 m
Rate of penetration:	3–30 m/h
Interval between levels:	10–20 m

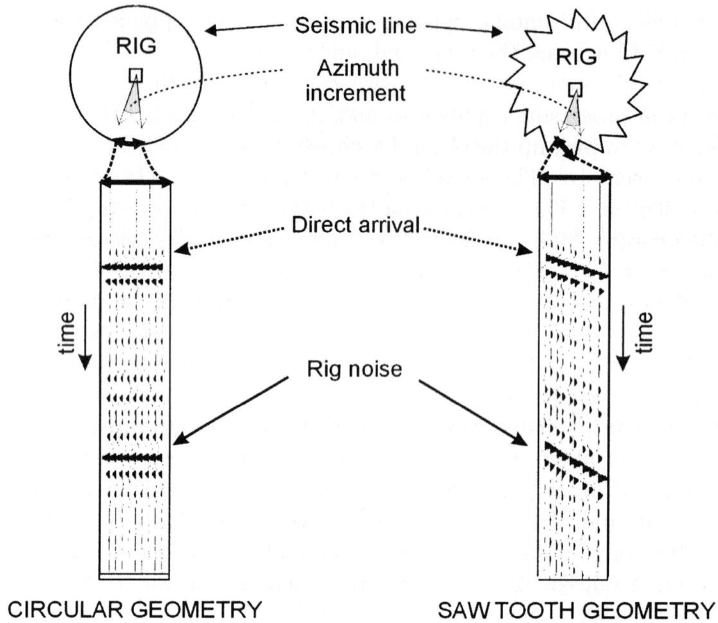

Figure 8.38: Synthetic example showing the use of irregular geometry (saw tooth vs circular azimuth) to discriminate signal and coherent noise with different moveout.

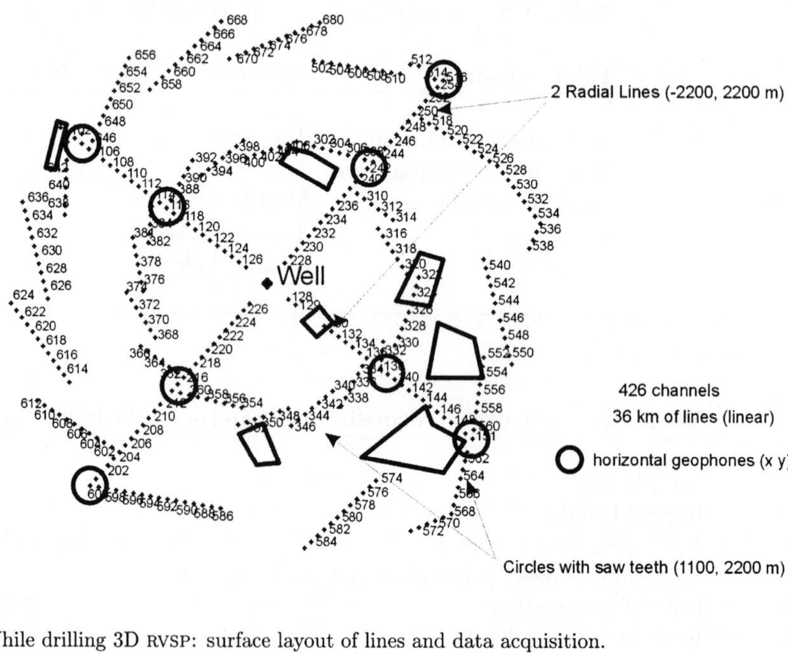

Figure 8.39: While drilling 3D RVSP: surface layout of lines and data acquisition.

The data quality was analyzed while drilling, by performing the pre-processing on the field, together with the data-quality control activity. The data quality both of shallow and deep data is good and, according to the radiation pattern of the drill-bit source, the signal is detectable in the radial and circular seismic lines. The good signal-to-noise ratio was obtained due to the good geological conditions of the area, characterized by undisturbed calcareous and calcareous-marly sequences, and use of roller-cone bits. While-drilling analysis was made on data obtained by crosscorrelating the receivers laid out in the field with the vertical-axial accelerometer pilot signal. Figures 8.40 and 8.41 shows an example of common-source gathers acquired at a depth of 2081 m on the radial receiver lines and on the circular lines, respectively.

8.10.3 Azimuthal analysis of rig-radiated noise

The disposition of the geophone lines in circles around the well makes it possible to analyze the radiation of the noise from the rig site in azimuth. The radiation shows an azimuthal pattern in agreement with the radiation pattern of a horizontal-force model for the vibrating rig (Section 4.8). This analysis is done with the receiver traces of the outer circle.

In Figure 8.42 we see the synthetic common-shots calculated by 3D-full-wave elastic modeling before drilling of the well (Petronio, Poletto, Carcione, Seriani, Luca and Miranda, 2001). Figure 8.43 shows the stationary noise obtained by vertical stacking of the outer-circle traces taken in real amplitude, without pilot-delay correction, for all the available bit-depth levels. The result shows a distribution of noise energy that is in agreement with the result of the synthetic simulation of Figure 8.42, calculated with the same trace geometry and assuming a surface-horizontal force as the source. Hence, the result is interpreted as a confirmation that the rig acts as a source of horizontal vibrations at the surface.

8.10.4 While-drilling analysis of 3D data

The synthetic data calculated on the 3D model were used for in-field data-quality control and as an aid in the signal interpretation in the areas where the signal-to-noise ratio is low. Synthetic seismograms as an aid in the first-arrival picking were, in some occasions, used. Figure 8.44 shows the comparison between the drill-bit data of a radial seismic line with the corresponding synthetic traces and the comparison of the data of a circular seismic line with the corresponding synthetic traces. That is, for VSP processing it is necessary to identify the direct arrivals and perform first-arrival picking. The availability of an accurate 3D model has provided synthetic data useful in the identification of the direct and reflected arrivals in both the shot and common-receiver-VSP domains, analyzed with and without stationary noise. The while-drilling 3D-RVSP processing was performed in the common-receiver domain. The processing sequence is shown in Table 8.5.

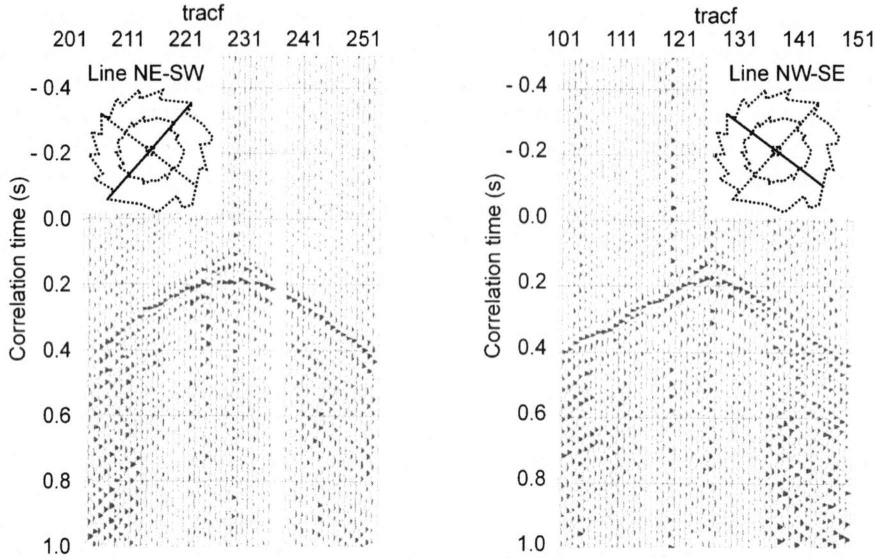

Figure 8.40: Roller-cone common-source level on the radial receiver lines (common-source depth 2081 m).

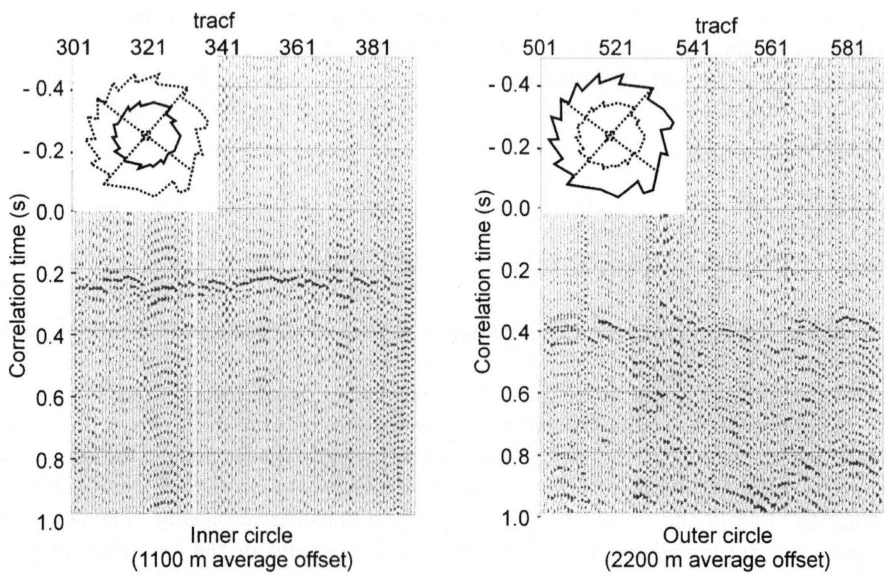

Figure 8.41: Roller cone common-source level on the circular receiver lines (common-source depth 2081 m).

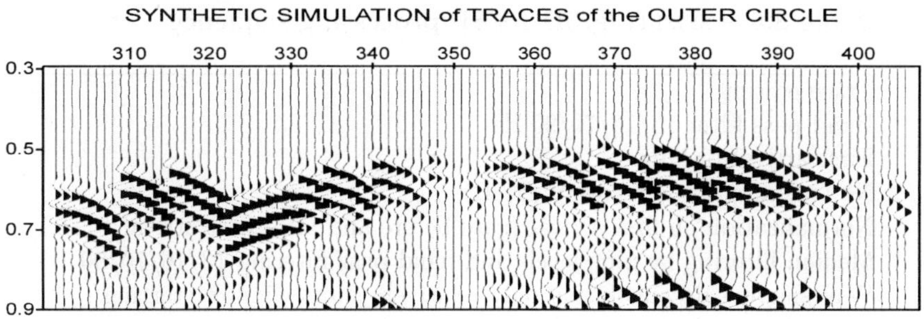

Figure 8.42: Synthetic simulation of the seismograms measured at the traces of the outer circle.

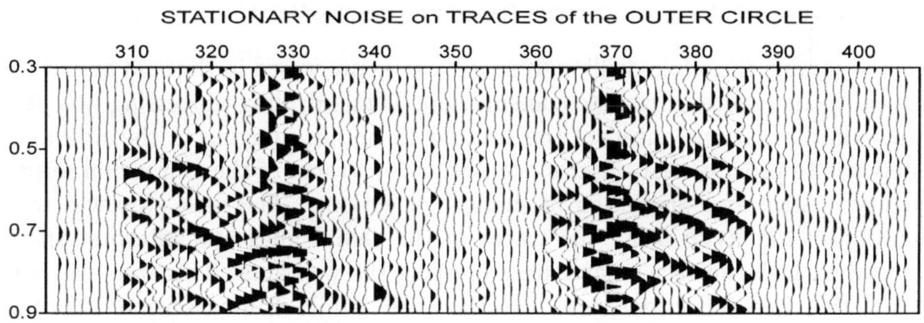

Figure 8.43: Stationary noise on traces of the outer circle.

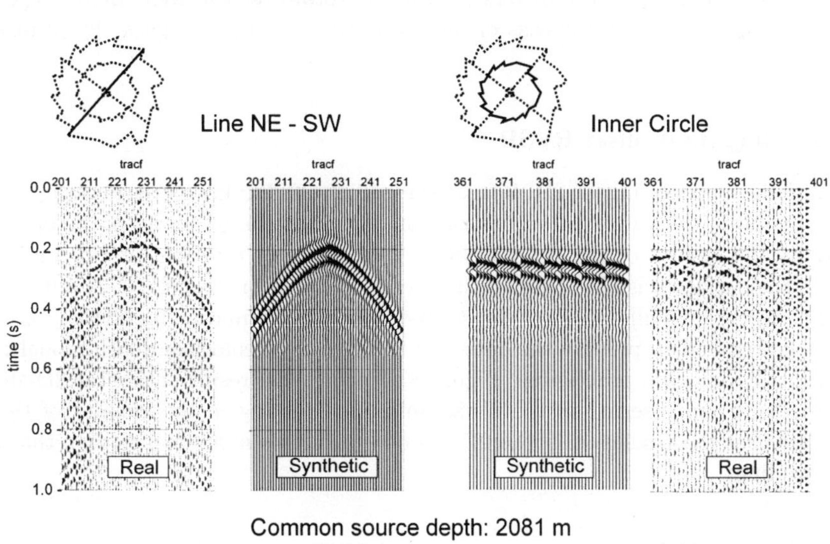

Figure 8.44: Data-quality control and signal interpretation aid with synthetic seismograms; comparison of the radial seismic line with the corresponding synthetic traces and of the circular seismic line with the corresponding synthetic traces.

Table 8.5 – SEISBIT ® **3D RVSP processing sequence**

Trace editing
Band pass filter 13/16/70/90 Hz
First arrival picking
No true-amplitude recovery (TAR)
Normalization on first arrival (gate before and after first arrival)
Median filtering by nine levels and band pass filtering 13/16/70/90
Waveshaping deconvolution only relatively short window starting before first arrival (no predictive deconvolution)
Static correction only datum plane correction with a 2000 m/s velocity

Prediction of reflecting interfaces by near-offset RVSP

During the survey, the while-drilling RVSP data were processed in the field and at the office connected to the rigsite by remote transmission. While-drilling processing was performed to detect the formations ahead of the bit and to obtain imaging in the well area. The first step was to process near-offset RVSPs, using few traces of the four radial seismic lines for quality control and prediction of shallow reflections. The next step was to process the multioffset 2D RVSP using the traces of the radial lines and interpret the seismic results tied with the surface seismic lines crossing the well area. This step was performed to predict the deep reflections, and investigate the while-drilling seismic information in correspondence with the surface seismic lines. Thanks to the good data quality achieved in the survey, it was possible to identify reflections ahead of the bit with a prediction distance of several hundreds of meters even in the total-raw-field RVSP before separation of the reflected waves as in the case of the top of the volcanic sequence shown in Figure 8.45.

While-drilling multioffset RVSP

The multioffset processing of 2D data of the radial seismic lines was performed while-drilling, to obtain the structural reconstruction of the well target ahead of the bit. Figure 8.46 shows the results of the 2D-multioffset processing of the data of two radial lines tied with the surface seismic. These results were obtained in the field. The data analysis allowed us to obtain while-drilling velocity and imaging information in different directions in the well area and to prepare the data and the velocity/geological depth model for the subsequent 3D-imaging processing. Figure 8.47 shows the result of several 2D-MORVSP lines processed and inserted in the 2D-seismic lines passing in the vicinity of the well. We can see that the resolution of the 2D-MORVSP results is much higher than that of the 2D-surface lines.

Comparison of SWD results, conventional VSP and well-log data

One of the uses of SWD, as outlined at the beginning of this chapter, is the checkshot while drilling. The validity of this result is shown in Figure 8.48, which shows the comparison

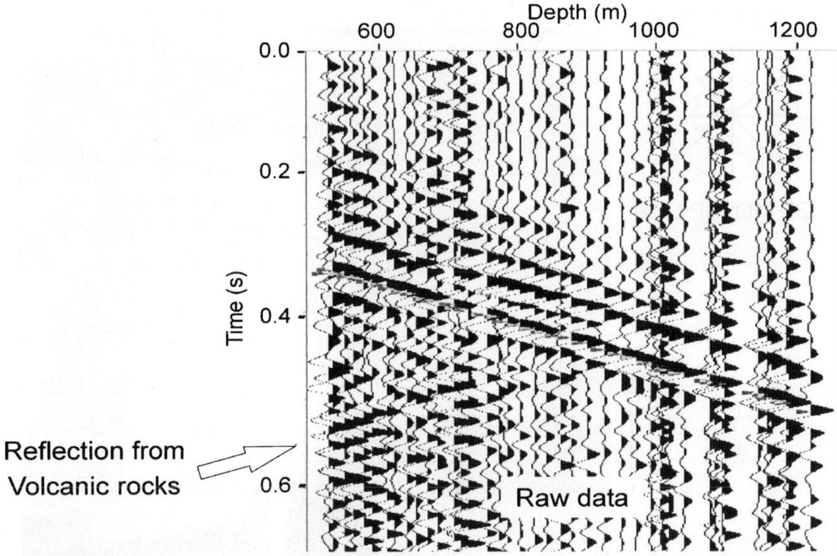

Figure 8.45: Reflection given by volcanic rocks readily visible also on unprocessed data.

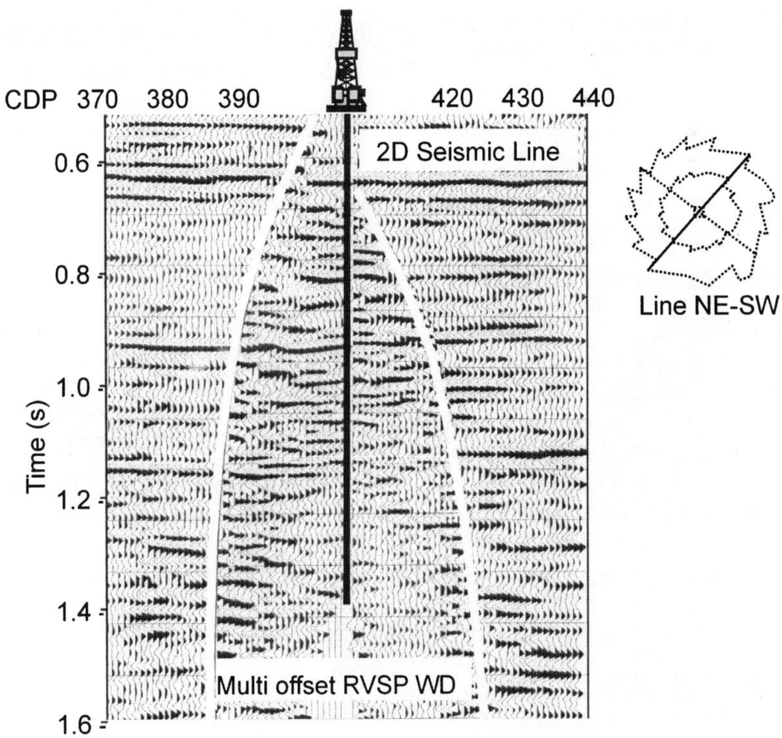

Figure 8.46: In-field while-drilling 2D-MORVSP processing for prediction and imaging.

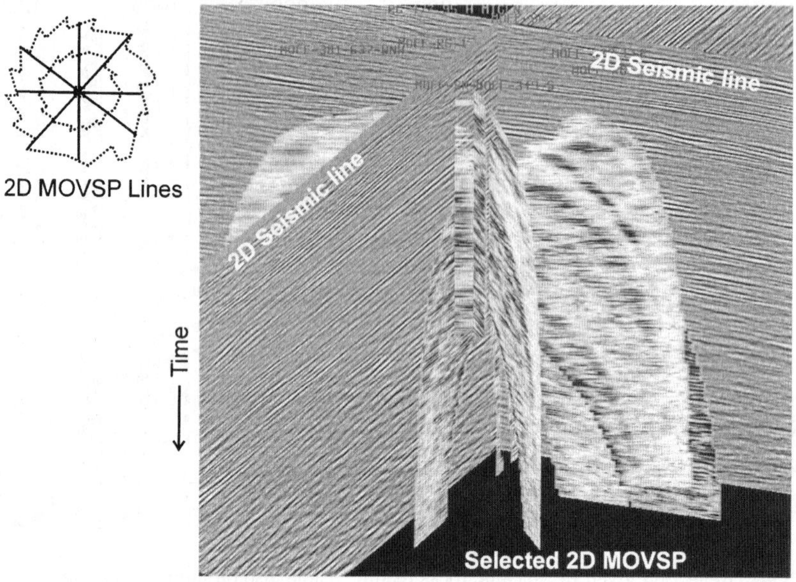

Figure 8.47: 3D-display of while-drilling 3D-RVSP preliminary processing, with a fence display of two surface lines for comparison.

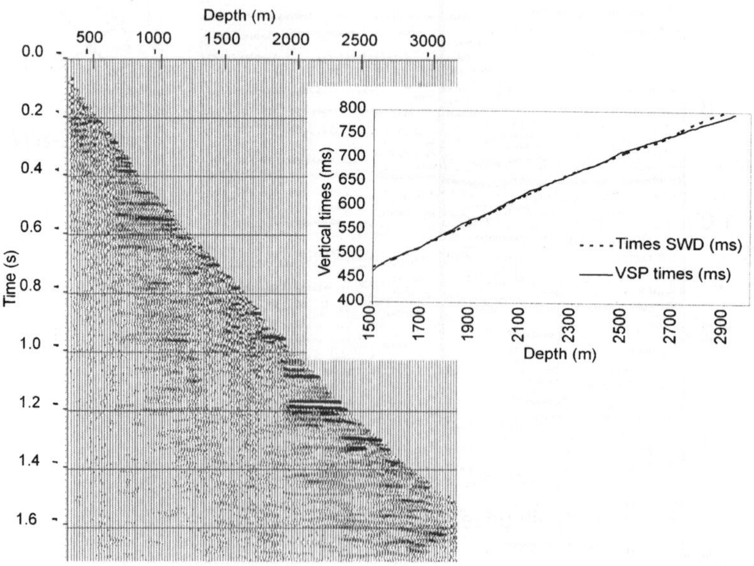

Figure 8.48: Comparison of zero-offset conventional VSP and SWD first-arrival times.

between the time-depth curves obtained using the first-arrival times of a zero-offset wireline VSP acquired after the drilling completion and the SWD first arrivals picked on a near-offset trace. The good matching between the two curves can be seen on the whole section.

8.10.5 Analysis of shear-wave data

The analysis of S and P waves can improve the knowledge of the reservoir and the caprock. In addition, the observation of shear-wave-splitting effects can make it possible to identify the orientation of discontinuities due to rock layering or fractures. In the 3D acquisition, in addition to the vertical geophones, 16 horizontal geophones (radial and transverse direction) were deployed at the crossing points of the radial and saw-toothed lines. The S arrivals can be observed also in the common-receiver gathers of the vertical geophones. That is, the direct SH arrivals are typically detected in the seismograms of the horizontal-transversal (tangential-to-circle) geophone channels. SV arrivals are detected both by the vertical and horizontal-radial geophones. In general, the quality of the S-wave arrivals decreases with increasing drill-bit depths. This effect is due to the S-wave radiation pattern of the drill-bit source. Furthermore, the interference with the stationary noise increases with the seismic time on the traces, as shown also by the synthetic seismograms. Figure 8.49 shows the common-receiver-gathers (offset 1199 m) of the radial (x) and transverse (y) components. The S-wave direct arrivals have different arrival times in the radial- and transverse-horizontal components. This effect may be attributed to layering or fracturing which produces anisotropy and shear-wave splitting ("*birefringence*" effects). S-arrival picking in records of radial (x) and transverse (y) components are also shown on Figure 8.49, where a very clear splitting effect can be recognized.

Anisotropy analysis

The geometry of a 3D-RVSP experiment is ideally suited to the task of measuring azimuth dependent phenomena associated with the elastic wave propagation. This analysis has been done on the data recorded by the radial lines. Figure 8.50 shows the difference between synthetic and observed travel times. Synthetic first arrival times are computed by the 3D-isotropic model built during the feasibility study. In the near-offset traces with 880 m drill-bit depth, the synthetic and measured travel times differ less than 10–15 ms. At increasing offsets, the difference increases. This discrepancy can have different causes (Table 8.6).

Table 8.6 – Causes of travel-time anomaly

3D-model errors (velocity and/or geometry)
Topographic and/or near-surface effects
Near-surface geology
Source-deepening and receiver-array filtering effects
Anisotropy

The potential causes of this time anomaly have been analyzed. The receiver elevations along the radial lines are smooth and the near-surface velocity model computed using the rig-stationary noise and surface-seismic data is smooth also. In order to preserve the high-frequency content of the drill-bit signal, geophone strings where deployed along linear arrays (radial direction) with limited length so that strong receiver-array filtering effects can be excluded (Section 5.14). At increasing offsets, the drill-bit radiation pattern can cause a change in the direct-arrival waveform. In this case, a difference of some milliseconds between first-arrival picking due to non-stationary wavelet can be expected. The imprecision in the 3D model building and/or anisotropy effects can explain the anomalies observed. Direct-arrival inversion by tomography will be performed to update the initial 3D model and to obtain new synthetic data. A new comparison between measured and computed data may give us evidence of anisotropy.

8.10.6 3D-RVSP migration of SWD data

Model computation for 3D static correction

A model computation for static correction was performed by surface and downhole seismic data. To improve the field-static correction computed in the preliminary processing phase, both the stationary-rig-noise arrivals and the first breaks of the surface reflection data acquired in the well area were used (Section 5.16.1). An average velocity of 2000 m/s was obtained for the near-surface layer. With this value, the static correction to the regional datum plane was computed. The difference in the results obtained using this method and using the geological model generated from the surface-seismic-refraction survey in the area gives values lower than 4 ms for all the field traces.

3D-SWD seismic tomography

An accurate velocity field is necessary before we can proceed to process the final 3D migration; we calculate this velocity field using 3D-seismic tomography. Seismic tomography is a basic key for estimating the seismic velocity distribution in the subsurface. Before 3D-seismic tomography inversion was calculated, an accurate revision of the first-break picking was performed both in the common-shot-gather and common-receiver-gather domains. About 20 % of the overall first-break dataset (approximately 12 000 traces), corresponding to data acquired at larger offsets by the radial lines and the external saw-toothed circle, was re-examined and adjusted. The tomographic analysis was performed after constructing the initial 3D-velocity model, ray tracing and data inversion. At the end, the final velocity model – including anisotropy – was obtained and used for the migration of the data. The velocity values on the well location were constrained to those obtained from the first-arrival-time picking of the SWD and from conventional VSP acquired on the well after drilling ended.

3D-RVSP migration

At the end of the data acquisition and before migration, a more sophisticated processing was performed on the data to enhance their signal-to-noise ratio. This included a better

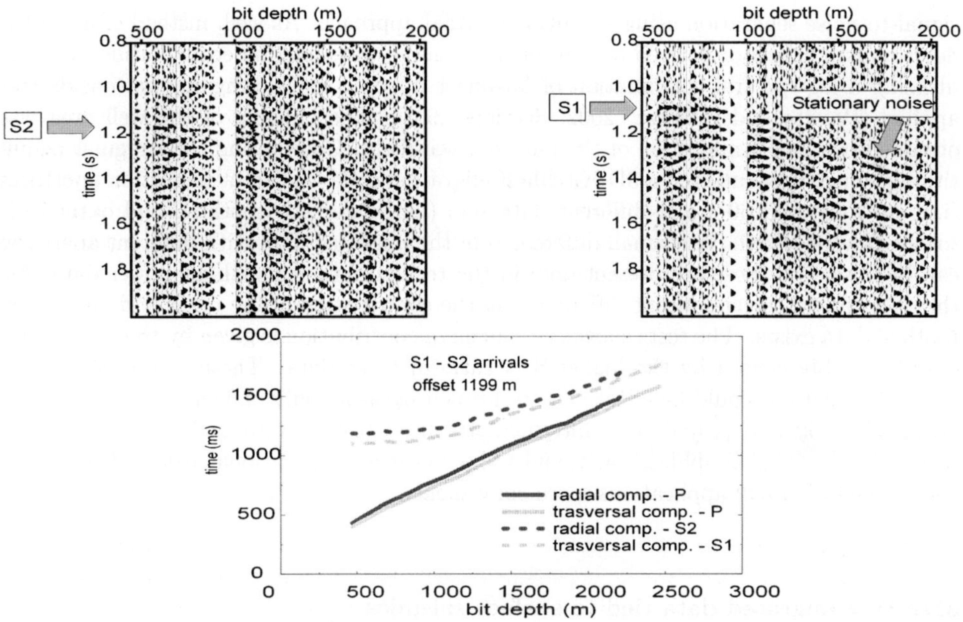

Figure 8.49: Radial (left) and transversal (right) component of the channel located at 1200 m offset from wellhead. Arrows indicate fast- (S_1), slow- (S_2) shear arrivals and stationary noise. Picking of the interpreted S arrival recorded by horizontal components.

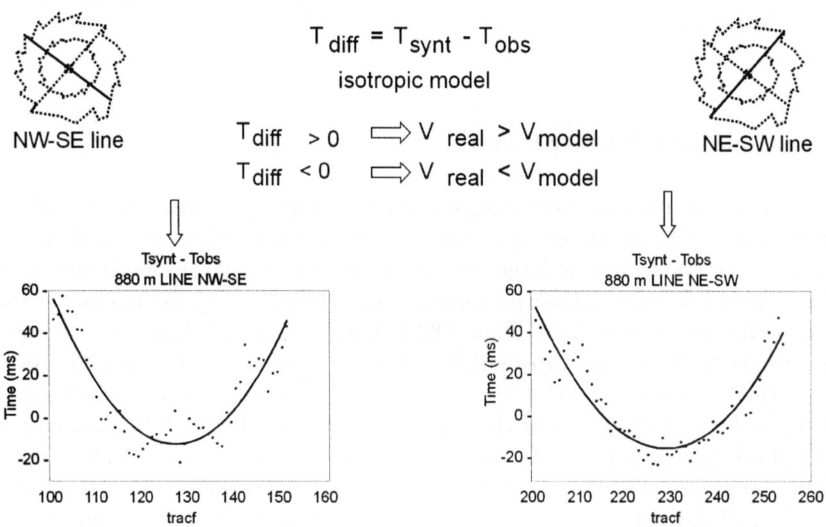

Figure 8.50: Difference between direct arrivals (drill bit to surface receiver) computed by synthetic isotropic model and observed data.

signal-to-noise separation using a multi-azimuthal approach. As well, instead of using the deterministic operator derived only from the upgoing wavefield, a deconvolution operator utilizing a log-spectral combination of far-offset upgoing and downgoing wavefields was applied (Poletto and Bellezza, 2002; Section 7.9.4). The imaging of the well area was obtained by the 3D migration of the reflected wavefield. Considering the irregular acquisition geometry, we utilized a 3D-Kirchhoff migration code. Different migration apertures (i.e., 30°, 45° and 60°) and different data sets (i.e., inner saw-toothed, and outer saw-toothed) were tested. Very small differences in the migrated data with different apertures can be observed; these are present only in the results obtained with shallower data. On the other hand, the significant differences in the contribution of the inner and outer saw-toothed data exists. The tests show that the main contribution is given by the inner-circle data, probably caused by the higher S/N ratio in these data. These results show that future acquisitions could be effectively performed by using only one circular saw-toothed layout of receivers deployed at an intermediate offset, to be determined in relation to the target depth. This simplification would reduce cost and operational risk and make the commercial 3D-RVSP applications more convenient.

3D-SWD migrated data tied to surface seismics

2D sections extracted from the cube of the 3D migrated SWD data are compared to 2D-surface reflection seismic line. Figure 8.51 shows the 2D-migrated surface reflection seismic data (SW-NE direction) compared with the SWD data. The principal interfaces are indicated by arrows. The SW-NE line surface seismic data were acquired by Hydrapulse. The target of the well, at about 1.45 s TWT, is clearly detectable on the surface data as well as on the SWD data.

Interpretation of 3D-RVSP cube

The 3D migration results were interpreted and compared to other seismic data, sections and 2D lines. The anticline trap, representing the well target, was identified. The continuity of the seal/reservoir interface located at about 2700 m depth was better in the drill-bit-SWD images obtained by tomographic inversion and 3D migration than in the surface-reflection-seismic data. Figure 8.52 shows, on the left side, an E-W line extracted from the 3D-RVSP SWD migrated volume. The correlation with the logs and corridor stack and synthetic trace processed with data acquired in the conventional wireline mode shows a very good match especially at the top of the reservoir. Figure 8.52 shows on the right a detail of the interpretation of the top of the reservoir done on the 3D-SWD-migrated data volume. It shows that the top of the reservoir is slightly offset from the well, meaning that the well did not cross the reservoir in its highest position. This interpretation is in agreement with the results of dip-meter borehole information. The 3D RVSP SWD is a good tool to plan side tracking of the well when the top of the structure interpreted on the 2D-lines is not reached.

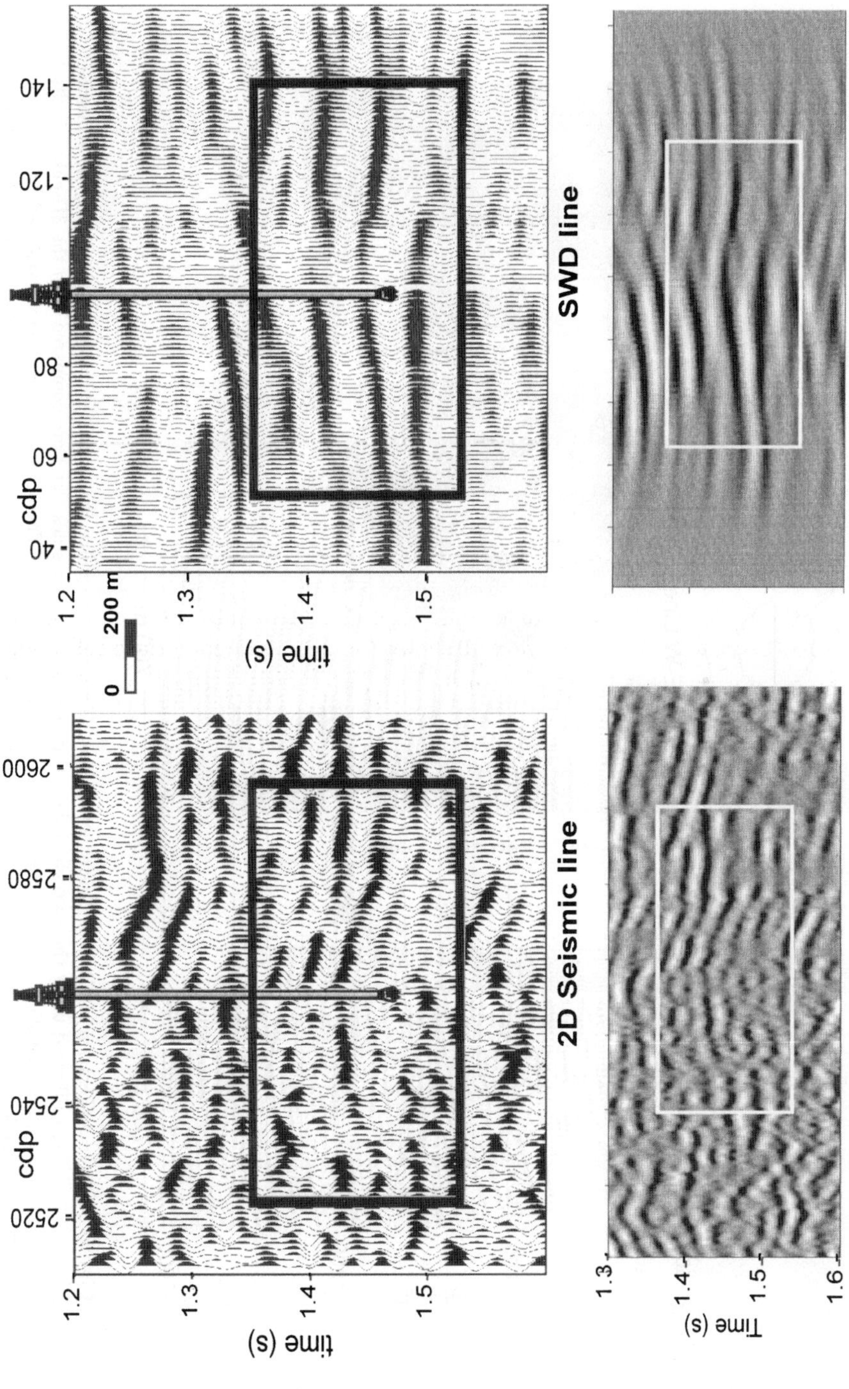

Figure 8.51: SW-NE direction: surface reflection seismic (left) and 3D-SWD (right) data comparison. The bottom panel shows surface reflection seismic and SWD data along the NW-SE direction. These surface data were acquired by Hydrapulse seismic source. The target can be compared in the two sections. The white rectangle indicates the target area.

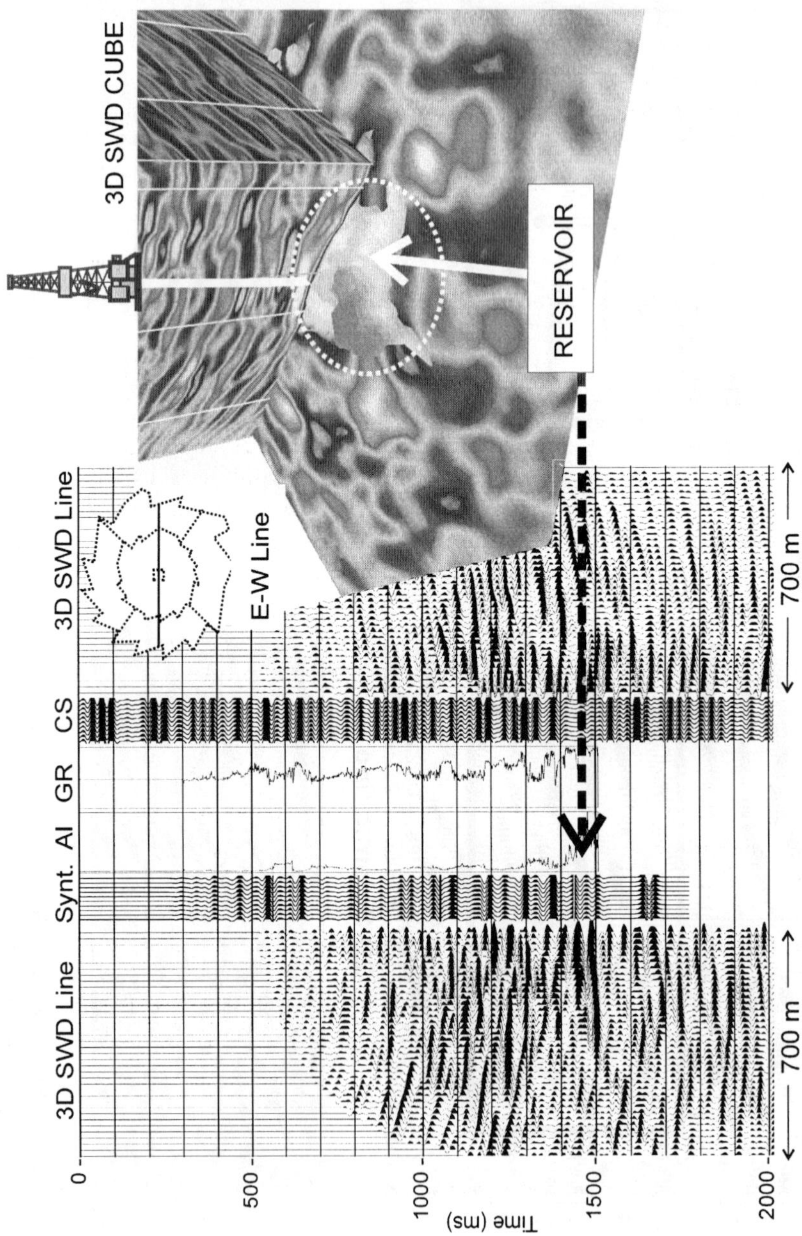

Figure 8.52: Interpretation of 3D RVSP Seisbit SWD. The 2D multioffset results, the synthetic trace obtained using the acoustic log calibrated using the SWD checkshot, and the SWD corridor stack (left) are compared with the 3D-cube results (with a time slice) in which the reservoir is interpreted (right). The 3D-SWD-RVSP imaging was useful to improve the knowledge of the reservoir and the overburden. The anticline trap, representing the well target, was identified.

8.11 While-drilling application of 3D-RVSP imaging

3D-imaging by drill-bit SWD data was successfully demonstrated in the first survey performed in Sicily (Italy). However, in this first survey the final 3D-imaging results were achieved only after the well was completed. Following the experience of this pilot survey, a second 3D RVSP was acquired in another exploration well drilled in the Southern Apennines (Italy) in 2002. In this work, Poletto, Petronio, Malusa, Schleifer, Corubolo, Miandro, Bellezza and Gressetvold (2003) obtained and used 3D-imaging results for prediction while drilling during the new 3D-drill-bit-RVSP application. The goal of this application was achieved by a careful organization and day-by-day tuning of the resources and team work. The main steps of the 3D-while-drilling survey were: initial- and near-surface-model definition; data acquisition and day-by-day processing; while-drilling tomography and model update; while-drilling Kirchhoff migration of 3D data; interpretation and prediction. Results have been used to support decisions made in the well-program variation as well as for the revision of the geological interpretation.

Geological setting and exploration problems

The well site is located in the Southern Apennines chain, an east-verging accretionary wedge resulting from the deformation of the Apulian continental margin (Section 8.8.1). The investigated area is characterized by a complicated overthrust tectonic style where several nappes overlay the Apulia Platform carbonates. Structural complexity and rough topography determine a poor surface seismic data quality (Bertelli, Abramo and Gatti, 1998) so "a priori" definition of the target depth and geological prognosis are difficult.

Work flow and 3D-while-drilling seismic imaging

About 300 channels (20 km total length of seismic lines) were used in this 3D-SWD survey. With this receiver configuration in a mountain area, a strong effort to acquire, process and manage the whole dataset was required. In Figure 8.53, the block diagram representing SWD-data work flow at the well site and at headquarters during well drilling is shown. During the drilling, in correspondence with two drill-bit levels (2442 and 2778 m), tomographic inversion of the drill-bit direct arrivals was performed to update the 3D-velocity model. On the base of this model, the imaging of the well area was obtained. A first 3D-Kirchhoff migration of the drill-bit reflected wavefields was performed while drilling with data up to 2442 m depth. A numerical mesh of $N_X = N_Y = 161$ and $N_Z = 251$ grid points, spaced with intervals of $dx = dy = 25$ m and $dz = 20$ m, was adopted as the input model, while the size of the output seismic cube was 97×97 traces with 25 m trace spacing.

While-drilling 3D results

3D RVSP and multioffset 2D-VSP-CDP mapping was utilized to predict the geological interfaces ahead of the bit. This information helped drillers in well-plan execution as well as geologists in geological re-interpretation. At a depth of 2610 m, hard-quartz sandstone caused a slow penetration rate. 3D-SWD migration processed at a drill-bit depth of 2442 m shows the top and the thickness of this unexpected layer. In the 3D-migrated data, at a depth of about 2980 m a strong reflection was observed. This predicted discontinuity was apparently not confirmed by changes of the main drilling data, such as lithology in cuttings samples and penetration rate (ROP). A sonic log acquired at the end of the drilling phase indicated an abrupt *P*-velocity decrease in correspondence with the strong reflection detected in the SWD data later related to a high-porosity zone with gas (Figure 8.54).

In summary, 3D-drill-bit-imaging results were achieved while-drilling and demonstrated to be beneficial for drilling and geological interpretation.

8.12 New trends for SWD

New trends of seismic while drilling are aimed at enlarging the field of application of the methodology and may follow from the application of new-emerging technologies. This goal involves overcoming limitations of the method with respect to the following.

Technical-recording capabilities – This aspect is related to the improvement of the downhole-recording capabilities and the increase of the allowed data-transmission rates – e.g., up to thousands of bits per second – from downhole to surface and vice versa.

Environmental and drilling conditions – The second point is related to solving the problems of acquiring and using SWD data in critical conditions, such as with difficult (for SWD) drill-bit types (in particular, PDC and diamond bits), downhole motor drilling, sliding drilling mode, highly deviated and horizontal well drilling, as well as drilling in deep water. This point is linked to the previous one.

Geophysical information – This aspect is related to the improvement in the signal-to-noise ratio for obtaining better reflection data, to the use of SWD for pressure assessment and prediction, as well as to borehole geosteering with near-well imaging and high-resolution while-drilling information and, eventually, to the integration of downhole/borehole technologies for smart-well drilling.

8.12.1 SWD in deep water

The importance of SWD is related to the critical environmental conditions for deep-water drilling, e.g., water depth > 1500 m. High daily cost of the rig and explorative conditions with uncertainty make the checkshot product very important. In this environment, technical problems are the following.

- ◇ The possible loss of the pilot signal transmitted in the drill string, for dispersion of the acoustic energy in the traveling through the riser (see Sections 2.4 and 4.3).
- ◇ the necessity to obtain quick information.

Normal seismic while drilling with receivers in the pipes and surface sources (Section 3.3.1) is a method that may be used in these conditions. Moreover, the use of downhole

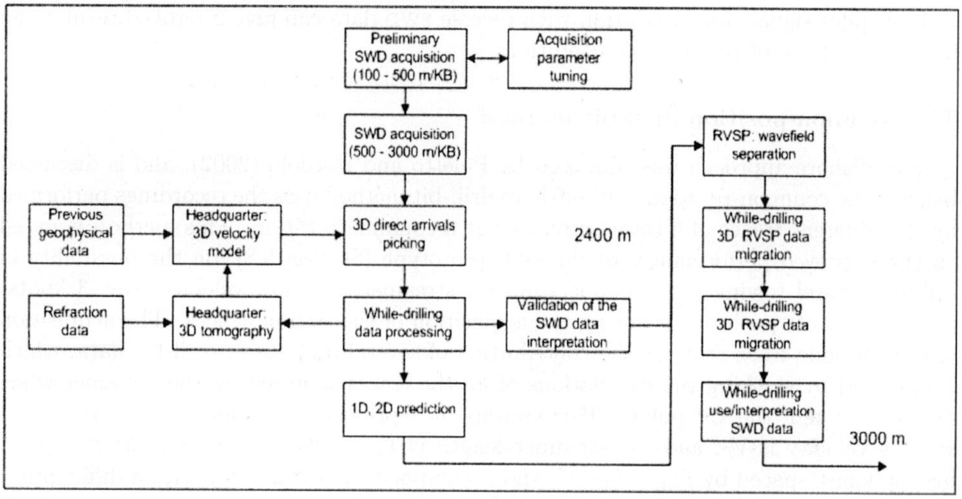

Figure 8.53: 3D RVSP survey: data and work flow.

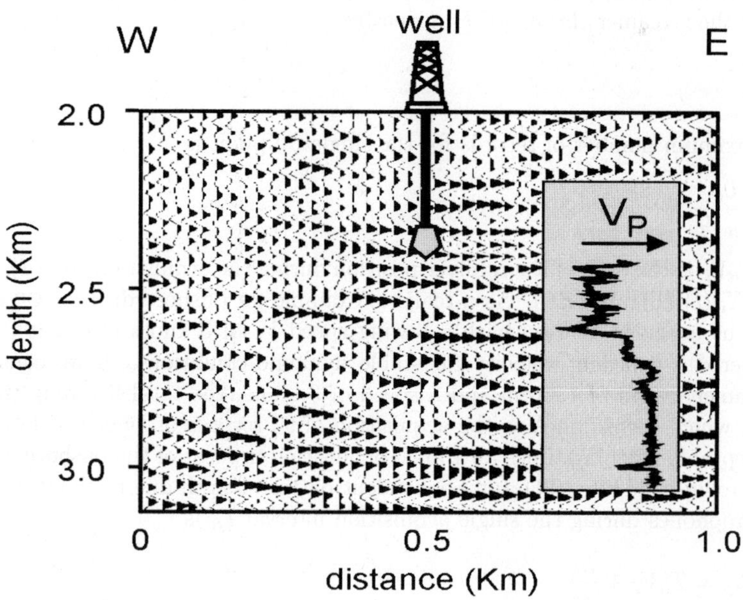

Figure 8.54: Prevision with SWD-migrated data of the interval up to 2442 m (bit position in figure). The depth image is compared with the P-velocity function obtained by sonic log.

drill-bit pilot signals for correlation with reverse SWD data can give improved results and has not yet been tested in these conditions.

The common-position drill-bit method

A new offshore approach was proposed by Poletto and Dordolo (2002), and is discussed below. The common-position (CP) offshore drill-bit method uses the recordings performed by the different traces of a towed streamer at a common position. This method is based on the improved performances of the SWD prototype (Section 5.8), on the possibility of sailing a vessel towing one (or more) marine streamer(s) at low velocity (say, 3 knots) and on accurate positioning during the acquisition of the seismic traces. The acquisition of each seismic trace requires the contribution of several minutes of drill-bit data, which is obtained by stacking the correlations of all the traces acquired by the streamer when passing through the CP point. This concept is represented in Figure 8.55, where the vessel's velocity is V_T, and the streamer length is L_C. The streamer has M groups of hydrophones, spaced by $\triangle_R = L_C/M$, and is equipped with a tailbuoy with a differential-global-positioning system (DGPS), while a reference global-positioning system (GPS) is located on the rig. To position the streamer, remote-reading compasses, an acoustic location system, a laser (MARS), and depth controllers are also used. We indicate the generic CP point, i.e., a fixed point on the sea surface, by CP_i. When sailing the streamer through it, and after a trigger signal, the vessel starts the acquisition of a sequence of N_R records $X_{ij}(t)$ $(j = 1, \ldots, N_R)$ with time-length T_R (Figure 8.56). This number is limited by the maximum listening time of the single CP point CP_i, which is equal to the total sailing time of the streamer through CP_i, given by

$$T_S = \frac{L_C}{V_T}. \tag{8.17}$$

Hence, the maximum number of records for the single CP point is

$$N_R = \frac{T_S}{T_R} = \frac{L_C}{T_R V_T} = \frac{L_C}{\triangle_x}, \tag{8.18}$$

where $\triangle_x = T_R V_T$ is the drift of the receiver position during the acquisition time T_R of the single record $X_{ij}(t)$ (Figure 8.56). As a realistic example for offshore drill-bit CP-gather feasibility, let us consider a vessel with a towing speed V_T of 3 knots (1.6 m/s), which can be considered a sufficient velocity for stable dynamic conditions. If we consider a reference streamer length of 1500 m and a speed of 3 knots, the total listening time T_S is 18 minutes, which corresponds to the acquisition of 90 records of length of 12 s each. These are acceptable listening times for robust acquisition in most of the onshore drill-bit surveys using roller-cone bits. The total equivalent pattern-length due to the drift of the groups of hydrophones during the single acquisition interval T_R is

$$\triangle_P = \triangle_x + \triangle_R = T_R V_T + \frac{L_C}{M}. \tag{8.19}$$

Under stationary signal conditions in the time-interval T_R, the shape of this equivalent pattern accumulating the continuously shifted contribution of the hydrophone array is a trapezoid, or a triangle in the particular case when $\triangle_x = \triangle_R$.

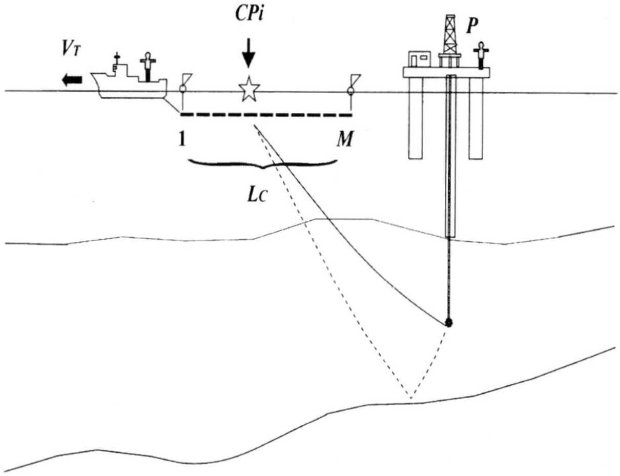

Figure 8.55: The drill-bit signal is recorded with two systems. A system on the vessel records the seismic traces as the streamer passes through the common-position point CP_i, marked by a star on the sea surface. The vessel tows the streamer with M traces through CP_i with velocity V_T. Simultaneously, a system on the rig records the pilot reference traces.

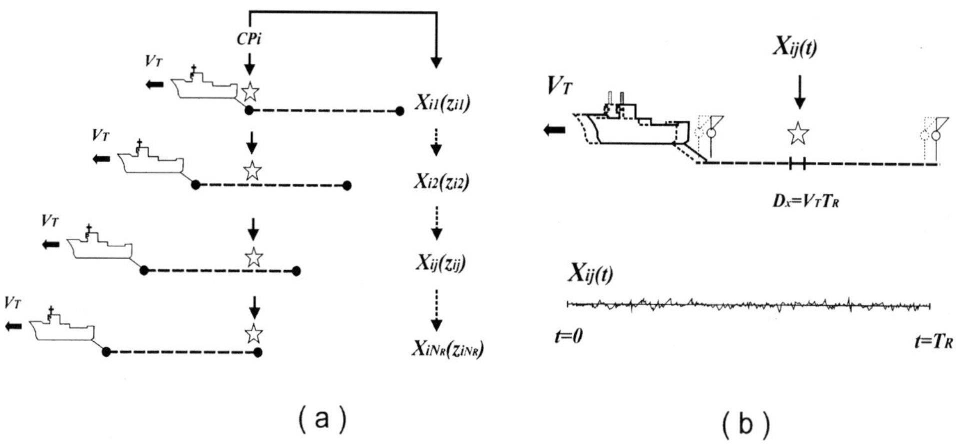

Figure 8.56: a) The traces of the common-position gather are acquired by different hydrophone groups when sailing the streamer through CP_i. Each trace corresponds to a different depth of the bit z_{ij}. b) The vessel system records N_R records when the streamer sails through CP_i. The recording time of each record is T_R, and X_{ij} indicates the trace acquired in CP_i with the j-th record ($j = 1, \ldots, N_R$). During the acquisition interval of the single record X_{ij}, the drift of the streamer is $\triangle_x$ (after Poletto and Dordolo, 2002).

Figure 8.56 shows the concept of CP gathering. In the common-position gather CP_i, we select all the traces X_{ij} acquired in correspondence with the same i-th position. These traces belong to different field records which are denoted by the index $j = 1, \ldots, N_R$, and recorded in different time intervals. In general, the traces of the CP gather correspond to different, but close, bit depths. Hence, this gather has an average acquisition time, an average position and an average bit depth. If we consider a neighboring CP point, the recording-times and bit-depths change because of the steamer drift, with slightly different average times and bit-depths. If z_{ij} is the bit-depth of j-th record, the average depth of the i-th CP-gather is

$$\bar{z}_i = \frac{1}{N_R} \sum_{j=1}^{N_R} z_{ij}. \tag{8.20}$$

Let P_j, with $j = 1, \ldots, N_R$, be the Fourier transform of the corresponding rig-pilot signal acquired with the synchronized SWD system. We obtain the seismic signal in the i-th position by correlating the traces of the CP_i gather with the pilot signal, and stacking these correlations. In the frequency domain, the correlated trace of the i-th position is given by

$$C_i = \sum_{j=1}^{N_R} P_j^* X_{ij}, \tag{8.21}$$

where '$*$' denotes complex conjugate. However, in the presence of steeply dipping events (such as diffraction flanks), sailing-Doppler effects due to relative velocity of receivers with respect to the fixed source and medium may be not negligible (Dragoset, 1988). The Doppler effect, if not compensated, may produce phase distortion and signal attenuation in the correlated data, and introduce limitations for the total time-length of the traces correlated in equation (8.21).

The errors introduced by streamer towing.

The effectiveness of the sum in equation (8.21) depends on the stationarity of the conditions in which the signal is acquired. In offshore drill-bit seismic while drilling, which uses fixed ocean-bottom cables (OBC) and vertical cables, the non-stationarity is mainly related to the source variability. Here, we have to take into account the sailing effects. These may introduce acquisition errors due to several reasons. These include, for instance, the buoying instability and the vertical fluctuations of the streamer, the lateral bending of the streamer, the sea waves, and the errors in the positions of the traces while sailing, including the tides. We distinguish between vertical and lateral effects.

Lateral streamer effects – When the drift due to feather exceeds the equivalent-array length of equation (8.19), correction due to the streamer feathering should be taken into account (Carcione, Padoan and Cavallini, 2000). In onshore drill-bit seismic while drilling, receiver-group-array lengths of several tens of meters are in standard use. Comparable values can be acceptable both for the in-line apparent pattern effect and for lateral feather, which must be controlled in order not to induce significant errors. For a streamer length of about 1.5 km, this feather should be limited to about 5° if the feather is uniform along the streamer. In case of higher feather, the processing should separate the groups of traces in the partial stack of more common-position gathers. In addition, in the presence of significant feather, the differential-NMO correction before the stack of the correlated traces of equation (8.21) must be applied to preserve resolution (Wardell, Diviacco and Sinceri, 2000).

Frequency and far-offset effects – The CP method may introduce long-array effects for the bit signal at far offset traces in radial lines. At large offsets and for shallow depths, the apparent velocity of the bit signal has its lowest asymptotic $V_{\rm RMS}$ value and the apparent radial-array length (equation (8.19)) filters the signal above the minimum wavelength $\lambda_{\rm min}$, which is small when using long recording times T_R. If we assume that the array-response notch for $\lambda_{\rm min}$ corresponds to the array length, the frequency notch is related to the sailing and acquisition parameters by

$$f_{\rm max} = V_{\rm RMS}(T_R V_T + \triangle_R)^{-1}, \quad (8.22)$$

where we have used equation (8.19). If $V_{\rm RMS} = 2400$ m/s, $\triangle_R = 12.5$ m, $T_R = 10$ s, and $V_T = 1.56$ m/s, we obtain $f_{\rm max} = 85$ Hz. This value is higher for non-radial sailing lines and deep bit signals.

In general, the average area of acquisition determined by the in-line apparent array effect and by the lateral feather should be taken into account while sailing. This area depends on the angle between the sailing direction and the direction of the radius from the well, and on the lateral-offset of the sailing line from the well. This analysis allows us to determine the acquisition and sailing configuration, taking into account the sailing conditions, to preserve the resolution of far-offset signals, and to give the correct average positioning of the CP point.

Doppler error – Dragoset (1988) showed that a Doppler factor δ can be used to relate the frequency f_s of a marine-vibrator source to the frequency $f_r = f_s(1+\delta)$ detected by a receiver. The Doppler factor δ can be calculated as $\delta = (V_T/V_W)\sin\alpha_r$, where V_T, V_W, and α_r, are the boat velocity, the speed of sound in water, and the in-line inclination angle of the emerging ray, respectively (Figure 8.57). This factor can be of the order of $\delta = 0.5{\cdot}10^{-3}$ for a ray with inclination of 30°. Considering a 10 s trace of 5000 samples, this corresponds to a relative difference of 2.5 samples. Hence, acquisition geometry (such as an apparent dip of events and sailing direction at large offsets) and time length used for correlation have to be carefully evaluated during acquisition and processing.

Vertical effects – The vertical variations of the streamer and the platform may introduce timing errors. As a reference value, a depth difference of 1.5 m between any two traces of the streamer corresponds to a 1 ms delay for vertical arrivals. In general, we assume that these vertical fluctuations correspond to a single time error in the arrival time of the signal. Let $\tau_{p,j}$ be the time-error associated with the pilot signal P_j, due to possible vertical fluctuations of the rig, and $\tau_{x,ij}$ the time-error

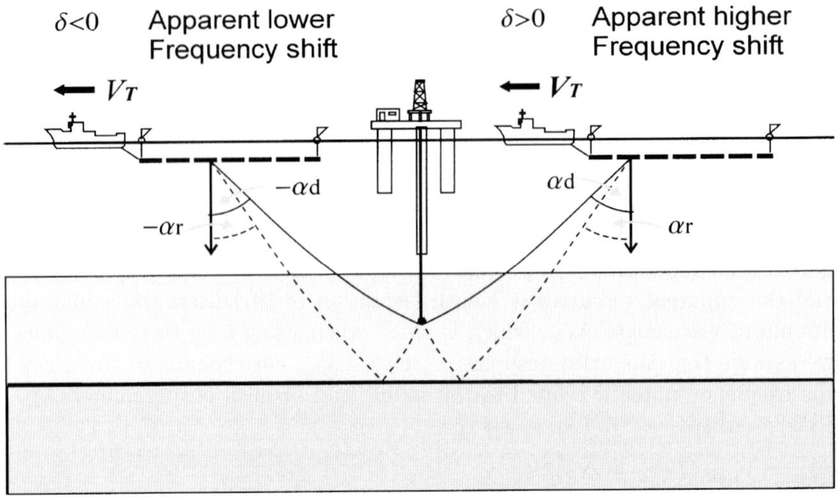

Figure 8.57: Doppler effect due to sailing velocity. The approaching angle is positive, the moving-away angle is negative. Inclination angles of direct arrivals (α_{d}) and reflections (α_{r}) are different. Reflections are subject to relatively smaller Doppler effects.

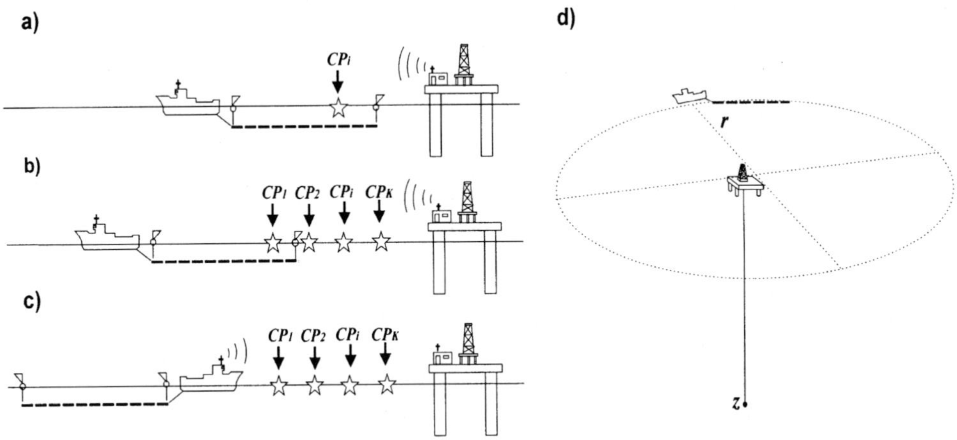

Figure 8.58: a) Layout of the acquisition phases. At the beginning of the acquisition, the rig-system controls the drilling parameters, interrogates the vessel system about the status of the line and sends a trigger signal to start acquisition (a). During sailing, many neighbor acquisition points are collected, with regular sampling of the seismic line. For each acquisition gather, the rig system transmits the pilot traces to the vessel system (b). After correlation and VSP seismic processing, c) the vessel system transmits the geophysical results to the rig laboratory. d) The offshore common-position method makes it possible to acquire many traces with large offset, radial, transverse and circular geometry, suitable also for 3D RVSP purposes.

of the trace X_{ij}. Taking into account these time-errors and the correction for the differential-NMO due to the variation in the trace location, the correlated traces are

$$C_i = \sum_{j=1}^{N_R} \left[P_j^* X_{ij}\, e^{\imath\omega(\tau_{x,ij} - \tau_{p,j})} \right]_{\mathrm{NMO}} , \tag{8.23}$$

where the subscript 'NMO' denotes the differential NMO correction of the traces of the CP gather with respect to the NMO of the CP point. The automatic while-drilling estimation of the time-fluctuations $\tau_{x,ij} - \tau_{p,j}$ and of the NMO variation in the CP gather is important to assess the quality of the correlated signal of equation (8.23).

Acquisition technology and data flow of offshore CP-SWD

The offshore CP-point acquisition method is based on a vessel towing one (or more) streamer(s) and two SWD laboratories (on the rig and vessel). The technology and the data flow are depicted in Figure 8.58. The data acquisition unit on the rig acquires the pilot traces, controls/records the drilling parameters, acquires a GPS synchronization trace, and stores the pilot traces on a local disk. A workstation synchronizes both a local acquisition PC and the remote recording on the vessel by sending a trigger signal via radio (Figure 8.58a).

On the vessel, the other acquisition unit acquires the seismic signals using a conventional (or solid-state) marine streamer, which is continuously balanced and positioned by the tailbuoy with DGPS, and compasses located along the streamer, e.g., every 100 m of streamer length. The streamer has also "birds" with sea-level sensors to better define and control the depth and the location of the hydrophone groups. These data are collected in real time by the vessel workstation, which evaluates the acquisition configuration. This workstation communicates via radio with the rig workstation and receives the trigger signal. The workstation is interfaced with the streamer-control system. At the triggering signal, the acquisition starts on the rig and on the vessel, synchronized by the GPS signal. Then, a sequence of records is acquired with close regular intervals, to cover the entire listening time of the CP point. The acquired raw data are stored on a local disk with large storage capabilities.

When the acquisition of the CP point is completed, the rig workstation transmits, via radio or cellular telephone, the corresponding pilot traces to the vessel workstation (Figure 8.58b), together with the drilling information. In the subsequent step, the vessel workstation correlates the traces, which are then processed with VSP programs. Finally, the near-offset, walkaway or multioffset reverse VSP results are sent, via radio or cellular telephone, to the rig laboratory for a first geological interpretation (Figure 8.58c).

CP-SWD acquisition geometry: offshore 2D versus 3D

Figure 8.58d shows an example of an acquisition scheme with circular-sailing geometry. The radius from the well is $r = 2$ km, and the circumference is 12.5 km. With a tangential-sailing velocity $V_T = 7.5$ km/h (4 knots), the sailing time around the well is about 1 hour and 40 minutes. More than 200 CP points can be gathered in close positions by acquiring continuously during this time interval. The difference between the average source-bit-depths for the first and last CP points after an entire turn depends on the rate-of-penetration (ROP) of the bit. If the revolution-period for completing a turn is $\Delta T_{\mathrm{rev}} = \frac{2\pi r}{V_T}$, the corresponding drill-bit deepening is $\Delta \bar{z}_{\mathrm{rev}} = \mathrm{ROP}\, \frac{2\pi r}{V_T}$. This means that after a complete turn of the vessel, the correlated traces at the same position correspond to different depth-levels of the source. If ROP = 10 m/h, we have $\Delta \bar{z}_{rev} = 16.7$ m. This correction must be taken into account in the processing sequence.

If the radius is sufficiently large, the CP technique can be safely applied to a circular geometry acquisition without corrections due to lateral feather effects. The approximation of the circular geometry with a series of straight lines is also considered by taking into account the acquisition delays of the maneuvers. Since the circular geometry can be repeated for many drill-bit depth levels, it becomes theoretically possible to collect an offshore 3D-RVSP dataset with many CP points at the surface and many drill-bit source depth levels. The expected results of the method are a sequence of large-offset walkaway RVSP shot gathers, acquired to extend the investigation range around the well, to improve the velocity analysis (coverage in tomography), and to better discriminate signal and noise wavefields in the offset-time domain. The aim of this technology is to provide a complementary tool to integrate the existing offshore-while-drilling methods, i.e., to be compared with other technologies based on the use of OBC and vertical cables. The advantages of this approach are the reduction of costs for acquiring extended-offset data by using simpler offshore operations with 3D-RVSP configurations and in deep water. Limitations are foreseen for the continuous acquisition of RVSP levels, since this method is not usable in stormy weather conditions. However, the design of the acquisition geometry will be aimed at compensating any loss of information caused by missing depth levels, with the improved offset information obtained during the reliable acquisition phases. This application will be more effective with the integrated use of downhole pilot measurements to improve the overall signal-to-noise ratio for the bit signal acquired in different sailing conditions.

8.12.2 SWD in highly-deviated wells

The SWD results are well demonstrated and proved when roller-cone bits are employed both in soft and in consolidated lithotypes (Section 8.5.1) and in wells with deviation up to 50 degrees. Highly deviated and horizontal wells are more problematic for drill-bit seismic based on surface pilot correlation (see also Section 8.4). This is due to loss of pilot signal transmission through the drill string for increased friction and drill-string borehole contacts, and reduction of drill-bit forces partially loaded on the borehole wall. In the formation, the radiation properties of the source change the amount of compressional and shear waves. Results with horizontal drilling were shown by Yu and Schuster (2002) with only crosscorrelogram migration of drill-bit surface-geophone data (Section 7.12.1). Downhole pilots may be of great help in this problem (see Section 7.8.1).

8.12.3 Geopressure prediction and assessment

The hydrostatic pressure of a column of fluid of density ρ and height z is

$$p = \rho g z, \tag{8.24}$$

where g is the acceleration of gravity. The pressure gradient is

$$\frac{\partial p}{\partial z} = \rho g, \tag{8.25}$$

that is the fluid weight per unit volume, which is often used to define borehole pressure conditions. Borehole fluid pressures different from the hydrostatic pressure (gradient $\approx$ 1 kg/lit) are defined as anomalous. Pressure detection while drilling is one of the most important targets of future SWD applications. In Section 2.2 we have defined high-pressure wells (HP) as wells where the pressure is higher than 690 bar. Another definition, equivalent for a well depth of about 4000 m, is that the pressure gradient – or the corresponding mud weight – is higher than 1.8 kg/lit. Detection of anomalous borehole pore pressures higher than hydrostatic is often referred to as the problem of detection of "overpressures". Overpressure effects reduce the difference between confining and pore-fluid pressures, resulting in a weakening of the overpressured rock (Poston and Berg, 1997). Mud weight is used to balance overpressure and avoid kick risks. However, underpressure (lower than hydrostatic) can also be of critical importance for drilling. For example, low pressures may follow the gas gradient or be related to hydrodynamic flow of aquifers (Poston and Berg, 1997). Predicting the expected pressure zones, the assessment of overburden matrix, the pore and fracture pressures is a fundamental part of any drilling plan to drill in a "balanced mode" and avoid blow out risks by properly setting mud plan and casing points (Section 2.6).

There are direct and indirect evidences that can be used to detect borehole pressures. Some of these methods are discussed in Section 2.6.1 and include use of borehole-log and drilling parameter measurements (Eaton, 1975). Seismic methods can be useful before and during drilling. In the prediction analysis, it is important to understand the origin of overpressures, especially in new exploration areas. Overpressures may occur either at small (few hundred meters) and large depths, at more than 8000 m (Poston and Berg, 1997). Among other mechanisms of generation, the major settings are shale undercompaction in areas of rapid deposition, oil-to-gas conversion, and recently active tectonic zones. An important requirement for pressure confining and anomalous pressure generation is that rocks have low permeability, so that no fluid flow can occur. For this reason overpressures commonly occur in shale formations. Undercompaction in young shale formations may occur when the increasing weight of the overburden caused by burial is faster than the compacting porous rock capability to bear the stress.

Oil/gas conversion starts at a given depth in a reservoir volume sealed with faults whose permeability is sufficiently low so that the increase in pressure due to gas generation greatly exceeds the dissipation of pressure by flow (Carcione, Helle, Pham and Toverud, 2003). The effects of abnormal fluid pressures on the seismic properties are investigated by Carcione and Gangi (2000) who considered disequilibrium compaction and gas generation. Other mechanisms are sediment loading (absence of drainage), diagenesis, such as smectite dehydratation, organic material maturation, and aquathermal expansion due to high temperature (Mouchet and Mitchell, 1989).

Reservoir rocks and hydrocarbon accumulations are frequently found in close association with abnormal pressures. As well pressure compartments are common in sedimentary basins (Carcione and Helle, 2002). The method is applied to the Tune field in the Viking Graben sedimentary basin of North Sea. The velocity map of the Trabert reservoir and the inverted pressure distribution agree with the structural features of the Trabert formation and its known pressure compartments (Carcione, Helle, Pham and Toverud, 2003). Physical properties such as compressibility, porosity and permeability in relation to in situ stress conditions determine the impedance of the rocks to seismic waves. Design aspects in reflection seismology aiming at petrophysical rock properties detection for stratigraphic purposes are discussed by Bilgeri and Carlini (1988).

Effective and differential pressures (or differential stress) play an important role in rock properties (Carcione and Helle, 2002), because they control the sedimentary compaction process. Differential pressure $p_{\rm d}$ is defined as the rock confining pressure $p_{\rm c}$ minus the pore fluid pressure $p_{\rm f}$ (Terzaghi, 1943)

$$p_{\rm d} = p_{\rm c} - p_{\rm f}. \tag{8.26}$$

The effective pressure is obtained by introducing the effective stress coefficient n in equation (8.26)

$$p_{\rm e} = p_{\rm c} - np_{\rm f}. \tag{8.27}$$

In static experiments $n \approx 1$ and for the purposes of this discussion we can assume that differential and effective pressures are equal. A more refined analysis may be undertaken to put in relation seismic measures of $V_{\rm P}$, $V_{\rm S}$, Q-factor and AVO effects to pressure conditions, rock compressibility, porosity and permeability properties (Carcione and Helle, 2002). This analysis needs the support of laboratory tests and/or logging data.

A basic concept in pressure prediction and detection is that in shale undercompaction conditions rock velocity is proportional to the logarithm of the effective pressure, which is essentially determined by the grain compaction local mechanism of a given provincial area. A deviation in effective pressure from the normal trend corresponds to a change of seismic velocity. Therefore, a plot of the measured velocity versus effective pressure, compared to that of velocity with the normal hydrostatic pressure gives a measure of the pressure variation from hydrostatic pressure. The point is to determine the so called "normal trend" of velocity in a shale formation – that is, an empiric law – and to estimate the real stress conditions, which in tectonic areas can be produced not only by the vertical overburden lithostatic component but also by non-negligible horizontal stress components. The pressure variation results in a variation of seismic velocity or interval travel times which may be detected and predicted by seismic methods (we may cite Bilgeri and Adameno, 1982; Weakley, 1989; Bowers, 1995; Sayers, Johnson and Denyer, 2000; Kan and Swan, 2001).

Overpressure, related to high porosity and undercompaction, corresponds to a decrease in seismic velocity. Reliability of seismic measurements for pressure prediction depends on signal-to-noise ratio of measured velocity data, of variable quality from stacking to tomographic-inversion velocity and interval VSP and log velocities (Sayers, Johnson and Denyer, 2000; Dutta, 2002; Carcione, Helle, Pham and Toverud, 2003).

Seismic while drilling can be used to assess and predict geopressures in two ways. It has to be emphasized that while-drilling assessment of borehole pressures is an important result – routinely obtained by using the empirical relationship based on drilling parameters – and even more important is the prediction ahead of the bit. The while-drilling assessment is by integrating drilling data, the prediction by velocity analysis.

Overpressure analysis using drilling data – Overpressure is related to higher ROP in rocks of depositional environments overpressured because they are undercompacted. Moreover, overpressure lowers the differential pressure between drilling mud – assumed to be of a given weight in slightly overbalanced drilling – and formation fluid (note that here the meaning of "differential" is between pore fluid and mud). This condition helps cutting cleaning in rotary drilling. Hence, overpressures are typically related to higher drilling rates (Warren and Smith, 1985; Warren, 1987; Peltier and Atkinson, 1987). In addition, the drill-bit seismic signal is influenced by differential drilling pressure conditions. See the discussion on drill-bit forces in rotary drilling in Section 3.12.3 (Maurer, 1962; Maurer, 1965). A while-drilling analysis of drill-bit source amplitude may be of help for use in conjunction with other indirect pressure detection methods based on drilling parameters, e.g., "*d*-exponent" and "sigma-log" empirical methods, related to rate of penetration normalized by drilling conditions (see Section 2.6.1; Jorden and Shirley, 1978; Mouchet and Mitchell, 1989).

Analysis using seismic velocity – While-drilling seismic velocity can be used with empirical relationships relating velocity to effective pressure in shales (Terzaghi, 1943). This method may provide while-drilling seismic information about the pressure of the drilled shale rock by analysis of SWD direct arrivals. Analysis of reflected wavefields (tomography, full-wave inversion, imaging) may provide seismic velocity, and consequently pressure information, ahead of the bit if the "normal velocity trend" in the well area is known.

Indirect detection – Indirect detection is also a while-drilling product. As discussed in Section 8.8.5, prediction of interfaces ahead of the bit identifies areas of potential risk. These formations are usually of known pressure properties and the prediction of their correct location in depth is an important task for which SWD is already used in areas of complex geology.

S and *P* SWD waves are commonly used to detect anisotropy and may provide information about pressured rock properties. The use of SWD data for *Q*-factor analysis may be of more difficulty due to S/N and variability of the drill-bit source.

8.12.4 The road ahead: SWD by downhole technology

The main technical difficulties of the drill-bit SWD methodology are in the applications with PDC bits, horizontal and highly deviated wells and with downhole motor in sliding mode, i.e., without pipe rotation. Many aspects related to the improvement of the signals acquired in these conditions have been solved. Others are still in a research phase. The PDC bit acts prevalently as a shear tool and its signal with lower axial vibrations of quasi-random nature, is more difficult to detect. The particular lithological environments in which PDC bits are used (Hanson and Hansen, 1995) and the nature of PDC forces are discussed in Sections 2.6.7 and 3.14. Help in the processing of the PDC data is given when some runs of the roller-cone bit can be included in a main run of the PDC bit. Refined processing by focusing on the multioffset data may also be of help in reinforcing the

PDC seismic signal, thus producing PDC seismic sections to tie with the surface seismic. However, this may be time consuming and not effective for while-drilling use of data.

Encouraging results with downhole motor in sliding and PDC bits have been obtained by using downhole recordings (see Section 7.8.1 and Miranda, Poletto, Abramo, Craglietto and Bernasconi, 1999). However, downhole instrumented tools are still in a test phase for routine SWD applications. Some important contributions in this field have already been given. Tests with use of SWD downhole technology in wells – in this case drilled with roller bits – were performed by using cabled pipes by Naville, Layotte, Pignard, and Guesnon (1994). Also, in geothermal exploration, tests were performed by using downhole memory for local data storage by Naville, Serbutoviez, Throo, Batini, Dini and Cecconi (2000).

Nowadays, "*smart well*" technology is a new challenge – a road ahead – for drilling which is essentially based on non-occasional downhole diagnostics technology, high-speed transmission (Finger, Mansure, Knudsen and Jacobsen, 2003) and the capability to power downhole instrumentation from the surface. Diagnostic while drilling (DWD) is based on the development of high-bandwidth continuous-transmission technology that can be used on a routine basis to communicate data sampled with rates of several kHz. The smart-well technology is essentially based on wireline systems in wired pipes with tool-joint and surface connections. There are different solutions to realize the tool-joint connections, which can be sealed (Laurie, Head and Smith, 2003) and/or based on inductive coiled coupling (Jellison, Hall, Howard, Hall, Long, Chandler and Pixton, 2003). Available speed rates are of the order of several kilo bit per second (KBPS) or mega bit per second (MBPS), while current mud-pulse telemetry is limited to about 10 bit/sec and electromagnetic telemetry is of the order of 100 bit/sec. The potentials of these new technologies for transmission – even with improved downhole electronic, processing and compression capabilities – and power supplying may also lead to a general improvement in SWD and geosteering technologies.

8.12.5 Drilling diagnostics and geosteering

Geosteering defines those activities that try to get the wellbore to a predetermined location defined both by its spatial coordinates in three dimensions, and by its position in the geological column. Geosteering methods use advanced downhole formation-evaluation (FE) – MWD sensors in conjunction with directional data to steer the drilling assembly into the target horizon and maintain it within its stratigraphic position to maximize the drilled section in productive reservoir rocks (Martin, Philo, Decker and Burgess, 1994; Meyer, Jian-Qun Wu, Macune and Harvey, 1995; Allan, Hess and Todd, 1996; Lesso and Kashikar, 1996; Phillips, Paulk and Constant, 2000; Robison, 2002).

With the possible exception of rank wildcat drilling in virgin territory, all wells are "geosteered" in the sense that they have both a spatial and geological target. In the past, the traditional geosteering team consisted of the combined talents and knowledge of the driller and the wellsite geologist. The results of their efforts were variable. They had little data to work with, and limited capability to get the drill bit to go in the desired direction.

Modern drill string components like steerable motors, better drill-bits, plus new drill-fluids, now enable the driller to actually steer the bottom hole assembly up and down. The geologist can call on new tools and data sources: centralized data collection, and graphical displays of drilling progress in modern mudlogging units; sophisticated biostratigraphy for detailed rock sequencing; and the development of MWD tools providing real-time petrophysical data to the surface. It is now possible to figure out where the drill string is, where it should be, and how to get there!

Geosteering is undertaken in distinct phases. Pre-job preparation integrates multiple offset well and digital map data into a 3-dimensional spatial model of the anticipated stratigraphy. The geosteering engineer works with the model to extract expected MWD tool response and to create a "most likely" log. Additional models are built to test alternative geological interpretations, and all data are consolidated into a pre-drill package.

During the well-drilling phase, survey and MWD data are captured and entered into geosteering software. A variety of custom plot formats are used to present the data in its geological context in both measured and stratigraphic depths. The geosteering engineer and well-site geologist work together with such visualization tools to correlate offset wells (and the model), to accurately land the well for the target section. Once in the reservoir, MWD tool measurements are continuously compared to the expected responses. When indications are seen of an approaching change in lithology or fluid content, the geosteering engineer will advise on corrective action that can be taken before the bit leaves the target horizon.

Most measurement configurations applied in MWD geosteering assemblies utilize dedicated formation-evaluation sensors like resistivity, azimuthal and non-azimuthal gamma-ray, neutron and density systems for formation parameter determination as well as acoustic sonic for single-well imaging (SWI) (Chabot, Henley, Brown and Bancroft, 2001). The main information source, providing the deepest look into the formation, is the resistivity measurement. Due to its used frequency range (kHz to MHz) and its non-azimuthal character, it provides only an integral circumferential outlook in a vicinity of less than 5 m around the wellbore. Additional drawbacks are a poor response in low conductive formations, e.g., sand versus shale, no directivity in the direction of the drill string axis (only normal to it), and no direct correlation to the surface seismic section, the basic road map of every well planning. So resistivity based geosteering systems provide only an averaged overview in the nearest vicinity of the wellbore, indicating changes in formation parameters while passing them. Thus, the best place for these tools would be as close to the bit as possible, a position not always compliant with other BHA purposes.

8.13 Geosteering

Optimum well placement is always a challenge, but especially important in an oil and gas market situation where financial investments for hydrocarbon recovery must be minimized in order to meet economic targets. Geosteering the borehole trajectory is a sub-set of many other efforts to direct the well in the minimum time with the highest possible precision into one or more targets.

Today geosteering has great promise in terms of optimizing the effective hydrocarbon production out of existing reservoirs. To determine the pay characteristics of potential oil and gas layers while drilling offers several advantages versus doing the same thing on a wireline after the drilling process. Primarily, the while-drilling method allows to react to unexpected or unknown formation situations by directing the trajectory to where the biggest recovery potential is met.

Traditionally the downhole evaluation of formation characteristics has been done by means of a wireline or test string after the section of hole was drilled. However, during recent years, a number of formation evaluation technologies have been made available for application while drilling (LWD and FEWD). Among these are the gamma ray, resistivity, nuclear density, lithology, etc. To install the sensors in the lower part of the bottom hole assembly allows for in-situ measurements, the results of which can be stored in the downhole tool for download after "pull out of hole" (POOH) or may be transmitted to the surface in real time by means of the standard MWD transmitter technology. This latter concept forms the basis for the geosteering approach.

Bibliography

Aarrestad, T. V., and Kyllingstad, Å., 1988, An Experimental and Theoretical Study of a Coupling Mechanism Between Longitudinal and Torsional Drillstring Vibrations at the Bit: SPE Drilling Engineering, (3), 12–18, Paper No. 15563.

Aarrestad, T. V., and Kyllingstad, Å., 1993, Rig suspension measurements and theoretical models and the effect on drill string vibrations: SPE Drilling & Completion, September, 201–206.

Aarrestad, T. V., Tønnesen, H. A., and Kyllingstad, Å., 1986, Drillstring Vibrations: Comparison Between Theory and Experiments on a Full-Scale Research Drilling Rig: IADC/SPE Paper No. 14760.

Adams, N. J., and Charrier, T., 1985, Drilling engineering – A complete well planning approach: PennWell Publishing Co., Tulsa, Oklahoma.

Aki, K., and Richards, P. G., 1980, Quantitative Seismology - Theory and methods: W.H. Freeman and Company, Vol. I.

Aldridge, D. F., 1989, Statistically perturbed geophone array response: Geophysics, **54**, 1306–1317.

Aleotti, L., Poletto, F., Miranda, F., Corubolo, P., Abramo, F., and Craglietto, A., 1999, Seismic while-drilling technology: use and analysis of the drill-bit seismic source in a cross-hole survey, Geophys. Prosp., **47**, 25–39.

Allan, D., Hess, M., and Todd, S. P., 1996, Integrated technologies exceed well placement challenge: SPE Paper No. 36917.

Angeleri, G. P., Persoglia, S., and Poletto, F., 1990, Drill bit noise as a source in geophysical surveys: Proceedings of the European Oil and Gas Conference, Altavilla Milicia, Sicily, Pre prints, 77–86. Published by Graham and Trotman for the CEC, 422–431.

Angeleri, G. P., Persoglia, S., Poletto, F., and Rocca, F., 1996, Process and device for detecting seismic signals in order to obtain Vertical Seismic Profiles during bore drilling operations: U.S. Patent No. 5,511,038.

Arntsen, B., and Carcione, J. M., 2000, A new insight into the reciprocity principle: Geophysics, **65**, 1604–1612.

Asanuma, H., and Niitsuma, H., 1992, Triaxial Inverse VSP Uses Drill Bit as a Downhole Seismic Source: 62nd Ann. Internat. Mtg., Soc. of Expl. Geophys., Expanded Abstracts, Session BG3.3, 108–111.

Backus, G. E., 1962, Long–Wave elastic anisotropy produced by horizontal layering: J. Geophys. Res., **67**, 4427–4440.

Baeten, G., and Ziolkowski, A., 1990, The Vibroseis source: Elsevier.

Bailey, J. R., 1974, Continuous Bit Positioning System: U.S. Patent No. 3,817,345.

Balch, A. H., and Lee, M. W., 1984, Vertical Seismic Profiling – *Technique, Applications, and Case Histories*: D. Reidel Publishing Co., Houston.

Barnes, T. G., and Kirkwood, B. R., 1972, Passbands for acoustic transmission in an idealized drill string: J. Acoust. Soc. Am., **52**, 1606–1608.

Bedford, A., and Drumheller, D. S., 1994, Elastic wave propagation: John Wiley & Sons, New York.

Behr, S. M., Warren, T. M., Sinor, L. A., and Brett, J. F., 1993, 3D PDC Bit Model Predicts Higher Cutter Loads: SPE Drilling & Completion, (12), 253–258, SPE Paper No. 21928.

Bellezza, C., and Poletto, F., 2002, Test di deconvoluzione con operatori misti downgoing-upgoing: Report OGS, OGS 01/2002 GDL/2.

Berkhout, A. J., 1984, Seismic resolution – Resolving power of acoustical echo techniques: Geophysical press, London.

Bernabe, Y., and Brace, W. F., 1990, Deformation and Fracture of Berea Sandstone: The Brittle/Ductile Transition in Rock: The Heard Volume of The Geophysical Monographs (#56, AGU), 91–101, American Geophysical Union.

Bernasconi, G., and Carraro, M., 1988, Identificazione delle onde emesse da uno scalpello in fase di perforazione e uso per prospezioni sismiche: Tesi di laurea, Politecnico di Milano, December.

Bernasconi, G., and Vassallo, M., 2001, Compression of downhole data for SWD application Drilling vibrations: 71st Ann. Internat. Mtg., Soc. of Expl. Geophys., Expanded Abstracts, 2001–2004.

Bernasconi, G., and Vassallo, M., 2001, Estimation of drill-bit rotation velocity with surface and downhole dynamics measurements: 71st Ann. Internat. Mtg., Soc. of Expl. Geophys., Expanded Abstracts, 2017–2020.

Bernasconi, G., and Vassallo, M., 2002, Drilling vibrations modeling and drillstring imaging: 72nd Ann. Internat. Mtg., Soc. of Expl. Geophys., Expanded Abstracts, 2401–2404.

Berni, A. J., and Roever, W. L., 1989, Field array performance: Theoretical study of spatially correlated variations in amplitude coupling and static shift and case study in the Paris Basin: Geophysics, **54**, 451–459.

Bertelli, L., Abramo, F., Gatti, W., 1998, Seismic while drilling and geophysical monitoring in the Southern Appennine Range: 60th Mtg. Eur. Assn. Geosci. Eng., Extended Abstracts, Session 10–50.

Bertelli, L., and di Cesare, F., 1999, Improving the subsurface geological model while drilling: First Break, **17**, No. 6, 223–228.

Bertelli, L., Miranda, F., Poletto, F., and Rocca, F., 1998, Design of SWD 3D RVSP Using SEISBIT ® Technology: 60th Mtg. Eur. Assn. Geosci. Eng., Extended Abstracts, Session: 10–49.

Besaisow, A. A., Jan, Y. M., and Shun, F. J., 1985, Development of a Surface Drillstring Vibration Measurement System: SPE Paper No. 14327.

Bickel, S. H., 1982, The effects of noise on minimum-phase Vibroseis deconvolution: Geophysics, **47**, 1174–1184.

Biggs, M. D., Cheatham, J. B. Jr., and Rice, U., 1969, Theoretical Forces for Prescribed Motion of a Roller Bit: SPE Paper No. 2391.

Bilgeri, D., and Adameno, E. B., 1982, Predicting abnormally pressured sedimentary rocks: Geophys. Prosp., **30**, 608–621.

Bilgeri, D., and Carlini, A., 1988, Design Aspects in Reflection Seismology Aiming at Rock Properties Detection: SPE Paper No. 17559.

Booer, A. K., and Meehan, R. J., 1993, Drilling Imaging - An Interpretation of Surface Drilling Vibrations: SPE Drilling & Completion, (6), 93–98.

Bourgoyne Jr., A. T., Millheim, K. K., Chenevert, M. E., and Young Jr., F. S., 1991, Applied drilling engineering: SPE Textbook series, Vol. 2, Richardson, TX, USA.

Bowers, G. L., 1995, Pore Pressure Estimation From Velocity Data: Accounting for Overpressure Mechanisms Besides Undercompaction: SPE Drilling & Completion, June, 89–95.

Bracewell, R., 1965, The Fourier transform and its applications: McGraw-Hill Book Co.

Brett, J. F., Warren, T. M., and Behr, S., 1990, Bit Whirl – A New Theory of PDC Bit Failure: SPE Drilling Engineering, December, SPE Paper No. 19571, 275–281.

Brittle, K. F., Lines, L. R., and Day, A. K., 2001, Vibroseis deconvolution: a comparison of cross-correlation and frequency-domain sweep deconvolution: Geophys. Prosp., **49**, 675–686.

Bruggeman, D. A. G., 1937, Berechnungen der verschiedenen physikalischen Konstanten von heterogenen Substanzen. Teil 3: Die elastischen Konstanten der quasi-isotropen Mischkörper aus aus isotropen Substanzen: Annalen dr Physik, (5) **29**, 160 ...

Burgess, T. M., and Lesso, Jr. W. G., 1985, Measuring the wear of milled tooth bits using MWD torque and weight-on-bit: SPE/IADC Paper No. 13475.

Burns, W. R., 1968, A statistically optimized deconvolution: Geophysics, **33**, 255–263.

Cambois, G., and Stoffa, P., 1993, Surface-consistent phase decomposition in the log/Fourier domain: Geophysics, **58**, 1099–1111.

Carcione, J. M., 2001, Wave Fields in Real Media: Wave Propagation in Anisotropic, Anelastic and Porous Media: Seismic Exploration, **31**, Klaus Helbig and Sven Treitel, Editors, Pergamon, Elsevier Science, Amsterdam.

Carcione, J. M., and Carrion, P. M., 1992, 3-D radiation pattern of the drilling bit source in finely stratified media: Geophysical Research Letters, **19**, No. 7, 717–720.

Carcione, J. M., and Gangi, A., 2000, Gas generation and overpressure: effects on seismic attributes: Geophysics, **65**, 1769–1779.

Carcione J. M., and Helle, H. B., 1999, Numerical solution of the poroviscoelastic wave equation on a staggered mesh: J. Comput. Phys., **154** (2), 520–527.

Carcione, J. M., and Helle, H. B., 2002, Rock physics of geopressure and prediction of abnormal pore fluid pressures using seismic data: *CSEG Recorder*, **27** (7), 8–32.

Carcione, J. M., Helle, H. B., Pham, N. P., and Toverud, T., 2003, Pore pressure estimation in reservoir rocks from seismic reflection data: Geophysics, **68**, 1569–1579.

Carcione, J. M., Kosloff, D., and Behle, A., 1991, Long wave anisotropy in stratified media: A numerical test: Geophysics, **56**, 245–254.

Carcione, J. M., Kosloff, D., Behle, A., and Seriani, G., 1992, A spectral scheme for wave propagation simulation in 3-D elastic-anisotropic media: Geophysics, **57**, 1593–1607.

Carcione, J. M., Padoan, G., and Cavallini, F., 2000, Synthetic seismograms of the sea bottom under different streamer conditions: Boll. Geof. Teor. Appl., **41** (1), 21–29.

Carcione, J. M., and Poletto, F., 2000, Simulation of stress waves in attenuating drill strings, including piezoelectric sources and sensors: J. Acoust. Soc. Am., **108** (1), 53–64.

Carcione, J. M., and Poletto, F., 2000, Sound velocity of drilling mud saturated with reservoir gas: Geophysics, **65**, 646–651.

Cardoso, J. F., 1990, Eigen-structure of the fourth-order cumulant tensor with application to the blind source separation problem: IEEE, 2655–2658.

Carrion, P., Persoglia, S., and Poletto, F., 1995, High resolution traveltime imaging: The Leading Edge, **14** (1), 26–28.

Carswell, A., and Moon, W. M., 1989, Application of multioffset vertical seismic profiling in fracture mapping: Geophysics, **54**, 737–746.

Cassand, J., and Lavergne, M., 1971, Seismic Emission by Vibrators: Seismic Filtering, Society of Exploration Geophysicists, *Trans. of Editions Technip – Paris.*

Cassano, E., and Rocca, F., 1973, Multichannel linear filters for optimal rejection of multiple reflections: Geophysics, **38**, 1053–1061.

Casserly, R., and Marion, B. P., 1992, Drill-bit source salt-proximity surveys for real-time determination of the bit position relative to the salt face: The Leading Edge, **11** (8), 37–41.

Chabot, L., Henley, D. C., Brown, R. J., and Bancroft, J.C., 2001, Single-well imaging using the full waveform of an acoustic sonic: 71st Ann. Internat. Mtg., Soc. of Expl. Geophys., Expanded Abstracts, 420–423.

Chameau, J. J., Batini, F., and Omnes, G., 1991, The seismic signals generated by the bit and the drillstring in a horizontal well in the Lardarello geothermal field: 53rd Mtg. Eur. Assn. Geosci. Eng., Extended Abstracts, Session B019, 140–141.

Cheatham, J .B., and Pattillo, P. D., 1984, Helical Post-Buckling Configuration of a Weightless Column Under the Action of an Axial Load: SPE Paper No. 10854.

Chen, S. L., Blackwood, K., and Lamine, E., 1999, Field Investigation of the Effects of Stick-Slip, Lateral, and Whirl Vibrations on Roller Cone Bit Performance: SPE Paper No. 56439.

Cheng, C. H., and Toksöz, M. N., 1981, Elastic wave propagation in fluid-filled boreholes and synthetic acoustic logs: Geophysics, **46**, 1042–1053.

Chin, W. C., 1988, Why Drill Strings Fail At The Neutral Point: PETROLEUM ENGINEERING International, (5), 62–67.

Chin, W. C., 1994, Wave Propagation in Petroleum Engineering: Gulf Publishing Company.

Chu, A. S., Eller, E. E., and Whittier, R. M., 1995, Vibration Transducers: Shock and Vibration Handbook, Chapter 12, McGraw-Hill Book Co., Fourth Edition.

Chur, C., and Oppelt, J., 1993, Vertical drilling technology: a milestone in directional drilling: SPE/IADC Paper No. 25759.

Claerbout, J. F., 1968, Synthesis of a layered medium from its acoustic transmission response: Geophysics, **33**, 264–269.

Claerbout, J. F., 1976, Fundamentals of geophysical data processing: New York, McGraw-Hill Book Co., Inc.

Claerbout, J. F., 1992, Earth Soundings Analysis. Processing versus Inversion: Boston, Blackwell Scientific Publications.

Clayer, F., Vandiver, J. K., and Lee, H. Y., 1990, The Effect of Surface and Downhole Boundary Conditions on the Vibration of Drillstrings: SPE Paper No. 20447.

Close, D. A., Owens, S. C., and MacPherson, J. D., 1988, Measurement of BHA Vibration Using MWD: IADC/SPE Paper No. 17273.

Cole, S., 1988, Examination of a passive seismic data set using beam-steering: *SEP-57*.

Comon, P., 1989, Separation of stochastic processes: See Proc. of Workshop on Higher Order Spectral Analysis, Vail, 174–179.

Cooley, C. H., Pastusek, P. E., and Sinor, L. A., 1992, The Design and Testing of Anti-Whirl Bits: SPE Paper No. 24586.

Cunningham, R. A., 1968, Analysis of Downhole Measurements of Drill String Forces and Motions: Journal of Engineering for Industry, Transactions of the ASME, (5), 209–216.

Dareing, D. W., 1982, Drill Collar Lengths is a Major Factor in Vibration Control: SPE Paper No. 11228.

Deily, F. H., Dareing, D. W., Paff, G. H., Ortloff, J. E., and Lynn, R. D., 1968, Downhole Measurements of Drill String Forces and Motions: Journal of Engineering for Industry, Trans. ASME, Series B, **90** (5), 217–225.

DeLucia, F. V., and Herbert, R. P., 1984, PDM vs. Turbodrill: A Drilling Comparison: SPE Paper No. 13026.

Devereux, S., 1998, Practical well planning and drilling manual: PennWell Publishing Company, Tulsa, Oklahoma.

Devereux, S., 1999, Drilling Technology in nontechnical language: PennWell Publishing Company, Tulsa, Oklahoma.

Dillon, P. B., and Thomson, R. C., 1984, Offset source VSP surveys and their image reconstruction: Geophys. Prosp., **32**, 790–811.

Dimri, V., 1992, DECONVOLUTION AND INVERSE THEORY Application to Geophysical Problems: Elsevier, Amsterdam.

Dobrin, M. B., 1981, Introduction to Geophysical Prospecting: Third Edition, International Student Edition.

Dordolo, G., and Calligaris, D., 1993, Progetto per la realizzazione degli impianti per i sensori pilota su una torre di perforazione ...:- REL. 10/93 - GDL/08, (4).

Dragoset, W. H., 1988, Marine vibrators and the Doppler effect; Geophysics, **53**, 1388–1398.

Drumheller, D. S., 1989, Acoustic properties of drill strings: J. Acoust. Soc. Am., **85**, 1048–1064.

Drumheller, D. S., 1992, Extensional stress waves in one-dimensional elastic waveguides: J. Acoust. Soc. Am., **92**, 3389–3402.

Drumheller, D. S., 1993, Attenuation of sound wave in drill strings: J. Acoust. Soc. Am., **94**, 2387–2396.

Drumheller, D. S., 1993, Coupled extensional and bending motion in elastic waveguides: Wave Motion, **17**, 319–327.

Drumheller, D. S., and Knudsen, S. D., 1995, The propagation of sound waves in drill strings: J. Acoust. Soc. Am., **97**, 2116–2125.

Dubinsky, V. S. H., Henneuse, H. P., and Kirkman, M. A., 1992, Surface monitoring of downhole vibrations: Russian, European, and American Approaches: SPE Paper No. 24969.

Dutta, N. C., 2002, Geopressure prediction using seismic data: Current status and the road ahead: Geophysics, **67**, 2012–2041.

Dykstra, M. W., Chen, D. C.-K., Warren, T. M., and Azar, J. J., 1996, Drillstring Component Mass Imbalance: A Major Source of Downhole Vibrations: SPE Paper No. 29350-P, SPE Drilling & Completion, (12), 234–241.

Eaton, B. A., 1975, The Equation for Geopressure Prediction from Well Logs: SPE Paper No. 5544.

Eronini, I. E., Somerton, W. H., and Auslander, D. M., 1982, A Dynamic Model for Rotary Rock Drilling: ASME Journal of Energy Resources Technology, **104**, 108–120.

Evans, B. J., 1997, A Handbook For Seismic Data Acquisition In Exploration: Geophysical monograph series, Vol. 7, Society of Exploration Geophysicists, Tulsa.

Falconer, I. G., Belaskie, J. P., and Varlava, F., 1989, Application of a Real Time Wellbore Friction Analysis: SPE Paper No. 18649.

Falconer, I. G., Burgess, T. M., and Sheppard, M. C., 1988, Separating Bit and Lithology Effects from Drilling Mechanics Data: IADC/SPE Paper No. 17191.

Fear, M. J., and Abbassian, F., 1994, Experience in the Detection and Suppression of Torsional Vibration From Mud Logging Data: SPE Paper No. 28908.

Finger, J. T., Mansure, A. J., Knudsen, S. D., and Jacobsen, R. D., 2003, Development of a System for Diagnostic-While-Drilling (DWD): SPE/IADC Paper No. 79884.

Fokkema, J. T., and Ziolkowski, A., 1987, The critical reflection theorem: Geophysics, **52**, 965–972.

Gabolde, G., and Nguyen, J. P., 1999, Drilling data handbook: Seventh edition, Editions Technip, Paris.

Galbraith, J. N., and Wiggins, R. A., 1968, Characteristics of optimum multichannel stacking filters: Geophysics, **33**, 36–48.

Gal'perin, E. I., 1974, Vertical Seismic Profiling: Society of Exploration Geophysicists, Special Pubblication No. 12, Tulsa, Oklahoma.

Gangi, A. F., 1987, A theoretical study of the radiation patterns of a tunnel-boring machine (infinite medium): Final Report GL-87-20, Texas A&M University.

Gangi, A. F., and Fairborn, J. W., 1968, Accurate determination of seismic-array steering delays by an adaptive computer program: Supp. al Nuovo Cim., **6**, 105–115.

Gardner, G. H. F., Gardner, L. W., and Gregory, A. R., 1974, Formation velocity and density—The diagnostic basics for stratigraphyc traps: Geophysics, **39**, 770–780.

Gibson, R. L., and Peng, C., 1994, Low- and high-frequency radiation from seismic sources in cased boreholes: Geophysics, **59**, 1780–1785.

Gilbert, P. F. C., 1972, Iterative methods for the three dimensional reconstruction of an object from projection: J. Theor. Biol., **36**, 105–117.

Glangeaud, F., and Mari, J. L., 1994, Wave separation: Editions Technip, Paris.

Glowka, D. A., 1986, The Use of Single-Cutter Data in the Analysis of PDC Bit Design: SPE Paper No. 15619.

Guy, M. M., 1961, Nouvelle methode de controle geologique de forage: Brevet d'Invention N. 1.584.951, Ministere du Developpent Industriel et Scientifique, Republique Francaise.

Guyen Minh Duc, N., Cholet, H., and Tricot, P., 1972, Stress Concentration Induced in Elastic Rocks Under a Diamond Bit Tooth: SPE Paper No. 3987.

Haldorsen, J. B. U., Esmersoy, C., Hawthorn, A., Krasovec, M., Raikes, S., Harrold, T., Day, D. N., and Clippard, J. D., 2003, Full Waveform Seismic Measurements While Drilling from the Caspian Sea: SPE/IADC Paper No. 79844.

Haldorsen, J. B. U., Miller, D. E., and Walsh, J. J., 1995, Walk-away VSP using drill noise as a source: Geophysics, **60**, 978–997.

Haldorsen, J. B. U., Miller, D. E., Walsh, J. J., and Zoch, H. J., 1992, A multichannel approach to signature estimation and deconvolution for drill-bit imaging: 62nd Ann. Internat. Mtg., Soc. of Expl. Geophys., Expanded Abstracts, 181–184.

Hamon, B. V., and Hannon, E. J., 1974, Spectral Estimation of Time Delay for Dispersive and Non-dispersive Systems: J. Roy. Statist. Soc. Ser. C (Applied Statistics), **23**, 134–142.

Hanson, J. M., and Hansen, W. R., 1995, Dynamics Modeling of PDC Bits: Drilling Conference, SPE/IADC Paper No. 29401.

Hardage, B. A., 1987, Seismic stratigraphy: Geophysical press, London.

Hardage, B. A., 1992, Crosswell Seismology and Reverse VSP: Geophysical press, London.

Hardage, B. A., 2000, Vertical Seismic Profiling: Principles: Third Updated and Revised Edition, Klaus Helbig and Sven Treitel, Editors, PERGAMON, Elsevier, Amsterdam.

Harris, C. M., 1995, Shock and Vibration Handbook: McGraw-Hill Book Co., Fourth Edition.

Hasley, G. W., Kyllingstad, Å., Aarrestad, T. V., and Lysne, D., 1986, Drillstring Torsional Vibrations: Comparison Between Theory and Experiment on Full-Scale Research Drilling Rig: SPE Paper No. 15564.

He, X., and Kyllingstad, Å., 1995, Helical Buckling and Lock-Up Conditions for Coiled Tubing in Curved Wells: SPE Drilling & Completion, SPE paper No. 25370, 10–15.

He, X., Hasley, G. W., and Kyllingstad, Å., 1995, Interactions between Torque and Helical Buckling in Drilling: SPE Paper No. 30521.

Heelan, P. A., 1953, Radiation from a cylindrical source of finite length: Geophysics, **18**, 685–696.

Helbig, K., 1958, Elastische Wellen in anisotropen Medien: Gerlands Beiträge zur Geophysik, **67**, 177–211; 256–288.

Helbig, K., 1977, Unit and dimensions in scientific publications: Geophys. Prosp., **25**, 382–384.

Hoffe, B. H., Bland, H. C., Margrave, G. F., Manning, P. M., and Foltinek, D. S., 1999, Analysis of the effectiveness of 3-C receiver arrays for converted wave imaging: 69th Ann. Internat. Mtg., Soc. of Expl. Geophys., Expanded Abstracts, 649–652.

Hokstad, K., Sollie, R., and Carlsen, I. M., 2001, VSP While Drilling for Geosteering Applications: 71st Ann. Internat. Mtg., Soc. of Expl. Geophys., Expanded Abstracts, 452–455.

Holster, J. L., and Kipp, R. J., 1984, Effect of Bit Hydraulic Horsepower on the Drilling Rate of a Polycrystalline Diamond Compact Bit: Journal of Petroleum Technology, SPE Paper No. 11949, 2110–2118.

Holzman, M., 1963, Chebyshev optimized geophone arrays: Geophysics, **28**, 145–155.

Hunter, S. C., 1957, Energy absorbed by elastic waves during impact: Journal of the Mechanics and Physics of Solids, **5**, 162–171.

Hutchinson, M., Dubinsky, V., and Henneuse, H., 1995, An MWD Downhole Assistant Driller: SPE Paper No. 30523.

IADC Fixed Cutter Drill Bit Committee, 1992a, First revision of the IADC fixed cutter dull grading system: SPE/IADC Paper No. 23939.

IADC Fixed Cutter Drill Bit Committee, 1992b, Development of a new IADC fixed cutter drill bit classification system: SPE/IADC Paper No. 23940.

Input/Output.INC, 2001, 3-Component Mega-Level VSP^{TM} Array: Technical specifications, USA, Stafford, TX.

Jaeger, J. C., 1969, Elasticity, Fracture and Flow with Engineering and Geological Applications: Science Paperbacks, 55, London.

Jahn, F., Cook, M., and Graham, M., 1998, Hydrocarbon Exploration and Production: Developments in petroleum science, 46, Elsevier, Amsterdam.

Jansen, J. D., van den Steen, L., and Zachariasen, L., 1995, Active Damping of Torsional Drillstring Vibrations With a Hydraulic Top Drive: SPE Drilling & Completion, (12), 250–254.

Jardine, S. I., Lesage, M. L., and McCann, D. P., 1990, Estimating tooth wear from roller cone bit vibration: IADC/SPE Paper No. 19966.

Jellison, M. J., Hall, D. R., Howard, D. C., Hall, H. T., Jr. Long, R. C., Chandler, R. B., and Pixton, D. S., 2003, Telemetry Drill Pipe: Enabling Technology for the Downhole Internet: SPE/IADC Paper No. 79885.

Jorden, J. R., and Shirley, O. J., 1978, Application of drilling performance data to overpressure detection: J. Pet. Tech., (7), 987–991.

Jutten, C., and Herault, J., 1988, Independent components analysis versus PCA: Proc. of EUSIPCO, Grenoble, France, 643–646.

Kan, T. K., and Swan, H. W., 2001, Geopressure prediction from automatically-derived seismic velocities: Geophysics, **66**, 1937–1946.

Kanasewich, E. R., 1981, Time sequence analysis in geophysics: The University of Alberta Press.

Kårstad, E., and Aadnøy, B. S., 1997, Analysis of temperature measurements during drilling: SPE Paper No. 38603.

Kaufman, A. A., and Levshin, A. L., 2000, Acoustic and elastic wave fields in geophysics, I: Elsevier, Amsterdam.

Katz, L. J., 1984, Method and System for Seismic Continuous Bit Positioning: U.S. Patent No. 4,460,059.

Katz, L. J., 1990, Inverse seismic profiling while drilling: U.S. Patent No. 5,012,453.

Klaveness, A., 1980, Seismic Well Logging System and Method: U.S. Patent No. 4,207,619.

Knopoff, L., and Gangi, A. F., 1959, Seismic reciprocity: Geophysics, **24**, 681–691.

Kolsky, H., 1953, Stress waves in solids: Oxford University Press.

Kostov, C., 1988, Drill-bit signal: *SEP-59.*

Kostov, C., 1990, Multichannel seismic experiment with a drill-bit source: Ph.D. Thesis, Stanford University.

Kostov, C., and Zanzi, L., 1988, Analysis of drill bit data: Preliminary results: Stanford Exploration Reports, *SEP-57.*

Kragh, E., and Christie, P., 2002, Seismic repeatability, normalized rms, and predictability: The Leading Edge, **21** (7), 640–647.

Krebs, J. R., Fara, D. R., and Berlin, A. E., 1995, Accurate migration using offset-checkshot surveys: 65th Mtg. Eur. Assn. Geosci. Eng., Extended Abstracts, 1186–1188.

Kreisle, L. F., and Vance, J. M., 1969, Mathematical Analysis of the Effect of a Shock Sub on the Longitudinal Vibrations of an Oilwell Drill String: SPE Paper No. 2778.

Kriesels, P. C., Keultjes, W. J. G., Dumont, P., Huneidi, I., Owoeye, O. O., and Hartmann, R. A., 1999, Cost Savings through an Integrated Approach to Drillstring Vibration Control: SPE/IADC Paper No. 57555.

Kyllingstad, Å., and Hasley, G. W., 1987, A Study of Slip-Stick Motion of the Bit: SPE Paper No. 16659.

Labo, J., 1987, *A Practical Introduction to Borehole Geophysics*: Society of Exploration Geophysicists, Tulsa, Oklahoma.

Lamine, E., Leseutre, A., Roberts, D., Walker, C., and Jonsson, A., 1998, Using downhole dynamics to better understand the drilling process with positive displacement motors and PDC bits: SPE Paper No. 49207.

Landrum, R. A., Brook, R. A., and Sallas, J., 1994, Polarity convention for vibratory Source/Recording system: SEG special report: Geophysics, **59**, 315–322.

Langenkamp, R. D., 1994, The Illustrated Petroleum Reference Dictionary: 4th edition, PennWell Books, Tulsa, Oklahoma.

Langeveld, C. J., 1992a, PDC Bit Dynamics: IADC/SPE Paper No. 23867.

Langeveld, C. J., 1992b, PDC Bit Dynamics (Supplement to IADC/SPE 23867): IADC/SPE Paper No. 23873.

Laurie, P., Head, P., and Smith, J. E., 2003, Smart Drilling with ELECTRIC DRILLSTRING™: SPE/IADC Paper No. 79886.

Lavely, E., Gibson, R. L., and Tzimeas, C., 1997, 3-D Seismic Survey Design for Optimal Resolution: 67th Ann. Internat. Mtg., Soc. of Expl. Geophys., Expanded Abstracts, 31–34.

Lea, L. H., and Kyllingstad, Å., 1996, Propagation of coupled pressure waves in borehole with drillstring: SPE Paper No. 37156, 963–972.

Leaney, W. S., 1994, Anisotropy and AVO from walkaways: 64th Ann. Internat. Mtg., Soc. of Expl. Geophys., Expanded Abstracts, 105–109.

Lee, H. Y., 1991, Drillstring Axial Vibration and Wave Propagation in Boreholes: Ph.D. Thesis, Massachusetts Institute of Technology.

Lee, M. W., and Balch, A. H., 1982, Theoretical seismic wave radiation from a fluid-filled borehole: Geophysics, **47**, 1308–1314.

Lee, M. W., Balch, A. H., and Parrott, K. R., 1984, Radiation from a downhole air gun source: Geophysics, **49**, 27–36.

Lesage, M., Falconer, I. G., and Wick, C. J., 1988, Evaluating Drilling Practice in Deviated Wells With Torque add Weight Data: SPE Drilling Engineering, (9), 248–252, SPE Paper No. 16114.

Lesso, W. G., and Kashikar, S. V., 1996, The Principles and Procedures of Geosteering: IADC/SPE Paper No. 35051.

Levin, F. K., 1989, The effect of geophone arrays on random noise: Geophysics, **54**, 1466–1473.

Levy, S., and Oldenburg, D. W., 1987, Automatic phase correction of common-midpoint stacked data: Geophysics, **52**, 51–59.

Ligrone, A., Oppelt, J., and Calderoni, A., 1996, The fastest way to the bottom: straight-hole drilling device - drilling concept, design considerations, and field experience: SPE Paper No. 36826.

Loewenthal, D., and Robinson, E. A., 2000, On unified dual fields and Einstein deconvolution: Geophysics, **65**, 293–303.

Lous, N. J. C., Rienstra, S. W., and Adan, I. J. B. F., 1998, Sound transmission through a periodic cascade with application to drill pipes: J. Acoust. Soc. Am., **103** (5), 2302–2311.

Love, A. E. H., 1952, A treatise on the mathematical theory of elasticity: Cambridge University Press, Fourth Edition.

Lutz, J., Raynaud, M., Gastalder, S., Quichaud, C., Raynald, J., and Muckerloy, J. A., 1972, Instantaneous logging based on a dynamic theory of drilling: Journal of Petroleum Technology, (6), 750–758.

Ma, D., and Azar, J. J., 1885, Dynamics of Roller Cone Bits: Journal of Energy Resources Technology, Trans. of AIME, **107** (12), 543–548.

Ma, D., Zhou, D., and Deng, R., 1995, The Computer Simulation of the Interaction Between Roller Bit and Rock: SPE Paper No. 29922.

MacBeth, C., 2002, Multi-Component VSP Analysis for Applied Seismic Anisotropy: Klaus Helbig and Sven Treitel, Editors, PERGAMON, Elsevier, Amsterdam.

MacBeth, C., Zeng, X., Li, X. Y., and Queen, J., 1995, Multicomponent near-surface correction for land VSP data: Geophys. J. Int., **121**, 301–305.

MacPherson, J. D., Jogi, P. N., and Kingman, J. E. E., 2001, Application and Analysis of Simultaneous Near Bit and Surface Dynamics Measurements: SPE Drilling & Completion, 230-238.

MacPherson, J. D., Mason, J. S., and Kingman, J. E. E., 1993, Surface measurement and analysis of drill string vibrations while drilling: SPE/IADC Paper No. 25777.

Magarini, P., and Monaci, E., 1999, Drilling design manual: ENI S.p.A. – Agip Division.

Malusa, M., 2001, Personal communication.

Malusa, M., and Poletto, F., 2001, Analisi del segnale pilota SEISBIT ® in funzione della posizione angolare dello scalpello: Internal Report OGS, No. Rel. 1/01 OGS 2/01 GDL.

Malusa, M., Poletto, F., and Miranda, F., 2002, Prediction ahead of the bit by using drill-bit pilot signals and reverse VSP (RVSP): Geophysics, **67**, 1169–1176.

Malusa, M., Poletto, F., and Miranda, F., 2003, SWD using lateral vibrations of PDC bits: 73rd Ann. Internat. Mtg., Soc. of Expl. Geophys., Expanded Abstract, 2223–2226.

Manner, 1996, Manual Sensor-Telemetry: Manner Sensortelemetrie Eschenwasen, Spaichingen, Germany.

Martin, C. A., Philo, R. M., Decker, D. P., and Burgess, T. M., 1994, Innovative Advances in MWD: IADC/SPE Paper No. 27516.

Marzetta, T. L., and Schoenberg, M., 1985, Tube waves in cased boreholes: 55th Ann. Internat. Mtg., Soc. of Expl. Geophys., Session BHG 2.1.

Maurer, W. C., 1962, The "Perfect - Cleaning" Theory of Rotary Drilling: SPE Paper No. 408.

Maurer, W. C., 1965, Shear Failure of Rock Under Compression: SPE Paper No. 1054.

Maurer, W. C., 1965, Bit-Tooth Penetration Under Simulated Borehole Conditions: SPE Paper No. 1260.

Maurer, W. C., 1980, Advanced Drilling Techniques: Petroleum Publishing Co., Tulsa, OK.

Maurer, W. C., and Heilhecker, J. K., 1969, Hydraulic Jet Drilling: SPE Paper No. 2434.

Mavko, G., Mukerji, T., and Dvorkin, J., 1998, The rock physics handbook: tools for seismic analysis in porous media: Cambridge University Press, Cambridge, UK.

Mazzucchelli, P., and Rocca, F., 2001, Antialiasing for n-D Kirchhoff operators: 63rd Mtg. Eur. Assn. Geosci. Eng., Extended Abstracts, Session P-75.

McGehee, D. Y., Dahlem, J. S., Gieck, J. C., Kost, B., Lafuze, D., Reinsvold, C. H., and Steinke, S. C., 1992a, The IADC roller bit classification system: IADC/SPE Paper No. 23937.

McGehee, D. Y., Dahlem, J. S., Gieck, J. C., Kost, B., Lafuze, D., Reinsvold, C. H., and Steinke, S. C., 1992b, The IADC roller bit dull grading system: IADC/SPE Paper No. 23938.

Meehan, R., Miller, D., Haldorsen, J., Kamata, M., and Underhill, B., 1993, Rekindling Interest in Seismic While Drilling: Oilfield Review, January, 4–13.

Meehan, R., Nutt, L., Dutta, N., and Menzies, J., 1998, Drill Bit Seismic: A Drilling Optimization Tool: IADC/SPE Paper No. 39312.

Menand, S., Sellami, H., Simon, C., Besson, A., and Da Silva, N., 2002, How the Bit Profile and Gages Affect the Well Trajectory: IADC/SPE Paper No. 74459.

Meredith, J. A., 1990, Numerical and Analytical Modelling of Downhole Seismic Sources: The Near and Far Field: Ph.D. Thesis, Massachusetts Institute of Technology.

Meredith, J. A., Cheng, C. H., Toksöz, M. N., and Bouchon, M., 1990, Near-Field Modeling of Radiation from a Downhole Point Seismic Source: 60th Ann. Internat. Mtg., Soc. of Expl. Geophys., Expanded Abstracts, 4–7.

Meyer, W. H., Jian-Qun Wu, Macune, D. T., and Harvey, P.R., 1995, Brief: Geosteering With Near-Bit Formation Evaluation Sensors: JPT, (2), 117.

Miller, D., Haldorsen, J., and Kostov, C., 1990, Methods for deconvolution of unknown source signatures from unknown waveform data: U.S. Patent No. 4,922,362.

Miller, G. F., and Pursey, H., 1954, The field and radiation impedance of mechanical radiators on the free surface of semi-infinite isotropic solid: Proc. Roy. Soc., **223**, 521–541.

Miranda, F., Abramo, F., Poletto, F., and Comelli, P., 2000, Process for improving the bit seismic signal using drilling parameters: European Patent Application 00201332.4-2213.

Miranda, F., Aleotti, L., Abramo, F., Poletto, F., Craglietto, A., Persoglia, S., and Rocca, F., 1996, Impact of seismic while drilling technique on exploration wells: First Break, **14**, 55–68.

Miranda, F., Poletto, F., Abramo, F., Craglietto, F., and Bernasconi, G., 1999, SWD surface and downhole pilot recording to improve PDC bit signal: 61st Mtg. Eur. Assn. Geosci. Eng., Extended Abstracts, Session 20.43.

Miska, S., and Cunha, J. C., 1995, An Analysis of Helical Buckling of Tubulars Subjected to axial and Torsional Loading in Inclined Wellbores: SPE Paper No. 29460.

Mitchell, R. F., 1995, Comprehensive Analysis of Buckling with Friction: SPE Paper No. 29475.

Moon, W., Carswell, A., Tang, R., and Dilliston, C., 1986, Radon transform wave field separation for vertical seismic profiling data: Geophysics, **51**, 940–947.

Moore, P. L., 1986, Drilling practices manual: Second edition, PennWell Publishing Co., Tulsa, Oklahoma.

Mouchet, J. P., and Mitchell, A., 1989, Abnormal pressures while drilling. Origins - Prediction - Detection - Evaluation: Elf Aquitanie, Manuels Techniques 2, Boussens.

Moufarrej, M., and Thompson, J., 2002, Real-rime borehole seismic runs on drillpipe connection schedules. LWD tool gathers for drilling ahead: Offshore magazine, (4), 82–150.

Myhre, K., 1991, Application of Bicenter Bits in Well-Deepening Operations: SPE Drilling Engineering, June, 105–110, SPE Paper No. 19921.

Naganawa, S., Tanaka, S., and Sugaya N., 1995, Experimental and theoretical analysis of roller cone bit vibrations: Proceedings of OTCE'95, 1995 ASME Energy-Sources Technology Conference & Exhibition, Houston, Texas, USA.

Naville, C., Layotte, P. C., Pignard, G., and Guesnon, J., 1994, Well seismic - application of the TRAFOR MWD system to drill-bit seismic profiling: 56th Mtg. Eur. Assn. Geosci. Eng., Extended Abstracts, Paper No. G045.

Naville, C.,Serbutoviez, S.,Throo, A., Batini, F.,Dini, I., and Cecconi, F., 2000, Drillbit SWD reverse walkaway using downhole reference signal: SPE Paper No. 65115.

Neill, W. M., Aminzadeh, F., Sarem, A. M. S., and Quintana, J. M., 1993, Guided oscillatory well path drilling by seismic imaging: U.S. Patent No. 5,242,025.

Ng, F. W., DiSiena, J. P., and Bseisu, A. A., 1990, Method and system for Vertical Seismic Profiling by measuring drilling vibrations: U.S. Patent No. 4,965,774.

Ochi, M. K., 1990, Applied Probability and Stochastic Processes: Wiley, New York.

Olgaard, D. L., and Brace, W. F., 1983, The Microstructure of Gouge from a Mining-induced Seismic Shear Zone: Int. J. Rock Mech. Min. Sci. & Geomech. Abstr., **20**, 11–19.

Oppelt, J., Chur, C., Feld, D., and Juergens, R., 1991, New concepts for vertical drilling of boreholes: SPE/IADC Paper No. 21905.

Oppenheim, A. V., and Shafer, R. W., 1975, Digital Signal Processing: Prentice-Hall, Inc., N.J., USA.

Papoulis, A., 1991, Probability, random variables, and stochastic processes: McGraw-Hill Co.

Parfitt, S. H. L., and Abbassian, F., 1995, A Model for Shock Sub Performance Qualification: SPE/IADC Paper No. 29354.

Paslay, P. R., and Bogy, D. B., 1963, Drill String Vibrations Due to Intermittent Contact of Bit Teeth: Journal of Engineering for Industry, Transactions of ASME, 187–194.

PCB, 2002, Technical Support – General Theory: PCB Piezotronics, website http://www.pcb.com/tech_gen.html.

Peacock, K. L., and Treitel, S., 1978, Predictive deconvolution: Theory and practice: Geophysics, **34**, 221–235.

Peltier, B., and Atkinson, C., 1987, Dynamic Pore Pressure Ahead of the Bit: SPE Drilling Engineering, December, 351–357.

Peng, C., Cheng, C. H., and Toksöz, M. N., 1994, Cased borehole effects on downhole seismic measurements: Geophys. Prosp., **42**, 777–811.

Pessier, R. C., and Fear, M. J., 1992, Quantifying Common Drilling Problems With Mechanical Specific Energy and a Bit-Specific Coefficient of Sliding Friction: SPE Paper No. 24584.

Petersen, S., and Heggernes, R., 1996, Method and apparatus for performing measurements while drilling for oil and gas: U.S. Patent No. 5,585,556.

Petronio, L., and Poletto, F., 2002, Seismic-while-drilling by using tunnel boring machine noise: Geophysics, **67**, 1798–1809.

Petronio, L., Poletto, F., Carcione, J. M., Seriani, G., Luca, A., and Miranda, F., 2001, Seismic-wave modelling of a 3D SWD survey: 63rd Mtg. Assn. Geosci. Eng., Extended Abstracts, Session PO10.

Petronio, L., Poletto, F., Miranda, F., and Dordolo, G., 1999, Optimization of receiver pattern in seismic-while-drilling: 69th Ann. Internat. Mtg., Sec. Expl. Geophys., Expanded Abstracts, 164–167.

Phillips, I. C., Paulk, M. D., and Constant, A., 2000, Real Real-Time Geosteering: SPE Paper No. 65141.

Pieuchot, M., 1984, Seismic instrumentation, Handbook of geophysical exploration, Section I. Seismic exploration, Vol. 2, GEOPHYSICAL PRESS, Amsterdam.

Pilant, W., 1979, Elastic waves in the Earth: Elsevier, Amsterdam.

Poletto, F., 2000, A blind interpretation of drill-bit signals: Geophysics, **65**, 970–978.

Poletto, F., 2000b, Optimum stack of drill-bit data by signal-to-noise analysis: Final Report THERMIE Project OG/136/95/IT/UK, Chapter 2, Rel. OGS 27/2000 GDL 25/2000, (7), 35–39.

Poletto, F., 2002, Use of dual waves for the elimination of reverberation in drill string: J. Acoust. Soc. Am., **111** (1), 37–40.

Poletto, F., 2002a, Energy balance of the drill-bit seismic source. Part A: rotary energy and radiation properties: submitted to Geophysics.

Poletto, F., 2002b, Energy balance of the drill-bit seismic source. Part B: drill-bit versus conventional seismic sources: submitted to Geophysics.

Poletto, F., and Carcione, J. M., 2001, On the group velocity of guided waves in drill strings: J. Acoust. Soc. Am., **109** (4), 1743–1746.

Poletto, F., Carcione, J. M., Lovo, M., and Miranda, F., 2002, Acoustic velocity of SWD borehole guided waves: Geophysics, **67**, 921–927.

Poletto, F., and Craglietto, A., 1991, Orthogonalized noise subtraction: 53rd Mtg. Assn. Geosci. Eng., Extended Abstracts, 566–567.

Poletto, F., and Dordolo, G., 2002, A new approach to offshore drill-bit RVSP: Geophysics, **67**, 1071–1075.

Poletto, F., Lovo, M., and Carcione, J. M., 2000, Acoustic velocity of drilling mud and SWD borehole guided waves: 70th Ann. Internat. Mtg., Soc. of Expl. Geophys., Expanded Abstracts, 1763–1766.

Poletto, F., and Malusa, M., 2002, Reflection of drill-string extensional waves at the bit-rock contact: J. Acoust. Soc. Am., **111** (6), 2561–2565.

Poletto, F., Malusa, M. and Miranda, F., 2000, Reflection of drill-bit extensional waves at the bit-rock interface: 70th Ann. Internat. Mtg., Soc. of Expl. Geophys., Expanded Abstracts, 1767–1770.

Poletto, F., Malusa, M., and Miranda, F., 2001, Numerical modeling and interpretation of drillstring waves: Geophysics, **66**, 1569–1581.

Poletto, F., Malusa, M., and Miranda, F., 2003, SWD using rig-pilot arrays: 73rd Ann. Internat. Mtg., Soc. of Expl. Geophys., Expanded Abstracts, 2227–2230.

Poletto, F., Malusa, M., Miranda F., and Tinivella, U., 2002, Seismic while drilling by using dual sensors in drill strings: submitted to Geophysics.

Poletto, F., Malusa, M., Petronio, L., Lovo, M., and Miranda, F., 2002, First results of the SEISBIT ® 3D RVSP project: Boll. Geof. Teor. Appl., **43**, 119–130.

Poletto, F., and Miranda, F., 1998, Seismic while drilling use of pilot signals with downhole motor drilling: 68th Ann. Internat. Mtg., Soc. of Expl. Geophys., Expanded Abstracts, 147–150.

Poletto, F., Miranda, F., Corubolo, P., and Abramo, F., 1997, Seismic while drilling using PDC signals - Seisbit experience and perspectives: 59th Mtg. Eur. Assn. Geosci. Eng., Extended Abstracts, Session E-053.

Poletto, F., Miranda, F., Corubolo, P., and Abramo, F., 1997, Monitoring lithology variations in complex exploration areas by the SWD technology. SEISBIT ® experience and perspectives: Abstr. AAPG Internat. Conference and Exhibition, Vienna, 7–10 September.

Poletto, F., Miranda, F., Petronio, L., Bertelli, L., Malusa, M., Luca, A. and Schleifer, A., 2001, 3D RVSP Drill-Bit Survey - Preliminary Results: 63rd Mtg. Eur. Assn. Geosci. Eng., Extended Abstracts, Session M-12.

Poletto, F., Petronio, L., Malusa, M., Schleifer, A., Corubolo, P., Miandro, R., Bellezza, C., and Gressetvold, B., 2003, Prediction and 3D imaging while drilling by drill-bit 3D RVSP: 65rd Mtg. Eur. Assn. Geosci. Eng., Extended Abstracts, Session P221.

Poletto, F., Rocca, F., and Bertelli, L., 2000, Drill-bit signal separation for RVSP using statistical independence: Geophysics, **65**, 1654–1659.

Poletto, F., and Savini, L., 1997, The SEISBIT ® seismic while drilling technology: a selected case history and new trends: The first GCC - EU Conference on advanced Oil and Gas technologies, Preprints, Bahrain 13–15 October.

Poli, S., Donati, F., Donati, J., and Ragnitz, D., 1996, Advanced tools for advanced wells: rotary closed loop drilling system - Results of prototype field testing: SPE Paper No. 36884.

Postma, G. W., 1955, Wave propagation in a stratified medium: Geophysics, **20**, 780–806.

Poston, S. W., and Berg, R. R., 1997, Overpressured gas reservoirs: Society of Petroleum Engineers, Inc., Richardson, Texas, U.S.A.

Product Review, 2000, Marine seismic acquisition: First Break, **18**.8, xii-xiv.

Rabia, H., 1985, Specific Energy as a Criterion for Bit Selection: Journal of Petroleum Technology, (7), 1225-1229, SPE Paper No. 12355.

Radtke, B., Smith, M., Riedel, R., and Daniels, B., 1999, New High Strength and Faster Drilling TSP Diamond Cutters: FETC, Oil and Gas Conference, 28–30 June, Dallas, Texas.

Rama Rao, V. N., and Vandiver, J. K., 1999, Acoustics of fluid filled boreholes with pipe: guided propagation and radiation: J. Acoust. Soc. Am., **105** (6), 3057–3066.

Rector, III J. W., 1989, System for reducing drill string multiples in field signals: U.S. Patent No. 4,862,423.

Rector, III J. W., 1990, Method for reducing noise in drill string signals: U.S. Patent No. 4,954,998.

Rector, III J. W., 1992, Drill String Wave Modes Produced by a Working Bit: 62nd Ann. Internat. Mtg., Soc. of Expl. Geophys., Expanded Abstracts, 155–158.

Rector, III J. W., 1992, Noise characterization and attenuation in drill bit recordings: J. Seis. Expl., **1**, 379–393.

Rector III, J. W., and Hardage, B. A., 1992, Radiation pattern and seismic waves generated by a working roller-cone drill bit: Geophysics, **57**, 1319–1333.

Rector, III, J. W., and Marion, B. P., 1990, Signal processing to enable utilization of a rig reference sensor with a drill bit seismic source: U.S. Patent No. 4,926,391.

Rector, III J. W., and Marion, B. P., 1991, The use of drill-bit energy as a downhole seismic source: Geophysics, **56**, 628–634.

Rector, III J. W., and Weiss, D. S., 1989, Real-Time Drill-Bit VSP from an Offshore Platform: 59th Ann. Internat. Mtg., Soc. of Expl. Geophys., Expanded Abstracts, 5–8.

Rickett, J., and Claerbout, J. F., 1999, Acoustic daylight imaging via spectral transformation: Helioseismology and reservoir monitoring: 69th Ann. Internat. Mtg., Soc. of Expl. Geophys., Expanded Abstracts, 1675–1678.

Ristow, D., and Jurczyk, D., 1975, Vibroseis deconvolution: Geophys. Prosp., **23**, 363–379.

Robello, G. S., Miska, S., 1997, Analytical Study of the Performance of Positive Displacement Motor (PDM): Modeling for Incompressible Fluid: SPE Paper No. 39026.

Robinson, E. A., 1984, Seismic inversion and deconvolution. Part A: Classical methods: Vol. **4A**, GEOPHYSICAL PRESS, London - Amsterdam.

Robinson, E. A., 1998, Model-driven predictive deconvolution: Geophysics, **63**, 713–722.

Robinson, J. C., 1970, Statistically optimal stacking of seismic data: Geophysics, **35**, 436–446.

Robinson, E. A., and Durrani, T. S., 1986, Geophysical Signal Processing: Englewood Cliffs, Prentice-Hall.

Robison, C. E., How Intelligent Should Well Completions' Flow Controls Be? – A Review and Comparison with More Mature Technologies: SPE Paper No. 74494.

Rocca, F., Persoglia, S., Poletto, F., and Craglietto, A., 1990, Use of drilling noise to generate a reciprocal VSP: CEE Contract DGXII/TH-0113-87, Final Report of the Project.

Rossi, G., 2002, Tomographic analysis of the Seisbit data of a Southern-Italy Well: Rep. GDL 20/2002-02-18, OGS 14/2002.

Rossi, G., Corubolo, P., Bohm, G., Ceraggioli, E., Dell'Aversana, P., Morandi, S., Poletto, F., and Vesnaver, A., 2001, Joint 3D inversion of SWD and surface seismic data: First Break, **19**, 453–459.

Saleh, S. T., and Mitchell, B. J., 1989, Wellbore drillstring mechanical and hydraulic interaction: SPE Paper No. 18792.

Sallas, J. J., 1984, Seismic vibrator control and the downgoing P-wave: Geophysics, **49**, 732–740.

Santos, H., Plácido, J. C. R., Wolter, C., 1999, Consequences and Relevances of Drillstring Vibration on Wellbore Stability: SPE/IADC Paper No. 52820.

Sayers, C. M., Johnson, G. M., and Denyer, G., 2000, Predrill Pore Pressure Prediction Using Seismic Data: SPE Paper No. 59122.

Schechtman, G. A., and Schneerson, M. B., 1983, Inventor's Certificate No. 1035549: Russian Patent, 1983.

Scheuer, T. E., and Wagner, D. E., 1985, Deconvolution by autocepstral windowing: Geophysics, **50**, 1533–1540.

Schleifer, A., 1998, Seisbit EXT CEE Sismica While Drilling: OGS Report 13/98, Trieste, Italy.

Schuster, G. T., Followill, F., Katz, L. J., Yu, J., and Liu, Z., 2003, Autocorrelogram migration: Theory: Geophysics, **68**, 1685–1694.

Seto, W. W., 1994, Acoustics: McGraw-Hill, New York.

Sheppard, M. C., and Lesage, M., 1988, The forces at the teeth of a drilling rollercone bit: Theory and experiment: SPE Paper No. 18042.

Sheriff, R. E., 1999, Encyclopedic Dictionary of Exploration Geophysics: Society of Exploration Geophysicists.

Schoenberger, M., 1970, Optimization and implementation of marine seismic arrays: Geophysics, **35**, 1038–1053.

Schoenberger, M., 1996, Optimum Weighted Stack for Multiple Suppression: Geophysics, **61**, 891–901.

Simon, R., 1963, Energy Balance in Rock Drilling: Society of Petroleum Engineers Journal, December, 298–306, SPE Paper No. 499.

Sinceri, R., 1994, About use of rigsite refracted noise: Personal communication.

Skaugen, E., 1987, The Effects of Quasi-random Drill Bit Vibrations Upon Drillstring Dynamic Behavior: SPE Paper No. 16660.

Skaugen, E., and Kyllingstad, Å., 1986, Performance Testing of Shock Absorbers: SPE Paper No. 15561.

Somerton, W. H., Timur, A., and Gray, D. H., 1961, Stress Behavior of Rock Under Drilling Loading Conditions: SPE Paper No. 166.

Soulier, L., 1999, Device and method for transmitting information by electromagnetic waves: U.S. Patent No. 5,945,923.

Soulier, L., and Lemaitre, M., 1993, E.M. MWD data transmission status and perspectives: SPE/IADC Paper No. 25686.

Spagnolini, U., 1993, 2-D phase unwrapping and phase aliasing: Geophysics, **58** (9), 1324–1334.

Squire, W. D., and Whitenhouse, H. J., 1979, A new approach to drill-string acoustic telemetry: SPE Paper, presented the 5th SPE Fall Meeting, (9), Las Vegas, Nevada.

Staron, P., Arens, G., and Gros, P., 1988, Method of instantaneous acoustic logging within a wellbore: U.S. Patent No. 4,718,048.

Stuart, J., 1989, Method for determining drill bit wear: European Patent Application, EP 0 336 477 A1.

Tang, X.-M., and Cheng, A., 2004, Quantitative Borehole Acoustic Methods: Seismic Exploration, **24**, Klaus Helbig and Sven Treitel, Editors, Elsevier, Amsterdam.

Terzaghi, K., 1943, Theoretical soil mechanics, John Wiley & Sons.

Thigpen, B. B., Dalby, A. E., and Landrum, R., 1975, Special report of the subcommitee on polarity standards: Geophysics, **40**, 694–699.

Thomsen, L., 1986, Weak elastic anisotropy: Geophysics, **51**, 1954–1966.

Tinivella, U., Poletto, F., and Carcione, J. M., 2002, Numerical simulation of borehole drillstring waves for geosteering: 64th Mtg. Eur. Assn. Geosci. Eng., Extended Abstract, Session P011.

Toksöz, M. N., and Stewart, R., 1984, Vertical seismic profiling – Part B: Advanced concepts: Geophysical press, London.

Ulrych, T. J., 1971, Application of homomorphic deconvolution to seismology: Geophysics, **36**, 650–660.

Vesnaver, A., 1996, The contribution of reflected, refracted and transmitted waves to seismic tomography: a tutorial. First Break, **14**, 159–168.

Walden, A. T., 1985, Non-Gaussian reflectivity, entropy and deconvolution: Geophysics, **50**, 2862–2888.

Wang, R. J., 1969, The determination of optimum gate lengths for time-varying Wiener filtering: Geophysics, **34**, 683–695.

Wang, R. J., 1977, Adaptive predictive deconvolution of seismic data: Geophys. Prosp., **25**, 342–381.

Wardell, N., Diviacco, P., and Sinceri, R., 2000, Pre-processing corrections of very high resolution 3D marine seismic data: 62nd Mtg. Eur. Assn. Geosci. Eng., Extended Abstracts, Paper P130.

Warren, T. M., 1987, Penetration-Rate Performance of Roller-Cone Bits: SPE Drilling Engineering, March, 9–17.

Warren, T. M., and Smith, M. B., 1985, Bottomhole Stress Factors Affecting Drilling Rate at Depth: SPE Paper No. 13391.

Warring, R. H, 1985, Fundamentals of transducers: TAB BOOKS Inc.

Weakley, R. R., 1989, Use of Surface Seismic Data To Predict Formation Pore Pressures (Sand Shale Depositional Environments): SPE/IADC Paper No. 18713.

White, J. E., 1960, Use of reciprocity theorem for computation of low-frequency radiation patterns: Geophysics, **25** (3), 613–624.

White, J. E., 1965, Seismic waves: Radiation, Transmission, and Attenuation: McGraw-Hill Book Company, New York.

White, D. B., and Curry, D. A., 1988, Effects of Nozzle Configuration on Roller Cone Bit performance: IADC/SPE Paper No. 17188.

White, R. E., and Simm, R., 2003, The importance of bandwidth in seismic cross equalisation: well tie and other examples: 65th Mtg. Eur. Assn. Geosci. Eng., Extended Abstracts, P071.

Widrow, B., 1990, Seismic processing and imaging with a drill-bit source: U.S. Patent No. 4,964,087.

Widrow, B., Glover, J. R., Jr., McCool, J. M, Kaunitz, J. Williams, C. S., Hearn, R. H., Zeidler, J. R., Dong, E., and Goodlin, R. C., 1975, Adaptive Noise Cancelling: Principles and Applications: Proceedings of the IEEE, Vol. 63, No. 12, 1692–1716.

Wiggins, R. A., 1978, Minimum entropy deconvolution: Geoexploration, **16**, 21–35.

Wiggins, R., 1985, Entropy guided deconvolution: Geophysics, **50**, 2720–2726.

Williamson, P. R., 1991, A guide to limits of resolution imposed by scattering in ray tomography: Geophysics, **56**, 202–207.

Williamson, J. S., and Lubinski, A., 1987, Predicting Bottomhole Assembly Performance: SPE Drilling Engineering, March, 37–45, SPE Paper No. 14746.

Wilson, E. J., 1995, Strain-gage instrumentation: Shock and Vibration Handbook, Chapter 17, McGraw-Hill Book Co., Fourth Edition.

Winters, W. J., Warren, T. M., and Onyia, E. C., 1987, Roller bit model with rock ductility and cone offset: SPE Paper No. 16696.

Wojtanowicz, A. K., and Kuru, E., 1987, Dynamic Drilling Strategy for PDC Bits: SPE/IADC Paper No. 16118.

Wolf, S. F., Zacksenhouse, M., and Arian A., 1985, Field Measurements of Downhole Drillstring Vibrations: SPE Paper No. 14330.

Wood, A. B., 1955, A texbook of sound: The MacMillan Co., New York.

Yang, L. Z., 1990, The Analysis of Relations between a Tricone Bit and Lobed Bottom in Low Frequency Axial Vibration: SPE Paper No. 20325.

Yilmaz, Ö, 2001, *Seismic Data Analysis*: Series: Investigations in Geophysics, Vol. I, Society of Exploration Geophysicists.

Yu, J., Followill, L., Katz, L., and Schuster, G. T., 2001, Autocorrelogram migration of drill-bit data: 63rd Mtg. Eur. Assn. Geosci. Eng., Extended Abstracts, Session M13.

Yu, J., and Schuster, G., 2001, Crosscorrelogram migration of IVSPWD data: 71st Ann. Internat. Mtg., Soc. of Expl. Geophys., Expanded Abstracts, 456–459.

Zannoni, S. A., Cheatham, C. A., Chen, C.-K. D., and Golla, C. A., 1993, Development and Field Testing of New Downhole MWD Drillstring Dynamics Sensor: SPE Paper No. 26341.

Zijsling, D. H., 1987, Single Cutter Testing - A Key for PDC Bit Development: SPE Paper No. 16529/1.

Name index

Subject index